Mental Health Matters: From Cross-Cultural Perspectives

Mental Health Matters: From Cross-Cultural Perspectives

Editors

Yasuhiro Kotera
Elaina Taylor

MDPI • Basel • Beijing • Wuhan • Barcelona • Belgrade • Manchester • Tokyo • Cluj • Tianjin

Editors
Yasuhiro Kotera
Faculty of Medicine and Health Sciences
University of Nottingham
Nottingham
United Kingdom

Elaina Taylor
College of Health, Psychology and Social Care
University of Derby
Derby
United Kingdom

Editorial Office
MDPI
St. Alban-Anlage 66
4052 Basel, Switzerland

This is a reprint of articles from the Special Issue published online in the open access journal *Healthcare* (ISSN 2227-9032) (available at: www.mdpi.com/journal/healthcare/special_issues/mentalhealth_matters).

For citation purposes, cite each article independently as indicated on the article page online and as indicated below:

LastName, A.A.; LastName, B.B.; LastName, C.C. Article Title. *Journal Name* **Year**, *Volume Number*, Page Range.

ISBN 978-3-0365-8481-2 (Hbk)
ISBN 978-3-0365-8480-5 (PDF)

Contents

About the Editors

Yasuhiro Kotera

Dr Yasuhiro Kotera is an Associate Professor in Mental Health at the University of Nottingham in the UK. His research focuses on cross-cultural perspectives and mental health recovery. He is an Accredited Psychotherapist by the British Association for Counselling and Psychotherapy and a Fellow of the Advance HE.

Elaina Taylor

Dr Elaina Taylor is a Lecturer at the University of Derby in the UK, and Chartered Psychologist. She is a Fellow of the Advance HE and a full member of the Division of Health Psychology at the British Psychological Society.

Preface to "Mental Health Matters: From Cross-Cultural Perspectives"

Awareness of mental health has been increasing rapidly worldwide in recent years, and even more so since the outbreak of COVID-19. Depression is now regarded as one of the most debilitating diseases, and wellbeing is incorporated into the United Nations' Sustainable Development Goals. In order for all of us to have a happy life, mental health cannot be ignored. As announced by the UK government, our health cannot be achieved without good mental health. Likewise, in Asia, the word 'health' in Chinese and Japanese encompasses both a healthy body and a calm mind. The Japanese government has implemented a work-style reform to protect employees'mental health. While these movements suggest the importance of mental health worldwide, a universal definition of mental health remains to be defined. This is partly attributed to a lack of understanding of mental health from different cultures. How an individual regards mental health can differ significantly according to their culture. Therefore, this Special Issue aims to address this problem by introducing alternative views to mental health through discussion of cross-cultural psychiatric matters.

Yasuhiro Kotera and Elaina Taylor

Editors

healthcare

MDPI

Article

Mental Well-Being of Czech University Students: Academic Motivation, Self-Compassion, and Self-Criticism

Yasuhiro Kotera [1,*], Sarah Maybury [2], Gillian Liu [2], Rory Colman [2], Jenai Lieu [2] and Jaroslava Dosedlová [3]

[1] School of Health Sciences, University of Nottingham, Nottingham NG7 2HA, UK
[2] College of Health, Psychology and Social Care, University of Derby, Derby DE22 1GB, UK
[3] Department of Psychology, Masaryk University, 602 00 Brno, Czech Republic
* Correspondence: yasuhiro.kotera@nottingham.ac.uk

Citation: Kotera, Y.; Maybury, S.; Liu, G.; Colman, R.; Lieu, J.; Dosedlová, J. Mental Well-Being of Czech University Students: Academic Motivation, Self-Compassion, and Self-Criticism. *Healthcare* **2022**, *10*, 2135. https://doi.org/10.3390/healthcare10112135

Academic Editor: Alyx Taylor

Received: 17 September 2022
Accepted: 24 October 2022
Published: 27 October 2022

Abstract: University students in the Czech Republic suffer from a low level of mental well-being. Research in other university student populations suggests that academic motivation, self-compassion, and self-criticism are strongly related to mental well-being. Students who are motivated to study, are kind toward themselves, and are less judgmental of themselves tend to have a high level of mental well-being. These relationships had not been evaluated in Czech students. Accordingly, this cross-sectional study aimed to evaluate the relationships between mental well-being, academic motivation (intrinsic motivation, extrinsic motivation, and amotivation), self-compassion (self-reassurance) and self-criticism (self-inadequacy and self-hate). Of 130 students approached, a convenience sampling of 119 psychology students at a university in the Czech Republic completed a survey regarding these constructs. Correlation, regression, and path analyses were conducted. Mental well-being was positively associated with intrinsic motivation and self-compassion, and negatively associated with amotivation and self-criticism. Self-compassion was identified as the strongest predictor of mental well-being. Lastly, intrinsic motivation mediated the pathway from self-compassion to mental well-being, but not the one from self-inadequacy to mental well-being, and the one from self-hate to mental well-being. Our findings can help educators to identify effective means to protect students' mental well-being. Cultivating students' self-compassion may be helpful to protect their mental well-being. University staff and educators in the Czech Republic need to consider ways to embed self-compassion training into their students' programmes or university life.

Keywords: Czech university students; self-compassion; intrinsic motivation; self-criticism; amotivation; self-hate; self-inadequacy

1. Introduction

1.1. Poor Mental Well-Being of Czech Students

The promotion of mental well-being is at the forefront of many government agendas, due to the social, economic, and human burden that accompanies poor mental well-being [1]. Although mental well-being is a global health concern, the approach toward and experience of mental well-being, as well as the reception of mental health interventions, differ between countries, due to unique historical and political considerations [2]. For example, in ex-Soviet countries, the treatment of individuals with mental health problems was highly stigmatised, discriminatory, and often abusive [3]. With the fall of the Soviet Union, came a shift in the classification of mental health, from models established by Soviet health institutions to Western psychiatric diagnostic categories [4]. Although this shift brought a change in practices, the negative attitudes toward mental health persist, jeopardising effective interventions, and leading to worsened mental health outcomes [5]. This lasting negative perception towards mental health has resulted in a treatment gap and insufficient resources to aid those living with mental disorders [6]. Moreover, these attitudes have been internalised by those suffering from mental illness, who demonstrate feelings of shame, resulting in additional barriers to treatment-seeking behaviours [7].

In the Czech Republic, mental health expenditure amounts to approximately 4% of total health expenditure, resulting in insufficient provision of services and psychiatrists with extremely unmanageable patient loads [8]. In a country where an estimated one in five person suffers from an affective, anxiety, or alcohol- or substance-abuse disorder, the absence of adequate care for these individuals often results in further deterioration in already neglected members of society [6]. Rates of alcohol dependency in the Czech Republic are almost double those in Central Europe, with the highest prevalence found in individuals aged 18 to 29, the common age of undergraduate university students [9]. University students are at risk for further mental health problems, which reach their peak in young adulthood, often exacerbated by the academic and social pressures associated with the transition to university [10].

Additionally, the COVID-19 pandemic has presented additional stressors to Czech university students. The way the Czech government disseminated information in the initial stages of the pandemic led to mass psychological traumatisation of the general population [11]. Combined, these factors highlight the need for attention and research into ways to improve mental well-being for university students in the Czech Republic [12].

1.2. Academic Motivation and Mental Well-Being

One well-established avenue for this improvement has been through enhancing academic motivation, as understood through the lens of the self-determination theory (SDT) [13]. SDT proposes that all humans possess three primary needs: relatedness (feelings of belonging and connection to others), autonomy (self-directed action to exercise one's will), and competence (the effective interaction with one's surroundings to build capabilities). If these needs are unmet, a person may feel their life lacks meaning, which can result in depressive symptoms and suicidal feelings [14].

A fundamental aspect of SDT is motivation, which exists on a continuum and includes intrinsic motivation, extrinsic motivation and amotivation. Intrinsic motivation encourages individuals to seek opportunities to learn, explore and be challenged based on a task itself being meaningful, and results in positive outcomes in university students when it comes to stress, psychological health, and academic performance [15]. Extrinsic motivation is driven by the need to meet an external demand or obtain rewards, and amotivation refers to the absence of an intention to act [13], which are both associated with poorer academic and mental well-being outcomes [16].

Although SDT claims universal applicability, most studies about motivation have been performed in the United States and other Western, educated, industrialised, rich, and democratic (WEIRD) nations [17]. Some argue that it employs an individualist conception of people due to the focus on autonomy that is not relevant across cultures [18]. However it is argued that SDT acknowledges autonomous interdependence, which aligns with more collectivist-based value systems [19]. SDT has been shown to be an effective model to investigate motivation across different societies, given that cultural nuances are considered and applied to the population being studied [20,21].

Although SDT has demonstrated its suitability to study non-WEIRD cultures, there remains a need to grow the body of research into the specificities of ex-Soviet states, including the Czech Republic, as cultural differences are important to understanding psychological outcomes [22]. Rather than a static set of shared practices that precedes psychological phenomena, culture cultivates a person's worldview, which is integrated into their psyche and is fundamental to the understanding of processes such as motivation [23].

1.3. Self-Compassion and Self-Criticism

Studies across cultures provide a wealth of evidence that self-compassion and self-criticism are also related to mental well-being. Comparative studies between a range of countries have shown self-compassion to be a valid construct [24] benefiting diverse populations either measured as a trait [25] or following intervention to enhance self-compassion [26]. Evidence of negative mental well-being implications from self-criticism are comparably robust [27], with students at risk from its psychopathological effects [28]. To date, there is little information on the role of self-compassion and self-criticism in students within the Czech Republic, indicating a requirement for research into how these psychological factors contribute to their mental well-being.

Gilbert's biopsychosocial theory conceptualises self-compassion and self-criticism as originating in distinct organisations of motivation, emotion, and behaviour corresponding to evolutionary social relating modes, referred to as social mentalities [29]. Self-compassion is a caring social mentality, corresponding with neurophysiological systems promoting rest and digest activities [30]. Self-criticism on the other hand is a competitive social mentality, corresponding with threat-based monitoring and fight–flight neurophysiological systems [29]. At times of stress or setback, individuals experience emotion and exhibit behaviour related to how these systems are activated. Those able to behave self-compassionately and with self-reassurance, are afforded protection from negative psychological consequences, whereas self-critical individuals lack such protection and may even exacerbate the negative consequences by self-attacking [31]. Self-criticism can take the forms of self-inadequacy, a sense of personal inadequacy and disappointment with self, and self-hate, a sense of self-disgust and self-hatred with intentions of self-harm [31]. Within non-Czech student populations, each of these forms of self-criticism are shown to be associated with poorer mental well-being [31–35].

1.4. Intrinsic Motivation as Mediator for Self-Compassion to Mental Well-Being

Gilbert's model regards motivation as determining the social mentality one adopts to cope with challenges. For instance, functioning to protect from possible harm to social standing, the self-critical process is motivated by underlying fears of social disconnection and shame [29]. Self-criticism has been shown to be associated with extrinsic motivation [36–38], hindering academic goal progress [39]. Driven by perceived threat, self-criticism stimulates defensive and defeatist behaviours with corresponding negative effects [40] which can manifest in amotivation. A self-reassuring process on the other hand is motivated by caring for self, rather than social comparison, fostering a sense of self-efficacy and courage [30,41].

Based on attachment research findings, self-reassurance is heorized to develop through experiences of a secure and reassuring environment. Indeed, students with secure attachment styles have been found most able to self-reassure [42]. Leak and Cooney found intrinsic motivation to mediate between a secure attachment style and psychological well-being when investigating adult student romantic relationships [43]. What is not yet established, however, is how self-reassurance and intrinsic motivation interact to influence mental well-being. As findings in university students of different cultures indicate self-compassion to have positive association with intrinsic motivation [44–48], it is hypothesised in this study that intrinsic motivation may mediate the relationship between self-compassion and mental well-being in Czech students. Because interventions to increase self-compassion in students have provided evidence of enduring mental well-being effects up to six months following intervention [49], self-compassion has promise as an ability that can be developed for Czech students. By ascertaining the relationships between self-compassion, self-criticism, motivation, and mental well-being, recommendations can be made on the most appropriate means to improve Czech students' mental well-being.

1.5. Study Aims

This study aimed to understand the mental well-being of Czech university students. Therefore, the objectives were (1) to evaluate the relationship among mental well-being, academic motivation (intrinsic motivation, extrinsic motivation, and amotivation), self-compassion, and self-criticism; (2) to identify predictors of mental well-being; and (3) assess whether intrinsic motivation mediates the pathway from self-compassion/self-criticism to mental well-being.

2. Materials and Methods

2.1. Participants

Opportunity sampling was used to recruit 130 undergraduate students in psychology at a Czech university in Brno, of which 119 completed the measures. The students were Czech (n = 98) and Slovakian (n = 21), had an age range between 19 and 44 years (M = 21.87, SD = 3.32), and consisted of 93 females and 20 males. This sample reflects the demographic of the general population of psychology students in the Czech Republic [50]. Students were recruited through announcements given by programme tutors. Overall, the number of participants reached the required sample size of 115 according to statistical power calculations (84: two tails, pH1 (r) = 0.30 medium [51], α = 0.05, Power = 0.80, pH0 = 0 [52]). Participants provided consent to their engagement in the study.

2.2. Materials

The study evaluated the relationship between mental well-being, academic motivation (intrinsic motivation, extrinsic motivation and amotivation), self-compassion (self-reassurance), and self-criticism (self-inadequacy and self-hate) through a survey regarding these constructs.

The Self-Criticising/Attacking and Self-Reassuring Scale (FSCSR) was used to assess self-criticism and self-reassurance in participants, by looking at how they think of themselves in adverse circumstances [31]. The FSCSR consists of 22 items scored on a five-point Likert scale (0 = "Not at all like me" to 4 = "Extremely like me"). The scale that contains three dimensions: two forms of self-criticism (self-inadequacy and self-hate) and one form of self-reassurance. Self-inadequacy refers to a belief of personal deficiency (e.g., "I am easily disappointed in myself"), self-hate refers to a desire to punish the self (e.g., "I have a sense of disgust toward myself"), and self-reassurance refers to the ability to soothe oneself when things go wrong (e.g., "I am able to remind myself of positive things about myself"). The FSCSR subscales are known for their reliability (α = 0.90 for self-inadequacy, α = 0.86 for self-hate, and α = 0.86 for self-reassurance) and validity ($|r|$ = 0.45–0.77) [31].

The Academic Motivation Scale (AMS) [53] measures three variations of motivation that are categorised into seven subtypes: (i) amotivation, (ii) extrinsic motivation, and (iii) intrinsic motivation. Amotivation refers to the lack of self-determination (e.g., "I can't understand what I am doing in school"), extrinsic motivation refers to external and identified factors (e.g., "Because I want to have 'the good life' later on), and intrinsic motivation is the sense of knowing or accomplishment from the action (e.g., "Because I experience pleasure and satisfaction while learning new things"). The scale consists of 28 items on a seven-point Likert scale (1 = "Does not correspond at all" to 7 = "Corresponds exactly"). The AMS is demonstrated to have adequate to high internal consistency (α = 0.62–0.91) [53].

The Short Warwick–Edinburgh Mental Well-Being Scale (SWEMWBS) [54] evaluated mental well-being by using the shortened version of the original 14-item scale to the seven-item scale [55]. The SWEMWBS is measured on a five-point Likert scale (1 = "None of the time" to 5 = "All of the time"). The scale has been demonstrated to be mostly free of item bias [54] and is considered well-established to appraise mental well-being [56]. Participants completed the survey based on their experiences within the past two weeks regarding hedonic and eudaimonic well-being (e.g., "I've been dealing with problems well"). The SWEMWBS is regarded to have high internal consistency (α = 0.85) [54].

2.3. Analysis

A cross-sectional design was used to conduct correlation, regression, and path analyses in the study. The data was screened for outliers and the assumptions of parametric tests. Afterward, correlations between mental well-being, academic motivation, self-compassion, and self-criticism were evaluated by using IBM SPSS version 27. Lastly, path analyses were conducted to analyse how self-criticism and self-reassurance may mediate in the relationship between mental well-being and academic motivation by using the Process Macro 3 for

SPSS [57] with 5000 bootstrapping re-samples and bias-corrected 95% confidence intervals (CIs) for indirect effects were applied.

2.4. Ethics

The research ethics were approved by the researcher, JD's university research ethics committee. Participants provided their consent and were informed of their rights including their right to withdraw from the study.

3. Results

Descriptive statistics of all variables are presented in Table 1. High internal consistency was demonstrated for all variables (α = 0.73–0.90).

Table 1. Descriptive statistics.

	M	SD	α
Gender	93 females, 20 males, 6 unanswered		
Age	21.87	3.32	
Mental well-being	25.21	3.80	0.73
Intrinsic motivation	4.55	1.10	0.90
Extrinsic motivation	4.97	1.02	0.83
Amotivation	1.52	0.88	0.80
Self-reassurance	20.35	6.08	0.85
Self-inadequacy	17.35	8.12	0.88
Self-hate	3.92	4.20	0.82

3.1. Correlation (Objective 1)

Pearson's correlation was performed to evaluate the relationship between mental well-being, academic motivation (intrinsic motivation, extrinsic motivation, and amotivation), self-reassurance, and self-criticism (self-inadequacy and self-hate) (Table 2). Point biserial correlations were calculated for gender (0 = female, 1 = male).

Table 2. Correlations among mental well-being, academic motivation, self-compassion, and self-criticism.

		1	2	3	4	5	6	7	8	9
1	Gender (0 = F, 1 = M)	-								
2	Age	0.25 **	-							
3	Mental Well-being	0.05	0.13	-						
4	Intrinsic motivation	0.12	−0.09	0.33 **	-					
5	Extrinsic motivation	−0.27 **	−0.11	−0.01	0.18	-				
6	Amotivation	0.05	−0.03	−0.41 **	−0.17	0.03	-			
7	Self-reassurance	0.03	−0.06	0.61 **	0.21 *	−0.03	−0.33 **	-		
8	Self-inadequacy	−0.10	−0.10	−0.52 **	−0.06	0.22 *	0.49 **	−0.63 **	-	
9	Self-hate	−0.02	−0.01	−0.37 **	−0.14	0.05	0.51 **	−0.60 **	0.73 **	-

* $p < 0.05$, ** $p < 0.01$.

Mental well-being was positively associated with intrinsic motivation and self-reassurance, and negatively associated with amotivation, self-inadequacy and self-hate. None of the three motivation types were interrelated. Self-reassurance was positively associated with intrinsic motivation and negatively associated with amotivation, self-inadequacy, and self-hate. Self-inadequacy and self-hate were positively related to each other.

3.2. Regression (Objective 2)

Multiple regression analyses were conducted to identify predictors of mental well-being (outcome variable). Significant correlates with mental well-being, namely intrinsic motivation, amotivation, self-reassurance, self-inadequacy, and self-hate were entered as a predictor variable. The adjusted coefficient of determination (Adj. R^2) was calculated to determine the degree of variance in the population (Table 3). Multicollinearity was not a concern (variance inflation factors < 10).

Table 3. Multiple regression analyses for mental well-being.

	B	**SE_B**	**β**	**95% CI (Low, Up)**	
Intrinsic motivation	0.76	0.24	0.22 **	0.27	1.23
Amotivation	−0.91	0.34	−0.21 **	−1.60	−0.23
Self-reassurance	0.29	0.06	0.46 ***	0.18	0.40
Self-inadequate	−0.15	0.05	−0.32 **	−0.25	−0.05
Self-hate	−0.26	0.09	−0.28 **	−0.07	−0.44
Adj. R^2			48%		

Outcome variable: Mental well-being. B = unstandardised coefficients; SE_B = standard error of the coefficient; β = standardised coefficients. ** $p < 0.01$; *** $p < 0.001$.

These five predictor variables were all significant, and predicted 48% of the variance in mental well-being altogether, indicating a large effect size [51]. Self-reassurance had the greatest impact on mental well-being (β = 0.46).

3.3. Mediation (Objective 3)

Three models of path analyses were conducted to assess whether intrinsic motivation would mediate the pathways from self-compassion/self-criticism to mental well-being. Model 4 in the process macro (parallel mediation model) [57] was used. Mental well-being was entered as an outcome variable, intrinsic motivation as a mediator variable, and self-reassurance, self-inadequacy, and self-hate as a predictor variable.

First, intrinsic motivation partially mediated the pathway from self-reassurance to mental well-being (Figure 1a). All pathways including the direct effect from self-reassurance to mental well-being, and total effect that includes the impact of intrinsic motivation, were significant (* $p < 0.05$; ** $p < 0.01$; *** $p < 0.001$). When self-reassurance increases mental well-being, it also impacts intrinsic motivation, which in turn improves mental well-being.

Second, intrinsic motivation did not mediate the pathway from self-inadequacy to mental well-being (Figure 1b). Although the direct effect from self-inadequacy to well-being, the total effect that includes the impact of intrinsic motivation on mental well-being, and intrinsic motivation's effect on mental well-being were significant, a pathway from self-inadequacy to intrinsic motivation was not significant.

Lastly, intrinsic motivation did not mediate the pathway from self-hate to mental well-being either (Figure 1c). Although the direct effect of self-hate on well-being, the total effect that includes the impact of intrinsic motivation on mental well-being, and intrinsic motivation's effect on mental well-being were significant, a pathway from self-hate to intrinsic motivation was not significant.

Figure 1. Three models of path analyses with intrinsic motivation as mediator. * $p < 0.05$; ** $p < 0.01$; *** $p < 0.001$, Direct = direct effect; Total = total effect. Mediation model of intrinsic motivation on the (**a**) self-reassurance to mental wellbeing pathway, (**b**) self-inadequacy to mental health pathway, and (**c**) self-hate to mental wellbeing pathway.

4. Discussion

This study evaluated the relationship between mental well-being, academic motivation, self-compassion (self-reassurance) and self-criticism (self-inadequacy and self-hate) in university students in the Czech Republic. Mental well-being was positively associated with intrinsic motivation and self-compassion, and negatively associated with amotivation and self-criticism. Self-compassion was identified as the strongest predictor of all of mental well-being. Lastly, intrinsic motivation mediated the pathway from self-compassion to mental well-being but did not mediate the pathways from self-criticism to mental well-being.

Consistent with previous research conducted in other cultures, intrinsic motivation and self-compassion were associated with mental well-being [15,44]. The significant correlation of mental well-being with intrinsic motivation and self-compassion in this study showed these may be universal regardless of culture differences. The positive association of mental well-being and intrinsic motivation in Czech students supports SDT as an explanatory theoretical model. Extending previous findings in Asia and Africa [46,58], the present study showed the applicability of SDT to non-WEIRD cultures that are located in Europe.

Self-compassion was also a strong predictor of mental well-being in the Czech students. This is consistent with previous findings from different populations [49,59,60]. Self-compassion interventions have been put into practice, and achieved desirable outcomes, such as increases in self-compassion, optimism, self-efficacy, and happiness and decreases in depression, anxiety, and stress [61–64]. Because of these outcomes, implementing self-compassion training appears to be a useful approach for better self-care and mental well-being [65,66]. Supporting self-compassion may be a less stigmatised approach for better mental well-being, instead of focusing on reduction of mental distress [67,68]. Because there is a negative perception toward people with mental health problems in Czech higher education [5], this significant relationship between self-compassion and mental well-being may help identify an effective means to support the mental well-being of Czech students.

Lastly, intrinsic motivation mediated the pathway from self-compassion to mental well-being, but did not mediate the pathways from self-criticism to mental well-being. When self-compassion impacts positively on mental well-being, there is an indirect positive impact of intrinsic motivation on mental well-being too. Associations between self-compassion and intrinsic motivation may play an important role in this pathway [69,70]. To be intrinsically motivated and highly autonomous, one needs to feel safe and content (i.e., self-compassionate) [71]. This mediation model suggests that by cultivating self-compassion, not only the Czech students' mental well-being, but also their intrinsic motivation can be enhanced. In other words, educators in the Czech universities can teach students to be kind and understanding toward themselves for mental well-being purposes, which can also help students to be more autonomous. The self-compassion training can be embedded in the curriculum. Alternatively, for students who feel it challenging to practice self-compassion exercises such as meditation and imageries [72,73], forest bathing may be recommended to support self-compassion [74,75].

Intrinsic motivation, however, did not mediate the self-criticism to mental well-being pathways. When self-criticism damages mental well-being, it does not impact intrinsic motivation positively nor negatively. This highlights the non-significant association between self-criticism and intrinsic motivation [76]. Furthermore, Powers et al. found no association between self-criticism and intrinsic motivation, suggesting that self-criticism and autonomy may independently contribute to mental well-being [39]. This may highlight the fact that self-criticism needs to be mitigated independent from motivation in order to protect mental well-being [77]. As noted earlier, self-compassion training may be helpful as it can reduce self-criticism [78].

Several limitations should be noted. First, the generalisability of our findings might have been compromised, as our recruitment was through opportunity sampling at one university. Furthermore, although the sample satisfied the required size, a larger sample could have enhanced the generalisability of the findings. Secondly, the two scales used in this study were rather long (FSCSR 22 items and AMS 28 items). Shorter versions [79,80] could have been used to reduce the workload of the students. Lastly, the causal directions of these relationships were not evaluated. Longitudinal data would help inform the causality, which may help develop more effective approaches.

5. Conclusions

This study evaluated the relationships among mental well-being, academic motivation, self-compassion, and self-criticism in university students in the Czech Republic. Consistent with findings from other student populations, strong relationships among mental well-being, self-compassion, and intrinsic motivation were identified in the Czech students. Cultivating self-compassion may enhance mental well-being as well as intrinsic motivation. Our findings can help educators to develop effective approaches to support the mental well-being among Czech university students.

Author Contributions: Conceptualization, Y.K. and J.D.; methodology, Y.K.; software, Y.K.; validation, all authors; formal analysis, Y.K.; investigation, Y.K. and J.D.; resources, J.D.; data curation, J.D.; writing—original draft preparation, Y.K., S.M., G.L., R.C. and J.L.; writing—review and editing, all authors; project administration, Y.K. All authors have read and agreed to the published version of the manuscript.

Funding: This research received no external funding.

Institutional Review Board Statement: The study was conducted according to the guidelines of the Declaration of Helsinki, and approved. Ethical approval was granted by the Ethics Panel of the Institute of Psychology, Faculty of Arts, Masaryk University (ID: 035/20; 28 January 2020).

Informed Consent Statement: Informed consent was obtained from all subjects involved in the study.

Data Availability Statement: The data presented in this study are available on request from the corresponding author. The data are not publicly available due to ethical restrictions.

Conflicts of Interest: The authors declare no conflict of interest.

References

1. Vigo, D.; Thornicroft, G.; Atun, R. Estimating the true global burden of mental illness. *Lancet Psychiatry* **2016**, *3*, 171–178. [CrossRef]
2. Choudhry, F.R.; Mani, V.; Ming, L.; Khan, T.M. Beliefs and perception about mental health issues: A meta-synthesis. *Neuropsychiatr. Dis. Treat.* **2016**, *12*, 2807–2818. [CrossRef] [PubMed]
3. Petrea, I. Mental Health in Former Soviet Countries:From Past Legacies to Modern Practices. *Public Health Rev.* **2012**, *34*, 5. [CrossRef]
4. Skultans, V. The Appropriation of Suffering. *Theory Cult. Soc.* **2007**, *24*, 27–48. [CrossRef]
5. Kotera, Y.; Andrzejewski, D.; Dosedlova, J.; Taylor, E.; Edwards, A.-M.; Blackmore, C. Mental Health of Czech University Psychology Students: Negative Mental Health Attitudes, Mental Health Shame and Self-Compassion. *Healthcare* **2022**, *10*, 676. [CrossRef]
6. Kagstrom, A.; Alexova, A.; Tuskova, E.; Csajbók, Z.; Schomerus, G.; Formanek, T.; Mladá, K.; Winkler, P.; Cermakova, P. The treatment gap for mental disorders and associated factors in the Czech Republic. *Eur. Psychiatry* **2019**, *59*, 37–43. [CrossRef]
7. Doblytė, S. Shame in a post-socialist society: A qualitative study of healthcare seeking and utilisation in common mental disorders. *Sociol. Health Illn.* **2020**, *42*, 1858–1872. [CrossRef]
8. Pec, O. Mental health reforms in the Czech Republic. *BJPsych Int.* **2019**, *16*, 4–6. [CrossRef]
9. Kotera, Y.; Dosedlova, J.; Andrzejewski, D. From Stress to Psychopathology: Relationship with Self-Reassurance and Self-Criticism in Czech University Students. *Int. J. Ment. Health Addict.* **2022**, *20*, 2321–2332. [CrossRef]
10. Pedrelli, P.; Nyer, M.; Yeung, A.; Zulauf, C.; Wilens, T. College Students: Mental Health Problems and Treatment Considerations. *Acad. Psychiatry* **2015**, *39*, 503–511. [CrossRef]
11. Trnka, R.; Lorencova, R. Fear, Anger, and Media-Induced Trauma During the Outbreak of COVID-19 in the Czech Republic. *Psychol. Trauma: Theory Res. Pract. Policy* **2020**, *12*, 546–549. [CrossRef] [PubMed]
12. Dosedlová, J.; Klimusová, H.; Burešová, I. Health-related Behavior over the Course of Life in the Czech Republic. *Procedia—Soc. Behav. Sci.* **2016**, *217*, 1167–1175. [CrossRef]
13. Ryan, R.M.; Deci, E.L. Self-determination theory and the facilitation of intrinsic motivation, social development, and well-being. *Am. Psychol.* **2000**, *55*, 68–78. [CrossRef]

14. Martela, F.; Riekki, T.J.J. Autonomy, Competence, Relatedness, and Beneficence: A Multicultural Comparison of the Four Pathways to Meaningful Work. *Front. Psychol.* **2018**, *9*, 1157. [CrossRef] [PubMed]
15. Bailey, T.H.; Phillips, L.J. The influence of motivation and adaptation on students' subjective well-being, meaning in life and academic performance. *High. Educ. Res. Dev.* **2016**, *35*, 201–216. [CrossRef]
16. Legault, L.; Green-Demers, I.; Pelletier, L. Why do high school students lack motivation in the classroom? Toward an understanding of academic amotivation and the role of social support. *J. Educ. Psychol.* **2006**, *98*, 567. [CrossRef]
17. Henrich, J. *The Weirdest People in the World: How the West Became Psychologically Peculiar and Particularly Prosperous*; Penguin Random House: Dublin, Ireland, 2020; p. 680.
18. Liu, X.; Flick, R. The relationship among psychological need satisfaction, class engagement, and academic performance: Evidence from China. *J. Educ. Bus.* **2019**, *94*, 408–417. [CrossRef]
19. Deci, E.L.; Ryan, R.M.; Gagné, M.; Leone, D.R.; Usunov, J.; Kornazheva, B.P. Need Satisfaction, Motivation, and Well-Being in the Work Organizations of a Former Eastern Bloc Country: A Cross-Cultural Study of Self-Determination. *Personal. Soc. Psychol. Bull.* **2001**, *27*, 930–942. [CrossRef]
20. Kotera, Y.; Ting, S.-H.; Neary, S. Mental health of Malaysian university students: UK comparison, and relationship between negative mental health attitudes, self-compassion, and resilience. *High. Educ.* **2021**, *81*, 403–419. [CrossRef]
21. Nalipay, M.J.N.; King, R.B.; Cai, Y. Autonomy is equally important across East and West: Testing the cross-cultural universality of self-determination theory. *J. Adolesc.* **2020**, *78*, 67–72. [CrossRef]
22. Muthukrishna, M.; Bell, A.V.; Henrich, J.; Curtin, C.M.; Gedranovich, A.; McInerney, J.; Thue, B. Beyond Western, Educated, Industrial, Rich, and Democratic (WEIRD) Psychology: Measuring and Mapping Scales of Cultural and Psychological Distance. *Psychol. Sci.* **2020**, *31*, 678–701. [CrossRef] [PubMed]
23. Zusho, A.; Clayton, K. Culturalizing Achievement Goal Theory and Research. *Educ. Psychol.* **2011**, *46*, 239–260. [CrossRef]
24. Tóth-Király, I.; Neff, K.D. Is Self-Compassion Universal? Support for the Measurement Invariance of the Self-Compassion Scale Across Populations. *Assessment* **2021**, *28*, 169–185. [CrossRef] [PubMed]
25. Zessin, U.; Dickhäuser, O.; Garbade, S. The Relationship Between Self-Compassion and Well-Being: A Meta-Analysis. *Appl. Psychol. Health Well-Being* **2015**, *7*, 340–364. [CrossRef] [PubMed]
26. Kirby, J.N.; Tellegen, C.L.; Steindl, S.R. A Meta-Analysis of Compassion-Based Interventions: Current State of Knowledge and Future Directions. *Behav. Ther.* **2017**, *48*, 778–792. [CrossRef]
27. Hermanto, N.; Zuroff, D.C.; Kopala-Sibley, D.C.; Kelly, A.C.; Matos, M.; Gilbert, P.; Koestner, R. Ability to receive compassion from others buffers the depressogenic effect of self-criticism: A cross-cultural multi-study analysis. *Personal. Individ. Differ.* **2016**, *98*, 324–332. [CrossRef]
28. McIntyre, R.; Smith, P.; Rimes, K.A. The role of self-criticism in common mental health difficulties in students: A systematic review of prospective studies. *Ment. Health Prev.* **2018**, *10*, 13–27. [CrossRef]
29. Gilbert, P.; Simos, G. *Compassion Focused Therapy: Clinical Practice and Applications*; Routledge: London, UK, 2022.
30. Gilbert, P. *The Compassionate Mind: A New Approach to Life's Challenges*; New Harbinger Publications: Oakland, CA, USA, 2010; 592p.
31. Gilbert, P.; Clarke, M.; Hempel, S.; Miles, J.N.V.; Irons, C. Criticizing and reassuring oneself: An exploration of forms, styles and reasons in female students. *Br. J. Clin. Psychol.* **2004**, *43*, 31–50. [CrossRef]
32. Kotera, Y.; Conway, E.; Van Gordon, W. Ethical Judgement in UK Business Students: Relationship with Motivation, Self-Compassion and Mental Health. *Int. J. Ment. Health Addict.* **2019**, *17*, 1132–1146. [CrossRef]
33. Kotera, Y.; Green, P.; Sheffield, D. Mental health of therapeutic students: Relationships with attitudes, self-criticism, self-compassion, and caregiver identity. *Br. J. Guid. Couns.* **2019**, *49*, 701–712. [CrossRef]
34. Kotera, Y.; Maughan, G. Mental health of Irish students: Self-criticism as a complete mediator in mental health attitudes and caregiver identity. *J. Concurr. Disord.* **2020**, *1*, 14–26. [CrossRef]
35. Kotera, Y.; Gilbert, P.; Asano, K.; Ishimura, I.; Sheffield, D. Self-criticism and self-reassurance as mediators between mental health attitudes and symptoms: Attitudes toward mental health problems in Japanese workers. *Asian J. Soc. Psychol.* **2019**, *22*, 183–192. [CrossRef]
36. Shahar, G.; Henrich, C.C.; Blatt, S.J.; Ryan, R.; Little, T.D. Interpersonal relatedness, self-definition, and their motivational orientation during adolescence: A theorical and empirical integration. *Dev. Psychol.* **2003**, *39*, 470–483. [CrossRef] [PubMed]
37. Shahar, G.; Kalnitzki, E.; Shulman, S.; Blatt, S.J. Personality, motivation, and the construction of goals during the transition to adulthood. *Personal. Individ. Differ.* **2006**, *40*, 53–63. [CrossRef]
38. Zuroff, D.C.; Koestner, R.; Moskowitz, D.S.; McBride, C.; Bagby, R.M. Therapist's Autonomy Support and Patient's Self-Criticism Predict Motivation during Brief Treatments for Depression. *J. Soc. Clin. Psychol.* **2012**, *31*, 903–932. [CrossRef]
39. Powers, T.A.; Koestner, R.; Zuroff, D.C. Self–Criticism, Goal Motivation, and Goal Progress. *J. Soc. Clin. Psychol.* **2007**, *26*, 826–840. [CrossRef]
40. Gilbert, P. The origins and nature of compassion focused therapy. *Br. J. Clin. Psychol.* **2014**, *53*, 6–41. [CrossRef]
41. Kotera, Y. De-stigmatising self-care: Impact of self-care webinar during COVID-19. *Int. J. Spa Wellness* **2021**, *4*, 213–217. [CrossRef]
42. Irons, C.; Gilbert, P.; Baldwin, M.W.; Baccus, J.R.; Palmer, M. Parental recall, attachment relating and self-attacking/self-reassurance: Their relationship with depression. *Br. J. Clin. Psychol.* **2006**, *45*, 297–308. [CrossRef]

43. Leak, G.K.; Cooney, R.R. Self-determination, attachment styles, and well-being in adult romantic relationships. *Represent. Res. Soc. Psychol.* **2001**, *25*, 55–62.
44. Kotera, Y.; Ting, S.H. Positive Psychology of Malaysian University Students: Impacts of Engagement, Motivation, Self-Compassion, and Well-being on Mental Health. *Int. J. Ment. Health Addict.* **2019**, *20*, 1611–1626. [CrossRef] [PubMed]
45. Kotera, Y.; Tsuda-McCaie, F.; Maughan, G.; Green, P. Cross-cultural comparison of mental health in social work students between UK and Ireland: Mental health shame and self-compassion. *Br. J. Soc. Work.* **2021**, *52*, 3247–3267. [CrossRef]
46. Kotera, Y.; Lieu, J.; Kirkman, A.; Barnes, K.; Liu, G.H.T.; Jackson, J.; Wilkes, J.; Riswani, R. Mental Wellbeing of Indonesian Students: Mean Comparison with UK Students and Relationships with Self-Compassion and Academic Engagement. *Healthcare* **2022**, *10*, 1439. [CrossRef]
47. Kotera, Y.; Taylor, E.; Fido, D.; Williams, D.; Tsuda-Mccaie, F. Motivation of UK graduate students in education: Self-compassion moderates pathway from extrinsic motivation to intrinsic motivation. *Curr. Psychol.* **2021**. [CrossRef] [PubMed]
48. Kotera, Y.; Young, H.; Maybury, S.; Aledeh, M. Mediation of Self-Compassion on Pathways from Stress to Psychopathologies among Japanese Workers. *Int. J. Environ. Res. Public Health* **2022**, *19*, 12423. [CrossRef] [PubMed]
49. Dundas, I.; Binder, P.E.; Hansen, T.G.B.; Stige, S.H. Does a short self-compassion intervention for students increase healthy self-regulation? A randomized control trial. *Scand. J. Psychol.* **2017**, *58*, 443–450. [CrossRef]
50. Czech Ministry of Education. *University Students by Groups Of Fields and Sex in 2018 [Studenti Vysokých Škol Podle Skupin Oborů a Pohlaví v Roce 2018]*; Czech Ministry of Education: Prague, Czech Republic, 2019.
51. Cohen, J. *Statistical Power Analysis for the Behavioral Sciences*; L. Erlbaum Associates: Mahwah, NJ, USA, 1988; 567p.
52. Faul, F.; Erdfelder, E.; Buchner, A.; Lang, A.-G. Statistical power analyses using G*Power 3.1: Tests for correlation and regression analyses. *Behav. Res. Methods* **2009**, *41*, 1149–1160. [CrossRef]
53. Vallerand, R.J.; Pelletier, L.G.; Blais, M.R.; Briere, N.M.; Senecal, C.; Vallieres, E.F. The academic motivation scale: A measure of intrinsic, extrinsic, and amotivation in education. *Educ. Psychol. Meas.* **1992**, *52*, 1003–1017. [CrossRef]
54. Stewart-Brown, S.; Tennant, A.; Tennant, R.; Platt, S.; Parkinson, J.; Weich, S. Internal construct validity of the Warwick-Edinburgh Mental Well-being Scale (WEMWBS): A Rasch analysis using data from the Scottish Health Education Population Survey. *Health Qual. Life Outcomes* **2009**, *7*, 15. [CrossRef]
55. Stewart-Brown, S.; Janmohamed, K. *Warwick-Edinburgh Mental Well-Being Scale (WEMWBS) User Guide Version 1*; Warwick Medical School: Coventry, UK, 2008.
56. Kotera, Y.; Green, P.; Sheffield, D. Positive Psychology for Mental Wellbeing of UK Therapeutic Students: Relationships with Engagement, Motivation, Resilience and Self-Compassion. *Int. J. Ment. Health Addict.* **2021**, *20*, 337–361. [CrossRef]
57. Hayes, A.F. *Introduction to Mediation, Moderation, and Conditional Process Analysis: A Regression-Based Approach*; Guilford Press: New York, NY, USA, 2013; 507p.
58. Kotera, Y.; Mayer, C.H.; Vanderheiden, E. Cross-Cultural Comparison of Mental Health Between German and South African Employees: Shame, Self-Compassion, Work Engagement, and Work Motivation. *Front. Psychol.* **2021**, *12*, 627851. [CrossRef] [PubMed]
59. Ehret, A.M.; Joormann, J.; Berking, M. Examining risk and resilience factors for depression: The role of self-criticism and self-compassion. *Cogn. Emot.* **2015**, *29*, 1496–1504. [CrossRef] [PubMed]
60. Neely, M.E.; Schallert, D.L.; Mohammed, S.S.; Roberts, R.M.; Chen, Y.-J. Self-kindness when facing stress: The role of self-compassion, goal regulation, and support in college students' well-being. *Motiv. Emot.* **2009**, *33*, 88–97. [CrossRef]
61. Kotera, Y.; Van Gordon, W. Effects of Self-Compassion Training on Work-Related Well-Being: A Systematic Review. *Front. Psychol.* **2021**, *12*, 630798. [CrossRef] [PubMed]
62. Neff, K.D.; Germer, C.K. A Pilot Study and Randomized Controlled Trial of the Mindful Self-Compassion Program. *J. Clin. Psychol.* **2013**, *69*, 28–44. [CrossRef]
63. Shapira, L.B.; Mongrain, M. The benefits of self-compassion and optimism exercises for individuals vulnerable to depression. *J. Posit. Psychol.* **2010**, *5*, 377–389. [CrossRef]
64. Smeets, E.; Neff, K.; Alberts, H.; Peters, M. Meeting suffering with kindness: Effects of a brief self-compassion intervention for female college students. *J. Clin. Psychol* **2014**, *70*, 794–807. [CrossRef]
65. Christopher, J.C.; Maris, J.A. Integrating mindfulness as self-care into counselling and psychotherapy training. *Couns. Psychother. Res.* **2010**, *10*, 114–125. [CrossRef]
66. Kotera, Y.; Ozaki, A.; Miyatake, H.; Tsunetoshi, C.; Nishikawa, Y.; Kosaka, M.; Tanimoto, T. Qualitative Investigation into the Mental Health of Healthcare Workers in Japan during the COVID-19 Pandemic. *Int. J. Environ. Res. Public Health* **2022**, *19*, 568. [CrossRef]
67. Kotera, Y.; Cockerill, V.; Chircop, J.G.E.; Forman, D. Mental health shame, self-compassion and sleep in UK nursing students: Complete mediation of self-compassion in sleep and mental health. *Nurs. Open* **2020**, *8*, 1325–1335. [CrossRef]
68. Kotera, Y.; Green, P.; Sheffield, D. Mental Health Shame of UK Construction Workers: Relationship with Masculinity, Work Motivation, and Self-Compassion. *J. Work Organ. Psychol.* **2019**, *35*, 135–143. [CrossRef]
69. Kotera, Y.; Conway, E.; Van Gordon, W. Mental health of UK university business students: Relationship with shame, motivation and self-compassion. *J. Educ. Bus.* **2018**, *94*, 11–20. [CrossRef]
70. Kotera, Y.; Rhodes, C. Pathways to Sex Addiction: Relationships with Adverse Childhood Experience, Attachment, Narcissism, Self-Compassion and Motivation in a Gender-Balanced Sample. *Sex. Addict. Compulsivity* **2019**, *26*, 54–76. [CrossRef]

71. Kotera, Y.; Asano, K.; Kotera, H.; Ohshima, R.; Rushforth, A. Mental Health of Japanese Workers: Amotivation Mediates Self-Compassion on Mental Health Problems. *Int. J. Environ. Res. Public Health* **2022**, *19*, 10497. [CrossRef]
72. Clarke, F.J.; Kotera, Y.; McEwan, K. A Qualitative Study Comparing Mindfulness and Shinrin-Yoku (Forest Bathing): Practitioners' Perspectives. *Sustainability* **2021**, *13*, 6761. [CrossRef]
73. Kotera, Y.; Sweet, M. Comparative evaluation of neuro-linguistic programming. *Br. J. Guid. Couns.* **2019**, *47*, 744–756. [CrossRef]
74. Kotera, Y.; Fido, D. Effects of Shinrin-Yoku Retreat on Mental Health: A Pilot Study in Fukushima, Japan. *Int. J. Ment. Health Addict.* **2022**, *20*, 2652–2664. [CrossRef]
75. McEwan, K.; Giles, D.; Clarke, F.J.; Kotera, Y.; Evans, G.; Terebenina, O.; Minou, L.; Teeling, C.; Basran, J.; Wood, W.; et al. A Pragmatic Controlled Trial of Forest Bathing Compared with Compassionate Mind Training in the UK: Impacts on Self-Reported Wellbeing and Heart Rate Variability. *Sustainability* **2021**, *13*, 1380. [CrossRef]
76. Deci, E.L.; Flaste, R. *Why We Do What We Do: Understanding Self-Motivation*; Penguins Books: London, UK, 1996; p. 230.
77. Shahar, G.; Soffer, N.; Gilboa-Shechtman, E. Sociotropy, autonomy, and self-criticism are three distinguishable dimensions of cognitive-personality vulnerability. *J. Cogn. Psychother.* **2008**, *22*, 219–227. [CrossRef]
78. Gilbert, P.; Procter, S. Compassionate Mind Training for People with High Shame and Self-Criticism: Overview and Pilot Study of a Group Therapy Approach. *Clin. Psychol. Psychother. Clin. Psychol. Psychother* **2006**, *13*, 353–379. [CrossRef]
79. Kotera, Y.; Conway, E.; Green, P. Construction And factorial validation of a short version of the Academic Motivation Scale. *Br. J. Guid. Couns.* **2021**, 1–10. [CrossRef]
80. Sommers-Spijkerman, M.; Trompetter, H.; ten Klooster, P.; Schreurs, K.; Gilbert, P.; Bohlmeijer, E. Development and validation of the forms of Self-Criticizing/Attacking and Self-Reassuring Scale—Short Form. *Psychol. Assess.* **2018**, *30*, 729–743. [CrossRef]

healthcare

MDPI

Article

Mental Wellbeing of Indonesian Students: Mean Comparison with UK Students and Relationships with Self-Compassion and Academic Engagement

Yasuhiro Kotera [1], Jenai Lieu [2], Ann Kirkman [2], Kristian Barnes [2], Gillian H. T. Liu [2], Jessica Jackson [2], Juliet Wilkes [2,*] and Riswani Riswani [3]

[1] School of Health Sciences, University of Nottingham, Nottingham NG7 2HA, UK; yasuhiro.kotera@nottingham.ac.uk

[2] College of Health, Psychology and Social Care, University of Derby, Derby DE22 1GB, UK; jenai.lieu@gmail.com (J.L.); a.kirkman2@derby.ac.uk (A.K.); kmjbarnes@gmail.com (K.B.); gillianlht@gmail.com (G.H.T.L.); j.jackson2@derby.ac.uk (J.J.)

[3] Faculty of Education and Teacher Training, State Islamic University of Sultan Syarif Kasim Riau, Pekanbaru 28293, Indonesia; riswani@uin-suska.ac.id

* Correspondence: julietlorettawilkes@gmail.com

Abstract: The number of Indonesian students in higher education has been increasing rapidly. However, many Indonesian university students report experiencing mental distress. Research on student wellbeing found that self-compassion (i.e., being kind towards oneself in challenging times) and academic engagement (i.e., a commitment and purposeful effort toward study) are essential to students' mental wellbeing. With the present study, we aimed to assess the mental wellbeing of Indonesian students. A convenience sample of 156 Indonesian students completed self-report measures on mental wellbeing, self-compassion and academic engagement (consisting of vigour, dedication and absorption). To contextualise their mental wellbeing, data from Indonesian students were compared with those from 145 UK students using Welch *t*-tests. Correlation, regression and mediation analyses were then performed to examine the relationships among these variables. Indonesian students had higher levels of self-compassion and absorption and a lower level of dedication than UK students. Self-compassion and academic engagement explained 36% of variance in mental wellbeing. Self-compassion and vigour were identified as significant predictors of mental wellbeing, whereas self-compassion was the strongest predictor. Lastly, vigour did not mediate the pathway from self-compassion to mental wellbeing. Interventions to support self-compassion and academic engagement, especially vigour, in Indonesian students are recommended to support their mental wellbeing.

Keywords: Indonesian students; mental wellbeing; self-compassion; academic engagement; vigour; dedication

Citation: Kotera, Y.; Lieu, J.; Kirkman, A.; Barnes, K.; Liu, G.H.T.; Jackson, J.; Wilkes, J.; Riswani, R. Mental Wellbeing of Indonesian Students: Mean Comparison with UK Students and Relationships with Self-Compassion and Academic Engagement. *Healthcare* **2022**, *10*, 1439. https://doi.org/10.3390/healthcare10081439

Academic Editors: Daniele Giansanti and John H. Foster

Received: 11 June 2022
Accepted: 26 July 2022
Published: 1 August 2022

1. Introduction

1.1. Rapid Growth of Indonesian Higher Education

Indonesia is the fourth most populated country in the world, with a population of approximately 276 million people [1]. Indonesia is viewed as having a promising economic future and is predicted to become the world's fourth largest economy by 2050 [2]. This rapid economic growth is positive; however, there are still socioeconomic problems within the country, with 27 million Indonesian people (more than 10% of the population) living on less than USD 0.75 a day [3]. Research has suggested that to allow Indonesia to experience full economic potential, the government needs to invest in key sectors, including education [4].

The Indonesian government has started to change their education policies to support its growing economy [5]. As a result, the number of Indonesian people attending university has grown rapidly. Decentralisation and increased government funding have supported a growth of 10 million Indonesian students since 2002 [6]. In 2016, 1.47 million students entered higher education (HE), almost doubling to 2.99 million by 2020 [7]. Additionally, the HE academic publication output has increased tenfold since 1996, with a 15% average annual growth (from 538 in 1996 to 5499 in 2014) [8]. The relationship between decentralisation and education development continues to be mutually reinforcing [9]. Investments in HE by the Indonesian government will ultimately contribute to increased work productivity, increased income and wealth and overall wellbeing in the general population [10].

1.2. Compromised Mental Wellbeing in Indonesian Students

Alongside a rapid increase in the number of students and scholarly output in Indonesian HE [11], the number of students claiming poor mental wellbeing has also increased. Consistent with findings in other countries [12–14], the mental wellbeing of Indonesian university students is compromised [15]. About a third to a half (37–53%) of undergraduate students in Indonesia reported high-stress symptoms [16], 25% experienced depression and 51% experienced anxiety [17]. In addition, high levels of loneliness have been reported [18]. Whereas the serious nature of depleted mental wellbeing among Indonesian university students has been clearly reported [15,17,19], information about the high financial cost of treatment, as well as the stigma surrounding mental health, support and treatment, is not readily available [20]. A lack of university mental health services and underutilisation of existing services is also common in non-Western countries [21], including Indonesia [22]. Overall, improved understanding of student mental wellbeing is needed in Indonesian HE.

1.3. Self-Compassion

Self-compassion is associated with improved mental wellbeing [23]. It has been found to help students actualise adaptive achievement goals by allowing them to embrace their shortcomings and prioritise productive learning strategies [24]. Self-compassion is rooted in Buddhist teachings, which include a willingness to face and alleviate suffering [23]. Self-compassion has three elements that affect one's wellbeing: treating oneself kindly during times of difficulty, recognising hardship and failure as a common human experience and using mindfulness to maintain a balanced awareness of painful thoughts and feelings [25,26]. Self-compassion acts as a mediator between positive psychological constructs and wellbeing [27] by enhancing resilience through the understanding and acceptance of one's inadequacies [28]. A systematic review demonstrated that self-compassion is connected to a decrease in stress and depressive symptoms [29], as well as an increase in happiness and life satisfaction [30]. Self-compassion helps regulate emotions through the reduction of self-criticism, rumination and avoidance of painful experiences [25] and reduces the negative effects of stress on wellbeing [31–33]. Self-compassion is therefore important for mental wellbeing.

1.4. Academic Engagement

Academic engagement is associated with healthy mental wellbeing in higher education [34]. It refers to the commitment and purposeful effort by students in regard to their learning and other scholastic activities, which is measured by its components of vigour, dedication and absorption [35,36]. Vigour is defined as the willingness to persevere through difficulties represented as energy and mental agility; dedication is defined as investment in one's work, resulting in inspiration and enthusiasm; and absorption is defined as the sense of immersion in one's work [37]. These components of engagement can positively impact students' cognitive appraisals of their academic experience through resource building and implementation [36], thus influencing their probability of increased motivation and academic performance [38]. Academic engagement in postsecondary education has been considered a determining factor for students' achievement and personal development [39]. A longitudinal survey of undergraduate students based in the UK showed that engagement and wellbeing were positively related, also suggesting a feedback loop of increased engagement, increased academic performance and increased wellbeing, leading again to increased engagement [40]. Overall, engagement leads to increased satisfaction in postsecondary learning, leading to improved wellbeing.

1.5. The Effect of Academic Engagement on Self-Compassion and Mental Wellbeing

Academic engagement and self-compassion are both positively associated with mental wellbeing [32,40] and share similarities despite little empirical evidence comparing the two. Research revealed that students who are passionate and engaged in their academic studies and who exhibit resilience toward challenges and compassion toward themselves have a high level of wellbeing [41]. Self-compassion is associated with student engagement and the ability to adapt achievement goals, which allows students to confront failure and adopt healthy and productive learning ethics [24]. Academic engagement and self-compassion both facilitate mental wellbeing through association with increased resilience and personal development [24,39,42]. Students who practice self-compassion and who are academically engaged are more resourceful when managing their goals and motivation [36], which promotes a positive effect on their wellbeing. This suggests that engagement and self-compassion share similarities in terms of facilitating positive mental wellbeing by enhancing students' ability to navigate difficult circumstance through positive cognitive appraisal [36]. Therefore, in the present study, we evaluate whether academic engagement could mediate the pathway from self-compassion to positive mental wellbeing.

1.6. Comparison with UK Students

In this study, the mean scores of mental wellbeing among Indonesian students were compared with those of UK students in order to (a) contextualise mental wellbeing in Indonesian students and (b) discuss how cultures may influence mental wellbeing. When comparing cross-cultural differences between postsecondary students in the UK and Eastern countries, such as Indonesia, the UK scored significantly high in individualism [43], relatively high in masculinity [43] and low for power distance in the school system [44]. Regarding Hofstede's cultural dimensions theory [43], the cultural dimension of individualism–collectivism involves societal orientations, masculinity–femininity identifies societal views and power distance explores unequal distributions of power between individuals. Hereafter, we will use the words 'success-driven' and 'quality-oriented' instead of 'masculine' and 'feminine' for accuracy and clarity [45]. UK culture is highly individualistic, and it emphasises personal achievement and independence [43], compared to collectivist cultures that emphasise group achievement and interdependence [46]. UK culture scored relatively high for the dimension of success-drive, indicating the society's tendency to prioritise accomplishment and achievement over caring for others and maintaining a high quality of life [43]. Lastly, UK culture scored low in power distance, suggesting that there is a value in equal treatment among individuals [43]. Low power distance in the school system is demonstrated through learner-centred environments with increased interaction between students and teachers, compared to the high levels of power distance common in Eastern culture school systems [44]. By comparing the mental wellbeing levels between these culturally contrasting groups, we aimed to inform how cultures may explain the level of difference in mental wellbeing. Culturally aware approaches to wellbeing are needed in modern higher education [47].

1.7. Study Aims

The aim of this study was to evaluate the mental wellbeing of Indonesian students, along with self-compassion and academic engagement, namely vigour, dedication and absorption. First, the levels of these five constructs (mental wellbeing, self-compassion, vigour, dedication and absorption) were assessed by comparing them with UK university students (Aim 1). Second, the extent of variance in mental wellbeing could be predicted by self-compassion, vigour, dedication and absorption, and significant predictors of mental wellbeing were identified (Aim 2). Lastly, we examined how engagement mediates the pathway from self-compassion to mental wellbeing (Aim 3).

2. Materials and Methods

2.1. Participants

Students >18 years studying a caring profession subject (counselling and education) full-time at a university in Indonesia were eligible for this study. Caring profession subjects relate to occupations in which humans take care of other humans [48], including the allied health professions, counselling and education [49]. All study materials were provided as hard copies, prepared in English and disseminated by an independent tutor to 162 students (October–December 2021), meaning the study had a 96% (n = 156) response rate. The majority of respondents were female (n = 128), with 25 males and 3 students who did not disclose their gender. Participants ranged in age from 18 to 22 years (19.07 ± 0.98). Five students were postgraduate students. Our sample was similar in age to the general Indonesian student population, but was more female-dominated (age: 20 years old, 49% female [50]). The study received ethical approval from the co-author, R.R.'s, university (State Islamic University of Sultan Syarif Kasim Riau: Ref KE/KEP-FPP/01/05/2022) and involved no deception or financial incentives.

Students were provided with supporting information about mental health during the study. Data were compared with data obtained in a study involving students studying a caring profession subject in a UK university (n = 145; entire sample), using the same participation criteria; the original studies were published previously [41,51] (University of Derby: Ref 011017YK). As with the Indonesian sample, a paper survey was disseminated by an independent tutor. No participation incentive was offered. Data were collected between April and May 2018. In this UK study, the majority of students were also female (n = 130), with 15 male students. They had a wider age range of 17–52 years (26.80 ± 8.64), with 133 undergraduates and 12 postgraduates. The respondent age range in the UK study reflects the wider trend that caring professions attract both young and mature students [52]. The younger age range of the Indonesian respondents reflects the wider student population of the country. The predominance of female participants in both study samples is consistent with the global trend within health and social care, which consists of a largely female workforce [53]. Table 1 summarises the demographic information of both samples.

Table 1. Demographic data for 156 Indonesian students and 144 UK students.

		Indonesian Students		UK Students	
Age		Years		Years	
	Mean	19.07		26.8	
	SD	0.98		8.64	
	Range	18–22		17–52	
Gender		n	%	n	%
	Female	128	82.05%	130	89.66%
	Male	25	16.03%	15	10.34%
	Unknown	3	1.92%	0	0.00%
Level of Study		N	%	N	%
	Undergraduate	151	96.79%	133	91.72%
	Postgraduate	5	3.21%	12	8.28%

2.2. Materials

The materials consisted of three validated scales to measure students' mental well-being, self-compassion and engagement in academia. The original English version of all three scales had not been previously used in Indonesian samples. Mental welling was measured using the established seven-item Short Warwick–Edinburgh Mental Wellbeing Scale (SWEMWBS; [54]). This scale was selected for its holistic evaluation of mental wellbeing and high internal consistency [55]. Respondents were asked to make both hedonic and eudemonic reflections on the past two weeks on a five-point Likert scale (e.g., 'I've been dealing with problems well'; 1 = 'None of the time' to 5 = 'All of the time'). The internal consistency of the SWEMWBS is high ($\alpha = 0.85$; [56]).

Self-compassion was measured using the 12-item Self-Compassion Scale–Short Form because of its high internal consistency (SCS-SF; [57]). Scale items are measured on a five-point Likert scale and include questions such as, 'I try to be understanding and patient towards those aspects of my personality I don't like' and measures whether, in difficult situations, a respondent is consistently kind to themselves (0 = 'Almost never' to 5 = 'Almost always'). SCS-SF demonstrates high internal consistency ($\alpha = 0.86$ [57]).

Engagement was measured to consider how confident and active students are in their academic workload, using the 17-item Utrecht Work Engagement Scale (UWES-S) because the scale (global score) and subscales demonstrate a good-to-high internal consistency (α = 0.63–0.81; [58]). The scale consists of three subscales—vigour, dedication and absorption—and each item is measured on a seven-point Likert scale. Vigour relates to the student's mental capacity, which leads to substantial effort in academia, for example, 'I am very resilient, mentally, as far as my studies are concerned'. Dedication relates to the student's commitment to academia, for example, 'My study inspires me'. Absorption relates to the student's immersion in academia, for example, 'When I am studying, I forget everything else around me' [37].

2.3. Analysis

Data were screened and tested for assumptions and outliers with parametric tests, correlations and a multiple regression model using SPSS (v25) and Process Macro version 3 [59]. An additional path analysis was performed in order to identify whether an engagement component could mediate the pathway from self-compassion to mental wellbeing.

2.4. Results

2.4.1. Levels of Mental Wellbeing, Self-Compassion and Academic Engagement (Aim 1)

An independent-samples t-test was conducted to compare the levels of mental wellbeing, self-compassion and academic engagement—namely, vigour, dedication and absorption—between Indonesian and UK students (Table 2). According to Levene's test for equality of variances, the assumption of homogeneity of variances was not maintained for self-compassion ($p < 0.001$) and absorption ($p < 0.001$); thus, Welch t-tests were used.

Indonesian students had higher levels of self-compassion (mean difference, −0.45; CI 95% [−0.59, −0.32], $t(261.07) = -6.55$, $p < 0.001$, $d = -0.77$) and absorption (mean difference, −0.66, CI 95% [−0.92, −0.39], $t(293.19) = -4.81$, $p < 0.001$, $d = 0.93$) and a lower level of dedication (mean difference, 0.88, CI 95% [0.67, 1.10], $t(293.19) = 8.06$, $p < 0.001$, $d = -0.56$) than UK students. There were no significant differences in mental wellbeing ($p = 0.17$) and vigour ($p = 0.28$).

Table 2. Comparing the levels of mental wellbeing, self-compassion and academic engagement (vigour, dedication and absorption) between 156 Indonesian students and 144 UK students.

	Indonesian Students			**UK Students**							
Variable	M	SD	α	M	SD	α	t	MD	CI 95% [L, U]		d
Mental Wellbeing	23.43	4.30	0.72	23.94	4.86	0.85	0.96	0.51	−0.54	1.55	0.11
Self-Compassion ***	3.32	0.50	0.67	2.87	0.67	0.84	−6.55	−0.45	−0.59	−0.32	−0.77
Vigour	3.63	0.94	0.72	3.70	1.08	0.76	0.60	0.07	−0.16	0.30	0.07
Dedication ***	3.87	0.93	0.60	4.75	0.97	0.63	8.06	0.88	0.67	1.10	0.93
Absorption ***	3.78	0.97	0.79	3.13	1.35	0.80	−4.81	−0.66	−0.92	−0.39	−0.56

*** $p < 0.001$ α = Cronbach's alpha significant difference between the two groups according to Welch t-tests.

2.4.2. Prediction of Mental Wellbeing (Aim 2)

Multiple regression analyses were conducted to identify predictors of mental wellbeing. After adjusting for age and gender (Step 1), self-compassion, vigour, dedication and absorption were entered as predictor variables (Step 2), and mental wellbeing was entered as an outcome variable (Table 3). Multicollinearity was of no concern (variance inflation factors < 10).

Table 3. Multiple regression: self-compassion, vigour, dedication and absorption for mental wellbeing in 156 Indonesian students.

	Outcome: Mental Wellbeing				
	B	**SE_B**	**b**	**95% CI [L, U]**	
Step 1					
Gender	−1.12	0.86	−0.11	−2.82	0.58
Age	−0.34	0.36	−0.08	−1.05	0.37
Adj. R^2			0.2%		
Step 2					
Gender	−1.30	0.69	−0.12	−2.67	0.06
Age	−0.38	0.29	−0.09	−0.96	0.20
Self-Compassion ***	2.14	0.59	0.25	0.98	3.30
Vigour *	1.23	0.54	0.27	0.17	2.29
Dedication	0.76	0.49	0.16	−0.22	1.73
Absorption	0.63	0.62	0.14	−0.60	1.86
Δ Adj. R^2			36%		

*** $p < 0.001$, * $p < 0.05$.

These four predictor variables accounted for 36% of mental wellbeing, indicating a large effect size [60]. Self-compassion and vigour were significant positive predictors of mental wellbeing, whereas self-compassion was the strongest predictor of mental wellbeing.

2.4.3. Mediation of the Self-Compassion–Mental Wellbeing Pathway (Aim 3)

Lastly, to evaluate whether academic engagement mediates the pathway from self-compassion to mental wellbeing, path analyses were conducted using model 4 in the Process macro (parallel mediation model; [59]) (see Figure 1). Among the three subscales of academic engagement, vigour was used, as it was found to be a significant predictor of mental wellbeing.

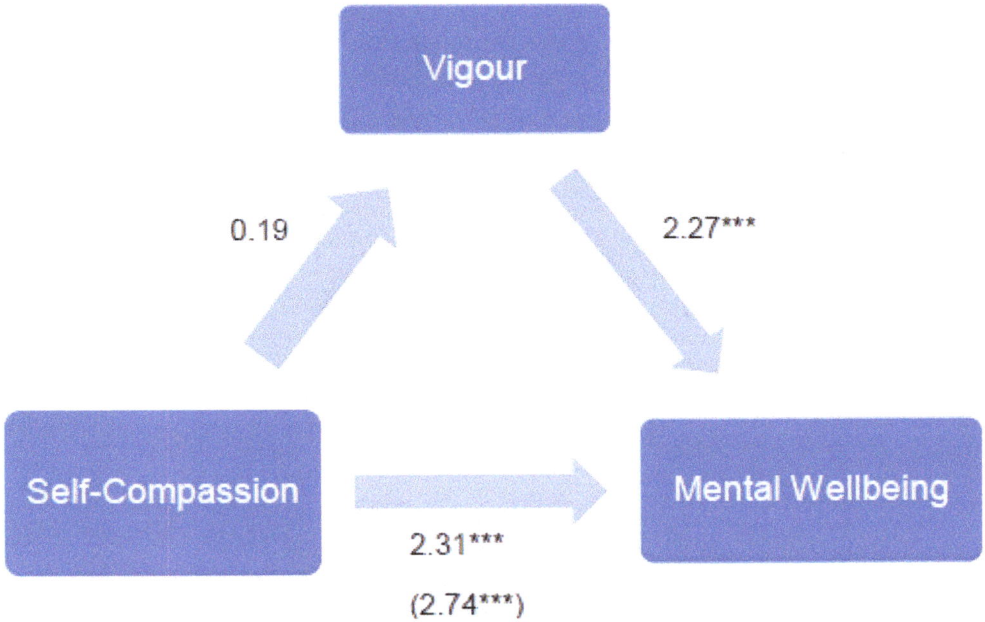

Figure 1. Evaluating whether vigour mediates the pathway from self-compassion to mental wellbeing. Parallel mediation model: self-compassion as a predictor of mental wellbeing, mediated by vigour. The confidence interval for the indirect effect is a BCa bootstrapped CI based on 1000 samples. *** $p < 0.001$; direct effect (total effect).

Vigour did not mediate the pathway from self-compassion to mental wellbeing, as the path from self-compassion to vigour was not significant ($b = 0.19$, $t(154) = 1.23$, $p = 0.22$). However, the total effect of self-compassion on mental wellbeing, including vigour, was significant ($b = 2.74$, $t(154) = 4.13$, $p < 0.001$). The direct effect of self-compassion on mental wellbeing, controlling for vigour, was also significant ($b = 2.31$, $t(153) = 4.05$, $p < 0.001$).

3. Discussion

This study assessed the mental wellbeing of Indonesian students. Indonesian students had higher levels of self-compassion and absorption and a lower level of dedication than UK students. Self-compassion and academic engagement explained 36% of variance in mental wellbeing, and self-compassion and vigour were identified as predictors of mental wellbeing, where self-compassion was the strongest predictor. Lastly, vigour did not mediate the pathway from self-compassion to mental wellbeing.

The significant difference in the levels of self-compassion between Indonesian and UK students observed in this study may be explained by cultural differences [61,62]. The higher self-compassion in Indonesian students can be attributed to their lower scores in Hofstede's cultural dimensions of individualism and success-drive [63]. Montero-Marin et al. [62] reported that individualism and success-drive have a moderate effect on negative items of self-compassion. More individualistic and success-driven cultural values create a more competitive system, evoking social comparison [64]. Because people from these cultural backgrounds are more self-referent to success, they try to avoid failures and are therefore more self-critical [62]. Indeed, the UK is highly individualistic and success-driven, and society emphasises personal success and achievement [43]. This characteristic of the UK's cultural value is aligned with the result of having a lower self-compassion score than that of Indonesian students. Moreover, it is noteworthy that despite displaying higher levels of self-compassion than UK students, the mental wellbeing of Indonesian students was similar to that of UK students. Previous research suggested that self-compassion mediates pathways from stress to psychopathologies [65]. It is possible that high stress in Indonesian society relating to the rapid growth in domains such as education and economy impacted students' mental wellbeing, despite the mediative effects of self-compassion. Future research is needed to understand the first-hand experience of Indonesian students. Interview-based research focusing on mental wellbeing constructs [66] is suitable to deeply appraise internal experiences.

Absorption scores were significantly higher in Indonesian students than UK students. This may also be explained by cultural differences between the two countries. The power distance in Indonesia is higher than that in the UK [63], with people in Indonesia, in general, being more accepting of social hierarchy. Their high-power distance may help Indonesian students to focus less on competing with each other and more on enjoying their studies. A high level of absorption in studies has been reported in Indonesian students [67,68].

Conversely, UK students' low power distance and high success-drive can explain their higher level of dedication relative to Indonesian students. Contrary to the inner quality of absorption, dedication relates more to the external; it is about the investment put into the work [37], relating to a sense of significance, challenge and pride taken in study [69]. Success-driven societies value personal accomplishment; therefore, students tend to commit more to their studies in order to achieve future success [43]. Low power distance can be another drive to enhance dedication. As students are able to interact with teachers more easily, they may find academic activities more meaningful, resulting in increased motivation to attain high achievement [44,70]. However, in higher power distance societies, instructions are usually given from an authority, causing students to follow passively [71]. The strong hierarchy may limit students' participation and motivation to commit, hence low dedication.

Of the four predictor variables, self-compassion and vigour were significant positive predictors of mental wellbeing, with self-compassion as the strongest predictor. This finding supports previous research in student populations [23,72] and suggests that self-compassion training is critical to students maintaining positive mental wellbeing. Self-compassion training helps students improve overall wellbeing by reducing academic stress and anxiety [10,73,74]. It does so by equipping students with techniques to help them acknowledge and accept their weaknesses and to develop more effective learning strategies [24]. Several studies support the positive impact of self-compassion training on mental wellbeing [75,76]. Self-compassion interventions have also been shown to have a sustained longer-term positive mental wellbeing effect [30,75,76]. In particular, the practice of mindfulness has been found to be an effective format for developing self-compassion [76,77]. Beaumont et al. [78] recommend that schools and universities develop a culture of compassion to protect students' mental wellbeing. They suggest that self-compassion training should not be an add on but an integral part of the taught curriculum. Implementation and evaluation of self-compassion training for Indonesian students are needed.

Lastly, vigour, although the second strongest predictor of wellbeing, did not mediate the pathway from self-compassion to wellbeing, suggesting that vigour's contribution to wellbeing occurs through a different pathway. One possible explanation for this finding is the mediation model we used, which assessed vigour separately from the other academic engagement components. Vigour is argued to predict high academic achievement [79] and is associated with efficacy [80]. However, vigour without dedication can lead to academic burnout [81] and a lack of motivation [82]. This suggests that dedication could be a pathway mediator from vigour to wellbeing, although further research is required to affirm this.

Vigour interventions focused on improving personal resources (e.g., self-efficacy) and physical movement have been shown to increase vigour [83] and are suggestive of effective types of vigour training. Vigour training has been suggested to help PhD students mediate burnout [84]. Importantly, research has indicated that first-year degree students have higher levels of vigour and dedication, which decrease in subsequent years [85].

This offers insight into the potential timing of vigour training in a student's life cycle for maximum effectiveness.

4. Limitations

Firstly, in the present study, we employed convenience sampling at one university, limiting the generalisability. Secondly, the comparisons were made between counselling and education students in Indonesia and counselling and occupational therapy students in the UK. Comparing students solely from a single discipline could contribute to a more accurate discussion of cultural differences (however, it can be difficult or impossible to do so in cross-national research). Additionally, in this study, we did not consider other aspects that could have enabled more accurate comparisons to be made, such as ethnicity, religion and socioeconomic status. Thirdly, the cross-cultural use of self-report measures might have limited the accuracy of our results [86]. Moreover, the accuracy of the SCS-SF scale used in this study has been questioned and is currently under debate [87]. However, as the scale has been used broadly among university student samples [88], we felt it was appropriate for use in the present study. Fourthly, the causal directions of the variables were not assessed. Lastly, although the study was conducted when COVID-19 cases were declining steadily in Indonesia [89], the impact of the COVID-19 pandemic was not discussed. Furthermore, the UK data were collected before the outbreak of the COVID-19 pandemic. Indonesian universities tried to mitigate the spread of the COVID-19 infection, including by limiting face-to-face teaching [90]. Resulting social isolation compromised the mental wellbeing of Indonesian students [91]. The impact of COVID-19 on the study variables needs to be evaluated.

5. Conclusions

As Indonesian higher education continues to develop rapidly, many Indonesian university students suffer from compromised mental wellbeing. In the present study, we identified that Indonesian students had higher levels of self-compassion than UK students. Self-compassion and vigour were identified as significant predictors of mental wellbeing. Our findings will help researchers and educators in Indonesia to effectively support the mental wellbeing of students in higher education.

Author Contributions: Conceptualization, Y.K.; methodology, Y.K. and J.J.; software, Y.K.; validation, Y.K., J.W. and R.R.; formal analysis, Y.K.; investigation, Y.K., J.L., A.K., K.B. and G.H.T.L.; resources, R.R.; data curation, R.R.; writing—original draft preparation, Y.K., J.L., A.K., K.B., G.H.T.L. and J.J.; writing—review and editing, Y.K. and J.W.; project administration, Y.K. All authors have read and agreed to the published version of the manuscript.

Funding: This research received no external funding.

Institutional Review Board Statement: The study received ethical approval from the co-author, R.R.'s, university (State Islamic University of Sultan Syarif Kasim Riau: Ref KE/KEP-FPP/01/05/2022).

Informed Consent Statement: Informed consent was obtained from all subjects involved in the study.

Data Availability Statement: The data presented in this study are available on reasonable request from the corresponding author. The data are not publicly available due to ethical restrictions.

Conflicts of Interest: The authors declare no conflict of interest.

References

1. The World Bank. Population, Total—Indonesia. 2022. Available online: https://data.worldbank.org/indicator/SP.POP.TOTL?locations=ID (accessed on 30 May 2022).
2. PwC. The Long View How Will the Global Economic Order Change by 2050? 2017. Available online: https://www.pwc.com/gx/en/world-2050/assets/pwc-the-world-in-2050-full-report-feb-2017.pdf (accessed on 1 May 2022).
3. Wardhani, E. The Equalization of English Language Skill in Indonesia in The Preparation for Indonesia Vision 2045. Sepuluh Nopember Institute of Technology, Faculty of Industrial Technology and Systems Engineering, Department of Industrial and

Systems Engineering. 2021. Available online: https://www.researchgate.net/publication/352367752_The_Equalization_of_English_Language_Skill_in_Indonesia_in_The_Preparation_for_Indonesia_Vision_2045 (accessed on 1 May 2022).
4. Narayan, S.; Narayan, P.K.; Mishra, S. Investigating the Relationship Between Health and Economic Growth: Empirical Evidence From a Panel of 5 Asian Countries. *J. Asian Econ.* **2010**, *21*, 404–411. [CrossRef]
5. Sukmayadi, V.; Yahya, A. Indonesian Education Landscape and the 21st Century Challenges. *J. Soc. Stud. Educ. Res.* **2020**, *11*, 219–234.
6. The World Bank. The Promise of Education in Indonesia. 2020. Available online: https://www.worldbank.org/en/country/indonesia/publication/the-promise-of-education-in-indonesia (accessed on 1 May 2022).
7. Statista Research Department. Education in Indonesia—Statistics and Facts, Statista. 2022. Available online: https://www.statista.com/topics/9229/education-in-indonesia/#topicHeader__wrapper (accessed on 1 May 2022).
8. Scimago. Indonesia. Scimago Institutions Rankings. 2021. Available online: https://www.scimagojr.com/countrysearch.php?country=id (accessed on 10 May 2022).
9. Shaturaev, J. Indonesia: Superior Policies And Management For Better Education (Community development through Education). *A a oa* **2021**, *1*, 1–10.
10. Kurniawan, R.; Managi, S. Economic Growth and Sustainable Development in Indonesia: An Assessment. *Bull. Indones. Econ. Stud.* **2018**, *54*, 339–361. [CrossRef]
11. Hill, H.; Wie, T.K. Indonesian Universities in Transition: Catching Up and Opening Up. *Bull. Indones. Econ. Stud.* **2012**, *48*, 229–251. [CrossRef]
12. Larcombe, W.; Finch, S.; Sore, R.; Murray, C.M.; Kentish, S.; Mulder, R.A.; Lee-Stecum, P.; Baik, C.; Tokatlidis, O.; Williams, D.A. Prevalence and Socio-demographic Correlates of Psychological Distress Among Students at an Australian University. *Stud. High. Educ.* **2016**, *41*, 1074–1091. [CrossRef]
13. Orygen. International Students and Their Mental Health and Physical Safety. 2020. Available online: https://internationaleducation.gov.au/International-network/Australia/InternationalStrategy/EGIProjects/Documents/ORYGEN%20-%20International%20Student%20Mental%20Health%20and%20Physical%20Safety%20June%202020.pdf (accessed on 10 May 2022).
14. Royal College of Psychiatrists. Mental Health of Students in Higher Education. College Report CR166. 2011. Available online: https://www.rcpsych.ac.uk/docs/default-source/improving-care/better-mh-policy/college-reports/college-report-cr166.pdf?sfvrsn=d5fa2c24_2 (accessed on 10 May 2022).
15. Kloping, N.A.; Citraningtyas, T.; Lili, R.; Farrell, S.M.; Molodynski, A. Mental Health and Wellbeing of Indonesian Medical Students: A Regional Comparison Study. *Int. J. Soc. Psychiatry* **2021**, 1–5. [CrossRef] [PubMed]
16. Mardea, N.A.; Widayanti, A.W.; Kristina, S.A. Stress Level Comparison Between Pharmacy Students and Non-Pharmacy Students in Indonesia. *Int. J. Pharm. Res.* **2021**, *12*, 204–209.
17. Astutik, E.; Sebayang, S.K.; Puspikawati, S.I.; Tama, T.D.; Sintha, D.M.; Dewi, K. Depression, Anxiety, and Stress Among Students in Newly Established Remote University Campus in Indonesia. *Malays. J. Med. Health Sci.* **2020**, *16*, 270–277.
18. Peltzer, K.; Pengpid, S. Loneliness and Health Risk Behaviors Among ASEAN Adolescents. *Iran J. Psychiatry Behav. Sci.* **2017**, *11*, e7691. [CrossRef]
19. Lili, R.; Molodynski, A.; Farrell, S.M.; Citraningtyas, T.; Kloping, N.A. Wellbeing and Mental Health Among Medical Students in Indonesia: A Descriptive Study. *Int. J. Soc. Pschiatry* **2021**, 00207640211057709. [CrossRef] [PubMed]
20. Barrera, I.; Longoria, D. Examining Cultural Mental Health Care Barriers Among Latinos. *J. Leadersh. Organ. Stud.* **2018**, *4*, 1–12.
21. Roberts, T.; Esponda, G.M.; Krupchanka, D.; Shidhaye, R.; Patel, V.; Rathod, S. Factors Associated with Health Service Utilisation for Common Mental Disorders: A Systematic Review. *BMC Psychiatry* **2018**, *18*, 262. [CrossRef] [PubMed]
22. Rahmadiana, M.; Karyotaki, E.; Schulte, M.; Ebert, D.D.; Passchier, J.; Cuijpers, P.; Berger, T.; van Ballegooijen, W.; Wimbarti, S.; Riper, H. Transdiagnostic Internet Intervention for Indonesian University Students with Depression and Anxiety: Evaluation of Feasibility and Acceptability. *JMIR Ment. Health* **2021**, *8*, e20036. [CrossRef] [PubMed]
23. Baer, R.A.; Lykins, E.L.B.; Peters, J.R. Mindfulness and Self-Compassion as Predictors of Psychological Wellbeing in Long-Term Meditators and Matched Nonmeditators. *J. Posit. Psychol.* **2012**, *7*, 230–238. [CrossRef]
24. Neff, K.D.; Hsieh, Y.-P.; Dejitterat, K. Self-Compassion, Achievement Goals, and Coping with Academic Failure. *Self Identity* **2005**, *4*, 263–287. [CrossRef]
25. Neff, K.D. The Development and Validation of a Scale to Measure Self-Compassion. *Self Identity* **2010**, *2*, 223–250. [CrossRef]
26. Neff, K.D.; Mcgehee, P. Self-Compassion and Psychological Resilience Among Adolescents and Young Adults. *Self Identity* **2010**, *9*, 225–240. [CrossRef]
27. Fong, M.; Loi, N.M. The Mediating Role of Self-Compassion in Student Psychological Health. *Aust. Psychol.* **2016**, *51*, 431–441. [CrossRef]
28. Gilbert, P. Introducing Compassion-Focused Therapy. *Adv. Psychiatr. Treat.* **2009**, *15*, 199–208. [CrossRef]
29. Macbeth, A.; Gumley, A. Exploring Compassion: A Meta-Analysis of the Association Between Self-Compassion and Psychopathology. *Clin. Psychol. Rev.* **2012**, *32*, 545–552. [CrossRef] [PubMed]
30. Neff, K.D.; Rude, S.S.; Kirkpatrick, K.L. An Examination of Self-Compassion in Relation to Positive Psychological Functioning and Personality Traits. *J. Res. Pers.* **2007**, *41*, 908–916. [CrossRef]

31. Muris, P.; van den Broek, M.; Otgaar, H.; Oudenhoven, I.; Lennartz, J. Good and Bad Sides of Self-Compassion: A Face Validity Check of the Self-Compassion Scale and an Investigation of Its Relations to Coping and Emotional Symptoms in Non-Clinical Adolescents. *J. Child. Fam Stud.* **2018**, *27*, 2411–2421. [CrossRef] [PubMed]
32. Kotera, Y.; Ting, S.-H. Positive Psychology of Malaysian University Students: Impacts of Engagement, Motivation, Self-Compassion, and Well-Being on Mental Health. *Int. J. Ment. Health Addict.* **2019**, *19*, 227–239. [CrossRef]
33. Poots, A.; Cassidy, T. Academic Expectation, Self-Compassion, Psychological Capital, Social Support and Student Wellbeing. *Int. J. Educ. Res.* **2020**, *99*, 101506. [CrossRef]
34. Rogers, A.A.; DeLay, D.; Martin, C.L. Traditional Masculinity During the Middle School Transition: Associations with Depressive Symptoms and Academic Engagement. *J. Youth Adolesc.* **2017**, *46*, 709–724. [CrossRef]
35. Krause, K.-L.; Coates, H. Students' Engagement in First-Year University. *Assess. Eval. High. Educ.* **2018**, *33*, 493–505. [CrossRef]
36. Martinez, I.M.; Youssef-Morgan, C.M.; Chambel, M.J.; Marques-Pinto, M. Antecedents of Academic Performance of University Students. *Educ. Psychol.* **2019**, *39*, 1047–1067. [CrossRef]
37. Schaufeli, W.B.; Salanova, M.; Gonzalez-roma, V.; Bakker, A.B. The Measurement of Engagement and Burnout: A Two Sample Confirmatory Factor Analytic Approach. *J. Happiness Stud.* **2002**, *3*, 71–92. [CrossRef]
38. Luthans, F.; Youssef, C.M. Emerging Positive Organizational Behavior. *J. Manag.* **2007**, *33*, 321–349. [CrossRef]
39. Pascarella, E.T.; Terenzini, P.T. *How College Affects Students: A Third Decade of Research*; Jossey-Bass: San Francisco, CA, USA, 2005.
40. Boulton, C.A.; Hughes, E.; Kent, C.; Smith, J.R.; Williams, H.T.P. Student Engagement and Wellbeing Over Time at a Higher Education Institution. *PLoS ONE* **2019**, *14*, e0225770. [CrossRef]
41. Kotera, Y.; Green, P.; Sheffield, D. Positive Psychology for Mental Wellbeing of UK Therapeutic Students: Relationships with Engagement, Motivation, Resilience and Self-Compassion. *Int. J. Ment. Health Addict.* **2021**, *20*, 1611–1626. [CrossRef]
42. Trompetter, H.R.; de Kleine, E.; Bohlmeijer, E.T. Why Does Positive Mental Health Buffer Against Psychopathology? An Exploratory Study on Self-Compassion as a Resilience Mechanism and Adaptive Emotion Regulation Strategy. *Cogn. Ther. Res.* **2017**, *41*, 459–468. [CrossRef] [PubMed]
43. Alheneidi, H.; Smith, A.P. Problematic Internet Use and Wellbeing: A Study of Kuwait and UK Students. *Int. J. Educ. Hum. Soc. Sci.* **2020**, *3*, 1–14. Available online: http://ijehss.com/ (accessed on 10 May 2022).
44. Cortina, K.S.; Arel, S.; Smith-Darden, J.P. School Belonging in Different Cultures: The Effects of Individualism and Power Distance. *Front. Educ.* **2017**, *2*, 56. [CrossRef]
45. Kotera, Y.; Ozaki, A.; Miyatake, H.; Tsunetoshi, C.; Nishikawa, Y.; Tanimoto, T. Mental health of medical workers in Japan during COVID-19: Relationships with loneliness, hope and self-compassion. *Curr. Psychol.* **2021**, *40*, 6271–6274. [CrossRef]
46. Gökçen, E.; Furnham, A.; Mavroveli, S.; Petrides, K.V. A Cross-Cultural Investigation of Trait Emotional Intelligence in Hong Kong and the UK. *Pers. Individ. Differ.* **2014**, *65*, 30–35. [CrossRef]
47. Gopalkrishnan, N. Cultural Diversity and Mental Health: Considerations for Policy and Practice. *Front. Public Health* **2018**, *6*, 179. [CrossRef] [PubMed]
48. Kotera, Y. De-stigmatising self-care: Impact of self-care webinar during COVID-19. *Int. J. Spa Wellness* **2021**, *4*, 213–217. [CrossRef]
49. Hugman, R. *New Approaches in Ethics for the Caring Professions*; Palgrave Macmillan: London, UK, 2005.
50. Indonesian Higher Education Data Base. College Profile, PDDikti—Higher Education Data Base. 2020. Available online: https://pddikti.kemdikbud.go.id/data_pt/QjJDQjRENDQtMTg3Qy00RTJELThFNUItRjRGMkFFMUE2REY4 (accessed on 15 May 2022).
51. Kotera, Y.; Green, P.; Sheffield, D. Mental health of therapeutic students: Relationships with attitudes, self-criticism, self-compassion, and caregiver identity. *Br. J. Guid. Couns.* **2021**, *49*, 701–712. [CrossRef]
52. Wilkes, L.; Cowin, L.; Johnson, M. The reasons students choose to undertake a nursing degree. *Collegian* **2015**, *22*, 259–265. [CrossRef] [PubMed]
53. World Health Organization. Delivered by Women, Led by Men: A Gender and Equity Analysis of the Global Health and Social Workforce, Human Resources for Health Observer Series No. 24. 2019. Available online: https://apps.who.int/iris/handle/10665/311322 (accessed on 15 May 2022).
54. Stewart-Brown, S.; Janmohamed, K. *Warwick-Edinburgh Mental Well-being Scale (WEMWBS) User Guide Version 1*; Warwick Medical School: Coventry, UK, 2008.
55. Cooke, P.; Melchert, T.; Connor, K. Measuring well-being: A review of instruments. *Couns. Psychol.* **2016**, *44*, 730–757. [CrossRef]
56. Tennant, R.; Hiller, L.; Fishwick, R.; Platt, S.; Joseph, S.; Weich, S.; Parkinson, J.; Secker, J.; Stewart-Brown, S. The Warwick-Edinburgh Mental Well-being Scale (WEMWBS): Development and UK validation. *Health Qual. Life Outcomes* **2007**, *5*, 63. [CrossRef] [PubMed]
57. Raes, F.; Pommier, E.; Neff, K.; Van Gucht, D. Construction and Factorial Validation of a Short Form of the Self-Compassion Scale. *Clin. Psychol. Psychother.* **2011**, *18*, 250–255. [CrossRef] [PubMed]
58. Schaufeli, W.; Bakker, A. Utrecht Work Engagement Scale: Preliminary Manual, Occupational Health Psychology Unit, Utrecht University. 2004. Available online: https://www.wilmarschaufeli.nl/publications/Schaufeli/Test%20Manuals/Test_manual_UWES_English.pdf (accessed on 15 May 2022).
59. Hayes, A. *Introduction to Mediation, Moderation, and Conditional Process Analysis*, 2nd ed.; Guildford Publications: Guildford, UK, 2018.
60. Cohen, J. *Statistical Power Analysis for the Behavioral Sciences*, 2nd ed.; Lawrence Erlbaum Associates: Hillsdale, NJ, USA, 1988.

61. Goetz, J.; Keltner, D.; Simon-Thomas, E. Compassion: An Evolutionary Analysis and Empirical Review. *Psychol. Bull.* **2010**, *136*, 351–374. [CrossRef] [PubMed]
62. Montero-Marin, J.; Kyuken, W.; Crane, C.; Gu, J.; Baer, R.; Al-Awamleh, A.A.; Akutsu, S.; Araya-Véliz, C.; Ghorbani, N.; Chen, Z.J.; et al. Self-Compassion and Cultural Values: A Cross-Cultural Study of Self-Compassion Using a Multitrait-Multimethod (MTMM) Analytical Procedure. *Front. Psychol.* **2018**, *9*, 2638. [CrossRef] [PubMed]
63. Hofstede, G.; Hofstede, G.J.; Minkov, M. *Cultures and Organizations: Software of the Mind*; McGraw-Hill: New York, NY, USA, 2010.
64. Gilbert, P. The Origins and Nature of Compassion Focused Therapy. *Br. J. Clin. Psychol.* **2014**, *53*, 6–41. [CrossRef]
65. Kotera, Y.; Dosedlova, J.; Andrzejewski, D.; Kaluzeviciute, G.; Sakai, M. From Stress to Psychopathology: Relationship with Self-Reassurance and Self-Criticism in Czech University Students. *Int. J. Ment. Health Addict.* **2021**, 1–12. [CrossRef]
66. Kotera, Y.; Chircop, J.; Hutchinson, L.; Rhodes, C.; Green, P.; Jones, R.-M.; Kaluzeviciute, G.; Garip, G. Loneliness in online students with disabilities: Qualitative investigation for experience, understanding and solutions. *Int. J. Educ. Technol. High. Educ.* **2021**, *18*, 64. [CrossRef]
67. Adiputra, S. Self-Compassion Profile in Lampung Culture Perspective. *JOMSIGN J. Multicult. Stud. Guid. Couns.* **2018**, *2*, 8–21. [CrossRef]
68. Aisha, S.; Mulyana, D. Indonesian Postgraduate Students' Intercultural Communication Experiences in the United Kingdom. *J. Ilmu. Komun.* **2019**, *7*, 1–13. [CrossRef]
69. Alrashidi, O.; Phan, H.; Ngu, B. Academic Engagement: An Overview of Its Definitions, Dimensions, and Major Conceptualisations. *Int. Educ. Stud.* **2016**, *9*, 41–52. [CrossRef]
70. Kotera, Y.; Taylor, E.; Fido, D.; Williams, D.; Tsuda-McCaie, F. Motivation of UK Graduate Students in Education: Self-Compassion Moderates Pathway from Extrinsic Motivation to Intrinsic Motivation. *Curr. Psychol.* **2021**, *22*, 1–14. [CrossRef] [PubMed]
71. Khatri, N. Consequences of Power Distance Orientation in Organisations. *Vis. J. Bus. Perspect.* **2009**, *13*, 1–9. [CrossRef]
72. Neely, M.E.; Schallert, D.L.; Mohammed, S.S.; Roberts, R.; Chen, Y. Self-Kindness When Facing Stress: The Role of Self-Compassion, Goal Regulation, and Support in College Students' Well-being. *Motiv. Emot.* **2009**, *33*, 88–97. [CrossRef]
73. Regehr, C.; Glancy, D.; Pitts, A. Interventions to Reduce Stress in University Students: Review and Meta-analysis. *J. Affect. Disord.* **2013**, *148*, 1–11. [CrossRef]
74. Law, D.W. A Measure of Burnout for Business Students. *J. Educ. Bus.* **2010**, *85*, 195–202. [CrossRef]
75. Gunnell, K.; Mosewich, A.; McEwen, C.; Eklund, R.; Crocker, P. Don't be so Hard on Yourself! Changes in Self-Compassion during the First Year of University are Associated with Changes in Well-being. *Pers. Individ. Differ.* **2017**, *107*, 43–48. [CrossRef]
76. Ko, C.; Grace, F.; Chavez, G.; Grimley, J.; Dalrymple, E.; Olson, L. Effect of Seminar on Compassion on student self-compassion, mindfulness and well-being: A randomized controlled trial. *J. Am. Coll. Health* **2018**, *66*, 537–545. [CrossRef]
77. Shapiro, S.L.; Brown, K.W.; Biegel, G.M. Teaching Selfcare to Caregivers: Effects of Mindfulness-based Stress Reduction on the Mental Health of Therapists in Training. *Train. Educ. Prof. Psychol.* **2007**, *1*, 105–115. [CrossRef]
78. Beaumont, E.; Durkin, M.; Hollins, M.C.; Carson, J. Compassion for Others, Self-Compassion, Quality of Life and Mental Well-being Measures and Their Association with Compassion Fatigue and Burnout in Student Midwives: A Quantitative Survey. *Midwifery* **2016**, *34*, 239–244. [CrossRef] [PubMed]
79. Cadime, I.; Pinto, A.M.; Lima, S.; Rego, S.; Pereira, J.; Ribeiro, I. Well-being and Academic Achievement in Secondary School Pupils: The Unique Effects of Burnout and Engagement. *J. Adolesc.* **2016**, *53*, 169–179. [CrossRef] [PubMed]
80. Merhi, R.; Paniagua, Á.S.-E.; Descals, F.J.P. The Role of Psychological Strengths, Coping Strategies and Well-being in the Prediction of Academic Engagement and Burnout in First-year University Students. *Acción Psicol.* **2018**, *15*, 51–68. [CrossRef]
81. Stoeber, J.; Childs, J.H.; Hayward, J.A.; Feast, A.R. Passion and Motivation for Studying: Predicting Academic Engagement and Burnout in University Students. *J. Educ. Psychol.* **2011**, *31*, 513–528. [CrossRef]
82. Kotera, Y.; Cockerill, V.; Chircop, J.G.E.; Kaluzeviciute, G.; Dyson, S. Predicting Self-Compassion in UK Nursing Students: Relationships with Resilience, Engagement, Motivation, and Mental Wellbeing. *Nurse Educ. Pract.* **2021**, *51*, 102989. [CrossRef] [PubMed]
83. Levecque, K.; Rigolle, F.; De Beuckelaer, A.; Mortier, A. PhD Students and Vigour: On Bursting with Energy, Feeling Fit and Being Enthusiastic About Work. *Ecoom Briefs* **2019**, 1–5.
84. Kusurkar, R.A.; van der Burgt, S.M.E.; Isik, U.; Mak-van der Vossen, M.; Wilschut, J.; Wouter, A.; Koster, A. Burnout and engagement among PhD students in medicine: The BEeP study. *Perspect. Med. Educ.* **2021**, *10*, 110–117. [CrossRef] [PubMed]
85. Fernández-Martínez, E.; Andina-Díaz, E.; Fernández-Peña, R.; García-López, R.; Fulgueiras-Carril, I.; Liébana-Presa, C. Social Networks, Engagement and Resilience in University Students. *Int. J. Environ. Res. Public Health* **2017**, *14*, 1488. [CrossRef]
86. Kotera, Y.; Van Laethem, M.; Ohshima, R. Cross-cultural comparison of mental health between Japanese and Dutch workers: Relationships with mental health shame, self-compassion, work engagement and motivation. *Cross Cult. Strat. Manag.* **2020**, *27*, 511–530. [CrossRef]
87. Kotera, Y.; Sheffield, D. Revisiting the Self-compassion Scale-Short Form: Stronger Associations with Self-inadequacy and Resilience. *SN Compr. Clin. Med.* **2020**, *2*, 761–769. [CrossRef]
88. Hayes, J.A.; Lockard, A.J.; Janis, R.A.; Locke, B.D. Construct validity of the Self-Compassion Scale-Short Form among psychotherapy clients. *Couns. Psychol. Q.* **2016**, *29*, 405–422. [CrossRef]

89. UNICEF. Indonesia COVID-19 Situation Report. 2021. pp. 1–18. Available online: https://www.unicef.org/media/114246/file/Indonesia%20COVID-19%20Situation%20Report,%20Oct-%20Dec%202021.pdf (accessed on 15 May 2022).
90. Haris, I.; Haris, M.I. Response of Indonesian Universities to the COVID-19 Pandemic—Between Strategy and Implementation. *J. Public Health Res.* **2021**, *10*, jphr-2021. [CrossRef] [PubMed]
91. Rahiem, M.D.H.; Krauss, S.E.; Ersing, R. Perceived Consequences of Extended Social Isolation on Mental Well-Being: Narratives from Indonesian University Students during the COVID-19 Pandemic. *Int. J. Environ. Res. Public Health* **2021**, *18*, 10489. [CrossRef] [PubMed]

healthcare MDPI

Article

"Crying on the Bus": First Time Fathers' Experiences of Distress on Their Return to Work

Suzanne Hodgson [1,*], Jon Painter [2], Laura Kilby [3] and Julia Hirst [4]

1 Department of Nursing, Manukau Institute of Technology—Te Pūkenga, Manukau, Auckland 2104, New Zealand
2 Department of Nursing and Midwifery, Sheffield Hallam University, Sheffield S10 2BP, UK; j.painter@shu.ac.uk
3 Centre for Behavioural Science and Applied Psychology, Sheffield Hallam University, Sheffield S10 2BP, UK; l.kilby@shu.ac.uk
4 Department of Psychology, Sociology and Politics, Sheffield Hallam University, Sheffield S10 2BP, UK; j.hirst@shu.ac.uk
* Correspondence: suzanne.hodgson@manukau.ac.nz

Abstract: There is increasing research interest in the experiences of new fathers taking paternity leave, but less insight into men's experiences of returning to work after the birth of their first baby. For many men in the UK context, this could take place immediately after the birth or after one or two weeks of paternity leave. This paper utilizes data from a UK-based study whilst also drawing on international literature and policy contexts. A constructivist grounded theory method was adopted to generate theory from the data gathered. Twelve new fathers shared their experiences in this study by participating in audio-recorded, semi-structured interviews. This paper focuses on fathers' experiences of negotiating the workplace as part of an overall theoretical framework related to broader transitions to fatherhood and sheds light on the distress, guilt and psychological challenges that the participants experienced when they initially returned to work. Whether fathers did or did not explicitly describe distress at this time, they all described a change in their worker identity, which for some participants led to uncertainty in the workplace. Men returning to work at this time in the postnatal period are vulnerable to experiencing distress. Flexibility and support in the workplace could be protective of their mental health. Finally, policy and practice developments are offered to support men's transitions to fatherhood in the workplace context.

Keywords: distress; return to work; fathers; fatherhood; transitions; paternal; perinatal; identity

Citation: Hodgson, S.; Painter, J.; Kilby, L.; Hirst, J. "Crying on the Bus": First Time Fathers' Experiences of Distress on Their Return to Work. *Healthcare* **2023**, *11*, 1352. https://doi.org/10.3390/healthcare11091352

Academic Editors: Yasuhiro Kotera and Elaina Taylor

Received: 10 March 2023
Revised: 27 April 2023
Accepted: 27 April 2023
Published: 8 May 2023

1. Introduction

Fathering, and men's role in childcare, is continuously evolving [1] and there are now multiple different fathering roles and performances available to men; Cannito (p661) describes these as "several fatherhoods" [2]. The traditional role of the distant, disciplinarian provider is now less valued by men [3], with many aspiring to a more emotionally involved ideal [4], which is frequently different to the approaches of their own fathers [1]. Thus, some new fathers attempt to combine multiple parenting roles, including breadwinning, emotional support and practical care [5]. Positive paternal relationships (demonstrated by father–infant sensitivity and play) have numerous benefits including significant positive impacts on emotional and behavioural development [6–11]. Conversely, the absence of fathers can have negative consequences throughout a child's emotional and social development [12,13].

Men who are able to take longer parental leave are found to be more involved in their children's care, in developmental play and have closer emotional relationships than those who take less paternity leave [14–17]. Furthermore, studies have shown a positive correlation between the length of paternity leave and the extent of the positive influence on men's lives [18]. Internationally, paternity leave provision varies considerably. For example, paternity leave policy in the USA [14] differs starkly from countries such as Sweden, Finland and Norway, whose governments offer the most generous leave to new fathers [18–20]. Paternity leave in these Nordic territories is considerably more than the two weeks offered in the UK and is independent of maternal leave allowances [21].

There appear to be differences in fathers' perspectives on using and accessing leave and flexible working [14] depending on their socioeconomic or employment status. In a UK-based study, manual, semi-skilled and unskilled workers were typically found to have fewer flexible working options and access to leave than their professional counterparts [4]. Additionally, in the UK, men in higher occupational groups such as consultancy roles tend to have more access to shared parental leave, and this may be supplemented with full pay by their employers [22]. Moreover, due to the nature of their work, they also enjoy more flexible working options than fathers engaged in manual roles or the service industry, which cannot be performed from home [23]. Other researchers have found that Italian men's decisions on whether to take paternity leave were influenced strongly by social norms, the perception that care from the mother is best for the child and, whilst the concept of involved fathering was present, it appeared to be more theoretical than something regularly enacted [24].

Returning to work can be both challenging and rewarding. Fathers wishing to remain significantly involved with their children may still face barriers such as a lack of flexibility and autonomy in their working conditions. These experiences can increase levels of stress, unhappiness and anxiety [25]. On their return to work, some fathers report a sense of dread, often wishing to have more time at home with their baby [26] and feeling sad about how much time they will spend away from their family. In the same study, however, others found a new sense of purpose in work, feeling more useful in the workplace than at home in the early days and weeks of fatherhood [26].

Research has evidenced that men can experience mental health problems related to their transitions to fatherhood [27–30]. Perinatal distress, particularly postnatal depression (PND), was, in the past, considered to be an entirely maternal problem [27]. It is however now understood that men can experience different forms of perinatal distress such as anxiety, depression and obsessive compulsive disorder (OCD). Witnessing traumatic births can also lead to anxiety, stress and relationship problems between new fathers and their partners and babies [28]. It is suggested that men manifest perinatal distress differently to women [27] and may engage in substance misuse as a form of self-medication or demonstrate anger and avoidance behaviours [31,32].

Men experience various psychological reactions during their transition to fatherhood as they try to find a role for themselves relative to their partner and as they engage in paid work, whilst navigating early parenting [33,34]. Moreover, low parenting-related self-efficacy and lower self-esteem in new and prospective fathers has been associated with paternal postnatal depression [35]. Positive intimate partner relationships have been shown to support parenting self-efficacy and parental autonomy and may be protective factors against mental health problems in both parents [36]. When mothers experience perinatal mental health problems, the risk to fathers' mental health may increase [30,37,38]. Social support bolsters the wellbeing of new fathers, but new fatherhood has equally been shown to lead to friendship losses or changes and reductions in time for exercise, leisure and social activities [39,40].

Studies have shown that parental leave policy alone does not enable all fathers to co-parent their children, with deeper sociocultural and structural barriers also apparent [15,41,42]. These include traditional views on masculinity [43] and gendered norms regarding whose job it is to care for children [4,23,44]. Individually, some UK fathers are instigating changes for themselves in how they father, despite broader social and institutional norms lagging, particularly in workplace and health service policy and practice [14,45]. A recent study found that during the COVID-19 pandemic, new fathers were more able to enact involved fathering as they did not have to navigate the structural workplace barriers which consistently prevent it [46]. This same study demonstrated an association between poorer paternal mental health outcomes and what is deemed by many to be inadequate paternity leave policy in the UK [46].

Further research is required to better understand the factors which maximise the opportunities for new fathers to co-parent and be involved with all aspects of the care of their babies. Exploring this specifically when they return to work will help to understand the distress they experience during this specific point in their transitions to fatherhood. Furthermore, the mental health of "involved" fathers and men who take up parental leave and return to work is an under-researched topic.

This broad constructivist grounded theory (CGT) study was a doctoral research study examining all aspects of transitions to fatherhood. Different elements of the findings have applications for specific audiences (see [47] for findings relevant to healthcare practice). The current paper reports the findings with direct relevance to new fathers returning to the workplace after one to two weeks of paternity leave and is therefore likely to be of particular interest to occupational health practitioners, human resources professionals and organisational managers. The paper provides important insights into the experiences of new fathers as they navigate the UK's current paternity leave provision and return-to-work arrangements.

2. Materials and Methods

Whilst there has been an increase in the study of fatherhood internationally, a study which explores the interaction between different aspects of men's lives as they become fathers was identified as a gap in the knowledge base. To this end, a constructivist grounded theory method (CGTM) [48–50] underpinned by symbolic interactionism [51,52] was adopted to generate theory from the data gathered [53].

Approval for this study was granted by the university's research ethics committee (HEC 2015/137 and AM/KW/D&S-311). First-time fathers with a child under one year were recruited via social media, a university website and by third sector organisations sharing adverts with their service users. Participant information sheets were emailed to the 21 respondents, with 12 men ultimately consenting to participate. Audio-recorded, face-to-face, semi-structured interviews [49] ranging from 40 min to two hours (over two sessions) were conducted between November 2016 and October 2017. Interviews generally began by asking the participants to tell their story of becoming a father for the first time. Follow-up questions sought to understand their preparations for becoming a father, experiences in the workplace, and any other topics participants highlighted as being important.

Theoretical sampling and constant comparison processes [54] enabled the interview schedule to be adapted to explore emerging concepts and extend recruitment to include first-time fathers with a child under two years. Line-by-line coding, focused coding and theoretical coding led to the iterative formulation of the theoretical categories and final core category [55]. For further explanation of the methodological process, see [47]. CGTMs typically seek transferability rather than generalisability [48,56], hence the findings of this study may be applicable to men in different contexts or workplaces to those interviewed.

Participants

Twelve new fathers shared their experiences in this study, six aged 25–34 and six 35–44. Eleven identified as "White British" and one as "White Other". All lived with their partners; six married and six co-habiting. Eleven fathers had a baby under one year and one had a child under two years. All had completed formal education, one left school at 16, another at 18, and all worked 30+ hours per week. One had a further education certificate, four were graduates and five had post-graduate qualifications. Pseudonyms have been used to protect the anonymity of participants, partners and children.

3. Results

These findings focus on participants' perspectives of current paternity leave provision in the UK and, consequently, their experiences of returning to work after taking paternity leave. These experiences are often underpinned by their experiences in the healthcare setting, see [47]. Feelings such as guilt, distress and abandonment of their partner and child are shared explicitly and implicitly by the fathers, whilst worker identity is discussed in relation to feelings of uncertainty and potential insecurity.

Participants experienced distress in the immediate postnatal hours and days and carried this with them on their return to work. These feelings relate to traumatic experiences during birth and returning home alone on the night of the birth.

> *"It was scary, possibly the most scary experience that I've ever been through especially looking back now, it was so close to losing both of them."* Mark, pp. 26–27

Returning home from the hospital alone soon after the birth was particularly confronting for some participants.

> *"So, I'll go home then at four o'clock in the morning, and I think all that happened so quick and then you're almost left on your own to deal with it really ... Just sort of difficult really, feeling separated from your wife ... not seeing her and then to go home but then all of that stuff to happen and then to see the sights that you've seen and the emotions that flood through."* Alex, p. 12 and p. 15

Paternity leave was discussed at length with an overwhelming sense of dissatisfaction with current policy. There was a consensus that two weeks of paternity leave was inadequate and that this paid leave should be significantly longer:

> *"Two weeks of paternity leave ... is ridiculously short."* Ethan, p. 19

Iain expands on this with his perspective on this policy:

> *"I found that almost ridiculous that the man doesn't get any time off; a week, two weeks, but it's definitely not long enough to get used to it. Even if they did extend it to a couple of months maybe just to get used to this change in your life and then you could go back to work at least ... I get two weeks and I'm thrown straight back into work and the biggest change of my life, I don't know what I'm doing at all at that point and then I just go back to work ... I'll see you tonight for an hour."* Iain, pp. 29–30

On their return to work, participants expressed a feeling of abandoning their new family. There was a sense of not having had enough time to settle as a newly formed family and that inadequate paternity leave may be denying new families the opportunity to bond properly.

> *"Everything got on top of me, I had to go back to work and leave the baby crying ... she's absolutely screaming ... I can't just not go to work, as much as I'd love to stay here and help you, I can't, because I'm only allowed two weeks, that's all you're allowed."* William, pp. 24–25

Participants indicated that their role as fathers was not currently endorsed by policies related to paternity leave. They understood and articulated the consequences, including missing out on family time and on their child's developmental milestones, and not being able to adequately support their partners. The inequality felt in relation to paternity leave was unmistakable in participants' stories and a significant point of sadness and regret.

> *"I'd not really thought about coming back to work and kind of knowing that I will miss a big chunk of his life because I am at work, I'm out of the house most of the day and thinking about, is this what it will be like for me, not forever, but for the long-term future, and I'm out of his life for more time than I'm actually in it."* Iain, p. 27

Participants expressed difficulty in motivating themselves when they returned to work. They questioned their impact and usefulness as paid employees when they were exhausted and distracted in the early weeks of fatherhood.

> *"I remember . . . crying on the bus on the first day. So, a combination of just being sleep deprived and really not wanting to be there when I was at work, I felt pretty rotten."* Frank, p. 16

The challenges of the early weeks of new fatherhood appeared to be understood by some employers where expectations on participants' workplace performances were lowered, at least in the initial weeks.

> *"I think I've been lucky in that my boss, my manager at work is very understanding, kind of put me in a quiet corner for those first few weeks back."* Frank p. 2

In some cases, participants had very positive experiences with employers extending leave or providing compassionate leave if their babies had been unwell post birth.

> *"Certainly, in the last year my work has been pretty flexible, I think, in terms of allowing me a little bit more leeway of coming maybe slightly later and maybe not working, not exactly working all my hours in a week, things like that they've been quite good with."* Alex, p. 14

Participants aspired to fulfil their roles as co-parents. A significant aspect of this was a desire to attend to their babies overnight and share such responsibilities with their partners. However, tensions arose between workplace pressures and expectations at home during the early weeks of fatherhood which compromised their desired performances as worker, father and partner.

> *"I felt like I was abandoning my duties. If the baby's crying her eyes out at ten o'clock at night and I'm up at four I'm like 'will you see to the baby' and you can see in her eyes she's like kind of 'what'? And you feel so guilty but what can you do, you know what I mean, you're not given much choice but to live that kind of life . . . That feeling you can't help somebody even though it's part of your job as well, but you just can't help."* William, pp. 25–26

These tensions were more overtly illustrated in some workplaces by explicit demands to be job focused and not necessarily care-giving focused.

> *"No, you can't be a dad because we need you to be working here and obviously you shouldn't be trying to do this anyway, why bother with that kind of thing . . . yeah it's really frustrating when you want to do things to be a better dad but you're kind of impeded from doing that."* Ethan, p. 23

Tensions were present in the context of the competing responsibilities to their employer and partner. Moreover, participants described having little choice than to satisfy these demands because of their employers' expectations. As can be seen from the quotes above, attempting to enact a fathering identity within the workplace, for some participants, was challenging.

There was an appreciation of the demands of early parenting and the work that mothers were performing. New fathers were attempting to balance their fathering role alongside workplace demands whilst also trying to support their partners by sharing the parenting load.

> *"It wasn't literally like this, but I'd come home from work and she'd basically hand the baby to me and just be like can you just take her and fine, no problem, absolutely fine, like I'd go to work, it would be intense because it was Christmas, but I'd come home and my actual job would start. That's how I kind of see it now, it's significantly more tiring being at home than it is being at work."* Fred, p. 44

Participants attempted to find other ways to balance work expectations with fatherhood responsibilities in the early weeks of their return to work.

> *"It's about managing a little bit of work to still keep a job but most of your time devoting your energy towards him for those or supporting Ann for those twelve months and that's*

what my family has needed and, yeah that's the way I've viewed it essentially and so my career has probably been secondary." Alex, p. 18

There was also a reflection on changes in their values and attitudes towards work. Some participants changed their workplace behaviour because of becoming a father, re-evaluating their roles and responsibilities.

"I feel like if I did act stupid and got sacked or quit then that's having an effect on the baby because she's not going to be able to have the things she wants and that's not a good role model." William, p. 28

For other participants, their new responsibilities at home impacted upon their professional identity which made them quickly feel out of the loop.

"I dunno, losing a little bit of that identity about, you know, I'm kind of, kind of known at work as the film buff and someone to talk to about those things and that is kind of slowly going and people are saying, oh have you seen this film and I'm like, what's that?" Albert, p. 3

In light of their experiences of their return to work, participants provided insight into the nature of support they might find or did find helpful. They spoke about this in relation to gendered stereotypes of men's help seeking but also provided positive solutions for workplace support.

Some participants' perspectives on men talking about their feelings were embedded within stereotypical gendered norms.

"Don't know if that's just me with anxiety, or whether that's more normal ... what everyone goes through. It's hard to tell. Cos with blokes as well, a lot of blokes keep their feelings to themselves. It's very rare that you get the truth about what blokes are feeling, cos they like to put up a wall, as protection for themselves whereas women are a lot more emotional between each other." Neville, pp. 37–38

Peer support was valued by some participants as a means of gaining support from both informal and formal sources. The importance of being able to talk to others about being a new father was shared.

"Maybe just talking to other dads, maybe some erm, some kind of session for that, I dunno, that might be useful. Erm, yeah, new dads talking to existing dads." Albert, p. 25

Opportunities to talk about feelings, on confirmation of pregnancy through to the early weeks and months of fatherhood, would be valuable.

"I felt there were a couple of times specifically relating to some of the bad, the worries you know and the emergencies ... there were a few times, I think just after that, those occasions when I would have quite liked to have been able to talk." Frank, pp. 8–9

"Well from the moment you find out right through, you know somebody you could, even just pick up the phone or go and see, right from when you find out to a few months afterwards." Mark, p. 17

Overall, the participants spoke of challenges on their return to work which included tensions between home and work life. There was a strong desire to support their partners and to be involved fathers, whilst balancing their continuing role as a financial provider. Workplace behaviours changed and professional identity was challenged as a consequence of their new role as fathers being unsupported in the workplace. Participants spoke of peer support as a valuable approach to being able to talk more openly about the joys and challenges of being a new father and gaining insight and support both formally and informally.

4. Discussion

Negotiating the workplace during transitions to fatherhood appeared to be a source of distress for new fathers in this study. Whilst there were some examples of overt distress experienced by the participants in this study, the lack of workplace acknowledgement of men's transitions to fatherhood may lend itself to the internalisation of distress. These findings reflect the extant literature relating to new fathers navigating workplace policies and practices including flexible working [57] and the perception of inadequate paternity leave in the UK context [58]. Fathers in other studies have identified that returning to work is both challenging and rewarding [33]. Changes in worker identity for participants were a potential source of distress, particularly for men who previously (prior to fatherhood) felt valued by others in their organisation for their knowledge, commitment and skills. In this context, the workplace could be compared to a soap opera with multiple story lines. If you are unable to watch every episode you may quickly fall behind with some of the stories, thus potentially changing your perceived status within the organisation. Recent evidence suggests that fathers requesting or taking flexible working are subjected to a "fatherhood forfeit" [59]. In this evidence, men who were balancing caregiving with paid work were viewed with scepticism and deemed to be lazy by co-workers [59]. Workplace flexibility is likely to support men to be more engaged with both the pregnancy and as involved fathers in the early weeks of parenthood. However, participants in this study and men in other studies appear to have to rely on discretionary flexibility, which is uncertain, whereas formalised flexible working should be embedded across workplace policy to promote equity [60].

As mentioned above, participants in this study described returning to work in relation to feelings of abandonment, guilt and missing out on their baby's developmental milestones, with some indicating a lack of closeness as a consequence of this. This has been reported in other studies [25] where inflexibility in working conditions resulted in higher levels of stress, sadness and anxiety, and prohibited the enactment of caring masculinities and involved fatherhood. In the present study, when workplaces or individual managers provided a level of flexibility, new fathers were able to participate more in pregnancy-related appointments and thus felt more engaged in the process.

Traditional gendered workplaces, such as those which are male-dominated or involve manual jobs, may not afford fathers the opportunity to take time out to perform other aspects of fathering, as reflected in the current literature [15]. In these more traditional workplaces, men have been shown to be reluctant to take parental leave or to request flexible working [61,62] as they are concerned about the potential consequences on their careers and of being perceived as less masculine [57,63].

It is within the power of employers to support the development of a positive, involved father identity, by way of their respective policies and philosophies [59,60]. The pervasive presence of hegemonic masculine norms [43], a lack of flexible working opportunities and poor parental leave provision for fathers coupled with poor support for mental health creates an environment which has the potential to cause paternal distress. This is particularly concerning considering that this is a space where new fathers will spend much of their time in the early weeks of fatherhood.

As alluded to above, research suggests that gendered workplace norms heavily influence the uptake of paternity leave [15] and there is a growing understanding that parental leave policy alone is insufficient to facilitate a family's desire to co-parent and that factors such as socioeconomics and traditional gendered norms influence these opportunities [42].

Participants in this study were open to talking to someone about how they were feeling and felt that some form of peer support would be useful. This mirrors current evidence and also grassroots practices such as Andy's Man Club [64] and other charitable organisations such as Leeds Dads [65] and Dads Rock [66] in the UK. In previous studies, peer support has been demonstrated to be of value in supporting transitions to fatherhood [67]. Benefits described by participants in these studies include having feelings validated, reducing isolation, sharing experiences, enjoying other fathers' company and the acknowledgement of their roles as fathers [67–70]. Validation of feelings is especially pertinent as a previous study found that new fathers felt that their feelings of distress experienced in the perinatal period were not legitimate [45].

How and whether men choose to access support is heavily influenced by traditional concepts of masculinity, and men's help seeking takes on many forms [71–73]. It must not be assumed that men will access services in the same ways as women, nor that all men have the same health beliefs and behaviours [74]. Some men may prefer a male-only or male-led environment [75–77]. Creating spaces that are sensitive to the challenges of hegemonic masculinity and that seek to actively support new fathers in adjusting to their changing roles and identities would be a welcome start. That way, talking could be normalized, and furthermore, new fathers could be more valued and appreciated within workplaces for both their challenges and the joy and excitement of new fatherhood.

Strengths and Limitations

The strengths of this study are reflected in its attention to the specific experiences of men navigating the workplace as they become fathers for the first time. There is growing interest in men's experiences of paternity leave provision and flexible working, and this study offers insight into the potential distress experienced when navigating the workplace in early fatherhood. The interwoven findings of this grounded theory study have broad application to both policy and practice.

Participants' willingness and desire to share details of their emotional wellbeing counters the common perception that men are reluctant to discuss their feelings. The helping professions may need to consider whether services are positioned to receive men's stories and what the barriers to disclosure might be [78].

The sample of participants in this study were white, predominantly middle-class men in heterosexual relationships and over the age of 18 years. Therefore, their experiences may not be representative of fathers who identify differently from these demographic characteristics. Whilst in CGT studies, there is no requirement to be demographically representative, there is a limitation to how far these findings can be applied to other fathers, for example from non-cisgender or gay identities, different age groups, social classes, ethnicities or cultural backgrounds. Evidence does however indicate potential similarities between fathers regardless of ethnicity or cultural background [4,79–81] and so the findings may be somewhat applicable to other groups of fathers.

5. Conclusions and Implications for Workplaces and Workplace Policies

There continue to be calls for the extension and expansion of paternity leave in the UK [58] mirroring the offer in Scandinavian countries such as Sweden, Finland and Norway [21,81]. When men take up extended leave in these countries, there are benefits for the whole family [17], and fathers appear to change their perspective on work; this results in happier employees and increasing productivity and retention [81]. The benefits to infant development have been found to be significant when men are facilitated in taking parental leave or adopting shorter working hours [18]. Further research is needed to explore whether the length of paternity leave taken is associated with less distress and improved mental wellbeing in new fathers.

Whilst suggestions for policy and practice lend themselves to improved support for and acknowledgement of the needs of fathers, mothers too are likely to benefit from changes that improve workplace equity.

Since fathering is performed within the workplace, this seems an appropriate place to support the fathering roles that men want and to provide a platform for emotional support. Much more work is needed in this area of policy and development of inclusive and equal workplace philosophy; a father's quota of dedicated and protected paternity leave such as that seen in Nordic countries may help here [81].

Recent policy changes in England have been welcomed and include paid leave for parents who have a baby in a neonatal intensive care unit [82], providing relief for fathers concerned about balancing a return to work with the distressing experience of caring for an unwell baby. However, whilst further policy change would be welcome, there are a broad range of factors which influence the uptake of paternity leave, and these must be addressed alongside any policy implementation [83]. What is well known and clearly illustrated in the present study is that men returning to work at this time are vulnerable to experiencing postnatal distress, and that flexibility in employment could be supportive of paternal mental health [25]. With this evidence in mind, we suggest the following policy initiatives.

Policy Suggestions

- Actively promoting access to flexible working for both fathers and mothers.
- Education and awareness raising to promote parental wellbeing in the workplace and encourage an environment which is supportive and inclusive.
- Training for all levels of management and human resource departments in how to positively support new fathers in the early weeks of parenting.
- Dedicated and protected paternity leave to replace shared parental leave.
- Transparent workplace-based psychological support or mental health champions for new fathers, which aims to endorse and validate their feelings of distress and exhaustion but also their joy and pride in their new father roles.
- Workplace peer supporters for new fathers.

As part of a complete package of initiatives to support new fathers' workplace culture, change is warranted which acknowledges the importance of involved fathers on their children's lives, as much as mothers. Challenging workplace gendered norms may be difficult but is essential to promote equality with the potential to benefit all employees and their families.

Author Contributions: Conceptualization, S.H.; methodology, S.H.; software, S.H.; validation, S.H.; formal analysis, S.H. and L.K. and J.H.; investigation, S.H.; resources, S.H.; data curation, S.H.; writing—original draft preparation, S.H. and J.P.; writing—review and editing, S.H., J.P., L.K. and J.H.; visualization, S.H.; supervision, L.K. and J.H.; project administration, S.H. All authors have read and agreed to the published version of the manuscript.

Funding: This research received no external funding.

Institutional Review Board Statement: The study was conducted in accordance with the Declaration of Helsinki and approved by the Institutional Review Board (or Ethics Committee) of Sheffield Hallam University (HEC 2015/137 and 151 AM/KW/D&S-311 8 November 2016).

Informed Consent Statement: Informed consent was obtained from all subjects involved in the study.

Data Availability Statement: For access to data generated in this study, please contact the first author.

Acknowledgments: Gratitude is extended to the participants in this study who gave up their valuable time to share their stories.

Conflicts of Interest: The authors declare no conflict of interest.

References

1. Dermott, E.; Miller, T. More than the sum of its parts? Contemporary fatherhood policy, practice and discourse. *Fam. Relatsh. Soc.* **2015**, *4*, 183. [CrossRef]
2. Cannito, M. Beyond "traditional" and "new": An attempt of redefinition of contemporary fatherhoods through discursive practices and practices of care. *Men Masc.* **2020**, *23*, 661–679. [CrossRef]
3. Marsiglio, W.; Roy, K. *Nurturing Dads: Fatherhood Initiatives beyond the Wallet*; Russell Sage Foundation: New York, NY, USA, 2012.
4. Dex, S.; Ward, K. *Parental Care and Employment in Early Childhood*; Equal Opportunities Working Paper No. 57; Equal Opportunities Commission: London, UK, 2007.
5. Genesoni, L.; Tallandini, M.A. Men's psychological transition to fatherhood: An analysis of the literature, 1989–2008. *Birth Issues Perinat. Care* **2009**, *36*, 305–318. [CrossRef] [PubMed]
6. Cabrera, N.J.; Shannon, J.D.; Tamis-LeMonda, C. Fathers' influence on their children's cognitive and emotional development: From toddlers to pre-K. *Appl. Dev. Sci.* **2007**, *11*, 208–213. [CrossRef]
7. Ramchandani, P.G.; O'Connor, T.G.; Evans, J.; Heron, J.; Murray, L.; Stein, A. The effects of pre- and postnatal depression in fathers: A natural experiment comparing the effects of exposure to depression on offspring. *J. Child Psychol. Psychiatry Allied Discip.* **2008**, *49*, 1069–1078. [CrossRef]

8. Ramchandani, P.; Stein, A.; Evans, J.; O'Connor, T.G.; Team, A.S. Paternal depression in the postnatal period and child development: A prospective population study. *Lancet* **2005**, *365*, 2201–2205. [CrossRef]
9. Sarkadi, A.; Kristiansson, R.; Oberklaid, F.; Bremberg, S. Fathers' involvement and children's developmental outcomes: A systematic review of longitudinal studies. *Acta Paediatr.* **2008**, *97*, 153–158. [CrossRef]
10. St George, J.M.; Wroe, J.K.; Cashin, M.E. The concept and measurement of fathers' stimulating play: A review. *Attach. Hum. Dev.* **2018**, *20*, 634–658. [CrossRef]
11. Tamis-LeMonda, C.S.; Shannon, J.D.; Cabrera, N.J.; Lamb, M.E. Fathers and mothers at play with their 2-and 3-year-olds: Contributions to language and cognitive development. *Child Dev.* **2004**, *75*, 1806–1820. [CrossRef]
12. Cabrera, N.; Tamis-LeMonda, C.S.; Bradley, R.H.; Hofferth, S.; Lamb, M.E. Fatherhood in the twenty-first century. *Child Dev.* **2000**, *71*, 127–136. [CrossRef]
13. Cabrera, N.J.; Tamis-LeMonda, C.S. (Eds.) *Handbook of Father Involvement: Multidisciplinary Perspectives*; Routledge: England, UK, 2013.
14. Petts, R.J.; Knoester, C. Paternity leave-taking and father engagement. *J. Marriage Fam.* **2018**, *80*, 1144–1162. [CrossRef] [PubMed]
15. Haas, L.; Hwang, C.P. Policy is not enough—The influence of the gendered workplace on fathers' use of parental leave in Sweden. *Community Work Fam.* **2019**, *22*, 58–76. [CrossRef]
16. Huerta, M.C.; Adema, W.; Baxter, J.; Han, W.J.; Lausten, M.; Lee, R.; Waldfogel, J. Fathers' leave and fathers' involvement: Evidence from four OECD countries. *Eur. J. Soc. Secur.* **2014**, *16*, 308–346. [CrossRef] [PubMed]
17. Tanaka, S.; Waldfogel, J. Effects of parental leave and work hours on fathers' involvement with their babies: Evidence from the millennium cohort study. *Community Work Fam.* **2007**, *10*, 409–426. [CrossRef]
18. Brandth, B.; Kvande, E. Workplace support of fathers' parental leave use in Norway. *Community Work Fam.* **2019**, *22*, 43–57. [CrossRef]
19. Duvander, A.Z.; Johansson, M. Does Fathers' Care Spill Over? Evaluating Reforms in the Swedish Parental Leave Program. *Fem. Econ.* **2019**, *25*, 67–89. [CrossRef]
20. Johansson, T. Fatherhood in Transition: Paternity Leave and Changing Masculinities. *J. Fam. Commun.* **2011**, *11*, 165–180. [CrossRef]
21. Sigurdardottir, H.M.; Garoarsdottir, O. Backlash in gender equality? Fathers parental leave during a time of economic crisis. *J. Eur. Soc. Policy* **2018**, *28*, 342–356. [CrossRef]
22. Scott, K. Accenture Equalises Parental Policies to Encourage Male Take up. Available online: https://employeebenefits.co.uk/issues/april-2018/accenture-shared-parental-leave/ (accessed on 17 February 2023).
23. Birkett, H.; Forbes, S. Where's dad? Exploring the low take-up of inclusive parenting policies in the UK. *Policy Stud.* **2019**, *40*, 205–224. [CrossRef]
24. Cannito, M. The influence of partners on fathers' decision-making about parental leave in Italy: Rethinking maternal gatekeeping. *Curr. Sociol.* **2020**, *68*, 832–849. [CrossRef]
25. Bolzan, N.; Gale, F.; Dudley, M. Time to Father. *Soc. Work Health Care* **2005**, *39*, 67–88. [CrossRef] [PubMed]
26. Chin, R.; Daiches, A.; Hall, P. A qualitative exploration of first-time fathers' experiences of becoming a father. *Community Pract.* **2011**, *84*, 19–23. [PubMed]
27. Shafian, A.K.; Mohamed, S.; Raduan, N.J.N.; Ann, A.Y.H. A systematic review and meta-analysis of studies validating Edinburgh Postnatal Depression Scale in fathers. *Heliyon* **2022**, *8*, e09441. [CrossRef] [PubMed]
28. Elmir, R.; Schmied, V. A qualitative study of the impact of adverse birth experiences on fathers. *Women Birth* **2022**, *35*, 41–48. [CrossRef]
29. Scarton, D. *Postpartum Depression Strikes New Dads as Well as Moms*; U.S. News & World Report: New York, NY, USA, 2008; p. 83.
30. Baldwin, S.; Malone, M.; Sandall, J.; Bick, D. Mental health and wellbeing during the transition to fatherhood: A systematic review of first time fathers' experiences. *JBI Database Syst. Rev. Implement. Rep.* **2018**, *16*, 2118. [CrossRef]
31. Pérez, F.; Brahm, P.; Riquelme, S.; Rivera, C.; Jaramillo, K.; Eickhorst, A. Paternal post-partum depression: How has it been assessed? A literature review. *Ment. Health Prev.* **2017**, *7*, 28–36. [CrossRef]
32. O'Brien, A.P.; McNeil, K.A.; Fletcher, R.; Conrad, A.; Wilson, A.J.; Jones, D.; Chan, S.W. New fathers' perinatal depression and anxiety—Treatment options: An integrative review. *Am. J. Men's Health* **2017**, *11*, 863–876. [CrossRef]
33. Chin, R.; Hall, P.; Daiches, A. Fathers' experiences of their transition to fatherhood: A metasynthesis. *J. Reprod. Infant Psychol.* **2011**, *29*, 4–18. [CrossRef]
34. Morse, C.A.; Buist, A.; Durkin, S. First-time parenthood: Influences on pre-and postnatal adjustment in fathers and mothers. *J. Psychosom. Obstet. Gynecol.* **2000**, *21*, 109–120. [CrossRef]
35. Singley, D.B.; Edwards, L.M. Men's perinatal mental health in the transition to fatherhood. *Prof. Psychol. Res. Pract.* **2015**, *46*, 309. [CrossRef]
36. Bradley, R.; Slade, P. A review of mental health problems in fathers following the birth of a child. *J. Reprod. Infant Psychol.* **2011**, *29*, 19–42. [CrossRef]
37. Figueiredo, B.; Field, T.; Diego, M.; Hernandez-Reif, M.; Deeds, O.; Ascencio, A. Partner relationships during the transition to parenthood. *J. Reprod. Infant Psychol.* **2008**, *26*, 99–107. [CrossRef]
38. Wee, K.Y.; Skouteris, H.; Pier, C.; Richardson, B.; Milgrom, J. Correlates of ante-and postnatal depression in fathers: A systematic review. *J. Affect. Disord.* **2011**, *130*, 358–377. [CrossRef] [PubMed]

39. Castle, H.; Slade, P.; Barranco-Wadlow, M.; Rogers, M. Attitudes to emotional expression, social support and postnatal adjustment in new parents. *J. Reprod. Infant Psychol.* **2008**, *26*, 180–194. [CrossRef]
40. Hodgson, S. Transitions to Fatherhood: A Constructivist Grounded Theory Study. Ph.D. Thesis, Sheffield Hallam University, Sheffield, UK, 2021. [CrossRef]
41. Ma, L.; Andersson, G.; Duvander, A.Z.; Evertsson, M. Fathers' Uptake of Parental Leave: Forerunners and Laggards in Sweden, 1993–2010. *J. Soc. Policy* **2020**, *49*, 361–381. [CrossRef]
42. Plantin, L. Different classes, different fathers? On fatherhood, economic conditions and class in Sweden. *Community Work Fam.* **2007**, *10*, 93–110. [CrossRef]
43. Connell, R.W. Change among the gatekeepers: Men, masculinities, and gender equality in the global arena. *Signs J. Women Cult. Soc.* **2005**, *30*, 1801–1825. [CrossRef]
44. Sriyasak, A.; Almqvist, A.L.; Sridawruang, C.; Neamsakul, W.; Häggström-Nordin, E. The new generation of Thai Fathers: Breadwinners involved in parenting. *Am. J. Men's Health* **2018**, *12*, 1368–1378. [CrossRef]
45. Darwin, Z.; Galdas, P.; Hinchliff, S.; Littlewood, E.; McMillan, D.; McGowan, L.; Gilbody, S. Fathers' views and experiences of their own mental health during pregnancy and the first postnatal year: A qualitative interview study of men participating in the UK Born and Bred in Yorkshire (BaBY) cohort. *BMC Pregnancy Childbirth* **2017**, *17*, 45. [CrossRef]
46. Gheyoh Ndzi, E. Paternal Leave Entitlement and Workplace Culture: A Key Challenge to Paternal Mental Health. *Int. J. Environ. Res. Public Health* **2023**, *20*, 5454. [CrossRef]
47. Hodgson, S.; Painter, J.; Kilby, L.; Hirst, J. The experiences of first-time fathers in perinatal services: Present but invisible. *Healthcare* **2021**, *9*, 161. [CrossRef] [PubMed]
48. Charmaz, K. The power and potential of grounded theory. *Med. Sociol. Online* **2012**, *6*, 2–15.
49. Charmaz, K. *Constructing Grounded Theory*; Sage: Los Angeles, CA, USA, 2014.
50. Suddaby, R. From the editors: What grounded theory is not. *Acad. Manag. J.* **2006**, *49*, 633–642. [CrossRef]
51. Heath, H.; Cowley, S. Developing a grounded theory approach: A comparison of Glaser and Strauss. *Int. J. Nurs. Stud.* **2004**, *41*, 141–150. [CrossRef] [PubMed]
52. Coyne, I.; Cowley, S. Using grounded theory to research parent participation. *J. Res. Nurs.* **2006**, *11*, 501–515. [CrossRef]
53. Bryant, A. Re-grounding Grounded Theory. *J. Inf. Technol. Theory Appl.* **2002**, *4*, 25–42.
54. Starks, H.; Brown Trinidad, S.B. Choose your method: A comparison of phenomenology, discourse analysis, and grounded theory. *Qual. Health Res.* **2007**, *17*, 1372–1380. [CrossRef]
55. Bryant, A.; Charmaz, K. *The Sage Handbook of Grounded Theory*; Sage: Los Angeles, CA, USA, 2007.
56. Cooney, A. Rigour and grounded theory. *Nurse Res.* **2011**, *18*, 17–22. [CrossRef]
57. Rudman, L.A.; Mescher, K. Penalizing men who request a family leave: Is flexibility stigma a femininity stigma? *J. Soc. Issues* **2013**, *69*, 322–340. [CrossRef]
58. Jackson, S. Shared Parental Leave Take-up May Be as Low as 2%. BBC News 2018. Available online: http://www.bbc.co.uk/news/business-43026312 (accessed on 17 February 2023).
59. Kelland, J.; Lewis, D.; Fisher, V. Viewed with suspicion, considered idle and mocked-working caregiving fathers and fatherhood forfeits. *Gend. Work Organ.* **2022**, *29*, 1578–1593. [CrossRef]
60. Weldon-Johns, M. The Future of UK Work-Family Rights: The Case for More Flexible Working. In Proceedings of the 5th International Conference on Gender Research, Aveiro, Portugal, 28–29 April 2022; Academic Conferences International: Sonning Common, UK, 2022; pp. 259–265. Available online: https://papers.academic-conferences.org/index.php/icgr/article/view/89/109 (accessed on 13 January 2023).
61. Borgkvist, A.; Moore, V.; Crabb, S.; Eliott, J. Critical considerations of workplace flexibility 'for all' and gendered outcomes: Men being flexible about their flexibility. *Gend. Work Organ.* **2021**, *28*, 2076–2090. [CrossRef]
62. Harvey, V.; Tremblay, D.G. Paternity leave in Québec: Between social objectives and workplace challenges. *Community Work Fam.* **2020**, *23*, 253–269. [CrossRef]
63. Acker, J. Hierarchies, jobs, bodies: A theory of gendered organizations. *Gend. Soc.* **1990**, *4*, 139–158. [CrossRef]
64. Andy's Man Club. Available online: https://andysmanclub.co.uk/ (accessed on 17 February 2023).
65. Leeds Dads. Available online: https://leedsdads.org/ (accessed on 17 February 2023).
66. Dads Rock. Available online: https://www.dadsrock.org.uk/ (accessed on 17 February 2023).
67. Bragadóttir, H. Computer-mediated support group intervention for parents. *J. Nurs. Scholarsh.* **2008**, *40*, 32–38. [CrossRef] [PubMed]
68. Archibald, S.J. What about fathers? A review of a fathers' peer support group on a Neonatal Intensive Care Unit. *J. Neonatal Nurs.* **2019**, *25*, 272–276. [CrossRef]
69. Doucet, S.; Letourneau, N.; Blackmore, E.R. Support needs of mothers who experience postpartum psychosis and their partners. *J. Obstet. Gynecol. Neonatal Nurs.* **2012**, *41*, 236–245. [CrossRef]
70. Letourneau, N.; Tryphonopoulos, P.D.; Duffett-Leger, L.; Stewart, M.; Benzies, K.; Dennis, C.L.; Joschko, J. Support intervention needs and preferences of fathers affected by postpartum depression. *J. Perinat. Neonatal Nurs.* **2012**, *26*, 69–80. [CrossRef]
71. Addis, M.E.; Mahalik, J.R. Men, masculinity, and the contexts of help-seeking. *Am. Psychol.* **2003**, *58*, 5–14. [CrossRef]
72. Galdas, P.M.; Cheater, F.; Marshall, P. Men and help-seeking behaviour: A literature review. *J. Adv. Nurs.* **2005**, *49*, 616–623. [CrossRef]

73. Mahalik, J.R.; Good, G.E.; Englar-Carlson, M. Masculinity scripts, presenting concerns, and help-seeking: Implications for practice and training. *Prof. Psychol. Res. Pract.* **2003**, *34*, 123–131. [CrossRef]
74. White, A. Men's health in the 21st century. *Int. J. Men's Health* **2006**, *5*, 1–17. [CrossRef]
75. Friedewald, M.; Fletcher, R.; Fairbairn, H. All-male discussion forums for expectant fathers: Evaluation of a model. *J. Perinat. Educ.* **2005**, *14*, 8–18. [CrossRef] [PubMed]
76. Premberg, A.; Lundgren, I. Fathers' Experiences of Childbirth Education. *J. Perinat. Educ.* **2006**, *15*, 21–28. [CrossRef]
77. Symon, A.; Lee, J. Including men in antenatal education: Evaluating innovative practice. *Evid. Based Midwifery* **2003**, *1*, 12–20.
78. Malcher, G. Engaging men in health care. *Aust. Fam. Physician* **2009**, *38*, 92–95.
79. Hauari, H.; Hollingworth, K. Understanding Fathering: Masculinity, Diversity and Change. Joseph Rowntree Foundation. 2009. Available online: https://discovery.ucl.ac.uk/id/eprint/10005675/1/understanding-fathering-diversity-full.pdf (accessed on 17 February 2023).
80. Salway, S.; Chowbey, P.; Clarke, L. Parenting in Modern Britain: Understanding the Experiences of Asian Fathers. Joseph Rowntree Foundation. 2009. Available online: http://www.jrf.org.uk/sites/files/jrf/Asian-fathers-Britain-full.pdf (accessed on 17 February 2023).
81. Kvande, E. Individual Parental Leave for Fathers: Promoting Gender Equality in Norway. In *Engaged Fatherhood for Men, Families and Gender Equality*; Springer: Berlin/Heidelberg, Germany, 2022.
82. HM Treasury. Budget 2020: Delivering on Our Promise to the British People. 2020. Available online: https://assets.publishing.service.gov.uk/government/uploads/system/uploads/attachment_data/file/871802/Budget_2020_Print.pdf (accessed on 17 February 2023).
83. Kaufman, G. Barriers to equality: Why British fathers do not use parental leave. *Community Work Fam.* **2018**, *21*, 310–325. [CrossRef]

healthcare

MDPI

Brief Report

Brief COPE Short Version (Mini-COPE): A Proposal of Item and Factorial Reduction in Mexican Population

Aniel Jessica Leticia Brambila-Tapia [1,*], Reyna Jazmin Martínez-Arriaga [2], Joel Omar González-Cantero [3], Victor Hugo González-Becerra [3], Yesica Arlae Reyes-Domínguez [4], María Luisa Ramírez-García [4] and Fabiola Macías-Espinoza [5,*]

1 Departamento de Psicología Básica, Centro Universitario de Ciencias de la Salud (CUCS), Universidad de Guadalajara, Guadalajara 44340, Mexico
2 Departamento de Clínicas de Salud Mental, Centro Universitario de Ciencias de la Salud (CUCS), Universidad de Guadalajara, Guadalajara 44340, Mexico
3 Departamento de Ciencias del Comportamiento, Centro Universitario de los Valles (CUVALLES), Universidad de Guadalajara, Guadalajara 46600, Mexico
4 Maestría en Psicología de la Salud, Departamento de Psicología Básica, Centro Universitario de Ciencias de la Salud (CUCS), Universidad de Guadalajara, Guadalajara 44340, Mexico
5 Departamento de Psicología Aplicada, Centro Universitario de Ciencias de la Salud (CUCS), Universidad de Guadalajara, Guadalajara 44340, Mexico
* Correspondence: aniel.brambila@academicos.udg.mx (A.J.L.B.-T.); fabiola.macias@academicos.udg.mx (F.M.-E.)

Abstract: The factorial reduction of Brief COPE has not been successfully replicated by independent studies, and few have been performed in Spanish-speaking populations; therefore, the objective of this study was to perform a factorial reduction of the instrument in a large sample of the Mexican population and perform a convergent and divergent validity of the factors obtained. We distributed a questionnaire via social networks with sociodemographic and psychological variables, including the Brief COPE and the scales of the CPSS, GAD-7, and CES-D to measure stress, anxiety, and depression. A total of 1283 persons were included, most of whom (64.8%) were women and had a bachelor's degree (55.2%). After performing the exploratory factorial analysis, we did not find a model with an adequate fit and a reduced number of factors; therefore, we decided to reduce the number of items according to the most representative ones of adaptive, maladaptive, and emotional coping strategies. The resulting model with three factors showed good fit parameters and good internal consistency of the factors. In addition, the nature and naming of the factors were confirmed by convergent and divergent validity, with significant negative correlations between factor 1 (active/adaptive) and stress, depression, and anxiety, significant positive correlations between factor 2 (avoidant/maladaptive) and these three variables, and no significant correlation between factor 3 (emotional/neutral) and stress or depression. This shortened version of the brief COPE (Mini-COPE) is a good option to evaluate adaptive and maladaptive coping strategies in Spanish-speaking populations.

Keywords: brief COPE; factorial analysis; reduced version; Spanish-speaking populations; Mini-COPE

Citation: Brambila-Tapia, A.J.L.; Martínez-Arriaga, R.J.; González-Cantero, J.O.; González-Becerra, V.H.; Reyes-Domínguez, Y.A.; Ramírez-García, M.L.; Macías-Espinoza, F. Brief COPE Short Version (Mini-COPE): A Proposal of Item and Factorial Reduction in Mexican Population. *Healthcare* **2023**, *11*, 1070. https://doi.org/10.3390/healthcare11081070

Academic Editors: Yasuhiro Kotera and Elaina Taylor

Received: 20 February 2023
Revised: 31 March 2023
Accepted: 6 April 2023
Published: 8 April 2023

1. Introduction

Coping strategies are defined as an individual's attempts to use cognitive and behavioral strategies to manage and regulate pressures, demands, and emotions in response to stress [1]. Stress is considered a "particular relationship between the person and the environment that is appraised by the person as taxing or exceeding his or her resources and endangering his or her well-being" [2]. So far, coping strategies have been classified into different types according to authors and instruments employed; for instance, the most commonly used instruments are the coping strategies inventory (CSI) [3,4] and the Brief COPE [5,6]. Although these two instruments are similar and share the assessment (measurement) of common strategies, the Brief COPE is very commonly used in clinical and health contexts [7,8].

The Brief COPE measures a higher number of coping strategies (14 vs. 8). The Brief COPE consists of 28 items that measure 14 different subscales (including: self-blame, behavioral disengagement, self-distraction, denial, substance use, emotional support, instrumental support, active coping, planning, acceptance, positive reframing, religion, venting, and humor), with two items for each subscale. The Brief COPE has been reduced to different dimensions depending on the study. For example, Cooper et al. reduced the instrument into three dimensions: emotion-focused strategies, problem-focused strategies, and dysfunctional-focused strategies [9]; and Meyer et al. classified it into two big dimensions: adaptive and maladaptive coping strategies [10]. However, none of these reports has been validated with the factorial analysis. To date, many studies have tried to reduce the number of factors in the Brief COPE using exploratory factor analysis (EFA) and/or confirmatory factor analysis (CFA) in different clinical and general populations [8]. However, most of them have been performed with small sample sizes (less than 500 subjects) and have yielded a similar number of factors as the original instrument, making the factorial reduction useless in practice. In addition, most of the factorial reductions among the different populations studied are not similar. For instance, Su et al. [11] reduced the instrument with EFA and CFA to six factors in an HIV population, while Baumstarck et al. [12] reduced the instrument to four different factors in cancer patients and their caregivers in the French population, with a few coincidences in the item arrangement for each factorial structure in both studies. In addition, only four reports of factorial reduction of the Brief COPE have been proposed in Spanish-speaking populations: one in Argentina, two in Spain, and one in Mexico, where the authors used EFA or CFA and found great variability in the number of factors: from 2 to 14, showing discordances even within the same population. By way of example, the two different studies performed in Spain yielded 14 and 12 factors in the EFA [13,14], and the only study performed in a Mexican population of breast cancer patients [15] yielded 7 factors of the original 14, leaving intact 5 original subscales: humor, self-blame, substance use, self-distraction, and planning.

Among the variables that can modify the factorial structures observed in the different populations are: (a) The different frequencies and combinations of coping strategies used in each studied population, and (b) The type of coping being measure with the Brief COPE (situational or dispositional), this by considering that the instrument varies in its redaction, and in some studies it measures situational coping strategies (state and trait oriented, i.e., "I criticize myself") while in other studies it measures dispositional coping strategies (which identifies coping skills utilized during an specific period of time, i.e., "I have been criticizing myself"), which, although they are expected to be related, they represent different measurements [16]. In this study we utilized the situational Brief COPE.

Considering these discrepancies in the factorial reduction of the instrument, it is fundamental to perform more studies that intend to reduce the Brief COPE into fewer factors, which can be easily applicable in different health and general contexts, even if this reduction implies some item reduction. In this sense, it is important to mention that although the Brief COPE is already reduced from the original full COPE instrument of 60 items [17], its analysis is complex when considering that it has 14 subscales, many of them belonging to similar coping styles, like avoidant or active coping. Therefore, its reduction into fewer factors, which represent the type of coping they are measuring, could be very helpful in order to make the instrument more useful and practical. The objectives of this study are: (1) to reduce the brief COPE into functional dimensions, even if this implies an item reduction, in a sample of the Mexican general population; and (2) to corroborate the factorial structure obtained with convergent and divergent validity in order to identify the type of coping that each factor represents. For this last objective, we measured three psychological scales of stress, anxiety, and depression, expecting that a factor that represents maladaptive coping strategies would show significant positive correlations with them while a factor representing adaptive strategies would show a negative correlation with them.

2. Subjects and Methods

The target population was the Mexican adult population, which was reached with the snowball sampling method, in which the research team distributed an electronic questionnaire with sociodemographic and psychological instruments by social networks including WhatsApp, Facebook, and e-mail. In this sense, most participants were university students, which represents mainly young and educated people. The instrument was designed by the research team, and its understanding was tested on a small number of students (the first participants).

The study was approved by the ethics and research committee of the Health Sciences University Center of the University of Guadalajara (registration number: CI-06821), and the participants gave their consent to participate in the same questionnaire.

The socio-demographic data included sex, age, whether they have a romantic partner, schooling, whether they have a job, and socio-economic level, which was measured in five possible categories: from very low to very high.

The psychological measures included the coping strategies, measured with the brief-COPE scale [5,6] with a range of 0–3, and for this instrument, we used each item to perform the factorial reduction. In addition, we obtained the average of the punctuation obtained in the items included in each factor in order to perform the correlations with the other psychological scales (stress, depression, and anxiety). To identify the convergent and divergent validity of the instrument, we measured: stress with the Cohen Perceived Stress Scale (CPSS) [18,19], with a range of the instrument of 1–5; depression, measured with the CES-D Scale [20,21], with a range of the instrument of 0–3; and anxiety, measured with the GAD-7 Scale [22,23] with a range of the instrument of 0–3. These three instruments have also been validated in the Spanish-speaking population, and they only have one whole scale per instrument, which means that they are not divided into subscales. For analysis purposes, we also obtained the average of the punctuation obtained in the items conforming to each scale for each participant.

Statistical Analysis

To describe qualitative variables, we used frequencies and percentages, and for quantitative ones, means and standard deviation. To determine the EFA and CFA, we used the JASP software [24]. In the EFA, we obtained the factor load using the orthogonal varimax rotation analysis, assuming that factors are correlated, and the estimator method was the minimum residual. For the CFA, we obtained the comparative fit index (CFI) with the maximum likelihood (ML) estimator method; we also obtained the Tucker-Lewis index (TLI), the standardized root mean square residual (SMRS), and the root mean square error of approximation (RMSEA) values. Values of CFI and TLI ≥ 0.90 and SMRS and RMSEA ≤ 0.08 are considered good fits for the model [25]. In order to determine the convergent validity of the factors obtained with the confirmatory factorial analysis, we used the Spearman correlation test (considering the non-parametric distribution of the data) to compare each factor with stress, depression, and anxiety. We also obtained the Cronbach's alpha tests for all the instruments employed and factors obtained from the Brief COPE; a value ≥ 0.60 was considered acceptable [26]. These statistical analyses were performed with the SPSS v.25 software.

3. Results

We verified that all questionnaires were congruently filled out in order to discard false information. A total of 1283 participants older than 18 years old were included, of whom 64.8% were women. The mean $\pm$ SD of age was 31.42 $\pm$ 11.28 years. Most of the studied population had a romantic partner (61.6%), had a job (67.3%), had a bachelor's degree (55.2%), and had a medium socioeconomic level (81.2%); this refers to the social and economic position that a person has. The Cronbach's alpha test for the CPSS scale was 0.855; for GAD-7, 0.923; and for CES-D, 0.867.

3.1. EFA and CFA

The EFA analysis gave a result of six potential factors; however, three of these factors were made up of two items each (corresponding to the subscales of substance use, humor, and religion), and six items were not clearly integrated into any factor. Finally, when we performed the confirmatory factorial analysis with this item distribution, the CFI value was barely acceptable (<0.91). We additionally performed the confirmatory factorial analysis for the factor structure proposed by Cooper and Meyer [9,10] and by the reports in China and France [11,12], but the fitness of the model in neither case was acceptable (with CFIs of 0.490, 0.366, 0.795, and 0.613, respectively).

3.2. Item Reduction and CFA

We then performed an item reduction, leaving only the most representative subscales for active/adaptive strategies (4 items: active coping and planning), avoidant/maladaptive strategies (4 items: denial and behavioral disengagement), and emotional/neutral coping strategies (4 items: emotional and instrumental support), with 12 items in total. This selection was performed based on the results observed in the EFA, together with the theoretical analysis of these items as being the most representative of a specific coping style. The minimum item loading was 0.45, with no item overlap in two factors (Table 1). With this factor structure, we obtained a good fit for the model (CFI = 0.954) (Table 2). Additionally, the Cronbach's alpha values of the three factors were high for factors 1 and 3 (>0.80) and acceptable for factor 2 (>0.60) (Table 2). Considering the important item reduction, we propose a different name for this version: brief COPE short version (Mini-COPE). Very similar fit parameters were obtained if only factors 1 and 2 were included in the model, with a CFI of 0.954, TLI of 0.933, RMSEA of 0.067, and SRMS of 0.040 (Table 2).

Table 1. Items loaded in the proposed factors (English/Spanish).

	Items	Factor 1	Factor 2	Factor 3
1.	I've been concentrating my efforts on doing something about the situation I'm in./*"Concentro mis esfuerzos en hacer algo sobre la situación en la que estoy."*	0.654		
2.	I've been taking action to try to make the situation better./*"Tomo medidas para intentar que la situación mejore."*	0.714		
3.	I've been trying to come up with a strategy about what to do./*"Intento proponer una estrategia sobre qué hacer."*	0.771		
4.	I've been thinking hard about what steps to take./*"Pienso detenidamente sobre los pasos a seguir."*	0.631		
5.	I've been saying to myself "this isn't real"./*"Me digo a mí mismo: esto no es real"*		0.454	
6.	I've been giving up trying to deal with it./*"Renuncio a intentar ocuparme de ello."*		0.592	
7.	I've been refusing to believe that it has happened./*"Me niego a creer que haya sucedido."*		0.652	
8.	I've been giving up the attempt to cope./*"Renuncio al intento de hacer frente al problema."*		0.542	
9.	I've been getting emotional support from others./*"Consigo apoyo emocional de otros."*			0.770
10.	I've been getting help and advice from other people./*"Consigo que otras personas me ayuden o aconsejen."*			0.791
11.	I've been getting comfort and understanding from someone./*"Consigo el consuelo y la comprensión de alguien."*			0.762
12.	I've been trying to get advice or help from other people about what to do./*"Intento conseguir que alguien me ayude o aconseje sobre qué hacer."*			0.598

Table 2. A confirmatory factorial analysis of the factorial model and internal consistency.

	Xi^2	df	*p* Value	CFI	TLI	SRMR	RMSEA
3 factor model	266.27	51	<0.001	0.954	0.941	0.041	0.057
2 factor model	128.175	19	<0.001	0.954	0.933	0.067	0.040
Cronbach's alphas							
Factor 1 (Active/Adaptive)	0.805						
Factor 2 (Avoidant/Maladaptive)	0.645						
Factor 3 (Emotional/Neutral)	0.837						

Convergent and divergent validity.

Finally, we performed correlations between each of the three factors obtained and the values of stress, depression, and anxiety in the studied population. In these analyses, we confirmed the convergent and divergent validity of each factor (Table 3), finding significant negative correlations between factor 1 (active/adaptive coping strategies) and the three negative psychological variables; we also found significant positive correlations between factor 2 (avoidant/maladaptive coping strategies) and the three psychological variables; no significant correlations between factor 3 (emotional/neutral coping strategies) and stress or depression; and only a very low positive correlation was found between anxiety and factor 3 (Table 3). These correlations, along with theory, led us to name these coping types as adaptive, maladaptive, and neutral.

Table 3. Bivariate correlations between each factor and stress, depression, and anxiety (convergent and divergent validity).

Factors	Stress	Depression	Anxiety
Factor 1 (Active/Adaptive)	−0.323 **	−0.258 **	−0.127 **
Factor 2 (Avoidant/Maladaptive)	0.401 **	0.444 **	0.348 **
Factor 3 (Emotional/Neutral)	0.037	0.043	0.090 **

** p value < 0.01. p values obtained with Spearman correlation test.

4. Discussion

Brief COPE is a widely used instrument in the measurement of coping strategies; however, the large number of strategies that it measures and the wide variability of factorial reduction models [8–13] do not permit an easy and confident use of the instrument. This indicated the need for more studies with larger sample sizes and new proposals that can give rise to a useful and confident factorial model that is applicable to the Spanish-speaking population and that can even be replicated in different populations.

We observed after a careful factorial analysis, where EFA yielded a high number of factors similar to those observed in previous attempts to reduce this instrument, as reported in the review of Rodrigues et al. [8]; that it was possible to perform a considerable item reduction, giving three useful factors, and whose model showed an adequate fit. In addition, we could determine, by convergent and divergent validity, the names of each factor that better represent the items included. With this proposal, we reduced the number of items from 28, measuring 14 different coping styles, to 12, measuring three different coping styles: active/adaptive, avoidant/maladaptive, and emotional/neutral.

According to our second objective, we could name these factors based on the convergent and divergent validity obtained with the correlations between each factor and the values of stress, depression, and anxiety obtained in the studied population. In this sense, it is noteworthy that the factor of emotional/neutral strategies showed no significant correlations with stress or depression, and only a very low positive correlation was detected with anxiety; therefore, we decided to name this factor "neutral". Nevertheless, it is important to consider that the very low correlations observed with stress, depression, and anxiety, reaching significance only with anxiety, could indicate that these emotional strategies (emotional and instrumental support) are slightly maladaptive; however, this should be determined with more and larger studies. The other two factors were clearly adaptive (Factor 1) and maladaptive (Factor 2) according to theory and the correlations performed. In this sense, and considering that these are the more relevant factors for clinical and research usage, it is possible to use only these factors, reducing the instrument to eight items. The model fit values did not change when only factors 1 and 2 were included.

The naming of the three factors obtained with this item reduction is similar to the dimensions proposed by Cooper et al. [9], which were emotion-focused strategies, problem-focused strategies, and dysfunctional-focused strategies. So, the originally 14 subscales proposed by Carver [5] are reduced to three functional and practical factors or dimensions that correspond to a previously proposed one and are very similar to the two big dimensions proposed by Meyer (adaptive and maladaptive coping styles) [10].

Although with this proposal many coping strategies, most of them maladaptive, were not included, the main adaptive and maladaptive strategies are included in the model. We consider that this new reduced version of the instrument can be more useful in terms of instrument application (saving more time) and mainly for the data analysis, performing comparisons only with 3 factors (equivalent to subscales) instead of the 14 different subscales of the original instrument. According to the CFA performed, the instrument can be even more reduced by including only the main two factors (Factors 1 and 2), which represent adaptive and maladaptive coping strategies, and making even faster application, considering that these factors are measured with only eight items.

In addition, by considering the good fit of the model obtained and the congruence between the items within each factor (each one representing a different type of coping strategy), we think that this new version can be easily replicated in different populations.

The advantages of this study are the large sample size, considering that larger sample sizes approximate better to the evaluated population and give more reliable results; the evaluation of previous reported models by the CFI; and the measurement of stress, depression, and anxiety to determine the convergent and divergent validity. The limitations of the study are the use of a non-random sampling method, which diminishes the representativeness of the Mexican general population; the fact that we measured mainly young and educated people, which underrepresents the older people and those from lower socioeconomic levels; and, finally, the lack of follow-up, which would have permitted us to determine the intraclass correlation coefficients.

In conclusion, we performed a new proposal of item and factor reduction of the dispositional Brief COPE (Mini-COPE) instrument in the Mexican population, where we included 12 items corresponding to 3 factors, named active/adaptive, avoidant/maladaptive, and emotional/neutral, and which showed statistical parameters indicating a good fit of the model, an acceptable internal consistency of each factor, and expected correlations with stress, depression, and anxiety in the convergent and divergent validity. This reduction is expected to help the professional community by focusing on the clinically relevant strategies (adaptive and maladaptive ones), which can be measured with a higher degree of accuracy (considering that four items are included in each factor), and by saving time on its application.

In addition, the inclusion of convergent and divergent validities sheds more light on the comprehension of the relationship between coping strategies and mental health. The performance of further studies with this proposal, including the situational version of the Brief COPE, would determine its usefulness and confirm the fitness of the model.

Author Contributions: Conceptualization, A.J.L.B.-T. and F.M.-E.; methodology A.J.L.B.-T., Y.A.R.-D., M.L.R.-G. and F.M.-E.; validation, A.J.L.B.-T. and R.J.M.-A.; formal analysis, A.J.L.B.-T. and J.O.G.-C.; investigation, A.J.L.B.-T., F.M.-E., R.J.M.-A., V.H.G.-B., J.O.G.-C., Y.A.R.-D. and M.L.R.-G.; writing—review and editing, A.J.L.B.-T. and V.H.G.-B. All authors have read and agreed to the published version of the manuscript.

Funding: This research received no external funding.

Institutional Review Board Statement: The study was approved by the ethics and research committee of the Health Science University Center (CUCS) of the University of Guadalajara, number CI:06821.

Informed Consent Statement: Informed consent was obtained from all subjects.

Data Availability Statement: Data are available upon reasonable request.

Conflicts of Interest: The authors declare no conflict of interest.

References

1. McHugh, S.; Vallis, T.M. *Illness Behavior. A Multidisciplinary Model*; Plenum Press: New York, NY, USA, 1986; pp. 303–308.
2. Lazarus, R.S.; Folkman, S. *Stress, Appraisal and Coping*; Springer: New York, NY, USA, 1984; p. 19.
3. Tobin, D.L.; Holroyd, K.A.; Reynolds, R.V.; Wigal, J.K. The hierarchical factor structure of the coping strategies inventory. *Cogn. Ther. Res.* **1989**, *13*, 343–361. [CrossRef]
4. Cano García, F.J.; Rodríguez Franco, L.; García Martínez, J. Adaptación española del Inventario de Estrategias de Afrontamiento. *Actas Españolas Psiquiatr.* **2007**, *35*, 29–39.
5. Carver, C.S. You want to measure coping but your protocol' too long: Consider the brief cope. *Int. J. Behav. Med.* **1997**, *4*, 92–100. [CrossRef]
6. Morán, C.; Landero, R.; González, M.T. COPE-28: Un análisis psicométrico de la versión en español del BriefCOPE. *Univ. Psychol.* **2010**, *9*, 543–552. [CrossRef]
7. Owen, C.P.; Djukic, M.; Whisenant, M.; Lobiondo-Wood, G. Factors of maladaptive coping in emergency healthcare professionals: A systematic review. *J. Nurs. Scholarsh.* **2022**, *55*, 536–548. [CrossRef]
8. Rodrigues, F.; Figueiredo, N.; Rodrigues, J.; Ferreira, R.; Hernández-Mendo, A.; Monteiro, D. A Comprehensive Review and Bifactor Modeling Analysis of the Brief COPE. *INQUIRY J. Health Care Organ. Provis. Financ.* **2022**, *59*, 004695802211081. [CrossRef] [PubMed]
9. Cooper, C.; Katona, C.; Orrell, M.; Livingston, G. Coping strategies and anxiety in caregivers of people with Alzheimer's disease: The LASER-AD study. *J. Affect. Disord.* **2006**, *90*, 15–20. [CrossRef] [PubMed]
10. Meyer, B. Coping with Severe Mental Illness: Relations of the Brief COPE with Symptoms, Functioning, and Well-Being. *J. Psychopathol. Behav. Assess.* **2001**, *23*, 265–277. [CrossRef]
11. Su, X.Y.; Lau, J.T.; Mak, W.W.; Choi, K.; Feng, T.J.; Chen, X.; Liu, C.L.; Liu, J.; Liu, D.; Chen, L.; et al. A preliminary validation of the Brief COPE instrument for assessing coping strategies among people living with HIV in China. *Infect. Dis. Poverty* **2015**, *4*, 41. [CrossRef]
12. Baumstarck, K.; Alessandrini, M.; Hamidou, Z.; Auquier, P.; Leroy, T.; Boyer, L. Assessment of coping: A new french four-factor structure of the brief COPE inventory. *Health Qual. Life Outcomes* **2017**, *15*, 8. [CrossRef]
13. Mate, A.I.; Andreu, J.M.; Peña, M.E. Propiedades psicométricas del "Inventario breve de afrontamiento" (COPE-28) en una muestra de adolescentes [Psychometric properties of the Spanish version of the Brief COPE Inventory (COPE-28) in a sample of teenagers]. *Behav. Psychol.* **2016**, *24*, 305–318.
14. Perczek, R.E.; Carver, C.S.; Price, A.A.; Pozo-Kaderman, C. Coping, Mood, and Aspects of Personality in Spanish Translation and Evidence of Convergence With English Versions. *J. Personal. Assess.* **2000**, *74*, 63–87. [CrossRef] [PubMed]
15. Mejorada, R.; Tufiño, M.; Sierra, A.; Tena-Guerrero, O.; Riveros-Rosas, A.; Sánchez-Sosa, J.J. Afrontamiento en pacientes con cáncer de mama en radioterapia: Análisis de la Escala COPE Breve. *Psicol. Salud* **2013**, *23*, 55–62.
16. Solberg, M.A.; Gridley, M.K.; Peters, R.M. The Factor Structure of the Brief Cope: A Systematic Review. *West. J. Nurs. Res.* **2022**, *44*, 612–627. [CrossRef] [PubMed]
17. Carver, C.S.; Scheier, M.F.; Weintraub, J.K. Assessing coping strategies: A theoretically based approach. *J. Personal. Soc. Psychol.* **1989**, *56*, 267–283. [CrossRef]
18. Cohen, S.; Kamarck, T.; Mermelstein, R.A. Global measure of perceived stress. *J. Health Soc. Behav.* **1983**, *24*, 385–396. [CrossRef] [PubMed]
19. Remor, E. Psychometric properties of a European Spanish version of the perceived Stress Scale (PSS). *Span. J. Psychol.* **2006**, *9*, 86–93. [CrossRef] [PubMed]
20. Radloff, L. The CES-D Scale: A self-report depression scale for research in the general population. *Appl. Psychol. Meas.* **1977**, *1*, 385–401. [CrossRef]
21. González-Forteza, C.; Jiménez-Tapia, J.A.; Ramos-Lira, L.; Wagner, F.A. Aplicación de la Escala de Depresión del Center of Epidemiological Studies en adolescentes de la Ciudad de México. *Salud Publica Mex.* **2008**, *50*, 292–299. [CrossRef]
22. Spitzer, R.L.; Kroenke, K.; Williams, J.B.W.; Löwe, B. A Brief Measure for Assessing Generalized Anxiety Disorder. *Arch. Intern. Med.* **2006**, *166*, 1092. [CrossRef]
23. Garcia-Campayo, J.; Zamorano, E.; Ruiz, M.A.; Pardo, A.; Perez-Paramo, M.; Lopez-Gomez, V.; Freire, O.; Rejas, J. Cultural adaptation into Spanish of the generalized anxiety disorder-7 (GAD-7) scale as a screening tool. *Health Qual. Life Outcomes* **2010**, *8*, 8. [CrossRef] [PubMed]

24. JASP Team. *JASP*, Version 0.16.3, [Computer software]; Univeristy of Amsterdam: Amsterdam, The Neatherlands, 2022.
25. Baumgartner, H.; Homburg, C. Applications of structural equation modeling in marketing and consumer research: A review. *Int. J. Res. Mark.* **1996**, *13*, 139–161. [CrossRef]
26. Zhang, S.; Faries, D.; Vowles, M.; Michelson, D. ADHD rating scale IV: Psychometric properties from a multinational study as clinician-administered instrument. *Int. J. Methods Psychiatr. Res.* **2005**, *14*, 186–201. [CrossRef] [PubMed]

healthcare

MDPI

Article

Effect of Mindfulness Breathing Meditation on Depression, Anxiety, and Stress: A Randomized Controlled Trial among University Students

Maria Komariah [1,*], Kusman Ibrahim [2], Tuti Pahria [2], Laili Rahayuwati [3] and Irman Somantri [1]

[1] Department of Fundamental Nursing, Faculty of Nursing, Universitas Padjadjaran, Sumedang 45363, Indonesia
[2] Department of Medical-Surgical Nursing, Faculty of Nursing, Universitas Padjadjaran, Sumedang 45363, Indonesia
[3] Department of Community Health Nursing, Faculty of Nursing, Universitas Padjadjaran, Sumedang 45363, Indonesia
* Correspondence: maria.komariah@unpad.ac.id; Tel.: +62-896-6239-2294

Abstract: Background: The COVID-19 outbreak has caused various changes in all aspects of human life, including the educational system. These changes have forced students to undertake an adaptive process that has inevitably affected aspects of their life and psychological well-being. Adaptation of learning into online forms in universities, including nursing, triggers depression, stress, and anxiety. The high number of incidences of stress, anxiety, and depression in undergraduate students throughout the pandemic has made it important to prevent and deal with health approaches, such as mindfulness therapy. Objective: This research intended to examine whether an intervention based on mindfulness was effective and had the potential to become an interference to reduce anxiety, stress, and depression in Indonesian university students. Methods: This research applied a randomized controlled trial. One hundred and twenty-two students from Universitas Padjadjaran and other provinces in Indonesia participated in this study, with sixty-one students in each group. A pretest and a post-test were administered before and after the intervention using the Depression Anxiety Stress Scales (DASS-42). The intervention was carried out for 4 weeks with 15 min practice in each session. Results: In this study, there was an effect of mindfulness breathing meditation on decreasing the mean scores for depression, anxiety, and stress in the intervention group. However, only stress ($p = 0.007$) and anxiety ($p = 0.042$) showed a significant difference in the post-test results of the intervention and control groups. In addition, there was no difference in the scores of stress, anxiety, and depression for the pre–post-tests in each group based on religion. Conclusion: Mindfulness breathing meditation has an impact on reducing stress and anxiety in students, so it could be applied to all university students in order to develop psychosocial status and mindful attentiveness to one's needs.

Keywords: anxiety; depression; mindfulness; stress; students

Citation: Komariah, M.; Ibrahim, K.; Pahria, T.; Rahayuwati, L.; Somantri, I. Effect of Mindfulness Breathing Meditation on Depression, Anxiety, and Stress: A Randomized Controlled Trial among University Students. *Healthcare* **2023**, *11*, 26. https://doi.org/10.3390/healthcare11010026

Academic Editors: Yasuhiro Kotera and Elaina Taylor

Received: 15 November 2022
Revised: 17 December 2022
Accepted: 19 December 2022
Published: 22 December 2022

1. Introduction

The arrival of the COVID-19 pandemic has impacted the physical well-being and mental health of individuals around the world. Indonesia appealed to its people to carry out social distancing and physical distancing according to the WHO protocol by imposing large-scale social restrictions, including in the education sector, by enforcing all educational activities to be carried out online as an effort to reduce mass crowds and prevent the transmission of COVID-19. The transition to online learning poses special difficulties because the learning methodology requires adaptation [1], so the learning process that is experienced will be different from the face-to-face learning that is usually conducted. This may impact the psychological condition of students because many students struggle with mental health issues during their college years. The restrictions imposed by the COVID-19 pandemic and the shift to online learning environments may become even more apparent for students' emerging mental health issues [2]. Adapting learning to online university forms triggers depression, anxiety, and stress.

A survey conducted in China at the start of the outbreak showed that 53.8% of respondents experienced modest to extreme psychological effects, 16.5% had modest to extreme depressive symptoms, 28.8% experienced modest to extreme anxiety symptoms, and 8.1% experienced modest to extreme levels of stress [3]. Another study in Brazil showed that the majority of college students surveyed exhibited symptoms of mild to severe depression (60.5%), anxiety (52.5%), and stress (57.5%) [4]. A similar previous study in Lebanon concluded that the sudden change in learning methods during COVID-19 caused a stressful workload, which began to cause anxiety and depressive symptoms among undergraduate students [5].

Research in Korea reported that depression is the most critical cause of social problems or suicide. To determine efficacious depression anticipation schemes in the US, the American College Health Association (ACHA) actualized a national student mental health need and promoted it [6,7]. Depression also occurs in students in Indonesia. Research by Santoso et al. [8] on 148 nursing students in Central Java, Indonesia, found that students experienced mild mood disorders (25.7%), low depression (8.1%), moderate depression (0.7%), severe depression (12.2%), and extreme depression (0.7%).

Depression impacts around 67% of college students with anxiety, which becomes the main predictor of depression for nursing students [6,9]. Garcia-Gonzalez et al. [10] concluded that nursing students had high levels of anxiety during the first and fourth weeks of COVID-19 confinement. Besides that, research conducted on Chinese students throughout the COVID-19 pandemic discovered that around 25% of respondents experienced anxiety symptoms positively in proportion to increasing considerations about academic postponement, the economic impact of the pandemic, and the impact on everyday life [11]. Furthermore, a study by Young Minds showed that 83% of global young participants were positively certain that the pandemic exacerbated preceding mental health states, primarily because of the prolonged school time, loss of daily activities, and limited social contact [12]. Anxiety is influenced by emotion-oriented coping mechanisms, such as emotional responses. Therefore, it is suspected that an adaptive coping strategy, for instance, mindfulness-based intervention (MBI), can lessen depression, stress, and anxiety [13,14].

Apart from depression and anxiety, stress is a psychological problem that is prone to occur in students during a pandemic. Stress is an essential psychosocial aspect of learning activities that can affect students' academic accomplishment and well-being [15]. The high prevalence of stress among students throughout the COVID-19 pandemic was reported by Aristovnik et al. [16], who stated that the COVID-19 pandemic resulted in the closure of 1.59 billion educational institutions from 194 countries in the world, resulting in a dynamic alteration in the educational world of students, staff teachers, and families involved. Another study by Son et al. [17] found a rise in stress and anxiety levels throughout the COVID-19 pandemic social restrictions. A total of 173 out of a total of 195 students said that the stress experienced during the pandemic affected their concentration during study. Most of them, namely as much as 48%, thought that the situation at home was the biggest distraction that made it difficult for students to concentrate during online learning. In addition, another influencing factor was that the long period spent in front of a computer screen during online learning had side effects concerning vision and headaches and limited students' social interactions. Another study also stated that during the pandemic, society, including higher education students, experienced an increase in emotional disorders, such as stress, insomnia, frustration, and unstable irritability, causing psychological disorders, for instance, depression, anxiety, behavioral changes, and post-traumatic stress disorder (PTSD) [18].

The condition of psychological disorders experienced by students requires intervention that can minimize these disorders. Health experts have found various interventions to prevent and manage stress, depression, and anxiety events, one of which is therapy mindfulness [6,19]. Mindfulness-based intervention is currently becoming a popular therapy among the world community. Mindfulness is said to be a complete therapy that can minimize stress levels and improve psychological well-being because it includes various components of therapy, such as yoga, deep breathing techniques, focusing attention, and acceptance without judgment. Mindfulness is considered one of the core components of the meditation process, which includes two key elements, namely sustaining focus on the immediate experience and demonstrating acceptance of the experience [20]. Mindfulness therapy is effective and flexible because it can be applied to all ages, from children to the elderly, without any contraindications. In addition, mindfulness therapy is also very easy to do wherever and whenever according to the time desired by the individual [21].

In addition, in a study conducted by Spears [22], mindfulness is said to be a therapy that aims to decrease the level of stress, depression, and anxiety experienced by individuals universally. This means that mindfulness therapy can be given to individuals from various religions, beliefs, and ethnic and cultural backgrounds. Mindfulness is believed not to conflict with certain teachings or beliefs so that individuals can undergo this therapy without worry [22]. Barnes et al. [23] said that mindfulness is a therapy that involves the mind and body in an integrated manner to help individuals find calm when dealing with stressors, challenges, or illnesses. In the same study, it was also explained that mindfulness is not a health intervention but a skill that each individual should have [23]. Meanwhile, in Black's research, it was stated that mindfulness is a systematic procedure that sharpens and awakens the natural human capacity to be fully present in the present, welcomes and acknowledges, as well as possible, all the series of events that have occurred, whether pleasant, sad, or neutral events, and produce results regarding wisdom in behavior [24]. Therefore, it can be concluded that mindfulness is a therapy that involves the mind as well as the body to be fully present and accept the present experience to gain calm and wisdom in dealing with stressors.

Meanwhile, another study conducted on 131 students from several universities in America stated that mindfulness therapy did not significantly impact students' stress levels [25]. In addition, based on research on 288 students, mindfulness did not show a positive reaction to both reducing stress and student welfare [26]. This shows that there are quite contradictory differences in results.

A large number of events of stress, depression, and anxiety in global students, especially during the COVID-19 pandemic, make it important to prevent and deal with health approaches, such as mindfulness therapy. It is critical to consider the effects of these interventions in order for students to make evidence-based decisions concerning mental healthcare at university. While mindfulness is widely available, it is unknown whether it has a beneficial effect on the mental health of university students. Therefore, this study intended to examine whether an intervention based on mindfulness was effective and potentially an intervention to decrease stress, depression, and anxiety in university students.

2. Materials and Methods

2.1. Study Design

This research was a randomized controlled trial. Students from universities in various provinces in Indonesia, covering all provinces in Java, Sumatra, and Borneo island, took part in this study. Moreover, this study was registered and approved by TCTR with register number TCTR2022080800.

2.2. Recruiting and Sampling

A total of 122 students from 16 universities in Indonesia participated in this study. Recruitment of participants was carried out through open recruitment with information dissemination assisted by posting the flyer on social media. Each potential participant was reached through a cell phone communication application. A number of these students were selected based on the established criteria. The inclusion criteria in this study were: (1) Indonesian university students from all provinces in Java, Sumatra, and Borneo island who were willing to join the program and (2) adults (aged ≥18 years old) as the inclusion criteria. There was also an exclusion criterion including students with severe mental disorders. Before conducting the research, participants were given research information or informed consent, which was provided online through the Google Form. The informed consent was limited to one person, contained the respondent's willingness to participate, and participants could freely choose to participate or not in this research. This recruitment process was carried out in January 2022.

Participants in this study were divided into two groups, namely the intervention and control groups. The size of the sample uses the G* power formula by determining the significance level ($\alpha = 0.05$), effect size ($d = 0.80$), and power (80%). From the calculation based on the formula, the sample size was 26 people in each group. As for this study, there were 122 respondents, and none dropped out, so each group consisted of 61 students. Randomization with a simple random sampling approach was carried out to determine the distribution of groups for participants. This procedure was carried out by contactless enumerators of participants who were blindly selected by randomization minimization software 12.0. The results of this randomization then determined the placement of participants and whether they entered the intervention or control group.

2.3. Data Collection

Before the intervention was started, all participants' stress, anxiety, and depression were measured with the DASS-42, which was regulated through Google Forms. The DASS-42 includes 42 articles assessing the level of stress, anxiety, and depression [27]. The outcome of the study including the level of depression, anxiety, and stress was assessed using the DASS instrument. DASS is a psychological measurement instrument that measures depression, stress, and anxiety scale in individuals. Developed by Lovibond and Lovibond (1995), it consists of 42 question items in which each measurement scale consists of 14 items [28]. The instrument used in this study was DASS-42, which has been translated into an Indonesian version [29]. A validity test by Damanik and Rusli showed that the 42 items were valid. The reliability values obtained for the Indonesian language DASS instrument for the depression, anxiety, and stress scales were 0.872, 0.806, and 0.816, respectively [30]. The result indicated that the instrument is valid and reliable to use.

Research activities ranging from pretest measurements, carrying out the intervention, and the post-test were carried out from February to March 2022. The participants were then requested to join a WhatsApp group (WAG) initiated by researchers, which described the comprehensive protocol of mindfulness breathing meditation and provided support with video practice guidelines and tutorials. The intervention was carried out online through a Zoom meeting. Therefore, there were research assistants who assisted in the implementation of interventions during the research with blind group placement. The participants were required to exercise mindfulness breathing meditation every day for 4 weeks with 15 min duration on each day. In the first 2 weeks, the intervention was carried out with guidance through Zoom meetings. Meanwhile, in the following 2 weeks, the participants conducted their intervention without guidance from the research assistant. They were permitted to choose the most convenient times and places to exercise. To make sure that their exercise followed the guidelines, they were required to complete a reflective form provided by the researchers. Every week, participants were given reminders regarding the implementation of the intervention. The respondents practiced breathing meditation for 4 weeks, during which the researchers actively motivated and maintained contact with the participants through WAG chats. Unlike the intervention group, the control group did not carry out activities related to mindfulness breathing meditation during those 4 weeks. The researcher conveyed that the participants should continue to carry out activities as usual, both at home and elsewhere. Even so, the control group still received video practice guidelines and tutorials for mindfulness breathing meditation after a series of studies.

After 4 weeks, a follow-up was carried out, and the participants were required to fulfill the self-administered DASS-42 questionnaire via Google Forms as a post-test. The DASS-42 was directed by the researchers as a pretest survey before and as a post-test survey after the 4 weeks. During the administration of the intervention, no participants dropped out, so the number of students in each group was constant (n = 61) from the pretest to the post-test.

The data for analysis used in the first measuring is given in Figure 1, which shows the consort study flow chart.

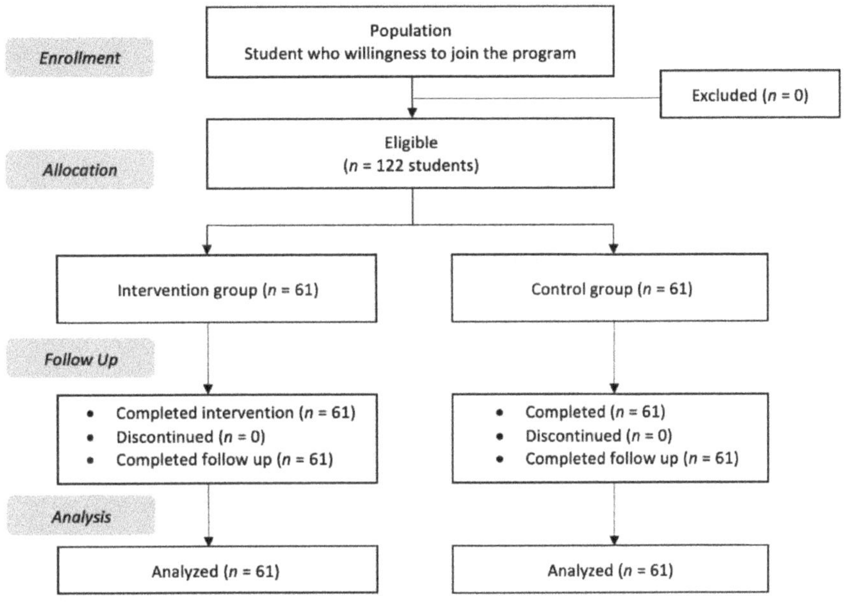

Figure 1. Consort study flow diagram.

2.4. Statistical Analysis

Data were collected from Google Forms and relocated to SPSS software version 24 (IBM Corp., Armonk, NY, USA). SPSS was used to analyze the demographic data and also baseline and intervention data for stress, anxiety, and depression. In this study, the analysis used univariate and bivariate analysis. A univariate analysis technique was conducted to determine the distribution, frequency, and percentage of respondents' characteristics, which included age, gender, and religion, besides anxiety, stress, and depression levels with the criteria: 0–14 = normal, 15–18 = mild stress level, 19–25 = moderate stress level, 26–33 = severe stress level, and >34 = very severe stress level; 0–9 = normal, 10–13 = mild depression level, 14–20 = moderate depression level, 21–27 = severe depression level, and >28 = very severe depression level; 0–7 = normal, 8–9 = mild anxiety level, 10–14 = moderate anxiety level, 15–19 = severe anxiety level, and >20 = very severe anxiety level, respectively.

The second analysis was a bivariate analysis. In this study, an independent *t*-test was performed to analyze the comparison of the intervention and control groups. We also used a one-way analysis of variance (ANOVA) to determine the effect of other variables on the effectiveness of mindfulness interventions on stress, anxiety, and depression.

2.5. Ethical Clearance

Ethical clearance for this research was granted by the ethics research committee of Universitas Padjadjaran with letter no. 1062/UN6.KEP/EC/2021. Participants involved in this study were instructed about the research purposes, and informed consent was acquired as a legal requisite.

3. Result

3.1. Characteristics of Participants

A total of 122 participants fulfilling the criteria were included in this study (mean age, 20.30 ± 1.116 in the intervention group and 22.42 ± 3.672 in the control group; most respondents were women, 51 (83.60%) in the control group and 42 (68.9%) in the intervention group; and majority religion was Islam with 54 respondents). Table 1 displays the university students' sociodemographic variables; there were no noteworthy differences between the control and intervention groups, except for the age variable.

Table 1. Demographic of participants.

Characteristic	Control Group	Intervention Group	*p*-Value
Gender			
Female	51 (83.60%)	42 (68.90%)	0.089
Male	10 (16.40%)	19 (31.15%)	
Religion			
Islam	54 (49.1%)	56 (50.9%)	0.361
Christian	5 (50%)	5 (50%)	
Buddhism	2 (100%)	0 (0%)	
Age	20.30 ± 1.116	22.42 ± 3.672	0.000

3.2. Baseline and Post-Intervention Stress, Anxiety, and Depression of Participants

Based on the analysis of stress, anxiety, and depression in the pre- and post-tests, in Table 2, it can be seen that there is a difference in mean and standard deviation in each group. From these data, it was also found that there was no significant differences in each group for the baseline data for stress (p = 0.862), anxiety (p = 0.664), and depression (p = 0.363). As for the post-test, the test results showed that there were significant differences between the post-test categories of the two groups in terms of stress (p = 0.007) and anxiety (p = 0.042). Whereas in depression, the results showed that there was no significant difference in the post-test data of the two groups. Even so, in the intervention group, there was a decrease in mean and SD for all variables. Thus, even though there was no significant change in the depression variable, the interventions were successful in reducing the average depression score in that group.

Table 2. Baseline and post-intervention stress, anxiety, and depression of participants.

Variable		Intervention Group M (SD)	Control Group M (SD)	*p*-Value
Stress	Baseline	16.31 (9.807)	16.00 (10.00)	0.862
	Post	10.07 (6.957)	14.03 (8.77)	0.007
Anxiety	Baseline	14.08 (8.737)	13.39 (8.74)	0.664
	Post	8.46 (5.448)	10.66 (6.32)	0.042
Depression	Baseline	13.93 (9.631)	12.36 (9.42)	0.363
	Post	7.54 (7.309)	9.46 (7.70)	0.161

3.3. Frequency Distribution of Participants' Levels of Stress, Anxiety, and Depression

Based on an analysis of the frequency distribution of stress, anxiety, and depression, it was found that respondents who participated in this study were divided into five categories, namely normal, mild, moderate, severe, and extremely severe, in each of the two groups' variables. Before giving the intervention, apart from being at a normal level, the other categories that were mostly found for stress occurred in the moderate category (60.0%)

in the intervention group and severe (55.6%) in the control group, anxiety occurred at a moderate level for the intervention group (55.9%) and control (44.1%), while depression occurred at a moderate level for the intervention group (46.2%) and control (53.8%). After giving the intervention, stress, anxiety, and depression categories were dominated by normal levels in the intervention group for stress (44.3%), anxiety (42.3%), and depression (57.7%), and in the control group for stress (55.7%), anxiety (57.7%) and depression (58%). These results are shown in Table 3.

Table 3. Frequency distribution of participants' levels of stress, anxiety, and depression.

Variable	Intervention Group		Control Group	
	Baseline	Post	Baseline	Post
Stress				
Normal	32 (51.6%)	44 (55.7%)	30 (48.4%)	35 (44.3%)
Mild	8 (50.0%)	8 (42.1%)	8 (50.0%)	11 (57.9%)
Moderate	8 (40.0%)	7 (46.7%)	12 (60.0%)	8 (53.3%)
Severe	10 (55.6%)	0 (0.0%)	8 (44.4%)	6 (100.0%)
Extremely severe	3 (50.0%)	2 (66.7%)	3 (50.0%)	1 (33.3%)
Anxiety				
Normal	15 (50.0%)	30 (57.7%)	15 (50.0%)	22 (42.3%)
Mild	6 (50.0%)	7 (46.7%)	6 (50.0%)	8 (53.3%)
Moderate	15 (44.1%)	17 (54.8%)	19 (55.9%)	14 (45.2%)
Severe	9 (60.0%)	4 (28.6%)	6 (40.0%)	10 (71.4%)
Extremely severe	16 (51.6%)	3 (30.0%)	15 (48.4%)	7 (70.0%)
Depression				
Normal	21 (43.8%)	29 (58.0%)	27 (56.2%)	21 (42.0%)
Mild	13 (52.0%)	18 (62.1%)	12 (48.0%)	11 (37.9%)
Moderate	14 (53.8%)	10 (37.0%)	12 (46.2%)	17 (63.0%)
Severe	6 (60.0%)	2 (22.2%)	4 (40.0%)	7 (77.8%)
Extremely severe	7 (53.8%)	2 (28.6%)	6 (46.2%)	5 (71.4%)

Note: data are expressed as frequency (n) and percentage (%) for each level of variables.

3.4. Age and Religion Factors on Participant's Interventions

Table 4 shows the results of testing the age and religion categories regarding their influence on successful interventions. These results are presented in the form of data on the average and standard deviation of each variable for each group, followed by the p-values for the post-test of the two groups with one-way ANOVA processing. Based on the test results, it was found that there was no difference in the scores for stress, anxiety, and depression for the pre- or -post-tests, either in the treatment group or the control group based on age (p = 0.63, 0.457, and 0.573) and religion (p = 0.568, 0.505, 0.534).

Table 4. Mean scores (standard deviations) based on age and religion of participants.

Variable	Baseline		Post		*p*-Value *
	IG (M, SD)	CG (M, SD)	IG (M, SD)	CG (M, SD)	
Age (years)					
Stress					
<21	14.66 (9.152)	21.83 (7.627)	10.00 (6.958)	14.83 (9.827)	0.63
21	18.82 (11.775)	15.19 (6.322)	11.12 (7.960)	13.05 (8.851)	
>21	18.00 (7.714)	15.47 (11.909)	8.33 (4.975)	14.50 (8.760)	
Anxiety					
<21	12.60 (8.388)	17.33 (5.007)	8.23 (5.180)	13.50 (6.504)	0.457
21	15.59 (10.381)	13.10 (6.503)	9.71 (6.659)	11.05 (7.619)	
>21	17.00 (5.831)	12.88 (10.304)	7.00 (3.742)	9.91 (5.384)	
Depression					
<21	12.66 (9.434)	16.67 (8.335)	7.74 (6.423)	11.83 (6.969)	0.573
21	16.76 (11.306)	11.67 (6.102)	8.35 (10.000)	11.10 (9.633)	
>21	13.56 (6.126)	12.03 (11.156)	5.22 (4.206)	8.03 (6.264)	
Religion					
Stress					
Islam	17.20 (9.727)	16.11 (10.101)	10.69 (7.012)	14.41 (8.891)	0.568
Christian	10.00 (8.337)	14.80 (9.731)	4.80 (2.683)	9.80 (6.496)	
Buddhism	8.00 (9.899)	-	6.50 (9.192)	-	
Anxiety					
Islam	14.89 (8.801)	13.43 (8.932)	9.00 (5.425)	10.71 (6.387)	0.505
Christian	8.20 (4.868)	13.00 (6.892)	3.80 (2.168)	10.00 (6.205)	
Buddhism	7.00 (8.485)	-	5.50 (7.778)	-	
Depression					
Islam	14.50 (9.743)	12.63 (9.680)	7.93 (7.581)	9.55 (7.865)	0.534
Christian	9.40 (7.436)	9.40 (5.550)	3.40 (1.140)	8.40 (6.025)	
Buddhism	10.00 (12.728)	-	7.50 (7.778)	-	

Note: * One-way ANOVA test; IG, = intervention group; CG = control group; M = mean; SD = standard deviation.

4. Discussion

4.1. Principal Finding

At the baseline data of the intervention and control groups, there was no significant difference between the two groups for stress, anxiety, and depression. Meanwhile, in the post-test data, it was found that there were significant differences between groups for stress and anxiety.

From the findings, the data showed that, compared with the control group, the intervention group that received the mindfulness therapy program resulted in a reduction in depression, anxiety, and stress. In this study, which is about the effectiveness of mindfulness therapy on depression, anxiety, and stress, it appears that the average score of the intervention group decreased more than the average score of the control group, in which the control group experienced a slight change in score. This finding is similar to a study by Ibrahim et al. [31], in which MBSR effectively increased psychological well-being. Additionally, a study by Nyklicek et al. [32] described that an MBSR program consisting of

90–120 min of instruction for only three sessions per week effectively reduced depression, anxiety, and stress in 107 percutaneous coronary patients.

Although it resulted in a higher mean score reduction in stress, anxiety, and depression in the intervention group, the post-test results for depression in the control and intervention groups showed no significant differences. Only stress and anxiety showed a significant difference in the post-test results between the intervention and control groups. Thus, the most significant change is indicated by stress, which is shown by a p-value that is smaller than the p-value of anxiety. The results are similar to Chiodelli et al.'s study that stress was the variable that had a significant difference. The use of nonclinical populations in the research has the potential to affect the effectiveness of mindfulness intervention. This could be because the reduction is significantly more likely to be experienced in populations who already experience high levels of depression and anxiety [33]. The same finding was also reported by Chen et al. [34]—decreasing levels in anxiety scores were significantly greater in the MBSR group than in the control group. However, there was not a significant result for depression levels. In connection with these matters, the respondents included in this study consisted of students of various ages and religions. However, the test results showed that these two aspects did not significantly influence the intervention for the two groups. Meanwhile, if viewed from a psychological condition, the respondents who were included in this study were not only respondents with high levels of stress, depression, and anxiety, but all levels were included, so this might also affect the effectiveness of the interventions carried out.

According to Chen et al. [34], the reason for this negative finding is that the MBSR program was only conducted as a short-term intervention, which was only for 7 consecutive days. In this study, the time of the intervention was carried out for 4 weeks. In a previous study on MBSR with 32 nursing students, an MBSR program was conducted for 8 weeks [35]. Moreover, in another study in which an MBSR was carried out for 2.5 h every week for 8 consecutive weeks, it showed a significant reduction in depression and stress in 83 patients with chronic disease [36]. Compared with other studies that show significant results on depression, stress, and anxiety, the intervention time in this research tends to be less. This is because some articles indicate an intervention duration of 8 weeks, while in this study, the intervention only lasted 4 weeks. Thus, the researchers suspect that the length of time an intervention is given can be one of the factors that increasingly influence the effectiveness of the mindfulness intervention.

Stress is a pressure that is felt by an individual as a result of an imbalance between the demands of their needs and the capacity they have, so it has an impact on the individual's biopsychological condition. Students belong to a group that is prone to experiencing stress related to academic and nonacademic responsibilities, such as relationships with family and students' social environment. Not only stress but anxiety and depression were also felt by students. Those psychological problems, if otherwise not handled properly, can result in various negative impacts that affect the physical and psychological well-being of students. Therefore, efforts to reduce stress, anxiety, and depression experienced by students need to be carried out so that it also has an impact on the optimization of the learning process during lectures. The results of this study indicate that mindfulness-based therapy can be an effective intervention to reduce student stress and anxiety. Mindfulness is a therapy that involves integrating the mind and body to fully focus, present oneself, and accept left-hand experiences to gain calm, wisdom, and peace in dealing with stressors.

In this study, the mindfulness therapy used is as a type of breathing meditation, which is a type of MBSR that is performed by focusing on abdominal breathing, along with feeling the sensations felt in the body, mind, and feelings without paying attention to things other than the sensations or emotions felt [37]. This exercise is performed in an upright sitting position but remaining relaxed, with the chest open for some time in a state of silence. According to Lutz (in Kropp and Sedlmeier) [38], Nagas meditation includes three basic abilities, namely monitoring the stimulus or sensation felt without distracting focus, the ability to separate oneself from possible distractions, and the ability to focus fully on the object or sensation that is felt.

From several findings, it was found that significant results in reducing stress levels were obtained from participants who participated in face-to-face and online MBSR therapy using the Mindful Skills for Student (MSS) application. Therefore, the combination of offline and online MBSR intervention was effective in helping participants manage the stress they experienced. Based on previous research, it was stated that internet-based mindfulness interventions have the potential to meet the needs of participants because they are more accessible and flexible so that they can be integrated with healthcare facilities to help manage stress, depression, and anxiety [39–42].

4.2. Limitations

The research findings obtained by several participants in the study conveyed the factors that became obstacles while carrying out mindfulness breathing meditation. The biggest obstacle that occurred during the intervention was the low level of public awareness and sociocultural influences. For instance, in this study, awareness was still very low among the campus community, and where participants lived resulted in mindfulness interventions not being widely known. This was indicated by a lack of public knowledge about the existence of mindfulness therapy and a lack of facilities and infrastructure in the campus environment and where participants live that provide information related to mindfulness. Other inhibiting factors are from a social and cultural perspective. Participants felt that the practice of mindfulness was not familiar in their social and cultural environment because, since childhood, there was no one that required them to behave mindfully. In this study, the age of the respondents was not homogeneous, but the practice of mindfulness can be carried out by various ages.

In addition, the respondents who were included in this study ranged from respondents with normal levels of stress, anxiety, and depression, so the possible symptoms they experienced were not too severe for some respondents. Besides that, the control group was only contacted individually for each respondent, and the researcher did not administer any treatment regarding mindfulness interventions that could be obtained with media or other methods that were different from the intervention group when the intervention group received the MBSR intervention.

4.3. Implications for Clinical Practice

The current result of our study indicates that mindfulness breathing meditation can be a promising strategy to efficiently reduce the levels of stress, depression, and anxiety. The fact that this meditation can be carried out at any time or any place also makes it a very cost-effective wellness intervention that can be accessed by everyone universally. Furthermore, this study provides a rationale for including this therapy in any nursing courses as one of the effective ways to make students relax. This intervention can be carried out using MP3 or MP4 video media files to guide the implementation of practical activities so that it can also be carried out by students independently. With the integration of an application for mindfulness implementation on campus, it is hoped that it will be able to increase students' abilities and concentration so that each intervention that is carried out can be more effective.

5. Conclusions

This study showed that mindfulness breathing meditation provides a reduction in scores of stress, depression, and anxiety among university students in Indonesia. However, an effective reduction occurred for stress and anxiety at the 4-week follow-up. Mindfulness breathing meditation is a nonpharmacological method, which means it can be implemented virtually at any time and any place. The use of mindfulness breathing meditation could be applied to all university students and so forth to develop psychosocial status and mindful attentiveness to one's needs.

From this study, we recommend a future study to conduct research about the factors that influence the effectiveness of giving mindfulness breathing meditation therapy. We also recommend that future researchers provide treatment to the control group related to the intervention during the research, for example, providing intervention information with poster media. This is intended so that all participants in the study receive the same treatment regarding the intervention. Additionally, a future study to assess the long effect of 6–12 months of follow-up is needed in the new normal era.

Author Contributions: Conceptualization, M.K., K.I., T.P., L.R. and I.S.; methodology, M.K. and I.S.; software, M.K., L.R. and I.S.; validation, K.I., T.P. and L.R.; formal analysis, M.K., K.I., T.P. and I.S.; investigation, M.K., K.I. and T.P.; resources, K.I., T.P. and I.S.; writing—original draft preparation, M.K., K.I., T.P., L.R. and I.S.; writing—review and editing, M.K., K.I., T.P., L.R. and I.S.; visualization, K.I. and I.S.; supervision, M.K., K.I., T.P. and L.R.; project administration, M.K.; funding acquisition, T.P. and L.R. All authors have read and agreed to the published version of the manuscript.

Funding: This research received funding from Universitas Padjadjaran with funding number no.1959/UN6.3.1/PT.00/2021. Moreover, this article processing charge (APC) was funded by Universitas Padjadjaran through a central invoice to the institution.

Institutional Review Board Statement: Not applicable.

Informed Consent Statement: Informed consent was obtained from all subjects involved in the study.

Data Availability Statement: Not applicable.

Acknowledgments: The authors acknowledge to all of the participants and the students of Universitas Padjadjaran who were involved in the study. Moreover, we would like to say thank you very much to Sidik Maulana, Syifa Fauziah, Elda Regita Dewi, and Hediati Hastuti for their assistance during the research.

Conflicts of Interest: The authors declare no conflict of interest.

References

1. Martinez, P.J.; Aguilar, F.J.; Ortiz, M. Transitioning from face-to-face to blended and full online learning engineering master's program. *IEEE Trans. Educ.* **2020**, *63*, 2–9. [CrossRef]
2. Pelucio, L.; Simões, P.; Cristina, M.; Dourado, N.; Quagliato, L.A.; Nardi, A.E. Depression and anxiety among online learning students during the COVID-19 pandemic: A cross-sectional survey in Rio de Janeiro, Brazil. *BMC* **2022**, *10*, 192. [CrossRef]
3. Wang, C.; Pan, R.; Wan, X.; Tan, Y.; Xu, L.; McIntyre, R.S.; Ho, C. A longitudinal study on the mental health of general population during the COVID-19 epidemic in China. *Brain Behav. Immun.* **2020**, *87*, 40–48. [CrossRef] [PubMed]
4. Lopes, A.R.; Nihei, O.K. Depression, anxiety and stress symptoms in Brazilian university students during the COVID-19 pandemic: Predictors and association with life satisfaction, psychological well-being and coping strategies. *PLoS ONE* **2021**, *61*, e0258493. [CrossRef]
5. Fawaz, M.; Samaha, A. E-learning: Depression, anxiety, and stress symptomatology among Lebanese university students during COVID-19 quarantine. *Nurs. Forum* **2021**, *56*, 52–58. [CrossRef] [PubMed]
6. Song, Y.; Lindquist, R. Effects of mindfulness-based stress reduction on depression, anxiety, stress and mindfulness in Korean nursing students. *Nurse Educ. Today* **2015**, *35*, 86–90. [CrossRef]
7. Buchanan, J.L. Prevention of depression in the college student population: A review of the literature. *Arch. Psychiatr. Nurs.* **2012**, *26*, 21–42. [CrossRef]
8. Santoso, A.; Ardi, W.R.; Prasetya, R.L.; Dwidiyanti, M.; Wijayanti, D.Y.; Mu'in, M.; Aisah, N.A. Tingkat depresi mahasiswa keperawatan di tengah wabah COVID-19. *Holist. Nurs. Health Sci.* **2020**, *3*, 1–8. [CrossRef]
9. Mahmoud, J.S.R.; Staten, R.T.; Hall, L.A.; Lennie, T.A. The relationship among young adult college students' depression, anxiety, stress, demographics, life satisfaction, and coping styles. *Issues Ment. Health Nurs.* **2012**, *33*, 149–156. [CrossRef]
10. García-González, J.; Ruqiong, W.; Alarcon-Rodriguez, R.; Requena-Mullor, M.; Ding, C.; Ventura-Miranda, M.I. Analysis of Anxiety Levels of Nursing Students Because of e-Learning during the COVID-19 Pandemic. *Healthcare* **2021**, *9*, 252. [CrossRef]
11. Cao, W.; Fang, Z.; Hou, G.; Han, M.; Xu, X.; Dong, J.; Zheng, J. The psychological impact of the COVID-19 epidemic on college students in China. *Psychiatry Res.* **2020**, *287*, 112934. [CrossRef] [PubMed]
12. COVID-19 Pandemic Has Worsened Pre-Existing Mental Health Conditions. News-Medical Net. 2020. Available online: https://www.news-medical.net/news/20200601/COVID-19-pandemic-has-worsened-pre-existing-mental-health-conditions.aspx (accessed on 14 April 2022).
13. Shikai, N.; Shono, M.; Kitamura, T. Effects of coping styles and stressful life events on depression and anxiety in Japanese nursing students: A longitudinal study. *Int. J. Nurs. Prac.* **2009**, *15*, 198–204. [CrossRef] [PubMed]

14. Warnecke, E.; Quinn, S.; Ogden, K.; Towle, N.; Nelson, M.R. A randomised controlled trial of the effects of mindfulness practice on medical student stress levels. *Med. Educ.* **2011**, *45*, 381–388. [CrossRef] [PubMed]
15. Jimenez, S.S.; Niles, B.L.; Park, C.L. A mindfulness model of affect regulation and depressive symptoms: Positive emotions, mood regulation expectancies, and self-acceptance as regulatory mechanisms. *Personal. Individ. Differ.* **2010**, *49*, 645–650. [CrossRef]
16. Aristovnik, A.; Keržič, D.; Ravšelj, D.; Tomaževič, N.; Umek, L. Impacts of the covid-19 pandemic on life of higher education students. *Sustainability* **2020**, *12*, 8438. [CrossRef]
17. Son, C.; Hegde, S.; Smith, A.; Wang, X.; Sasangohar, F. Effects of covid-19 on college students' mental health in the united states: Interview survey study. *J. Med. Internet Res.* **2020**, *22*, e21279. [CrossRef]
18. Limcaoco, R.S.G.; Mateos, E.M.; Fernandez, J.M.; Roncero, C. Anxiety, worry and perceived stress in the world due to the COVID-19 pandemic, March 2020. Preliminary results. *MedRiv* **2020**. [CrossRef]
19. Baer, R.A. Self-focused attention and mechanisms of change in mindfulness-based treatment. *Cogn. Behav. Ther.* **2009**, *38*, 15–20. [CrossRef] [PubMed]
20. Crowley, C.; Kapitula, L.R.; Munk, D. Mindfulness, happiness, and anxiety in a sample of college students before and after taking a meditation course. *J. Am. Coll. Health* **2020**, *70*, 493–500. [CrossRef] [PubMed]
21. Hidayati, N.; Aryanti, P.I.; Salim, S.; Zuliani Erniyawati, Y.; Yusuf, A. Penggunaan mindfulness meditation dalam manajemen stres: Sebuah systematic review. *J. Penelit. Kesehat. Suara Forikes* **2018**, *9*, 95–100.
22. Spears, C.A. Perceptions of mindfulness in a low-income, primarily african american treatment-seeking sample. *Physiol. Behav.* **2019**, *176*, 139–148. [CrossRef]
23. Barnes, N.; Hattan, P.; Black, D.S.; Schuman-Olivier, Z. An examination of mindfulness-based programs in US medical schools. *Mindfulness* **2017**, *8*, 489–494. [CrossRef]
24. Black, D.S. Mindfulness-based interventions: An antidote to suffering in the context of substance use, misuse, and addiction. *Subst. Use Misuse* **2014**, *49*, 487–491. [CrossRef] [PubMed]
25. Seppälä, E.M.; Bradley, C.; Moeller, J.; Harouni, L.; Nandamudi, D.; Brackett, M.A. Promoting mental health and psychological thriving in university students: A randomized controlled trial of three well-being interventions. *Front. Psychiatry.* **2020**, *11*, 590. [CrossRef] [PubMed]
26. De Vibe, M.; Solhaug, I.; Tyssen, R.; Friborg, O.; Rosenvinge, J.H.; Sørlie, T.; Bjørndal, A. Mindfulness training for stress management: A randomised controlled study of medical and psychology students. *BMC Med. Educ.* **2013**, *13*, 1–11. [CrossRef] [PubMed]
27. Psychology Foundation of Australia. 2013. Available online: http://www.psychologyfoundation.org.au (accessed on 12 July 2022).
28. Widyana, R.; Sumiharso; Safitri, R.M. Psychometric properties of internet-administered version of depression, anxiety and stress scales (dass-42) in sample Indonesian adult. *J. Talent Dev. Excell.* **2020**, *12*, 1422–1434.
29. Damanik, E.D.; Rusli, E. Pengujian Reabilitas, Validitas, Analisis Item, dan Pembuatan Norma Depression Anxiety Scale (DASS): Berdasarkan Penelitian pada Kelompok Sampel Yogyakarta dan Bantul yang Mengalami Gempa Bumi dan Kelompok Sampel Jakarta dan Sekitarnya yang Tidak Mengalami Gempa Bumi. Thesis, Faculty of Psychology, Universitas Indonesia, Depok, Indonesia, 2006.
30. Mutaqqin, D.; Ripa, S. Psychometric properties of the Indonesian version of the department anxiety stress scale: Factor structure, reability, gender and age measurements invariance. *Psikohumaniora J. Penelit. Psikol.* **2021**, *6*, 61–76. [CrossRef]
31. Ibrahim, K.; Komariah, M.; Herliani, Y.K. The effect of mindfulness breathing meditation on psychological well-being: A quasi-experimental study among nurses working for COVID-19 patients. *Holist. Nurs. Pract.* **2022**, *36*, 46–51. [CrossRef]
32. Nyklíček, I.; Dijksman, S.C.; Lenders, P.J.; Fonteijn, W.A.; Koolen, J.J. A brief mindfulness-based intervention for increase in emotional well-being and quality of life in percutaneous coronary intervention (pci) patients: The mindfulheart randomized controlled trial. *J. Behav. Med.* **2014**, *37*, 135–144. [CrossRef]
33. Chiodelli, R.; Jesus SN, D.; Mello LT, N.D.; Andretta, I.; Oliveira, D.F.; Costa ME, S.; Russell, T. Effects of the Interculturality and Mindfulness Program (PIM) on University Students: A Quasi-Experimental Study. *Eur. J. Investig. Health Psychol. Educ.* **2022**, *12*, 1500–1515. [CrossRef]
34. Chen, Y.; Yang, X.; Wang, L.; Zhang, X. A randomized controlled trial of the effects of brief mindfulness meditation on anxiety symptoms and systolic blood pressure in Chinese nursing students. *Nurse Educ. Today* **2013**, *33*, 1166–1172. [CrossRef] [PubMed]
35. Kang, Y.S.; Choi, S.Y.; Ryu, E. The effectiveness of a stress coping program based on mindfulness meditation on the stress, anxiety, and depression experienced by nursing students in Korea. *Nurse Educ. Today* **2009**, *29*, 538–543. [CrossRef] [PubMed]
36. Dobkin, P.L.; Zhao, Q. Increased mindfulness–The active component of the mindfulness-based stress reduction program? *Complement. Ther. Clin. Pract.* **2011**, *17*, 22–27. [CrossRef] [PubMed]
37. Anālayo, B. In the seen just the seen: Mindfulness and the construction of experience. *Mindfulness* **2019**, *10*, 179–184. [CrossRef]
38. Kropp, A.; Sedlmeier, P. What makes mindfulness-based interventions effective? An examination of common components. *Mindfulness* **2019**, *10*, 2060–2072. [CrossRef]
39. Mak, W.W.; Chio, F.H.; Chan, A.T.; Lui, W.W.; Wu, E.K. The efficacy of internet-based mindfulness training and cognitive-behavioral training with telephone support in the enhancement of mental health among college students and young working adults: Randomized controlled trial. *J. Med. Internet Res.* **2017**, *19*, e84. [CrossRef] [PubMed]

40. Widiasih, R.; Ermiati; Jayanti, T.N.; Rais, Y. Psychosocial Interventions for Improving the Quality of Life in Breast Cancer Survivors: A Literature Review. *IOP Conf. Ser. Earth Environ. Sci.* **2019**, *248*, 12056. [CrossRef]
41. Komariah, M.; Amirah, S.; Faisal, E.G.; Prayogo, S.A.; Maulana, S.; Platini, H.; Suryani, S.; Yosep, I.; Arifin, H. Efficacy of Internet-Based Cognitive Behavioral Therapy for Depression and Anxiety among Global Population during the COVID-19 Pandemic: A Systematic Review and Meta-Analysis of a Randomized Controlled Trial Study. *Healthcare* **2022**, *10*, 1224. [CrossRef]
42. Kurniawan, K.; Yosep, I.; Maulana, S.; Mulyana, A.M.; Amirah, S.; Abdurrahman, M.F.; Sugianti, A.; Putri, E.G.; Khoirunnisa, K.; Komariah, M.; et al. Efficacy of Online-Based Intervention for Anxiety during COVID-19: A Systematic Review and Meta-Analysis of Randomized Controlled Trials. *Sustainability* **2022**, *14*, 2866. [CrossRef]

healthcare

Article

Development of a Psychological Intervention to Improve Depressive Symptoms and Enhance Adherence to Antiretroviral Therapy among Adolescents and Young People Living with HIV in Dar es Salaam Tanzania

Tasiana Njau [1,2,3,*], Fileuka Ngakongwa [4], Bruno Sunguya [5], Sylvia Kaaya [1] and Abebaw Fekadu [2,3,6]

1 Department of Psychiatry and Mental Health, The Muhimbili University of Health and Allied Sciences, 9 United Nations Road, Upanga West 11103, Dar es Salaam 65001, Tanzania
2 Centre for Innovative Drug Development and Therapeutic Trials for Africa (CDT-Africa), Addis Ababa University, Addis Ababa 9086, Ethiopia
3 Department of Psychiatry, School of Medicine, College of Health Sciences, Addis Ababa University, Addis Ababa 9086, Ethiopia
4 Department of Psychiatry and Mental Health, Muhimbili National Hospital, Dar es Salaam 65000, Tanzania
5 Department of Community Health, Muhimbili University of Health and Allied Sciences, 9 United Nations Road, Upanga West 11103, Dar es Salaam 65001, Tanzania
6 Department of Global Health & Infection, Brighton and Sussex Medical School, Brighton BN1 9PX, UK
* Correspondence: tasiana2009@live.com; Tel.: +255-717547606

Abstract: Background: Interventions that simultaneously target depression and antiretroviral therapy (ART) medication adherence are recommended for improving HIV treatment outcomes and quality of life for adolescents living with HIV. However, evidence is scarce on culturally feasible and acceptable interventions that can be implemented for HIV-positive adolescents in Tanzania. We, therefore, developed a manualized brief psychological intervention that utilizes evidence-based strategies to address depression and ART adherence in adolescents living with HIV in Tanzania. **Methods**: We used the Theory of Change Enhanced Medical Research Council framework (TOCMRC) for developing complex interventions in health care to develop the intervention in five phases. First, the literature was reviewed to identify potential intervention components. Second, we conducted a situational analysis using qualitative interviews with adolescents living with HIV, health care providers, and caregivers. Third, we conducted a mental health expert workshop; and fourth, theory of change workshops with representatives from the Ministry of Health, mental health professionals, HIV implementing partners, adolescents, and healthcare providers. Lastly, we synthesized results to finalize the intervention and a theory of change map showing the causal pathway for how we expect the developed intervention to achieve its impact. **Results:** Adolescents living with HIV in Tanzania experience several unmet mental health needs ranging from overwhelming depressive symptoms to not feeling understood by healthcare providers who lack mental health knowledge. Participants perceived psychological intervention that utilizes a task-shifting approach to be acceptable and beneficial to addressing those problems. The novel components of the NITUE intervention included incorporating evidence-based intervention components, namely, cognitive–behavioral therapy, motivational interviewing, and problem solving. In addition, caregiver inclusion in the treatment was essential to ensure access to care, compliance, and improved outcomes. **Conclusions:** A culturally appropriate brief psychological intervention that utilizes a task-shifting approach to address depression and medication adherence for adolescents living with HIV in Dar es Salaam, Tanzania, was developed. The intervention will be piloted for appropriateness, feasibility, and acceptability and will provide material for a future trial to determine its effectiveness.

Keywords: psychological intervention; behavioral intervention; adolescent depression; ART adherence; individual therapy; theory of change; intervention development

Citation: Njau, T.; Ngakongwa, F.; Sunguya, B.; Kaaya, S.; Fekadu, A. Development of a Psychological Intervention to Improve Depressive Symptoms and Enhance Adherence to Antiretroviral Therapy among Adolescents and Young People Living with HIV in Dar es Salaam Tanzania. *Healthcare* **2022**, *10*, 2491. https://doi.org/10.3390/healthcare10122491

Academic Editor: Elaina Taylor

Received: 12 October 2022
Accepted: 14 November 2022
Published: 9 December 2022

1. Introduction

About 38.4 million people live with Human Immunodeficiency Virus (HIV) worldwide, and 18.8 million are adolescent girls and young women [1,2]. Although new HIV infections in sub-Saharan Africa have decreased from 1.9 million in 2010 to 1.7 million in 2019, the prevalence of HIV in adolescents remains high [3]. About 3.8 million adolescents between 15 and 24 years were living with HIV in sub-Saharan Africa, equating to 76% of the global burden of the disease among young people [4,5]. Like other countries in the region, Tanzania is also experiencing a high burden of HIV among adolescents and young adults [3,6]. With a third of its population comprising adolescents and young people between 10–24 years old [7], nearly half of new HIV infections occur in this group [5]. Hence, Tanzania carries 5% of the global burden of HIV among adolescents [1,5].

Adolescents living with HIV (ALWHIV) have an increased risk for depression, negatively affecting ART adherence [8–10]. In Tanzania, the prevalence of depression among ALWHIV is 47.1% [11,12], and despite such a high burden, depression remains underdiagnosed and undertreated. Management of depression ensures better adherence to ART medication and hence better treatment outcomes, as well as improved quality of life [12–15].

Evidence suggests that single-approach interventions such as medication alone are inadequate in resolving depression and its associated difficulties, such as adherence to ART medication for adolescents with co-morbid HIV and depression. Adhering to antidepressants has also proved challenging because of their interactions with ART and the additional tablet burden; hence, including psychological interventions alongside medical treatment is essential [16]. For adolescents, the mental health Gap Action Program (mhGAP) recommends brief psychological interventions as a first-line treatment for common mental disorders such as depression [17].

Studies in Tanzania have widely recommended integrating mental health services and adherence interventions into the care and treatment of adolescents living with HIV [11,13,18–20]. A cause for alarm is that, even with a significant amount of work being carried out to support adolescents living with HIV/AIDS [21], there is a scarcity of data and evidence-based interventions that address both depression and ART adherence in adolescents living with HIV in Tanzania and SSA [9,20].

Studies that have identified effective interventions to improve depression and ART adherence in well-resourced countries suggest that cognitive–behavioral therapy is an effective treatment for depression [18,22,23]. Moreover, the client-centered approach in motivational interviewing [24] is suitable for addressing the complex multidimensional features of ART adherence [24–26]. Research on brief psychological interventions that combined the two approaches (CBT and MI) to address depression and medication adherence in ALWHIV has recently increased in high-resourced countries and with positive findings [25,27–29]. Similar interventions are encouraged to be scaled up in Tanzania even when they have not been developed or adapted for use in this poor-resourced setting [30–32].

The issues on the global mental health agenda have raised concerns about the cultural appropriateness of interventions taken to scale in low and middle-income countries LMIC [33]. The arguments highlight the challenges and risks of applying Western approaches and interventions to other cultures prior to assessing if inherent are socially appropriate, feasible, or effective [26,34,35]; hence, intervention development or good adaptation work needs consideration before going to scale. The aim is to ensure that the interventions can deliver care of acceptable standard and that the strategies are replicable [36,37] in these settings where human resources are limited [38,39]. The current focus of Global Mental Health and the World Health Organization (WHO) in addressing the gap in treating mental health problems, especially for ALWHIV in LMIC, recommends evidence-based interventions that can be integrated and scaled up through task shifting to improve access to service [34].

To fill that gap, we developed a brief psychological intervention that utilizes a task-shifting approach to address depression and ART medication adherence for ALWHIV in Dar es Salaam, Tanzania.

2. Methods

2.1. Design

We used the Theory of Change enhanced Medical Research Council framework (ToCMRC) for developing complex interventions in healthcare [40,41] to guide the development process. Iteratively, we followed the four phases of the MRC framework (intervention development, feasibility and piloting, evaluation, and implementation). The ToC map provides a graphic representation of the causal pathways through which the developed intervention is expected to achieve its impact. The map includes (i) the final outcome, which is the primary intervention outcome (improvement in depressive symptoms and enhanced ART adherence); (ii) intermediate outcomes, which include things that are currently unavailable, and that need to be there to achieve the outcome; (iii) interventions which are the different components of the intervention that are needed to move from one outcome to another; (iv) assumptions which are the conditions that are beyond the intervention that needs to be there to move forward in achieving the outcome); (v) the rationale from evidence or experience that explains the pathway for each outcome; and (vi) indicators which are measurable things that determine whether the outcome has been achieved or that there is progress toward aching the outcome [40]. The ToCMRC framework was favored because it provides a basis for enhanced stakeholder engagement that ensures specific intervention components will be well set within the local Tanzania cultural context [40]. The framework also systematically identifies knowledge gaps and clearly describes pathways to the outcomes of interest, in this case, reducing depressive symptoms and enhancing ART adherence. Additionally, with this framework, information obtained during the piloting process can be used to improve the intervention [40,42,43].

2.2. Intervention Development Process

Intervention development work started with a review of literature that facilitated the identification of potential components of the current intervention and a description of how those components could improve depressive symptoms or enhance ART adherence [44–46]. We also conducted a situational analysis comprising public data on adolescents living with HIV, treatment coverage, and integration of mental health services in Dar es Salaam. These data were obtained from Management and Development Health (MDH), a non-profit organization that supports HIV care and treatment services delivery in Tanzania, the District Medical Officer, and observation of service provision in selected centers that provide adolescent HIV care and treatment in Dar es Salaam. Furthermore, qualitative surveys were carried out to determine the unmet mental health needs of adolescents living with HIV, and the barriers and opportunities for implementing an integrated psychological intervention described elsewhere [47,48]. The qualitative study's findings shed light on lived experiences and indicate the problems from the perspectives of adolescents, caregivers, and healthcare providers. Qualitative findings also facilitated the identification of psychosocial factors that could be considered in the current intervention.

The development process engaged mental health professionals such as psychiatrists and clinical Psychologists from the Muhimbili University of Health and Allied Sciences (MUHAS) and the Muhimbili National Hospital (MNH). We also collected views on developing the intervention from the adolescents, caregivers, and health care providers in HIV care and treatment centers. Furthermore, essential stakeholders from the government, such as the district medical officer, the ministry of health, and HIV implementing partners in Tanzania, were also involved as they directly affect the implementation of interventions. The process ensured these stakeholders had a picture of possible outcomes within the existing structure of CTCs in Dar es Salaam.

2.3. Setting

The study was conducted in Dar es Salaam, the business capital city of Tanzania, with five districts: Ilala, Kinondoni, Ubungo, Temeke, and Kigamboni. The qualitative study was carried out at the HIV care and treatment centers in the Kinondoni district because these centers provide HIV services for adolescents on a different day than adults and have providers who may be more likely to understand the dynamics of HIV and mental health in ALHIV. MDH facilitated the selection of HIV care and treatment centers. The expert meeting and ToC workshops were conducted in the ilala district, where the Muhimbili University of Health and Allied Sciences and the Muhimbili National Hospital is located.

3. Study Participants and Sampling

We included individuals aged 18 years and older, able to provide written, informed consent to study participation. Participants also had to be either a mental health professional, a key stakeholder from the government/Ministry of Health, an HIV implementing partner, or an HIV healthcare provider with experience working in the field or directly with ALWHIV. The study also included ALWHIV who were at least 18 years old and their caregivers (parents or guardians). We used a purposive sampling approach [49] to recruit a diverse sample of participants. We considered those with experience or expertise in adolescent HIV and mental health care. Adolescents were recruited from the three selected clinics with the help of a healthcare provider from the respective clinic who was familiar with adolescents who can communicate, are physically and mentally stable, and are aware of their HIV-positive status.

4. Mental Health Professionals' Workshop

Two mental health professional meetings were conducted. The first included twenty-nine [29] mental health professionals (four psychiatrists, five clinical psychologists, three social workers, four psychiatric nurses, and thirteen masters level residents in psychiatry and clinical psychology). First, findings from phase one of the study, which consisted of the literature review and the qualitative studies, were presented and discussed. Next, mental health professionals discussed and gave recommendations on the content and delivery of the intervention based on the proposed findings and their clinical and research experience. The workshop was conducted in Swahili and English at the Muhimbili National Hospital's Department of Psychiatry and Mental Health and lasted two hours. Three people took notes for the meeting. The second meeting was between two psychiatrists and three psychologists from the first meeting, who reviewed inputs submitted in writing by participants from the first meeting. Combined notes were taken to come up with one document.

5. Theory of Change Workshops

Four ToC meetings were conducted each for a half day with the Ministry of Health and HIV implementing partners and district mental health officer (four), HIV health care providers (twelve), adolescents living with HIV (eight), representatives from implementing partners (two), health care providers (four and adolescents living with HIV (four)) We purposefully recruited the participants to include those more likely to give their opinion and consider their experience in supporting or delivering HIV and mental healthcare. All ToC workshops were facilitated by TN, a master-level clinical psychologist with intensive training in ToC. Efforts were made to ensure all participants got an opportunity to participate in each component fully. Summaries of key points were projected on a PowerPoint presentation or laid out in flip charts and revised after each subsection to validate the recorded information. After the workshop, using notes recorded during the workshop by two research assistants and flipcharts from the seminar, the Theory of Change map was updated.

The three TOC meetings started with a presentation that described the objectives of the meeting, which were (a) to explore the feasibility and acceptability of a psychological intervention to address depression in ALWHIV, and (b) to develop a ToC map that will indicate the causal pathway through which the newly developed psychological intervention was expected to achieve its impact. The presentation also covered findings from the literature review and qualitative studies. English and or Kiswahili were used as the medium of communication. After the introductory presentation, a discussion was led by

TN encouraging participants to discuss the findings with a focus on possible social-cultural benefits of psychological intervention for adolescents living with HIV. Participants were also encouraged to consider the feasibility of such intervention within the existing structure and the delivery of HIV care in Tanzania, including human resources, time, and other possible barriers.

Discussion of the various components of ToC started by asking participants what they would like as the outcome of the psychological intervention for HIV-positive adolescents with depressive symptoms and challenges in ART medication adherence. We used sticky notes to map things participants discussed as unavailable but needed to be there to achieve the desired outcomes. Next, we used an example of growing a home shade tree to help participants discuss intermediate activities and conditions that need to be there to move from one intermediate outcome to the other, indicators of outcomes, and measurable things that determines whether the outcome has been achieved or there is progress toward achieving the outcome. The example included questions like; what are the desired results of planting a tree in the garden? What needs to be conducted before planting a tree, and why the preparatory process is necessary for obtaining the desired effect/outcome? The tree example was beneficial, especially for adolescent ToC meetings, as their response made it easy to relate the outcome assumptions and interventions used to develop the psychological intervention.

For each ToC meeting, the ToC map for each session was drafted by reviewing workshop minutes and notes. An integrated ToC map was then developed from the three maps. Finally, it was presented in the fourth Toc workshop that included representatives from each group in the previous ToC meetings to come up with a final agreement on the content, length, and number of sessions, and strategies to overcome potential barriers in the implementation of integrated psychological services for adolescents living with HIV. Next, TN and two other psychologists did a two-day intervention planning workshop to refine the ToC map and finalize the content of the psychological intervention. Then, before reviewing and approving the manual for pilot testing, the draft manual was translated into Swahili.

6. Analysis

Following Graneheim and Lundman [50], qualitative content analysis guided data analysis. Two researchers analyzed the data to ensure reliability [51]. All Toc meeting transcripts, flip charts, and notes were first to read and re-read by two authors (TN and FN). The qualitative data analysis with NVivo software was used to manage and organize data. Condensed meaning units related to participants' description of components of the ToC map, including outcomes, intervention, preconditions, assumptions and indicators, acceptability, and feasibility, were formed through data reduction. Since codes were predefined by the ToCMRC frameworks, the authors agreed on the main categories by checking the similarities and differences of sub-categories and reflecting upon the interpretations of the participant's descriptions. Quotes were selected to support the presented themes and categories. Although the description seems to be linear, the analysis process was iterative. Through discussion, ToC maps were developed. The authors then combined the ToC maps from each workshop and devised a single ToC map. The final draft was refined and approved by the last workshop, which consisted of representatives of participants from the previous three ToC workshops.

7. Data Triangulation

A triangulation of findings from different sources was conducted to ensure the trustworthiness of this study. Credibility was ensured by collecting data from adolescents, health care providers, and other essential stakeholders to provide comprehensive accounts of their experiences, establish buy-in of the intervention in the health care system and delivery, as well as to ensure its content and structure are well rooted in the local situation [52,53]. At least two authors conducted the initial analysis, and all authors reviewed and agreed on categories. Data accuracy was validated through summaries for participants' checks and the final meeting that reviewed and approved the last ToC map [54]. To increase the study's credibility, we included participants' quotes to support the results [55]. Two senior researchers (AF and SK) closely monitored, reviewed, and examined the research process and data analysis to ensure consistent findings to promote dependability and conformability. Finally, the Guidance for Reporting Intervention Development (GUIDED) checklist [56] was used to report the intervention development process.

8. Ethical Considerations

The Muhimbili University of Health and Allied Sciences IRB granted ethical approval for the study in Dar-es-Salaam, Tanzania (Ref. No. DA.282/298/01.C/053). Additionally, the study was approved by the Addis Ababa University Institutional Review Board (Ref. No. 051/20/CDT). The Kinondoni Reginal Medical officer permitted the conduct of the study. Before the study, written informed consent to participation, a record of discussion, and findings used for publication were obtained for all study participants at least 18 years. In total, 2 participants were below 18 years and written parental permission and their accent were also obtained.

9. Results

9.1. Mental Health Expert Workshop

The expert workshop suggested that the intervention be delivered in six individual sessions. Participants argued that the intervention should be designed to address underlying psychological problems/stressors specific to adolescents living with HIV, as identified in the qualitative study. Psychoeducation was proposed to be included briefly, focusing on illness/symptoms and available treatment options. Mental health professionals suggested that nurse counselors be trained in communication skills to help them burst common myths about mental health problems, change attitudes, and reduce stigma. Communication skills were thought to be necessary to improve and build therapeutic relationships and improve assessment and management skills that will enhance the acceptability of the intervention.

The mental health professionals further highlighted the importance of utilizing evidence-based approaches from cognitive behavioral therapy and motivational interviewing, as supported by the literature [10,45,57,58]. It was argued that the cognitive–behavioral component would be helpful for adolescents to understand and practice how thoughts affect emotional behaviors and change maladaptive thoughts. It was, however, discussed and agreed that homework might not be very acceptable to adolescents; hence they should be limited to more practical activities such as relaxation exercises, mood monitoring, and behavioral activation. In addition, motivational interviewing strategies were thought necessary to enhance and promote healthy behaviors and motivation to take care of their health, adapt to their HIV status, and adhere to the treatment regime.

Mental health professionals suggested that including problem solving will make the intervention more culturally acceptable. It was agreed that problems identified in the qualitative data that required generating solutions, such as dealing with stigma and improving academic performance, could be solved by improved problem-solving skills. Mental health professionals suggested that the Intervention sessions last for approximately 30–45 min. The mental health professionals further recommended that the intervention manual be used flexibly but should provide the provider using it with guidelines on the content and structure of the session. It was discussed and agreed that the provider can, however, modify a session based on their clinical judgment if the overall approach is consistent with the principles of the intervention.

The expert suggested that one or two additional sessions may be included in the manual and used throughout the interventions to address specific concerns of the adolescent receiving the intervention or the provider. The supplement areas were suggested to be designed to assist the counselor in applying core skills to common psychological problems and stressors for adolescents living with HIV, such as the loss of a significant other, stigma, and violence due to HIV status.

9.2. *ToC Workshops*

Three ToC workshops and one ToC workshop that included the representative of all participants were conducted, with a total of twenty-two participants, as indicated in Table 1.

Table 1. Description of study participants.

Participants	Male	Female	Total
ToC with Stake Holders			
HIV implementing partners	-	4	4
MoH and government officials	3	-	3
ToC with adolescent	4	4	8
ToC with healthcare providers	3	4	7
Total			22
Representative meeting			
Stake Holders	1	2	7
Adolescent	2	2	

The fourth ToC meeting included representatives from the three workshops purposefully selected to meet representation requirements. The synthesized findings were reviewed, and TN and FN drafted the intervention manual that all authors critically reviewed.

9.3. *Intervention Development Recommendations*

Participants perceived that a psychological intervention was needed to address depression and associated problems in adolescents living with HIV. They, however, indicated the importance of ensuring feasibility, acceptability, and applicability within the local context.

Caregivers of adolescents insisted on using local terminologies and other easy-to-understand and acceptable approaches to helping adolescents and caregivers understand depression. Local idioms were vital as they will facilitate understanding, health help-seeking behavior, coping strategies, and the intervention's probability of bringing about the intended effect.

> *"You will need to use terminologies that we understand, do not say depression because most will not understand. Better use a simple description like overthinking and give some more description of the symptom."*

The intervention was perceived as necessary because it would provide a solution to many adolescents' problems and reduce the time spent by caregivers looking for help.

> *"It is hard to imagine that somewhere in this area has the solution to the problem that we parents have been unsuccessfully finding. Availability of treatment in the CTC will save us the movement from church, traditional healers, and so many places that have not been helpful in this."*

The HCP thought that the intervention would be acceptable for nurse counselors targeted to deliver the intervention. In addition, they suggested that management manuals and screening tools for depression be presented simply, which would also assist in delivery.

Similarly, findings from the adolescent workshop indicated that for adolescents to be free to request the service, providers need to be accepting, understanding, and non-judgmental. In addition, they asked for depression awareness to be offered to them and their caregivers to improve their mental health knowledge, early detection, and the likelihood of the adolescents receiving support during depression treatment. Finally, caregivers' awareness was thought necessary to increase help-seeking and treatment adherence and ultimately lead to expected treatment outcomes.

> *"For someone like me (adolescent) to benefit from the intervention, my parents need to be aware first to understand my problem and then support me to receive the treatment."*

Participants in all groups suggested that the CTC intervention should be delivered in HIV-CTCs since its where adolescents receive their regular follow-ups and prescription refill. The Nurse counselor was seen as the most fitting person for the intervention delivery, and most participants believed counselors should be available every clinic day. Adolescents wanted the intervention to be called NITUE. This Swahili word means "Help me offload," reflecting the intervention's aim, which is to provide relief from carrying a heavy load of depressive symptoms.

9.4. Structure and Delivery of the Session

Adolescents and their caregivers suggested that the experts decide on the number, duration, and frequency of intervention sessions depending on the content and experience delivering similar kinds of help for adolescents living with HIV. However, they preferred that the intervention be provided when needed and not wait until the regular monthly appointments. Health care providers recommended from 30 min to 1 h, as they usually spend during counseling with adherence or behavioral problems. However, they believed that that time would be less if they were appropriately trained and knew what to do; hence, they reached a consensus of 30 min.

> *"We have been using so much time because we did not know what to do or say. You might spend an hour sometimes consulting with others or trying to console a crying youth. I think once one is trained and knows what to do, 30 min may be enough."*

9.5. ToC Map

The final Theory of Change map in Figure 1 summarizes a participant's identification of knowledge gaps and describes pathways to the outcome of interests.

The outcome indicates the primary effect of the intervention. Adolescents and health care providers mentioned improvement in depressive symptoms and enhanced adherence for adolescents living with HIV as the outcome of interest. In addition, all participants indicated other outcomes such as improved academic performance and quality of life as defined by good health, making it easy to enjoy life and attain individual life goals. Stakeholders noted that the final impact of the intervention would be to reduce the treatment gap for adolescent depression.

The intermediate outcome for the intervention included unavailable things and needs to be there to achieve the final prefeed effect. For example, stakeholder and HCP workshops suggested that adequate training, treatment manual, and awareness materials must be available within HIV-CTC. Other intermediate outcomes included (1) improved depression awareness among adolescents, caregivers, and health care providers, (2) decreased mental health stigma among healthcare providers, (3) healthcare providers' confidence in managing depression improved, (4) adolescents attend CTC regularly for HIV, and depression care, regular screening for depression in routine clinics, (5) enhanced communication skills for providers, and (6) appointments are given for counseling/mental health care.

The interventions discussed were the different components of the intervention that are needed to move from one outcome to another. For example, the unmeant mental health needs were required to establish the gap that the intervention would cover. Healthcare providers suggested that for the nurse counselors to deliver the intervention, there was a need for an intervention to ensure these providers acquire the necessary skills required for screening and managing depression in adolescents. They perceive that refresher, ongoing training, and supervision should enable them to deliver the intervention and give them the skills to provide mental health awareness and care for themselves while helping adolescents. Adolescents suggested that providers need to be understanding and less judgmental for them to be able to ask for and use the intervention. The need for improved communication skills necessitates training as an essential intermediate intervention. Adolescents also perceive that the intervention gives them skills to deal with everyday problems that may lead to depression. Adding problem-solving to the intervention components will make the intervention more desirable and increase help-seeking behavior.

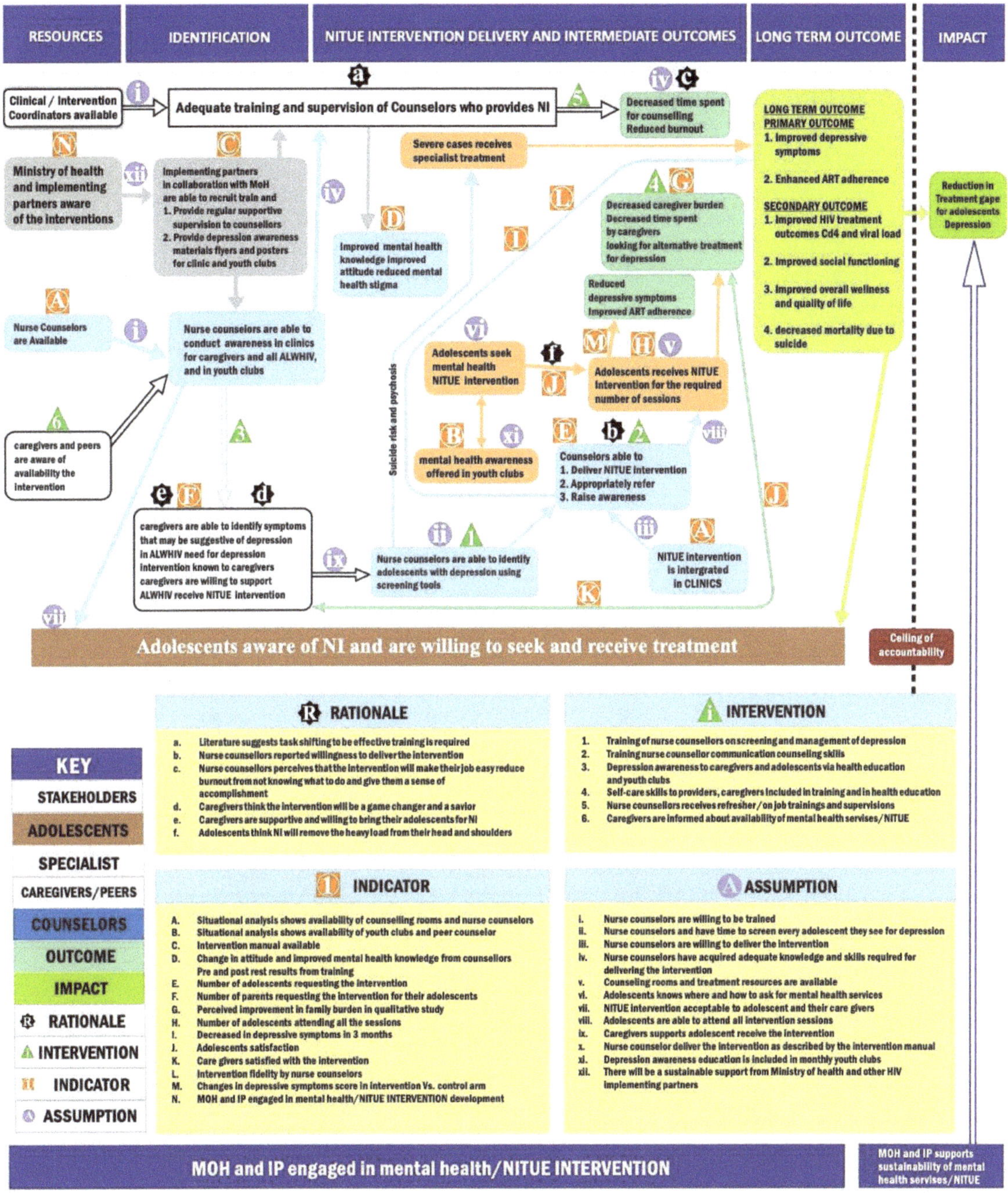

Figure 1. Theory of Change Map for the NITUE Intervention.

Assumption: Participants also discussed the conditions beyond the interventions that need to be there to move forward in achieving the outcome.

Stakeholders focused on the willingness of nurse counselors to be trained and their desire to deliver the intervention as the fundamental assumption for the intervention's success. Participants also mentioned sustainable support from the Ministry of Health and other HIV-implementing partners. The HCP workshop discussed issues related to the acceptability of the intervention for adolescents and their willingness to see and receive care. For adolescents, an important assumption was that all adolescents know where and how to ask for mental health services.

Participants were also asked to discuss the rationale from evidence or experience explaining each outcome's pathway. That is why one outcome is an outcome of another. For example, the available evidence was from a literature review that suggests that task shifting to be effective training is required. Other evidence was from the qualitative studies indicating the initial willingness of providers and positive attitude toward the intervention from adolescents, caregivers, and providers.

Indicators: Participants were asked to list the indicators. These measurable things determine whether the outcome has been achieved or whether there is progress toward achieving the result. For example, participants identified and agreed on the availability of structure, nurse counselors, and youth clubs to indicate that training and awareness education is possible.

A manual availability will determine whether the training and intervention will occur. Other indicators included (1) change in attitude and improved mental health knowledge from pre and post-test results from training will show if nurse counselors can deliver the intervention, (2) the number of adolescents attending all the intervention sessions, (3) the decrease in depressive symptoms in 3 months as measured by PHQ-A, (4) adolescents' satisfaction, (5) intervention fidelity by nurse counselors, and (6) changes in depressive symptoms score in intervention vs. Control arm. Figure 1 shows the final Theory of Change map for the NITUE intervention.

10. Description of the Intervention Manual

Information from the qualitative study and ToC workshops were merged to identify adolescents' unmet needs and priorities, which informed the intervention. During the intervention design, the identified needs and preferences were linked to the components of the intervention structured in six sessions.

Session 1: Psychoeducation and engagement to care: The first intervention sessions utilize motivational interviewing and counseling skills to (1) build rapport, (2) provide psychoeducation with a focus on depression, and HIV, as well as available treatment options for depression, (3) provide an explanation of the intervention model of treatment for depression and how it may help with depression and related problems, and (4) set treatment goals (including depression and adherence goals).

10.1. Sessions 2–4: Reducing Depressive Symptoms: CBT and Problem-Solving

The three sessions emphasize reducing depressive symptoms using mood monitoring, behavioral activation, cognitive restructuring, and problem-solving techniques. Behavioral techniques such as deep breathing exercises and progressive muscle relaxation are practiced, and adolescents are given homework at each session. Problem solving focuses on the problems adolescents currently face and on facilitating them to discover a way out.

10.2. Sessions 5–6: Enhancing Adherence to ART and Relapse Prevention: Taking Charge of One's Health

Application of brief motivational interviewing strategies to enhance and promote healthy behaviors and help adolescents adhere to ART treatment regimes. It also includes plans on stress management and coping skills.

11. Discussion

This study is a comprehensive, systematic, and theory-driven effort to develop the first manualized brief psychological intervention guide in an integrated care context for adolescents living with HIV in Tanzania. The participatory intervention development process was anchored in the theory of change, making it possible to show the causal pathway for how we expect the intervention to achieve the preferred long-term outcome and its impact, as well as the other necessary elements of care preconditions to its success.

We used qualitative exploration first to understand the unmet mental health needs and their associated consequences for adolescents living with HIV and incorporate the perspectives of their caregivers and healthcare providers. The findings of this qualitative data indicated areas that this brief psychological intervention could potentially address. Second, explore challenges to accessing mental health care and psychological intervention implementation opportunities. The understanding of adolescents, their caregivers, and HCP helped to ensure that the developed intervention is acceptable and addresses obstacles to help-seeking and utilization, adding the possibility that the interventions will prove effective. It also made it possible to consider barriers to mental health care access and include strategies to modify those barriers within the structure and content of the intervention [31].

Furthermore, the unmet needs and barriers identified by participants in the qualitative exploration are essential because psychological interventions work through specific factors necessary for the intervention to be effective. Therefore, these components, such as therapeutic alliance, the rationale for treatment, communication skills, positive emotional experience, and locally suitable and adolescent-friendly treatment components, are necessary considerations for intervention development [59].

Thirdly, we used ToC approaches involving different stakeholders, including adolescents living with HIV and their caregivers, HCP, the ministry of health, mental health professionals, and HIV implementing partners. This participatory approach helped the researchers to understand the context of HIV care delivery and make decisions that reflect scientific evidence and the views of essential stakeholders in policymaking, HIV care, and treatment in Tanzania [40,60]. Therefore, the ToC approach merged scientific evidence with stakeholders' contributions, thus establishing a local buy-in to ensure ownership, acceptance, and support for the intervention from these stakeholders, which are essential prerequisites for the implementation [40]. Furthermore, involving important stakeholders helped build trust, promoting the pooling of resources and knowledge while ensuring the intervention is rooted within the Tanzanian cultural context [60].

Adolescent participants highlighted the importance of involving caregivers in the intervention development process. In addition, literature from LMIC [61] and most mental health treatment guidelines recommend the participation of adolescents and their caregivers in treatment planning and decision-making [17].

Adolescents that live with HIV exhibit high levels of social stigma and isolation from society. Many maintain close contact with a caregiver, parents, or close relatives and perceive family support as critical. Caregivers, therefore, play an essential role in the recovery process. It is also noted that caregivers often establish the adolescent's initial access to mental health services and that caregiver involvement increases compliance and improves treatment outcomes.

The developed intervention utilizes a task-shifting approach [62] (and uses locally available resources that are accessible within HIV care and treatment facilities [31]. The World Health Organization has recommended integrating mental health services within PHC settings to address the unmet mental health needs of adolescents living with HIV [17,63]. All study participants agreed that nurse counselors in HIV care and treatment facilities to facilitate its future integration within routine HIV care. Although the participants were concerned about the willingness of HCP, nurse counselors who were the targeted providers of the intervention reported willingness and positive attitudes towards the intervention to simplify their work.

All study participants agreed that mental health training and communication for HCP were vital to improving their knowledge, changing negative attitudes about mental illness, and increasing the confidence of nurse counselors to deliver the intervention effectively.

Studies have shown that culturally adapted interventions delivered in a task-shared approach are feasible and acceptable in LMIC [64–66]. Findings from poorly resourced countries favor the possibility that HIV counselors in CTC facilities can be successfully trained and supervised to deliver brief psychological interventions that target depression in adolescents living with HIV [67–69].

12. Strengths and Limitations

We have developed an intervention that may be locally feasible, acceptable, and likely to be effective in reducing depressive symptoms and improving ART adherence for ALWHIV in Tanzania and possibly other low-income countries. In addition to the novel focus, an innovative element includes training resources and a manualized guide of the adapted intervention that targets (1) specific causes of mental health problems and issues facing adolescents living with HIV and (2) management strategies for depression and ART medication adherence problems for adolescents living with HIV and (3) using existing HIV care and treatment providers to deliver the intervention. The study also supports the attainment of Sustainable Development Goal three (Good health and well-being for people), Target 3.3, which requires countries to end AIDS epidemics by 2030, and Target 3.4 to reduce by one-third premature mortality from non-communicable diseases (suicide and depression included) through prevention, treatment, and promotion of mental health and well-being by the same year 2030.

13. Conclusions

Utilizing the TOC-MRC enhanced framework for complex intervention development, we developed a culturally appropriate brief psychological intervention that uses a task-shifting approach to address depression and medication adherence for ALWHIV in Dar es Salaam, Tanzania. The intervention will be piloted for appropriateness, feasibility, and acceptability. The study also provides the prerequisite materials, training, and infrastructure needed for a future trial to determine the effectiveness of this intervention in reducing depressive symptoms and enhancing ART adherence in adolescents living with HIV in Tanzania and possibly other countries in sub-Saharan Africa.

Author Contributions: T.N., A.F. and S.K. designed the study, with inputs from B.S. and F.N.; T.N. and F.N. were responsible for data collection and analysis with supervision from A.F. and S.K.; T.N. wrote the first draft of the manuscript. All authors have read and agreed to the published version of the manuscript.

Funding: This study was funded by the Centre for Innovative Drug Development and Therapeutics Trial for Africa (CDT-Africa), a World Bank Africa Centre of excellence at Addis Ababa University, as part of a Ph.D. scholarship for T.N.

Institutional Review Board Statement: Addis Ababa University Institutional Review Board (IRB), Ethiopia (Ref. No. 051/20/CDT) and from the Muhimbili University of Health and Allied Sciences IRB in Dar-es-Salaam Tanzania (Ref. No. DA.282/298/01.C/053) have approved the study.

Informed Consent Statement: The Kinondoni Regional Medical Officer permitted the conduct of the study. In addition, written parental or guardian permission (for those below 18 years) and informed consent were obtained for all study participants, and a record of discussion and findings were used for publication.

Data Availability Statement: All data generated in this study are available from the corresponding author upon reasonable request.

Acknowledgments: We are grateful to the mental health experts, HIV implementing partners, Government officials, healthcare providers, and adolescents living with HIV who generously contributed to this study.

Conflicts of Interest: The authors declare no conflict of interest.

References

1. UNAIDS. *Young People and HIV*; UNAIDS: Geneva, Switzerland, 2021; Volume 1–20.
2. Bezerra, L.M.D. *Global Report: UNAIDS Report on the Global AIDS Epidemic 2010*; UNAIDS: Geneva, Switzerland, 2010; Volume 3, ISBN 9789291738717.
3. UNAIDS. Young people are leading the HIV prevention revolution. *Outlook* **2010**, *7*, 3–7. Available online: http://www.unaids.org/sites/default/files/en/media/unaids/contentassets/documents/unaidspublication/2010/20100713_outlook_youngpeople_en.pdf (accessed on 8 July 2021).
4. UNAIDS. Data 2017. Program HIV/AIDS [Internet]. 2017. Available online: http://www.unaids.org/sites/default/files/media_asset/20170720_Data_book_2017_en.pdf (accessed on 10 February 2018).
5. UNICEF. Opportunity in Crisis. Preventing HIV from early adolescence to young adult. *Intern. Audit.* **2009**, *65*, 7.
6. United Republic of Tanzania. Global AIDS Monitoring Progress reports submitted by countries-United Republic of Tanzania. *Ctry. Prog. Rep.* **2020**, 5–48. Available online: https://www.unaids.org/sites/default/files/country/documents/TZA_2020_countryreport.pdf (accessed on 8 July 2021).
7. Tanzania, T.G. of the U.R. of Tanzania Nationwide Mills Census Report. 2022. Available online: https://docs.wfp.org/api/documents/WFP-0000142157/download/?_ga=2.181109345.792392857.1668754662-916564780.1668754662 (accessed on 12 November 2022).
8. Gonzalez, J.S.; Batchelder, A.W.; Psaros, C.; Safren, S.A. Depression and HIV/AIDS treatment nonadherence: A review and meta-analysis. *J. Acquir. Immune Defic. Syndr.* **2011**, *58*, 181–187. [CrossRef] [PubMed]
9. Adejumo, O.A.; Malee, K.M.; Ryscavage, P.; Hunter, S.J.; Taiwo, B.O. Contemporary issues on the epidemiology and antiretroviral adherence of HIV-infected adolescents in sub-Saharan Africa: A narrative review. *J. Int. AIDS Soc.* **2015**, *18*, 20049. [CrossRef]
10. Naar-King, S.; Parsons, J.T.; Murphy, D.; Kolmodin, K.; Harris, D.R. A Multisite Randomized Trial of a Motivational Intervention Targeting Multiple Risks in Youth Living With HIV: Initial Effects on Motivation, Self-Efficacy, and Depression. *J. Adolesc. Health* **2010**, *46*, 422–428. [CrossRef]
11. Dow, D.E.; Turner, E.L.; Shayo, A.M.; Mmbaga, B.; Cunningham, C.K.; O'donnell, K. Evaluating Mental Health Difficulties and Associated Outcomes Among HIV-Positive Adolescents in Tanzania. *AIDS Care* **2016**, *7*, 825–833. [CrossRef] [PubMed]
12. Traube, D.; Dukay, V.; Kaaya, S.; Reyes, H.; Mellins, C. Cross-cultural adaptation of the Child Depression Inventory for use in Tanzania with children affected by HIV. *Taylor Fr.* **2010**, *2*, 174–184. [CrossRef]
13. Antelman, G.; Kaaya, S.; Wei, R.; Mbwambo, J.; Msamanga, G.I.; Fawzi, W.W.; Smith Fawzi, M.C. Depressive Symptoms Increase Risk of HIV Disease Progression and Mortality Among Women in Tanzania. *JAIDS J. Acquir. Immune Defic. Syndr.* **2007**, *44*, 470–477. [CrossRef]
14. Gutmann, M.A.F. Mental Health and Hiv/Aids. Arlington, VA USAID/AIDSTAR ONE Proj. T. 2009. Available online: https://encompassworld.com/wp-content/uploads/2016/04/aidstaronementalhealthandhiv.pdf (accessed on 8 July 2021).
15. Prince, M.; Patel, V.; Saxena, S.; Maj, M.; Maselko, J.; Phillips, M.R.; Rahman, A. No health without mental health. *Lancet* **2007**, *370*, 859–877. [CrossRef] [PubMed]
16. Bärnighausen, T.; Chaiyachati, K.; Chimbindi, N.; Peoples, A.; Haberer, J.; Newell, M.-L. Interventions to increase antiretroviral adherence in sub- Saharan Africa: A systematic review of evaluation studies avenues for future work include intervention targeting and selection of interventions based on behavioural theories relevant to SSA. *Lancet Infect Dis.* **2011**, *11*, 942–951. [CrossRef]
17. World Health Organization mhGAP Intervention Guide. mhGAP Interv. Guid. Ment. Neurol. Subst. Use Disord. Non-Specialized Health Settings. 2010. Available online: http://www.drugsandalcohol.ie/13984/1/WHO_mhGAP_intervention_guide.pdf (accessed on 12 November 2022).
18. Marwick, K.F.M.; Kaaya, S.F. Prevalence of depression and anxiety disorders in HIV-positive outpatients in rural Tanzania. *AIDS Care* **2010**, *22*, 415–419. [CrossRef] [PubMed]
19. Kaaya, S.F.; Lee, B.; Mbwambo, J.K.; Smith-Fawzi, M.C.; Leshabari, M.T. Detecting depressive disorder with a 19-item local instrument in Tanzania. *Int. J. Soc. Psychiatry* **2008**, *54*, 21–33. [CrossRef]
20. Ramaiya, M.K.; Sullivan, K.A.; O'donnell, K.; Cunningham, C.K.; Shayo, A.M.; Mmbaga, B.T.; Dow, D.E. A Qualitative Exploration of the Mental Health and Psychosocial Contexts of HIV-Positive Adolescents in Tanzania. *PLoS ONE* **2016**, *11*, e0165936. [CrossRef] [PubMed]
21. Kim, M.H.; Mazenga, A.C.; Devandra, A.; Ahmed, S.; Kazembe, P.N.; Yu, X.; Nguyen, C.; Sharp, C. Prevalence of depression and validation of the Beck Depression Inventory-II and the Children's Depression Inventory-Short amongst HIV-positive adolescents in Malawi. *J. Int. AIDS Soc.* **2014**, *17*, 18965. [CrossRef] [PubMed]
22. Kennard, B.; Brown, L.; Hawkins, L.; Risi, A.; Radcliffe, J.; Emslie, G.; Mayes, T.; King, J.; Foxwell, A.; Buyukdura, J.; et al. Development and Implementation of Health and Wellness CBT for Individuals with Depression and HIV. *Cogn. Behav. Pract.* **2014**, *21*, 237–246. [CrossRef]
23. Lee, B.; Kaaya, S.F.; Mbwambo, J.K.; Smith-Fawzi, M.C.; Leshabari, M.T. Detecting depressive disorder with the Hopkins Symptom Checklist-25 in Tanzania. *Int. J. Soc. Psychiatry* **2008**, *54*, 7–20. [CrossRef]

24. Miller, W.R.; William, R.; Rollnick, S. *Motivational Interviewing: Preparing People for Change*; Guilford Press: New York, NY, USA, 2002; ISBN 1572305630.
25. Parsons, J.T.; Golub, S.A.; Rosof, E.; Holder, C. Motivational interviewing and cognitive-behavioral intervention to improve HIV medication adherence among hazardous drinkers: A randomized controlled trial. *J. Acquir. Immune Defic. Syndr.* **2007**, *46*, 443–450. [CrossRef]
26. Hanlon, C.; Luitel, N.P.; Kathree, T.; Murhar, V.; Shrivasta, S.; Medhin, G.; Ssebunnya, J.; Fekadu, A.; Shidhaye, R.; Petersen, I.; et al. Challenges and Opportunities for Implementing Integrated Mental Health Care: A District Level Situation Analysis from Five Low- and Middle-Income Countries. *PLoS ONE* **2014**, *9*, e88437. [CrossRef]
27. Stanley, B.; Brown, G.; Brent, D.A.; Wells, K.; Poling, K.; Curry, J.; Kennard, B.D.; Wagner, A.; Cwik, M.F.; Klomek, A.B.; et al. Cognitive-behavioral therapy for suicide prevention (CBT-SP): Treatment model, feasibility, and acceptability. *J. Am. Acad. Child Adolesc. Psychiatry* **2009**, *48*, 1005–1013. [CrossRef]
28. Golin, C.E.; Earp, J.; Tien, H.-C.; Stewart, P.; Porter, C.; Howie, L. A 2-arm, randomized, controlled trial of a motivational interviewing-based intervention to improve adherence to antiretroviral therapy (ART) among patients failing or initiating ART. *J. Acquir. Immune Defic. Syndr.* **2006**, *42*, 42–51. [CrossRef] [PubMed]
29. Gonzalez, J.S.; McCarl, L.A.; Wexler, D.J.; Cagliero, E.; Delahanty, L.; Soper, T.D.; Goldman, V.; Knauz, R.; Safren, S.A. Cognitive Behavioral Therapy for Adherence and Depression (CBT-AD) in Type 2 Diabetes. *J. Cogn. Psychother.* **2010**, *24*, 329–343. [CrossRef] [PubMed]
30. Kutcher, S.; Wei, Y.; Gilberds, H. Undefined Addressing Adolescent Depression in Tanzania: Positive Primary Care Workforce Outcomes Using a Training Cascade Model. 2017. Available online: hindawi.com (accessed on 10 February 2018).
31. Chisholm, D.; Flisher, A.; Lund, C.; Patel, V.; Saxena, S.; Thornicroft, G.; Tomlinson, M. Scale up services for mental disorders: A call for action. *Lancet* **2007**, *370*, 1241–1252. [CrossRef] [PubMed]
32. Mellins, C.A.; Malee, K.M. Understanding the mental health of youth living with perinatal HIV infection: Lessons learned and current challenges. *J. Int. AIDS Soc.* **2013**, *16*, 18593. [CrossRef]
33. Shidhaye, R.; Lund, C.; Chisholm, D. Closing the treatment gap for mental, neurological and substance use disorders by strengthening existing health care platforms: Strategies for delivery and integration of evidence-based interventions. *Int. J. Ment. Health Syst.* **2015**, *9*, 40. [CrossRef]
34. Kirmayer, L.J.; Pedersen, D. Toward a new architecture for global mental health. *Transcult. Psychiatry* **2014**, *51*, 759–776. [CrossRef] [PubMed]
35. Kirmayer, L.J. Cultural competence and evidence-based practice in mental health: Epistemic communities and the politics of pluralism. *Ocial Sci. Med.* **2012**, *75*, 249–256. [CrossRef]
36. Patel, V.; Chowdhary, N.; Rahman, A.; Verdeli, H. Improving access to psychological treatments: Lessons from developing countries. *Behav. Res. Ther.* **2011**, *49*, 523–528. [CrossRef]
37. Holmes, E.A.; Ghaderi, A.; Harmer, C.J.; Ramchandani, P.G.; Cuijpers, P.; Morrison, A.P.; Roiser, J.P.; Bockting, C.L.H.; O'Connor, R.C.; Shafran, R.; et al. The Lancet Psychiatry Commission on psychological treatments research in tomorrow's science. *Lancet Psychiatry* **2018**, *5*, 237–286. [CrossRef]
38. Abera, M.; Tesfaye, M.; Belachew, T.; Hanlon, C. Perceived challenges and opportunities arising from integration of mental health into primary care: A cross-sectional survey of primary health care workers in south-west Ethiopia. *BMC Health Serv. Res.* **2014**, *14*, 113. [CrossRef]
39. Kutcher, S.; Wei, Y.; Gilberds, H.; Brown, A.; Ubuguyu, O.; Njau, T.; Sabuni, N.; Magimba, A.; Perkins, K. Evaluating community health care providers knowledge and self-confidence in the identification, diagnosis and treatment of adolescent Depression in Tanzania. *Arch. Depress. Anxiety* **2016**, *2*, 026–030. [CrossRef]
40. De Silva, M.J.; Breuer, E.; Lee, L.; Asher, L.; Chowdhary, N.; Lund, C.; Patel, V. Theory of Change: A theory-driven approach to enhance the Medical Research Council's framework for complex interventions. *Trials* **2014**, *15*, 267. [CrossRef] [PubMed]
41. Craig, P.; Dieppe, P.; Macintyre, S.; Michie, S.; Nazareth, I.; Petticrew, M. Developing and evaluating complex interventions: The new Medical Research Council guidance. *Int. J. Nurs. Stud.* **2013**, *50*, 587–592. [CrossRef] [PubMed]
42. French, S.D.; Green, S.E.; O'Connor, D.A.; McKenzie, J.E.; Francis, J.J.; Michie, S.; Buchbinder, R.; Schattner, P.; Spike, N.; Grimshaw, J.M. Developing theory-informed behaviour change interventions to implement evidence into practice: A systematic approach using the Theoretical Domains Framework. *Implement. Sci.* **2012**, *7*, 38. [CrossRef]
43. Michie, S.; Johnston, M.; Abraham, C.; Lawton, R.; Parker, D.; Walker, A. "Psychological Theory" Group Making psychological theory useful for implementing evidence based practice: A consensus approach. *Qual. Saf. Health Care* **2005**, *14*, 26–33. [CrossRef]
44. Das, J.K.; Salam, R.A.; Lassi, Z.S.; Khan, M.N.; Mahmood, W.; Patel, V.; Bhutta, Z.A. Interventions for adolescent mental health: An overview of systematic reviews. *J. Adolesc. Health* **2016**, *59*, S49–S60. [CrossRef]
45. Bhana, A.; Kreniske, P.; Pather, A.; Abas, M.A.; Mellins, C.A. Interventions to address the mental health of adolescents and young adults living with or affected by HIV: State of the evidence. *J. Int. AIDS Soc.* **2021**, *24*, 25713. [CrossRef]
46. Weersing, V.R.; Jeffreys, M.; Do, M.-C.T.; Schwartz, K.T.G.; Bolano, C. Evidence-Base Update of Psychosocial Treatments for Child and Adolescent Depression. *J. Clin. Child Adolesc. Psychol.* **2017**, *46*, 11–43. [CrossRef]
47. Njau, T.; Sunguya, B. The Unmet Mental Health Needs of adolescents living With HIV in Eastern Tanzania: Experiences of healthcare providers, adolescents, and caregivers. *Res. Sq.* 2022. *in press*. [CrossRef]

48. Njau, T. Perceived barriers and opportunities for implementing integrated psychological intervention for depression in adolescents living with HIV in. *Res. Sq.* 2022. *in press*. [CrossRef]
49. Palinkas, L.A.; Horwitz, S.M.; Green, C.A.; Wisdom, J.P.; Duan, N.; Hoagwood, K. Purposeful Sampling for Qualitative Data Collection and Analysis in Mixed Method Implementation Research. *Adm. Policy Ment. Health Ment. Health Serv. Res.* **2015**, *42*, 533–544. [CrossRef]
50. Graneheim, U.H.; Lundman, B. Qualitative content analysis in nursing research: Concepts, procedures and measures to achieve trustworthiness. *Nurse Educ. Today* **2004**, *24*, 105–112. [CrossRef] [PubMed]
51. MacQueen, K.M.; McLellan, E.; Kay, K.; Milstein, B. Codebook Development for Team-Based Qualitative Analysis. *CAM J.* **1998**, *10*, 31–36. [CrossRef]
52. Polit, D.; Beck, C. Nursing Research. *Princ. Methods* **2012**, *66*, 37–39. [CrossRef]
53. Shenton, A.K. Strategies for ensuring trustworthiness in qualitative research projects. *SAGE Encycl. Action Res.* **2014**, *22*, 63–75. [CrossRef]
54. Creswell, J.W. *Qualitative Inquiry & Rerearch Design: Choosing Among Five Approaches*, 2nd ed.; SAGE Publications Inc.: New York, NY, USA, 2013; ISBN 9781412916066.
55. Noble, H.; Smith, J.; Noble, H.; Smith, J. Issues of validity and reliability in qualitative research. *Evid. Based Nurs.* **2015**, *18*, 34–35. [CrossRef]
56. Duncan, E.; O'Cathain, A.; Rousseau, N.; Croot, L.; Sworn, K.; Turner, K.M.; Yardley, L.; Hoddinott, P. Guidance for reporting intervention development studies in health research (GUIDED): An evidence-based consensus study. *BMJ Open* **2020**, *10*, 1–12. [CrossRef]
57. Watanabe, N.; Hunot, V.; Omori, I.M.; Churchill, R.; Furukawa, T.A. Psychotherapy for depression among children and adolescents: A systematic review. *Acta Psychiatr. Scand.* **2007**, *116*, 84–95. [CrossRef]
58. Palacio, A.; Garay, D.; Langer, B.; Taylor, J.; Wood, B.A.; Tamariz, L. Motivational Interviewing Improves Medication Adherence: A Systematic Review and Meta-analysis. *J. Gen. Intern. Med.* **2016**, *31*, 929–940. [CrossRef] [PubMed]
59. Cuijpers, P.; Reijnders, M.; Huibers, M.J.H. The Role of Common Factors in Psychotherapy Outcomes. *Annu. Rev. Clin. Psychol.* **2019**, *15*, 207–231. [CrossRef]
60. Oliveras Rodriguez, C.A. Engaging adolescents with HIV to ensure better health and more informed research. *J. Int. AIDS Soc.* **2017**, *20*, 110–111. [CrossRef] [PubMed]
61. Betancourt, T.S.; Ng, L.C.; Kirk, C.M.; Munyanah, M.; Mushashi, C.; Ingabire, C.; Teta, S.; Beardslee, W.R.; Brennan, R.T.; Zahn, I.; et al. Family-based prevention of mental health problems in children affected by HIV and AIDS: An open trial. *AIDS* **2014**, *28* (Suppl. 3), S359–S368. [CrossRef] [PubMed]
62. Buttorff, C.; Hock, R.; Weiss, H.; Naik, S.; Araya, R.; Kirkwood, B.; Chisholm, D.; Patel, V. Economic evaluation of a task-shifting intervention for common mental disorders in India. *Bull. World Health Organ.* **2012**, *90*, 813–821. [CrossRef]
63. Chatterjee, S.; Chowdhary, N.; Pednekar, S.; Cohen, A.; Andrew, G.; Araya, R.; Araya, R.; Simon, G.; King, M.; Telles, S.; et al. Integrating evidence-based treatments for common mental disorders in routine primary care: Feasibility and acceptability of the MANAS intervention in Goa, India. *World Psychiatry* **2008**, *7*, 39–46. [CrossRef]
64. Patel, V.; Weiss, H.A.; Chowdhary, N.; Naik, S.; Pednekar, S.; Chatterjee, S.; De Silva, M.J.; Bhat, B.; Araya, R.; King, M.; et al. Effectiveness of an intervention led by lay health counsellors for depressive and anxiety disorders in primary care in Goa, India (MANAS): A cluster randomised controlled trial. *Lancet* **2010**, *376*, 2086–2095. [CrossRef]
65. Mall, S.; Sorsdahl, K.; Swartz, L.; Joska, J. "I understand just a little ... " Perspectives of HIV/AIDS service providers in South Africa of providing mental health care for people living with HIV/AIDS. *AIDS Care* **2012**, *24*, 319–323. [CrossRef]
66. Patel, V.; Simon, G.; Chowdhary, N.; Kaaya, S.; Araya, R. Packages of care for depression in low- and middle-income countries. *PLoS Med.* **2009**, *6*, e1000159. [CrossRef] [PubMed]
67. Odokonyero, R.; Wagner, G.; Ngo, V.; Nakasujja, N.; Musisi, S.; Akena, D. Giving "Sadness" a Name: The Need for Integrating Depression Treatment into HIV Care in Uganda. *J. Int. Assoc. Provid. AIDS Care* **2015**, *14*, 108–111. [CrossRef]
68. Chibanda, D.; Bowers, T.; Verhey, R.; Rusakaniko, S.; Abas, M.; Weiss, H.A.; Araya, R. The Friendship Bench programme: A cluster randomised controlled trial of a brief psychological intervention for common mental disorders delivered by lay health workers in Zimbabwe. *Int. J. Ment. Health Syst.* **2015**, *9*, 21. [CrossRef] [PubMed]
69. Abas, M.; Nyamayaro, P.; Bere, T.; Saruchera, E. Feasibility and Acceptability of a Task-Shifted Intervention to Enhance Adherence to HIV Medication and Improve Depression in People Living with HIV in Zimbabwe, a Low Income Country in Sub-Saharan Africa. *AIDS Behav.* **2018**, *22*, 86–101. [CrossRef] [PubMed]

healthcare

MDPI

Article

A Social Emotional Learning Training Programme in a Poor Rural Primary School in Central China: A Pre-Post Intervention Study

Jiameng Li [1], Cuiling Ma [1], Qi Lu [1] and Therese Hesketh [1,2,*]

1 Centre for Global Health, Zhejiang University School of Medicine, 866 Yuhangtang Road, Xihu District, Hangzhou 310058, China
2 The Institute for Global Health, UCL, 30 Guilford Street, London WC1NEH, UK
* Correspondence: t.hesketh@ucl.ac.uk

Abstract: Introduction: Many universal school-based social and emotional learning (SEL) programmes in the U.S. and Europe have been found to improve social skills and reduce emotional distress and behaviour problems. The aim of this study is to determine whether an adapted version of the SEL can reduce social, emotional, and behavioural difficulties in children in mainland China, using a pre-post intervention design. Methods: The study was conducted in a primary school in an economically-disadvantaged rural area in Henan province in central China. The intervention consisted of 16 weekly 90-minunte classroom sessions involving all 190 children in the school. Social and emotional problems were assessed pre- and post- intervention using the Chinese version of the Strengths and Difficulties Questionnaire (SDQ). The results suggest that: (1) the programme can reduce children's peer relationship problems, and that the reduction was sustainable at the two post-intervention assessments; (2) the intervention effects on emotional symptoms or total difficulties in the overall population are very few, but children identified as high risk in the initial assessment benefited from the programme. Conclusions: This is the first published report on the effectiveness of a school-based SEL programme in mainland China. Although the improvement are limited, the programme does benefit some children.

Keywords: social and emotional learning (SEL); pre-post intervention study; school; children; mainland China

Citation: Li, J.; Ma, C.; Lu, Q.; Hesketh, T. A Social Emotional Learning Training Programme in a Poor Rural Primary School in Central China: A Pre-Post Intervention Study. *Healthcare* **2022**, *10*, 2332. https://doi.org/10.3390/healthcare10112332

Academic Editors: Yasuhiro Kotera and Elaina Taylor

Received: 3 October 2022
Accepted: 19 November 2022
Published: 21 November 2022

1. Introduction

Social-emotional learning (SEL) is the process of developing the self-awareness, self-control, and interpersonal skills that are vital for school, work, and life success. The idea of integrating SEL into the education system across the age range emerged in the early 1990s in the US. SEL was defined as the process of acquiring core abilities to recognise and manage emotions, set and achieve positive goals, feel and show empathy for others, establish and maintain supportive relationships, make responsible decisions, and handle interpersonal situations better [1]. The core goals of SEL programmes are to foster the development of five associated cognitive, affective, and behavioural competencies: self-awareness (e.g., recognising emotions, strengths and limitations), self-management (e.g., regulating emotions and behaviours), social awareness (e.g., taking the perspective of and empathising with others from diverse backgrounds and cultures), relationship skills (e.g., establishing and maintaining healthy relationships), and responsible decision making (e.g., making constructive choices across varied situations) [2]. The United States remains the hub for the development and dissemination of SEL programmes. SEL has been adopted in mainstream education in a number of countries, including Australia, the UK, other European countries, and parts of Asia such as Singapore [3,4].

The results from a meta-analysis of 213 school-based, universal SEL programmes (87% from United States) showed that participants demonstrated improved social skills, fewer behaviour problems, and less emotional distress [5]. A meta-analysis of the effects of universal, school-based social, emotional, and behavioural programmes in 75 studies (80% from North America and 15% from European countries) reported enhancement of social and emotional skills, positive self-image, reduction of antisocial behaviour and mental health problems [6]. A meta-analysis of follow-up effects of school-based, universal SEL interventions involving 97,406 kindergarten to high school students in 82 studies (54% from United States, and others from Australia and European countries) suggested that participants reported improvements in social-emotional skills, attitudes, and well-being regardless of their race or socioeconomic background [7]. This meta-analysis also found no significant difference in the effectiveness between interventions involving predominately low- and working-class students compared with those of middle- and upper-class. A review including 22 studies (mostly from Asian, African, and South American countries) indicated school-based SEL programmes could improve emotional and behavioural wellbeing among children from low- and middle-income backgrounds [8]. Some universal SEL interventions indicated that students from rural low socioeconomic status actually benefit more from the intervention [7]. Stronger intervention effects have been found for students from poor families in improving school achievement and reducing misbehaviour [9].

Very few universal school-based SEL programmes have been conducted in Chinese schools. Published reports on SEL programmes have only been published in Hong Kong. A study among first graders aged 6–7 years in Hong Kong found improvements in emotion regulation and prosocial behaviours after a universal classroom-led SEL programme [4]. A school-based universal SEL programme conducted among seventh-grade Hong Kong children aged 12–13 years reported that the intervention group had less emotional distress, fewer internalizing problems such as depression and anxiety, and attention difficulty [10]. In Chinese culture the pursuit of academic success is widely regarded as a high priority with children and adolescents forced to devote most of their waking hours to studying. Academic success is regarded as essential to higher income and social status [11]. As a result, very little attention is paid to the social and emotional well-being of children. This is despite strong evidence for the importance of confidence, compassion, good communication, good relationships, and emotional stability in order to function well in society and especially in the workplace [12].

In 2011, the Ministry of Education of China and UNICEF initiated a pilot social and emotional learning (SEL) programme in five Chinese provinces (Guizhou, Yunnan, Chongqing, Guangxi, and Xinjiang), based on the British Social and Emotional Aspects of Learning (SEAL) project. Resources were developed based on the pilot experience and were made available after the first COVID-19 lockdown in November 2020, in recognition of the particular need for social emotional training during the COVID pandemic [13]. However, the degree to which the programme was disseminated, as well as the outcomes, are unclear, and there have been no recommendations about incorporating it into the curriculum.

In this study, we piloted an adapted version of the MoE-UNICEF social and emotional learning programme with the aim of testing whether it can be used to reduce social, emotional, and behavioural difficulties in children. We aimed to explore: (1) whether there would be significant improvements in participants' social emotional wellbeing after the intervention; and (2) whether changes in social, emotional and behaviour problems were related to the initial risk levels of the participants.

2. Methods

2.1. Participants

The pilot intervention was conducted in a primary school located in an extremely poor village of Henan province in central China, with around 80% of residents with a household income per person of 10,000 yuan or less in 2021, which is much lower than the average level of rural China, i.e., 18,931 yuan [14]. More than 50% of the participants were "left-behind" children, with migrant workers as parents, and usually in the care of grandparents. There is only one class in each grade, a small school by Chinese standards. There are classes only in Chinese, mathematics, and English, with no classes in physical education, music, or painting, which are officially included in the school curriculum. Difficulties in recruiting local teachers to such poor rural schools mean that teachers are brought in on short-term contracts from the nearest city. During exam times, there are no class breaks because the whole day is spent revising for exams. This intensity is driven by competition with other schools for places at better secondary schools.

2.2. Procedures

In this pilot trial, a pre-post intervention design was adopted. All school attenders were included in the intervention condition with no control group because of the social emotional needs of the children, logistic challenges, and requirements of the school. All 243 students in the school participated in the intervention, but 43 (18%) first graders were not included in the pre-post analysis because of difficulties understanding the questionnaire; 7 (3%) from higher grades were not included because of missing one of the three assessments, and 3 (1.2%) were excluded because they missed more than two of the intervention sessions. Therefore, 190 (79%) students from 2nd to 6th grade were included in the analysis. Prior to the intervention, emotional and behaviour problems were assessed in January 2021 (Assessment 1). The intervention took place during regular school hours with the title "social skills training" for a whole 4-month semester from September 2021 to January 2022. After the programme was completed, emotional and behaviour difficulties were assessed in all participants in January 2022 (Assessment 2). After 5-month follow-up, the participants were once again assessed with the same questionnaire in June 2022 (Assessment 3).

2.3. Adaptation of the SEL Materials

The programme consisted of 16 weekly 90-min class sessions mostly adapted from the MoE-UNICEF social emotional learning resources [13]. The sessions were adapted on the basis of fieldwork and interviews with 30 students, evenly distributed across all grades, and 6 teachers, to fit the developmental, behavioural, and emotional needs of the local students. For example, teachers told us that almost all the children spoke rarely in class to ask or answer questions. To improve this, we used videos of popular cartoons, in which characters demonstrate confidence leading the children to discuss how to be confident. More than half of the children were "left-behind", by parents who were rural-urban migrant workers, returning typically once or twice per year. These children were mostly looked after by grandparents. At Interviews showed that communication with absent parents was often difficult, so to address this problem, we inserted several stories in the session "express myself" and "empathy training" to help with communicating with parents more comfortably. Several focus groups with children were conducted to identify which type of games from the SEL materials were most suited to the local context, as well as the "warm up" games children preferred in different sessions.

The 16 sessions covered five topics: (1) improvement of self-understanding and help with regulation of emotions; (2) building self-confidence and "feeling good to be me"; (3) help with addressing arguments, communicating effectively, and getting along with others; (4) saying NO to bullying; (5) setting targets and achieving them (Table 1). All sessions involved group discussion, role-play, art activities, storytelling, watching videos, handicrafts, and educational games. A handbook including the objectives and activities of the sessions was written specifically for the programme to guide the volunteers, and a separate booklet including the major content was given to each student.

Table 1. Content of the intervention Sessions.

Topic	Session Title
1. Learn about emotions	(1) Different and complex emotions
	(2) How emotions influence behaviours
	(3) Emotion management
2. Good to be me	(1) Be confident and aware of my strengths
	(2) Express myself bravely and stay true to myself
	(3) Relax and calm down
3. Get along with others	(1) Show kindness and care for others
	(2) Empathy training and put yourself in other's place to understand the differences
	(3) Learn to resolve conflicts
	(4) Take responsibility in collaboration
4. Say NO to bullying	(1) Bullying and being bullied
	(2) Empathise with bullying victims
	(3) What can we do if involved in bullying
	(4) How to stop bullying
5. Move toward your goals	(1) Make a goal and a plan
	(2) Strategies of overcoming difficulties such as boredom, tiredness, and procrastination

2.4. Selection of the School

We contacted the local education bureau to explain the aims and content of the programme and asked for their support and recommendations for a participating school. After getting approval from the school, a pamphlet was given to the students and their caregivers with detailed information about the programme and a request for their written consent for children's participation.

2.5. Training Volunteers and Implementing the Intervention

The SEL classes were led by a volunteer in each classroom. Six volunteers from the psychology department of a local university were recruited. They were third-year undergraduate students, three of which were majoring in applied psychology and the other three in educational psychology. They did not have experience in leading such intervention groups. They were trained over seven three-hour sessions by the coordinator. The training covered topics such as the general theory and main content of the curriculum, the adaptation of the sessions, and the introduction of the school and students. Volunteers were encouraged to think about daily examples in their experience that were relevant to the teaching of the sessions, and which would facilitate children's understanding of the topics. They also had the chance to practise and role-playing activities and discussed how to optimise delivery of the sessions in the classrooms.

When the sessions were underway in the classrooms, the coordinator observed and provided support if needed. The coordinator assessed the volunteers' implementation, including the way they conveyed the core ideas and concepts of the sessions. She gave feedback after each session and helped volunteers make adaptations if problems arose. Since the programme was included in the curriculum, the attendance rate was over 95% at each session.

3. Measurement Instruments

Sociodemographic and background information comprised gender, age, grade, number of siblings, the main caregiver, household composition (both parents, one parent, neither parent), family economy status (retrieved from school records and then categorised into three levels-good, fair, poor), and parents' occupations.

The Chinese version of the Strengths and Difficulties Questionnaire (SDQ) is a screening tool designed for the early detection of social, emotional, and behavioural problems in children and adolescents aged 4 to 16 years [15]. It is very widely used in Chinese research. The SDQ comprises 25 items, divided equally across five subscales: emotional symptoms, conduct problems, hyperactivity, peer relationship problems, and prosocial behaviour. The first four subscales measure potential difficulties and a combined score provide a child's total difficulties (TD) score with a higher score indicating greater difficulties. The fifth subscale, prosocial behaviour, is measured separately as a 'strength', with a higher score indicating better well-being.

Statistical Analyses

The intervention was carried out in all classes. In order to compare the effects of the intervention on children with different levels of emotional and behaviour difficulties, participants were stratified into low-, moderate- and high-risk groups based on the SDQ total difficulties scores at the first assessment. The upper quartile mean was 7.25 and lower quartile mean was 15, so the SDQ total difficulties score of less than 8 was classified as low risk, 8–14 moderate risk, and 15 and over high risk.

The impact of the programme on emotional and behaviour problems was evaluated by comparing scores on the SDQ subscales in the three risk groups at three time points: before the intervention (January 2021), immediately after completion of the intervention (January 2022), and 5-month follow-up (June 2022).

Data analysis was performed with SPSS 24.0. First, we generated descriptive statistics on the sociodemographic information. Second, we used Pearson's chi-square tests to examine the association between gender and sociodemographic information. Third, repeated measures analysis of variance (ANOVA) was conducted with three risk groups (low, moderate, high) as between-subjects factor and the SDQ components as repeated measures (pre-intervention, post-intervention, and follow-up assessment). Fourth, one-way repeated measures analysis of variance (ANOVA) was conducted in different risk groups separately with SDQ-difficulties as repeated measures (pre-intervention, post-intervention, and follow-up assessment).

4. Results

The sociodemographic characteristics of the participants who completed all three assessments are presented in Table 2. The chi-squared test for categorical variables revealed no significant differences between genders. Around 38% of the participants were mainly looked after by grandparents and around half did not live with both parents.

Table 2. Sociodemographic and background information of participants at Assessment 1.

	Total (n = 190)	Male (n = 81)	Female (n = 109)	Chi-Square	*p*
	Age (mean = 9.21, SD = 1.1)			0.124	0.725
7–9	120 (63.2)	50 (61.7)	70 (64.2)		
10–12	70 (36.8)	31 (38.3)	39 (35.8)		
	Number of siblings (mean = 1.91, SD = 1.038)			3.25	0.517
0	11 (5.8)	5 (6.3)	6 (5.6)		
1	76 (40)	36 (45)	40 (37)		
2	72 (37.9)	25 (31)	47 (43.5)		
3	22 (11.6)	10 (12.5)	12 (11)		
4	7 (3.7)	4 (5)	3 (2.8)		
	The main carer			0.83	0.66
Grandparents	72 (37.9)	33 (41.3)	39 (36.1)		
Father	8 (4.2)	4 (5)	4 (3.7)		
mother	108 (56.8)	43 (53.8)	65 (60.2)		
	Household composition			0.905	0.636
Both parents	93 (48.9)	37 (46.8)	56 (51.9)		
One parent	55 (28.9)	23 (29.1)	32 (29.6)		
Neither parent	39 (20.5)	19 (24.1)	20 (18.5)		
	Family economic status			2.542	0.281
Above average	60 (31.6)	26 (40.6)	34 (35.8)		
Average	89 (46.8)	32 (50)	57 (60)		
Below average	10 (5.3)	6 (9.4)	4 (4.2)		

4.1. Intervention Effects on SDQ Components

There was an effect of the time point on peer problem ($\eta^2 = 0.15$), which means there were significant differences among the peer problem scores in the three assessments (Tables 3 and 4). Peer problem scores lowered at Assessment 2 ($p < 0.001$) and Assessment 3 ($p < 0.001$) compared to Assessment 1, while there were no differences between Assessment 2 and Assessment 3. This suggested that peer problems decreased from pre- to post-assessments, and that the reduction was sustainable at the two post-intervention assessments. There was an effect of the time point on hyperactivity ($\eta^2 = 0.06$), which means there were significant differences in the scores of hyperactivity in the three assessments (Tables 3 and 4). There was no effect of the time point on SDQ total difficulties, conduct problems, emotional symptoms, or prosocial behaviours. Meanwhile, there was a time * risk group interaction effect of the SDQ-total difficulties ($\eta^2 = 0.18$), conduct problems ($\eta^2 = 0.1$), emotional symptoms ($\eta^2 = 0.11$) and hyperactivities ($\eta^2 = 0.11$). This suggested that the effect of the time point was different in the three risk groups of total difficulties, conduct problems, emotional symptoms and hyperactivities.

Table 3. Descriptive statistics for SDQ components, mean (SD).

	Total	Low Risk (n = 43)	Moderate Risk (n = 83)	High Risk (n = 46)
SDQ-total difficulties				
Assessment 1	11.9 (5.6)	5.1 (1.8)	11.5 (1.9)	19.0 (3.8)
Assessment 2	11.2 (6.0)	8.7 (5.2)	10.6 (5.6)	14.7 (6.0)
Assessment 3	12.1 (5.5)	10.0 (4.4)	11.7 (5.5)	14.8 (5.4)
SDQ-peer problem				
Assessment 1	3.7 (1.7)	2.2 (1.2)	3.9 (1.4)	4.9 (1.5)
Assessment 2	2.5 (2.0)	1.9 (1.8)	2.5 (1.7)	3.1 (2.4)
Assessment 3	2.6 (1.8)	1.8 (1.2)	2.7 (1.7)	3.2 (2.1)
SDQ-conduct problem				
Assessment 1	1.6 (1.5)	0.5 (0.7)	1.4 (1.0)	3.1 (1.7)
Assessment 2	1.6 (1.6)	1.2 (1.3)	1.5 (1.5)	2.2 (1.7)
Assessment 3	1.9 (1.4)	1.5 (1.2)	1.7 (1.4)	2.4 (1.6)
SDQ-emotional symptoms				
Assessment 1	3.6 (2.4)	1.4 (1.3)	3.4 (1.5)	6.1 (2.1)
Assessment 2	3.7 (2.5)	2.5 (2.1)	3.6 (2.5)	5.0 (2.4)
Assessment 3	3.9 (2.3)	3.2 (2.2)	3.8 (2.4)	4.7 (2.1)
SDQ-hyperactivity				
Assessment 1	3.0 (2.0)	1.1 (1.1)	2.9 (1.5)	4.9 (1.5)
Assessment 2	3.4 (2.3)	3.1 (2.3)	3.9 (2.2)	4.3 (2.2)
Assessment 3	3.8 (2.1)	3.4 (1.9)	3.6 (2.1)	4.6 (2.0)
SDQ-strength (prosocial behaviour)				
Assessment 1	7.9 (1.8)	8.7 (1.6)	7.8 (1.7)	7.3 (1.7)
Assessment 2	7.8 (2.0)	8.0 (1.9)	7.7 (2.1)	7.9 (1.9)
Assessment 3	7.5 (2.0)	7.9 (2.0)	7.6 (1.8)	7.1 (2.2)

Table 4. Two-way repeated measures ANOVA for assessing the differences of SDQ scores in three time points by risk groups.

Effect	MS	df	F	p	η^2
SDQ-total difficulties					
Time	34.4	1.8	2.14	0.124	0.01
Risk group	2335.8	2.0	65.11	<0.001	0.44
Time * risk group	296.7	3.7	18.48	<0.001	0.18
SDQ-peer problem					
Time	72.8	1.8	29.0	<0.001	0.15
Risk group	107.4	2.0	26.63	<0.001	0.24
Time * risk group	9.6	3.7	3.83	0.006	0.04
SDQ-conduct problem					
Time	3.1	1.9	2.31	0.10	0.01
Risk group	85.1	2.0	26.7	<0.001	0.24
Time * risk group	11.9	3.8	8.87	<0.001	0.1
SDQ-emotional symptoms					
Time	3.6	2.0	1.26	0.29	0.007
Risk group	283.3	2.0	36.15	<0.001	0.3
Time * risk group	30.6	4.0	10.66	<0.001	0.11
SDQ-hyperactivity					
Time	31.7	1.9	11.15	<0.001	0.06
Risk group	154.3	2.0	26.99	<0.001	0.24
Time * risk group	28.1	3.8	9.89	<0.001	0.11
SDQ-strength (prosocial behaviour)					
Time	7.6	2.0	3.15	0.04	0.02
Risk group	19.9	2.0	3.34	0.04	0.04
Time * risk group	4.4	4.0	1.83	0.12	0.02

Note. "*" refers to the interaction of time and risk group.

4.2. Intervention Effects on Participants of Different Risk Groups

The intervention effects of the SDQ-total difficulties, conduct problems, emotional symptoms, and hyperactivities in three risk groups are presented in Table 5.

Table 5. One-way repeated measures ANOVA of SDQ components in three risk groups.

Effect	MS	df	F	p	η^2
High risk group					
SDQ-total difficulties					
Time	318.4	1.8	14.76	<0.001	0.25
SDQ-conduct problem					
Time	12.8	1.6	5.92	0.007	0.12
SDQ-emotional symptoms					
Time	24.1	2.0	8.76	<0.001	0.16
SDQ-hyperactivity					
Time	4.26	2.0	1.66	0.196	0.04
Moderate risk group					
SDQ-total difficulties					
Time	36.55	1.8	2.28	0.112	0.03
SDQ-conduct problem					
Time	2.6	2.0	2.13	0.122	0.03
SDQ-emotional symptoms					
Time	4.1	2.0	1.36	0.261	0.02
SDQ-hyperactivity					
Time	10.5	1.8	3.1	0.053	0.04
Low risk group					
SDQ-total difficulties					
Time	272.2	2.0	24.68	<0.001	0.37
SDQ-conduct problem					
Time	12.7	2.0	15.1	<0.001	0.26
SDQ-emotional symptoms					
Time	37.0	2.0	14.0	<0.001	0.25
SDQ-hyperactivity					
Time	66.8	2.0	30.8	<0.001	0.42

In the high-risk group, there was an effect of the time point on SDQ-total difficulties (η^2 = 0.25) which lowered at Assessment 2 (p < 0.001) and Assessment 3 (p < 0.001) compared to Assessment 1, with no difference between Assessments 2 and 3. This suggested that in the high risk group, the total difficulties decreased from pre- to post-assessments, and that the reduction was sustainable at the two post-intervention assessments. In the moderate risk group, there was no effect of the time point on SDQ-total difficulties (η^2 = 0.027). In the low-risk group, there was an effect of the time point on SDQ-total difficulties (η^2 = 0.37), which increased at Assessment 2 (p < 0.001) and Assessment 3 (p < 0.001) compared to Assessment 1, with no difference between Assessments 2 and 3. This suggested that in the low risk group, total difficulties increased from pre- to post-assessments, and that the increase was sustainable at the two post-intervention assessments.

In the high-risk group, there was an effect of the time point on conduct problems (η^2 = 0.12), which lowered at Assessment 2 (p = 0.016) compared to Assessment 1, with no difference between Assessments 3 and 1 (p = 0.07). This suggested that in the high-risk group, conduct problems decreased from baseline to post-intervention, but increased at the 5-month follow-up. In the moderate risk group, there was no effect of the time point on conduct problems (η^2 = 0.026). In the low-risk group, there was an effect of the time point on conduct problems (η^2 = 0.26), which increased at Assessment 2 (p = 0.002) and Assessment 3 (p < 0.001) compared to Assessment 1, with no difference between Assessments 2 and 3. This suggested that in the low risk group conduct problems increased from pre- to post-assessments, and that the increase was sustainable at the two post-intervention assessments.

In the high-risk group, there was an effect of the time point on emotional symptoms ($\eta^2 = 0.16$), which lowered at Assessment 2 ($p = 0.03$) and Assessment 3 ($p < 0.001$) compared to Assessment 1, with no difference between Assessments 2 and 3. This suggested that in the high risk group, emotional symptoms decreased from pre- to post-assessments, and the reduction was sustainable at the two post-intervention assessments. In the moderate risk group, there was no effect of the time point on emotional symptoms ($\eta^2 = 0.016$). In the low-risk group, there was an effect of the time point on emotional symptoms ($\eta^2 = 0.25$), which increased at Assessment 2 ($p = 0.008$) and Assessment 3 ($p < 0.001$) compared to Assessment 1, with no difference between Assessments 2 and 3. This suggested that in the low risk group, emotional symptoms increased from pre- to post-assessments, and that the increase was sustainable at the two post-intervention assessments.

There was no effect of the time point on hyperactivities in the high-risk group ($\eta^2 = 0.036$) and moderate risk group ($\eta^2 = 0.037$). In the low risk group, there was an effect of the time point on hyperactivities ($\eta^2 = 0.42$), which increased at Assessment 2 ($p < 0.001$) and Assessment 3 ($p < 0.001$) compared to Assessment 1, with no difference between Assessment 2 and 3. This suggested that in the low risk group hyperactivities increased from pre- to post-assessments, and that the increase was sustainable at the two post-intervention assessments.

5. Discussion

To our knowledge, this is the first study to test the effectiveness of a universal school-based SEL programme for primary school children in mainland China. The key findings are: (1) the programme can reduce children's peer relationship problems and the reduction was sustainable at the two post-intervention assessments; (2) no intervention effect was found on emotional symptoms, conduct problems or hyperactivity in the whole participants, but the high-risk children benefited from the programme.

The peer problems reduced after the intervention and the reduction was sustainable at the two post-intervention assessments. This may have been because of the focus of our programme, which was on positive communication, collaboration, and problem-solving skills within the classroom context with the aim of enhancing children's ability to understand their differences, be more empathic, and resolve conflicts effectively. Children's peer communication skills can be improved with an increase in mutual understanding and acceptance [16]. The role plays are reported to be effective in fostering perspective taking and empathy [17]. In our study, the pre-prepared role plays scenarios relevant to the school setting are widely used through all sessions, for example, role plays dealing with how to communicate with students who have been making jokes on you that makes you embarrassed.

The intervention did not improve emotional symptoms, conduct problems, or hyperactivity in the whole participants. This may have been due to the relative brevity of the intervention, which spanned just 4 months. It is hard to acquire the skills that can have a consequential influence on emotional well-being or hyperactivity in such a short time [18]. There is also a problem that the programme put too much responsibility on the children to integrate the newly learned skills and strategies into their behaviours; for example, a common strategy in the programme involves teaching children how to modify unhelpful thoughts into helpful thoughts, which activate positive behaviours. An alternative approach would be to train schoolteachers to better recognise and reward helpful thoughts expressed by the children. Consequently, an approach where teachers are substantially trained in such core practice elements instead of only receiving a short introduction of the general idea of the programme may enhance the preventive effect [19]. Another possible explanation for this lack of significance is the delayed effect, which means the effectiveness cannot be found after a short period of the intervention, but can be seen after extended intervals for participants to practise, apply and consolidate skills [20]. So further assessments with longer follow-ups are required.

Since this is a universal study, the lack of effectiveness could also be attributed to the selection of the sample. In the general population of children, there will always be a number of subjects who do not present any risk of social emotional problems [21]. In our study, the emotional symptoms, conduct problems, and hyperactivity levels were quite low for the low-risk children at the pre-intervention assessment, and these difficulties even increased at post-intervention assessments but were still within the normal range. Room for improvement is quite small for such low-risk children [22]. However, the effectiveness of the reduction of total difficulties and emotional symptoms at post intervention and 5-month follow-up could be observed in high-risk children in our study. Similar results had been found from other studies, that is, the children most responsive to the effects of the social emotional training programmes were the ones with higher levels of emotional and behavioural difficulties at pre-intervention assessment [23,24].

There were a number of limitations in this study. A primary one was the lack of a control group. It is possible that the improvements in children's social and emotional outcomes were due to other factors and not the programme, for example, "natural" developmental progression in the 4-month period in which the intervention was implemented. Therefore, longer term follow-up is needed. Second, the results were based on children's self-reports and, ideally, future research should include data from other informants, such as parents, teachers, and caregivers. Third, given that the programme was conducted in just one school in a poor rural area of central China, the findings cannot be generalised.

6. Conclusions

This intervention was welcomed by the school, incorporated into the curriculum, and was cheap to implement, making it potentially sustainable. The programme is easy to implement with the guidance of the handbook. It can be delivered to school children in regular school hours and could be integrated into the classroom curriculum, ensuring a high attendance rate and retention.

To our knowledge, this is the first published report of the effectiveness of a universal school-based SEL programme in mainland China. Although the findings are preliminary, they support the evidence for the effectiveness of the programme to improve peer relationships among primary school children. With the support of the local Education Bureau, schools could incorporate the SEL programme into their formal curriculum, as has been the case in a number of countries [3,25]. Schools in Hong Kong have been urged by the governmental Education Bureau to regard the all-round development and social emotional wellbeing of their students as major educational aims, with the incorporation of SEL in the curricula [4]. Our study shows that the SEL programme is cheap to implement and potentially sustainable in poor rural areas in China. Local people who are interested in and capable of delivering the intervention sessions should be trained and employed to deliver the programme. Implementation should be the responsibility of the local government education department.

Our findings show that future studies should: (1) train schoolteachers and other school staff in SEL skills so that children's positive behaviours can be strengthened outside of SEL classes. SEL skills are also needed in playgrounds and in lunchrooms, and thus, school staff, who have interaction with children should be trained in such skills; (2) shorten the duration of each session to 40–50 min. Although there is a 10-min class break in the middle of the 90-min session, most children could not focus well in the second half, which might compromise the effectiveness of intervention sessions. (3) A more intense programme could be provided for high-risk children.

Author Contributions: J.L. and T.H. designed and conducted the research. J.L. conducted the data analysis and wrote the first draft of the manuscript. T.H. revised the manuscript with J.L., C.M. and Q.L. conducted the research and reviewed the manuscript. All authors have read and agreed to the published version of the manuscript.

Funding: This research received no external funding.

Institutional Review Board Statement: The study protocol was approved by the Ethics Committee of Zhejiang University School of Public Health (protocol number ZGL202101-3). All methods were performed in accordance with the study protocol and ethical guidelines and regulations.

Informed Consent Statement: Informed consent was obtained from all subjects involved in the study.

Acknowledgments: We would like to thank all participants and volunteers for their contribution to this study. We also thank the local education bureau, teachers and other school staff for their support.

Conflicts of Interest: The authors claim that there are no conflict of interest.

References

1. Ahlen, J.; Lenhard, F.; Ghaderi, A. Long-Term Outcome of a Cluster-Randomized Universal Prevention Trial Targeting Anxiety and Depression in School Children. *Behav. Ther.* **2019**, *50*, 200–213. [CrossRef] [PubMed]
2. Barrett, P.M.; Lock, S.; Farrell, L.J. Developmental Differences in Universal Preventive Intervention for Child Anxiety. *Clin. Child Psychol. Psychiatry* **2005**, *10*, 539–555. [CrossRef]
3. Barrett, P.; Turner, C. Prevention of anxiety symptoms in primary school children: Preliminary results from a universal school-based trial. *Br. J. Clin. Psychol.* **2001**, *40*, 399–410. [CrossRef] [PubMed]
4. Barry, M.M.; Clarke, A.M.; Jenkins, R.; Patel, V. A systematic review of the effectiveness of mental health promotion interventions for young people in low and middle income countries. *BMC Public Health* **2013**, *13*, 835. [CrossRef] [PubMed]
5. Beaumont, R.B.; Pearson, R.; Sofronoff, K. A Novel Intervention for Child Peer Relationship Difficulties: The Secret Agent Society. *J. Child Fam. Stud.* **2019**, *28*, 3075–3090. [CrossRef]
6. Durlak, J.A.; Weissberg, R.P.; Dymnicki, A.B.; Taylor, R.D.; Schellinger, K.B. The Impact of Enhancing Students' Social and Emotional Learning: A Meta-Analysis of School-Based Universal Interventions. *Child Dev.* **2011**, *82*, 405–432. [CrossRef]
7. Elias, M.J.; Zins, J.E.; Weissberg, R.P.; Frey, K.S.; Greenberg, M.T.; Haynes, N.M.; Kessler, R.; Schwab-Stone, M.E.; Shriver, T.P. Promoting Social and Emotional Learning: Guidelines for Educators. Association for Supervision and Curriculum Development. 1997. Available online: https://www.ascd.org/books/promoting-social-and-emotional-learning?variant=197157E4 (accessed on 3 May 2022).
8. Goodman, R. The Strengths and Difficulties Questionnaire: A Research Note. *J. Child Psychol. Psychiatry* **1997**, *38*, 581–586. [CrossRef] [PubMed]
9. Hawkins, J.D.; Catalano, R.F.; Kosterman, R.; Abbott, R.; Hill, K. Preventing Adolescent Health-Risk Behaviors by Strengthening Protection During Childhood. *Arch. Pediatr. Adolesc. Med.* **1999**, *153*, 226–234. [CrossRef] [PubMed]
10. Kam, C.-M.; Wong, L.W.-L.; Fung, K.M.-S. Promoting social-emotional learning in Chinese schools: A feasibility study of PATHS implementation in Hong Kong. *Int. J. Emot. Educ.* **2011**, *3*, 30–47. Available online: https://go.exlibris.link/1fKPHGGT (accessed on 3 May 2022).
11. Kozina, A. Can the "My FRIENDS" Anxiety Prevention Programme Also be Used to Prevent Aggression? A Six-Month Follow-Up in a School. *Sch. Ment. Health* **2018**, *10*, 500–509. [CrossRef]
12. Kozina, A. Can FRIENDS for Life social-emotional learning programme be used for preventing anxiety and aggression in a school environment: 6 months, 1-year and 1-and-a-half-year follow-up. *Eur. J. Dev. Psychol.* **2021**, *18*, 214–229. [CrossRef]
13. Lam, K.; Seiden, D. Effects of a Brief Mindfulness Curriculum on Self-reported Executive Functioning and Emotion Regulation in Hong Kong Adolescents. *Mindfulness* **2020**, *11*, 627–642. [CrossRef]
14. Li, J.; Sidibe, A.M.; Shen, X.; Hesketh, T. Incidence, risk factors and psychosomatic symptoms for traditional bullying and cyberbullying in Chinese adolescents. *Child. Youth Serv. Rev.* **2019**, *107*, 104511. [CrossRef]
15. Miller, C.F.; Kochel, K.P.; Wheeler, L.A.; Updegraff, K.A.; Fabes, R.A.; Martin, C.L.; Hanish, L.D. The efficacy of a relationship building intervention in 5th grade. *J. Sch. Psychol.* **2017**, *61*, 75–88. [CrossRef] [PubMed]
16. Resident Income and Consumption Expenditure in 2021. 2022. Available online: http://www.stats.gov.cn/tjsj/zxfb/202201/t20220117_1826403.html (accessed on 19 September 2022).
17. Rivet-Duval, E.; Heriot, S.; Hunt, C. Preventing Adolescent Depression in Mauritius: A Universal School-Based Program. *Child Adolesc. Ment. Health* **2011**, *16*, 86–91. [CrossRef] [PubMed]
18. A Series of Resources for the MoE-UNICEF Social and Emotional Learning Project. 2020. Available online: https://www.unicef.cn/documents/sel-resources (accessed on 3 May 2022).
19. Sheffield, J.K.; Spence, S.H.; Rapee, R.M.; Kowalenko, N.; Wignall, A.; Davis, A.; McLoone, J. Evaluation of universal, indicated, and combined cognitive-behavioral approaches to the prevention of depression among adolescents. *J. Consult. Clin. Psychol.* **2006**, *74*, 66–79. [CrossRef] [PubMed]
20. Singapore, G. Social and Emotional Learning in Singapore. 2022. Available online: https://www.moe.gov.sg/education-in-sg/our-programmes/social-and-emotional-learning (accessed on 5 May 2022).
21. Sklad, M.; Diekstra, R.; de Ritter, M.; Ben, J.; Gravesteijn, C. Effectiveness of school-based universal social, emotional, and behavioral programs: Do they enhance students' development in the area of skill, behavior, and adjustment? *Psychol. Sch.* **2012**, *49*, 892–909. [CrossRef]

22. Social and Emotional Aspects of Learning (SEAL). 2005. Available online: https://sealcommunity.org/node/356#five (accessed on 3 May 2022).
23. Taylor, R.D.; Oberle, E.; Durlak, J.A.; Weissberg, R.P. Promoting Positive Youth Development Through School-Based Social and Emotional Learning Interventions: A Meta-Analysis of Follow-Up Effects. *Child Dev.* **2017**, *88*, 1156–1171. [CrossRef] [PubMed]
24. Weissberg, R.P.; Durlak, J.A.; Domitrovich, C.E.; Gullotta, T.P. Social and emotional learning: Past, present, and future. In *Handbook of Social and Emotional Learning: Research and Practice*; Weissberg, R.P., Durlak, J.A., Domitrovich, C.E., Gullotta, T.P., Eds.; The Guilford Press: New York, NY, USA, 2015; pp. 3–19.
25. Wong, A.S.; Li-Tsang, C.W.; Siu, A.M.H. Effect of a Social Emotional Learning Programme for Primary School Students. *Hong Kong J. Occup. Ther.* **2014**, *24*, 56–63. [CrossRef]

healthcare

MDPI

Article

The Relationship between Psychological Disability and Religious Practice and Coping Strategies in Caregivers of Children with Traumatic Brain Injury in Pakistani Population

Ayesha [1,†], Saboor Ahmad [1,†], Shazadi Saba [1], Muhammad Kashif [1], Danish Ali Khan [1], Absarul Haque [2,3], Muhammad Imran Naseer [3,4], Adel M Abuzenadah [2,3], Anwar M. Hashem [3,5] and Shafiq Ur Rehman [1,*]

1 College of Physical Therapy, Northwest Institute of Health Sciences, Peshawar 25100, Pakistan
2 King Fahd Medical Research Center, King Abdulaziz University, Jeddah 21859, Saudi Arabia
3 Department of Medical Laboratory Sciences, Faculty of Applied Medical Sciences, King Abdulaziz University, Jeddah 21859, Saudi Arabia
4 Center of Excellence in Genomic Medicine Research, King Abdulaziz University, Jeddah 21589, Saudi Arabia
5 Vaccines and Immunotherapy Unit, King Fahad Medical Research Center, King, Jeddah 21589, Saudi Arabia
* Correspondence: shafiq.qau.edu@gmail.com or shafiqrehman@nwihs.edu.pk; Tel.: +92-333-9376758
† The authors contributed equally to this research work.

Citation: A.; Ahmad, S.; Saba, S.; Kashif, M.; Khan, D.A.; Haque, A.; Naseer, M.I.; Abuzenadah, A.M.; Hashem, A.M.; Rehman, S.U. The Relationship between Psychological Disability and Religious Practice and Coping Strategies in Caregivers of Children with Traumatic Brain Injury in Pakistani Population. *Healthcare* **2022**, *10*, 2158. https://doi.org/10.3390/healthcare10112158

Academic Editors: Yasuhiro Kotera and Elaina Taylor

Received: 29 September 2022
Accepted: 25 October 2022
Published: 28 October 2022

Abstract: Background: Traumatic brain injury (TBI) is a serious issue and a leading cause of death and disability worldwide. Caregivers of TBI patients experience psychological distress and a variety of social and financial issues. The present study aims to investigate the caregiver's burden and the factors that influence this burden. Furthermore, the present study will find out the association of religious practice, religious coping relations and psychological distress among caregivers of children affected with TBI. Methods: A cross-sectional survey was conducted on 302 caregivers of children with TBI using Duke University Religion Index (DURL) for religious practice. General Health Questionaire-12 (GHQ-12) was used for anxiety and depression and Brief Religious Coping Scale (RCOPE) was used for coping strategies. The caregivers were conveniently chosen from different regions of Khyber Pakhtunkhwa province and data was collected from different tertiary care hospitals in Peshawar. Results: Forty-nine (49) % of caregivers score ≥ 3 on GHQ suffer from psychological distress with a Mean of 20.957 ± 4.175). Positive coping methods were mostly used by caregivers than negative coping have a low level of distress with a Mean Positive Coping (P-COPE) of 6.93 ± 0.41, Mean of Negative Coping (N-COPE) 0.486 ± 1.023. In religious practice, caregivers mostly participate in Organized Reliogious Activities (ORA) or some Non-Organized Reliogious Activities (NORA) with a Mean ORA of 4.20 ± 1.27, and NORA Mean of 4.17 ± 1.37 used by the caregivers. Coping methods were related to Caregiver psychological distress (GHQ-12 and P-COPE co-relation scores are (ρ -0.022, p b 0.05); GHQ-12 scores and N-COPE (ρ + 0.221=, p b 0.001). There is a negative correlation between GHQ 12 and PCOPE, while GHQ12 is positively correlated with NCOPE. Conclusion: According to this study, there is a significant association between religious coping methods, religious practice, and psychological distress among caregivers of children with traumatic brain injury.

Keywords: traumatic brain injury; religious practice; psychological distress; coping strategies; caregiver; children

1. Introduction

Traumatic brain injury (TBI) is a brain injury, caused by out-sided force to the head that affects brain function [1]. TBI is caused by a focal gunshot wound or diffuse brain injuries like shaken baby syndrome or both types. Symptoms depend on the area of the lesion, and the extent of brain damage [2]. According to the Centers for Disease Control and Prevention (CDC), the most common causes of TBI in children aged 1 to 18 years old are falls (50.2%), struck by/against them (24.8%), automobile collisions (6.8%), assault on a person (2.9%) and other/unknown (15.3%) mechanisms of TBI in children TBI is a general health concern that can result in short- or long-term memory impairment. TBI causes psychological discomfort and financial burden all around the world [3]. Caregivers suffer from economic burdens as well as psychological discomfort during the TBI treatment process. The caregiver experiences feelings of hopelessness, frustration, impotence, despair, depression, grief, and a desire to quit. TBI has a significant impact on patients, their families, communities, and healthcare systems [4].

For both the survivor and the survivor's family, the healing process after a TBI can be lengthy and challenging. During a hospital stay, caregivers may neglect their home and work responsibilities, and they may have difficulty returning to their pre-injury activities [4]. They may feel a greater sense of responsibility to provide physical, financial, and emotional support than they are used to. TBI is the leading cause of death in children in the age group of 0–5 years and 15–19 years [5]. In addition, 145,000 children aged 1–19 years are living with long-term changes in behavior, physical disability and cognitive impairments following TBI [6]. The findings from a systematic review revealed that siblings experience psychological suffering and worse family functioning following TBI in children including behavioural problems in their brother and sisters and absence of perceived social support. The findings largely support that there is a need to grow and evaluate support interventions to improve siblings' adjustment to the effects of TBI [7]. Although evidence of efficacy is lacking, however, a number of tactics have been defined in the literature for siblings support in the context of TBI [8,9].

People react differently to different types of events and have varying levels of sensitivity. Caregivers who are experiencing psychological discomfort may have a different reaction. Cognitive evaluation and coping mechanisms, according to Lazarus and Folkman's model, have a major impact on the quality and severity of a person's stress reaction in a stressful event. Religion helps in stress management by increasing the occurrence of pleasant sensations and decreasing the likelihood of negative thoughts like depression, stress, and suicidal thoughts [10]. According to Kaplan et al, religion has a distinctive role in three ways when coping with various disorders: First people engage in a religious event, secondly, religion helps people to cope with stressful events, and lastly, religion is essential in people's lives to cope [11].

Numerous studies have been conducted to examine how religious participation influences caregivers' distress levels when their children suffer from long-lasting illnesses such as cancer, thalassemia, or cystic fibrosis. Caregivers, who pray, attend public religious services and study religious scriptures had lower stages of anxiety [12–14]. Some caregivers cope with their circumstances by seeking God's comfort [11]. As a result, religious beliefs and practices can alter people's perceptions of terrible occurrences and diminish unpleasant sensations, either directly or indirectly, while also encouraging pleasant emotions and the development of human values [15].

Religious coping stratagies are categorized into negative religious coping and positive religious coping [16]. There are more hopeful ways to deal with positive religious relationships that include spiritual help, forgiveness, shared religious conflicts, spiritual connections, and compassionate religious reconsideration Negative coping is a stressful approach to dealing with issues like spiritual misery and blaming God. Positive religious survival methods are more often used by people caring for children with chronic conditions than negative ones [17–19].

According to our knowledge, no study has been conducted on "the relationship between religious practice and psychological distress and coping strategies among caregivers of children affected with TBI. The diverse types of religious coping have had different effects on how people react to catastrophic life experiences. This study aims to look at different religious coping techniques in predicting coping strategies in a group of TBI child caregivers.

2. Materials and Methods

A cross-sectional survey was conducted at Northwest General Hospital & Research Centre (NWGH), Khyber Teaching Hospital (KTH), Hayatabad Medical Complex (HMC) Peshawar, the leading health care providing hospitals in Khyber Pakhtunkhwa province. The data was collected after approval from the ethical committee (Approval ID; NWIHS-EC/0028-0021) of the Northwest Institute of Health Sciences, Peshawar, Pakistan and the declaration of Helsinki was followed. The population for the study was caregivers of children with TBI at the above mention hospitals. Non-Probability convenience sampling was used. Inclusion criteria were considered (Caregivers of TBI patients, Patients aged 1 to 18 years). The data was collected through General Health Questionaire-12 (GHQ-12) (for behavioral changes and psychological distress), for religious practice the Duke University Religion Index (DUREL) and Brief Religious Coping Scale (Brief RCOPE) were used for coping. Participants meeting the inclusion criteria were selected for the study. To all participants, the purpose of the study was explained and informed consent was obtained in written form. All the willing participants were briefed about the purpose and procedure of this study. We provided a questionnaire in Urdu (translated through a faculty member of the department of Engish, University of Peshawar, who were native Urdu speakers) format for those who are not able to understand the questionnaire in English. For those participants who were illiterate, the questionnaires were administered by the researcher through a face-to-face interview, while others filled them through self-administration. The data collector verbally asked questions. The investigator was there to assist in case of difficulty in understanding. Prior to the survey, a short introduction was given to participants. Those participants who permitted us were screened by the inclusion (Caregivers of TBI patients, children aged 1 to 18 years) and exclusion criteria (known cases of Acute TBI patients, patients with any physical issues, diabetic neuropathy, amputation, mentally retarded patients and congenital brain disorder) in these tertiary care hospitals. To analyze the data SPSS (Statistical Package for Social Sciences) version 20 was used. Using descriptive statistics, Demographic data were described in tables. The links between, religious coping techniques, religious practice, and psychological distress were investigated using Spearman's rank-order correlation.

3. Results

The probable sample size was 302, 95% confidence interval, however, 310 questionnaires were given to Caregivers who met inclusion criteria throughout the sampling period. Three hundred and two caregivers agreed to take part and completed the questionnaires, in total, 79 % (n = 239) were male and 21% (n = 63) were female. In terms of religion, 99.4 % (n = 99) were Muslim, while only 0.6 % (n = 1) were Christian Table 1.

Table 1. Showing characteristics, frequencies, and percentages.

Variable	Characteristics	Frequency	Percentage
Gender	Male	239	79.1%
	Female	63	20.9%
	Total	302	100.0%

3.1. *Religious Practice, Religious Coping Methods and Psychological Distress*

The psychological distress Mean or average score was 20.957 ± 4.175. Almost all of the caregivers (n = 302) were depressed, with a GHQ score of 3 or above. The mean(M) score for Organized Religious activities (ORA), and caregivers' religious practice was 5.738 ± 0.7343, suggesting a high level of attendance at public religious services or meetings; the sample had a high ORA score. The mean for Non-Organized Religious Activities (NORA) was 4.410 ± 1.452, and the participants practiced it. The mean score for Intrinsic religiosity (IR) caregivers' religious practice was 14.367 ± 1.270, with IR ranging from 3 to 15; high IR for the majority, indicating strong religious motivation and faith. In terms of the coping sample score high on P-COPE, the mean P-COPE was 6.93 ± 0.418. The sample used negative religious coping methods N-COPE mean was 0.4868 ± 1.023. whereas the range show number of question in the questionnaire. Table 2 shows the mean, Std. Deviation, Ranges of religious practices, psychological distress and religious coping of the participants are presented in Table 2.

Table 2. Caregivers ' Mean, Std. Deviation, psychological distress, religious practices, and religious coping.

Variables	Total No	M ± SD	Ranges
Psychological distress (GHQ-12)			
≤3 scores		20.957 ± 4.175	0–12
≥3 scores	302		
Religious practice (DUREL)			
NORA		4.410 ± 1.452	1–6
ORA		5.738 ± 0.734	1–6
IR		14.367 ± 1.270	1–6
Religious coping Method			
N-COPE		0.486 ± 1.023	0–7
P-COPE		6.933 ± 0.418	0–7

3.2. Relationship between Psychological Distress and Religious Practice

The Spearman rank-order correlations between religious coping methods (N-COPE and P-COPE), religious practices sub-scales (NORA, ORA, and IR), and psychological distress (GHQ-12). ORA and GHQ-12 scores had a weak negative correlation (rs (df) = −0.020, p b 0.05), ORA is inversely related to GHQ 12 when a person deviates from ORA then his/her distress level will ultimately decrease. NORA and GHQ-12 scores had a weak positive connection (rs (df) = 0.094, p b 0.05), with an increase in NORA linked with high distress. IR and GHQ-12 had a negative connection (rs (df) = −0.045, p N 0.05). The connection between religious coping strategies and psychological discomfort-COPE and GHQ-12 scores had a weak negative correlation (rs (df) = −0.022, p b 0.05); increase P-COPE was linked with reduced distress level. The positive connection between GHQ-12 scores and N-COPE (rs (df) = 0.221, p b 0.05), with increased N-COPE scores being associated with increased distress (Table 3).

Table 3. Correlations among religious practice, religious coping methods, and psychological distress.

Variables	Values	GHQ-12 Scores
Religious Practices		
ORA	ρ	−0.020
	p	0.728
NORA	ρ	0.094
	p	0.102
IR	ρ	−0.045
	p	0.434
Religious Coping Method		
P-COPE	ρ	−0.022
	p	0.709
N-COPE	ρ	0.221
	p	0.000 *

* Correlation is significant at the 0.05 level (2-tailed). $p < 0.05$ was statistically significant.

3.3. Spearman Rank-Order Correlations between Psychological Distress and Religious Practice and Religious Coping Method

Table 4 shows the Spearman rank-order correlations between religious coping methods (N-COPE and P-COPE), religious practices sub-scales (NORA, ORA, and IR), and psychological distress (GHQ-12). The co-relation GHQ12 with NORA, ORA, IR, NCOPE and PCOPE has already been discussed in Table 3.

Table 4. Correlations among, religious coping methods, religious practice and psychological Distress.

			GHQ12	ORA	NORA	IR	PCOPE	NCOPE
Spearman's rho	GHQ12	Correlation Coefficient	1.000	−0.020	0.094	−0.045	−0.022	0.221
		Sig.(2 tailed) *p* value		0.728	0.102	0.434	0.709	0.000 *
		N	302	302	302	302	302	302
	ORA	Correlation Coefficient	−0.020	1.000	0.234	0.175	0.125 *	−0.363
		Sig. (2-tailed) *p* Value	0.728		0.000 *	0.002 *	0.030 *	0.000 *
		N	302	302	302	302	302	302
	NORA	Correlation Coefficient	0.094	0.234	1.000	0.298	0.049	−0.094
		Sig.(2 tailed) *p* value	0.102	0.000 *		0.000 *	0.399	0.104
		N	302	302	302	302	302	302
	IR	Correlation Coefficient	−0.045	0.175	0.298	1.000	0.131 *	−0.173
		Sig.(2 tailed) *p* value	0.434	0.002 *	0.000 *		0.022 *	0.003 *
		N	302	302	302	302	302	302
	PCOPE	Correlation Coefficient	−0.022	0.125 *	0.049	0.131 *	1.000	−0.296
		Sig.(2 tailed) *p* value	0.709	0.030 *	0.399	0.022 *		0.000 *
		N	302	302	302	302	302	302
	NCOPE	Correlation Coefficient	0.221	−0.363	−0.094	−0.173	−0.296	1.000
		Sig.(2 tailed) *p* value	0.000 *	0.000 *	0.104	0.003 *	0.000 *	.
		N	302	302	302	302	302	302

* Correlation is significant at the 0.05 level (2-tailed). $p < 0.05$ was statistically significant.

Whereas in Table 4, the second row shows the inter and intra-co-relation between questioners. There is a weak negative inter-co-relation between ORA and GHQ 12 (−0.020; $p = 0.728$). When a person deviates to ORA then his distress level will ultimately decrease. A moderate positive intra-co-relation between ORA and NORA (ρ 0.234) is observed with high significance ($p < 0.000$). There is a weak positive intra-co-relation between ORA and IR, which is highly significant (ρ = 0.175; $p = 0.002$). There is a weak positive intercorrelation between ORA and PCOPE (ρ = 0.125; $p = 0.030$). There is a moderate negative intercorrelation between ORA and NCOPE (ρ = −0.363; $p = 0.000$) which is highly significant. The third row shows the inter and intra-co-relation between questioners. There is a weak positive inter correlation between NORA and GHQ 12 (ρ = 0.094; $p = 0.102$) showing no significance. There is moderate positive intra-co-relation between NORA and ORA (ρ 0.234, $p = 0.000$) indicating highly significance. There is a moderate positive intra-co-relation between NORA and IR, which is highly significant (ρ = 0.298; $p = 0.000$). There is a weak positive inter-co-relation between NORA and PCOPE with no significance (ρ = 0.049; $p = 0.399$). There is a moderate negative intercorrelation between NORA and NCOPE with no significance (ρ = −0.094; $p = 0.104$). The fourth row shows the inter and intra-co-relation between IR, GHQ12, ORA, NORA PCOPE and NCOPE, There is a weak negative inter-co-relation between IR and GHQ 12 ($\rho = -0.045$; $p = 0.434$) which is non-significant. There is a weak positive intra-co-relation between IR and ORA (ρ 0.175). here, p-value 0.002 indicates high significance. There is moderate positive intra-co-relation between IR and NORA, which is highly significant ($\rho = 0.298$, $p = 0.000$). There is a weak positive inter co-relation between IR and PCOPE ($\rho = 0.131$, $p = 0.022$) indicating significance. There is a weak negative intercorrelation between IR and NCOPE ($\rho = -0.173$, $p = 0.003$) which is significant. The fifth row shows the inter and intra-co-relation between PCOPE, IR, GHQ12, ORA, NORA and NCOPE, There is a weak negative inter co-relation between PCOPE and GHQ 12 with no significance ($\rho = -0.022$; $p = 0.709$). There is a weak positive intercorrelation between PCOPE and ORA (ρ 0.125; $p = 0.709$) with no Significance. There is a moderate positive intercorrelation between PCOPE and NORA, which is non-significant ($\rho = 0.049$, $p = 0.399$). There is a weak positive intercorrelation between PCOPE and IR ($\rho = 0.131$, $p = 0.022$) which shows that the Correlation is significant at the 0.05 level (2-tailed). There is moderate negative intra-co-relation between PCOPE and NCOPE ($\rho = -0.296$, $p = 0.00$) which is highly significant. The last row shows the inter and intra-co-relation between NCOPE, IR, GHQ12, ORA, NORA and PCOPE, There is moderate negative inter co-relation between NCOPE and GHQ 12 with high significance ($\rho = -0.22$; $p = 0.000$). NCOPE and ORA showed moderate positive inter co-relation and high significance (ρ -0.363, $p = 0.000$). There is a weak negative inter co-relation between NCOPE and NORA, which is non–significant ($\rho = -0.049$, $p = 0.104$). There is a weak negative inter correlation between NCOPE and IR ($\rho = -0.173$, $p = 0.003$) which shows that the Correlation is significant at the 0.05 level (2-tailed). There is moderate negative intra-co-relation between NCOPE and PCOPE ($\rho = -0.296$, $p = 0.000$) which is highly significant.

4. Discussion

In this research, nearly all of the caregivers caring for a child suffering from traumatic brain injury ($n = 302$) reported psychological problems, with GHQ-12 ratings of 3 or higher. This finding is comparable to the findings of Shaligram et al. (2007), who found that about half (50%) of their observers had known psychiatric problems and scored > 3 on the GHQ-12 [20]. In a study conducted by Ali et al., 2012, out of 40, 27 (67.5%) parents of children who experience hypochromic anemia, reported psychological problems and scored > 3 on the GHQ-12 [21]. This discrepancy could be attributed to the short size of the sample or demographic variables. On the other side, the majority of parents (58%) scored < 3 on the GHQ-12, suggesting a decreased distress level or none at all, possibly due to acclimating to the standard therapies. The high ORA (M = 5.738, $n = 302$) and NORA (4.410, $n = 302$) scores suggested that caregivers were often engaged in religious practices. The average mean scores revealed an increased utilization of NORA and ORA, showing that caregivers take part in group religious activities and private religious practices. In contrast, Hexem et al. (2011) discovered that parents prayed and read religious scriptures more frequently, although not in a regular manner. Attending places of worship, according to Hexem et al is a sort of community aid in which friends offer support and strength to sick kids by praying for them. The parent's average intrinsic religiosity score indicates that they have a high level of IR, strong faith, and find inspiration in religious instruction. Parents who did not regularly join group religious services sensed a connection with God [13]. This is supported by previous studies conducted by Atkin and Ahmad et al. (2000) in which parents agreed that religion is important [14]. The P-COPE mean score of religious coping was 6933 indicating that parents used P-COPE more than negative methods. The finding is constant with earlier research demonstrating that parents select positive religious coping mechanisms [11,14,17]. Furthermore, the use of P-COPE may be due to their belief and faith in God which strengthens them to cope with it [12]. The connection between psychological distress and religious practice even though there is a weak negative connection between ORA and psychological distress in caregivers [rs (df) = -0.020, p b 0.05], the findings indicate that increased ORA will decrease psychological discomfort. This data suggests that those caregivers, who are experiencing a great level of emotional distress are trying to go to community religious services to connect and receive support from the community. Previous research has also indicated that religious activity is related to emotional support from God [13,14,22]. On the other hand, the study findings showed that caregiver psychological discomfort and NORA are negatively associated; caregivers' who regularly engaged in NORA (e.g reading religious scriptures, meditation, or prayer) had decreased distress. Because the link was weak, and the outcome may happen by coincidence. Longitudinal research would help to explain the association between psychological discomfort and religious activities in the future.

No statistically significant relationship between psychological distress and IR, implying that neither low nor high religious motivation scores were linked to caregivers' psychological discomfort or may be due to individual perceptions in the intensity of psychological suffering felt by caregivers who are subjected to the pressures [17]. Psychological distress is determined by how the individual assessed the circumstance based on religious instruction. An individual's stress response during stressful conditions is determined by their coping technique and intellectual evaluation [23]. The connection between psychological discomfort and religious coping strategies, Positive religious coping mechanisms were found to be associated with decreased levels of psychological discomfort in this study. Thombre et al. (2010) backed up this conclusion, claiming that religious coping was linked to higher levels of positive life-change perceptions. Caregivers who use negative coping methods have increased psychological distress than caregivers who used P-COPE. This is similar to the research of advanced cancer patients (aged 21 and older) and their non-related caregivers (N = 162) that specified that negative coping practices were connected with anxiety disorders and increased the risk of major depression [24,25]. Negative religious coping mechanisms show the effort that results from a damaged relationship with God. A negative view of life, and a sense of hostility from a religious community. As a result, nurses have a hard time paying attention to and being aware of their patients' religious coping styles during ordinary nursing care. On admission, nurses might conduct a cultural and spiritual assessment to gather previous information to assist caregivers who require religious support. Screening parents and caregivers for religious coping mechanisms could be useful in identifying people who are at risk of psychological distress and giving spiritual support to help them regain control of their circumstances [25,26].

4.1. Limitations

There are various restrictions. As convenience sampling was used and there were chances of occurrence of selection biases. There might have been some caregivers experiencing more psychological discomfort and they may have been overlooked. Random sampling can be used to avoid bias. Because this study relied on questionnaires and required participants to react at that moment, some caregivers may have struggled to focus due to the urge to comfort their distressed child. The investigators discovered that caregivers may answer quickly, to finish the questionnaire, leading to recall bias. As a result, the caregiver should be given enough time and space to complete the questionnaire, especially when their children are relaxing. There was also the possibility of self-report bias, which impeded the accuracy of the caregiver's information, reducing generalizability. A longitudinal study is recommended in the future to evaluate the relationship between religious coping methods and psychological suffering among caregivers caring for children with traumatic brain injury. Finally, because the researcher was in charge of all recruitment, the caregiver may have felt obligated to participate in the study. Note that the translation of Questionare from English to Urdu language was not subject to study verification.

4.2. Practice Implication

Caregivers in this study challenged a variety of problems that impacted their psychological well-being. Even though the data showed a modest association, there are several practical implications. When caregivers suffer from stress, religious coping approaches are offered to them as non-religious coping methods to cope with these stressful events. Nurses can connect parents in need with other religious colleagues. Caregivers gain control and confidence over their psychological well-being contributes to better child health outcomes. Religious coping might be an option for assisting caregivers in coping with their suffering. As a result, when delivering nursing care, nurses should respect each culture. Furthermore, continual nursing education on this topic is critical in strengthening the nurses' comprehension of their coping techniques. Moreover, nurses may work with a religious front-runner like an imam, holy man, pastor, or other religious crops to help careers manage stress and to offer nurses a better understanding of religious beliefs. Finally, caregivers who do not have religious beliefs should be taken into account. These caregivers may require therapy, coping skills training, or membership in a TBI support group.

5. Conclusions

In the present study, we for the first time found in Pakistani poulation that caregivers (n = 302) caring for children with traumatic brain injury endure psychological hardship. We noted an incidence of anxiety and depressive symptoms in caregivers. The caregivers take part in group religious activities as well as private religious practices which correlates with decreased psychological discomfort while managing the TBI children. This data also suggests that those caregivers who are experiencing a great level of emotional distress are trying to go to community religious services to connect and receive support from the community. They relied on positive religious coping strategies more than negative religious coping strategies. Non-Organized, and Organized Religious Activities, and negative and positive religious coping mechanisms had a strong relationship with caregivers' psychological suffering.

Author Contributions: Conceptualization, data collection manuscript writing, A. and S.A.; methodology, software, data collection, S.S. and M.K.; software, resources, D.A.K.; data curation, reviewing, funding resources, A.H., M.I.N., A.M.A. and A.M.H.; conceptualization, formal analysis, data curation, Reviewing, S.U.R. All authors have read and agreed to the published version of the manuscript.

Funding: This research work was funded by the Institutional Fund Projects under grant no. (IFPDP-49-22).

Institutional Review Board Statement: This study was conducted according to the guidelines of the Declaration of Helsinki and approved (approval ID: NWIHS-EC/0028-0021) by the Institutional Review Board of Northwest Institute of Health Sciences, Peshawar, Pakistan.

Informed Consent Statement: Informed consent was obtained from all subjects involved in the study.

Data Availability Statement: The data will be available upon the author's request to the corresponding author.

Acknowledgments: This research work was funded by the Institutional Fund Projects under grant no. (IFPDP-49-22). Therefore, the authors gratefully acknowledge technical and financial support from the Ministry of Education and Deanship of Scientific Research (DSR), King Abdulaziz University, Jeddah, Saudi Arabia.

Conflicts of Interest: The authors declare no conflict of interest.

References

1. Centers for Disease Control and Prevention. 2003. Available online: https://www.cdc.gov/ (accessed on 24 October 2022).
2. Pelatti, C.Y.; Bush, E.J.; Farquharson, K.; Schneider-Cline, W.; Harvey, J.; Carter, M.W. Speech-language pathologists' comfort providing intervention to children with traumatic brain injury: Results from a national survey. *Am. J. Speech Lang. Pathol.* **2019**, *28*, 1611–1624. [CrossRef]
3. Roudsari, B.S.; Anzai, Y. *1 Epidemiology of Traumatic Brain Injuries in the United States;* Thieme: New York, NY, USA, 2015; pp. 1–5.
4. Karoline, D.; Anne, N. Caregiver burden in Danish family members of patients with severe brain injury: The chronic phase. *Brain Inj.* **2016**, *30*, 334–342.
5. Frieden, T.R.; Houry, D.; Baldwin, G. *Report to Congress on Traumatic Brain Injury in the United States: Epidemiology and Rehabilitation;* The Centers for Diseases Control and Prevention (CDC): Atlanta, GA, USA, 2015; pp. 1–51.

6. Zaloshnja, E.; Miller, T.; Langlois, J.A.; Selassie, A.W. Prevalence of long-term disability from traumatic brain injury in the civilian population of the United States, 2005. *J. Head Trauma Rehabil.* **2008**, *23*, 394–400. [CrossRef] [PubMed]
7. Ownsworth, T.; Karlsson, L. A systematic review of siblings' psychosocial outcomes following traumatic brain injury. *Disabil. Rehabil.* **2022**, *44*, 496–508. [CrossRef] [PubMed]
8. Gan, C.; Gargaro, J.; Brandys, C.; Gerber, G.; Boschen, K. Family caregivers' support needs after brain injury: A synthesis of perspectives from caregivers, programs, and researchers. *NeuroRehabilitation* **2010**, *27*, 5–18. [CrossRef] [PubMed]
9. Geurtsen, G.J.; van Heugten, C.M.; Meijer, R.; Martina, J.D.; Geurts, A.C. Prospective study of a community reintegration programme for patients with acquired chronic brain injury: Effects on caregivers' emotional burden and family functioning. *Brain Inj.* **2011**, *25*, 691–697. [CrossRef]
10. Koenig, H.G. Religion, spirituality and psychotic disorders. *Arch. Clin. Psychiatr.* **2007**, *34*, 40–48.
11. Grossoehme, D.H.; Jacobson, C.J.; Cotton, S.; Ragsdale, J.R.; VanDyke, R.; Seid, M. Written prayers and religious coping in a paediatric hospital setting. *Ment. Health Relig. Cult.* **2011**, *14*, 423–432. [CrossRef]
12. Atkin, K.; Ahmad, W.I. Family care-giving and chronic illness: How parents cope with a child with a sickle cell disorder or thalassaemia. *Health Soc. Care Community* **2000**, *8*, 57–69. [CrossRef]
13. Hexem, K.R.; Mollen, C.J.; Carroll, K.; Lanctot, D.A.; Feudtner, C. How parents of children receiving pediatric palliative care use religion, spirituality, or life philosophy in tough times. *J. Palliat. Med.* **2011**, *14*, 39–44. [CrossRef]
14. Cotton, S.; Grossoehme, D.; Rosenthal, S.L.; McGrady, M.E.; Roberts, Y.H.; Hines, J.; Yi, M.S.; Tsevat, J. Religious/spiritual coping in adolescents with sickle cell disease: A pilot study. *J. Pediatr. Hematol. Oncol.* **2009**, *31*, 313. [CrossRef]
15. Wu, C.S.; Liu, Y.J. Religious coping in life difficulties: A case study of Taiwanese folk religion female believers. *Ment. Health Relig. Cult.* **2014**, *17*, 210–218. [CrossRef]
16. Panzini, R.G.; Bandeira, D.R. Coping (enfrentamento) religioso/espiritual. *Arch. Clin. Psychiatr.* **2007**, *34*, 126–135. [CrossRef]
17. Cardella, L.A.; Friedlander, M.L. The relationship between religious coping and psychological distress in parents of children with cancer. *J. Psychosoc. Oncol.* **2004**, *22*, 19–37. [CrossRef]
18. Aziz, K.; Sadaf, B.; Kanwal, S. Psychosocial problems of Pakistani parents of Thalassemic children: A cross sectional study done in Bahawalpur, Pakistan. *Biopsychosoc. Med.* **2012**, *6*, 1–7. [CrossRef] [PubMed]
19. Brown, B.; Okereke, J.; Lagunju, I.; Orimadegun, A.; Ohaeri, J.; Akinyinka, O. Burden of health-care of carers of children with sickle cell disease in Nigeria. *Health Soc. Care Community* **2010**, *18*, 289–295. [CrossRef]
20. Shaligram, D.; Girimaji, S.C.; Chaturvedi, S.K. Quality of life issues in caregivers of youngsters with thalassemia. *Indian J. Pediatr.* **2007**, *74*, 275–278. [CrossRef]
21. Ali, S.; Sabih, F.; Jehan, S.; Anwar, M.; Javed, S. Psychological distress and coping strategies among parents of beta-thalassemia major patients. *Int. Proc. Chem. Biol. Environ. Eng.* **2012**, *27*, 124–128.
22. Grossoehme, D.H.; Ragsdale, J.; Cotton, S.; Wooldridge, J.L.; Grimes, L.; Seid, M. Parents' religious coping styles in the first year after their child's cystic fibrosis diagnosis. *J. Health Care Chaplain.* **2010**, *16*, 109–122. [CrossRef]
23. Lazarus, R.S.; Folkman, S. *Stress, Appraisal, and Coping*; Springer Publishing Company: Berlin, Germany, 1984.
24. Pearce, M.J.; Singer, J.L.; Prigerson, H.G. Religious coping among caregivers of terminally ill cancer patients: Main effects and psychosocial mediators. *J. Health Psychol.* **2006**, *11*, 743–759. [CrossRef]
25. Thombre, A.; Sherman, A.C.; Simonton, S. Religious coping and posttraumatic growth among family caregivers of cancer patients in India. *J. Psychosoc. Oncol.* **2010**, *28*, 173–188. [CrossRef] [PubMed]
26. Nurasikin, M.S.; Khatijah, L.; Aini, A.; Ramli, M.; Aida, S.; Zainal, N.; Ng, C. Religiousness, religious coping methods and distress level among psychiatric patients in Malaysia. *Int. J. Soc. Psychiatry* **2013**, *59*, 332–338. [CrossRef] [PubMed]

healthcare MDPI

Article

Urban–Rural Distinction or Economic Segmentation: A Study on Fear and Inferiority in Poor Children's Peer Relationships

Shencheng Wang [1], Baochen Liu [2,*], Yongzheng Yang [3,*], Liangwei Yang [4] and Min Zhen [5]

1 Department of Social and Ecological Studies, Party School of the Central Committee of CPC (National Academy of Governance), Beijing 100089, China
2 School of Law, Shandong Jianzhu University, Jinan 250101, China
3 School of Public Administration and Policy, Renmin University of China, Beijing 100872, China
4 School of Public Policy and Management, Tsinghua University, Beijing 100190, China
5 School of Foreign Studies, South China Agricultural University, Guangzhou 510642, China
* Correspondence: liubaochen18@sdjzu.edu.cn (B.L.); yangyon@iu.edu (Y.Y.)

Abstract: Peer relationships play an important role in the growth of children. This study offers insights about feelings of fear and inferiority in children's peer relationships. Based on a national survey, the 2018 Construction for Social Policy Support System for Urban and Rural Poor Families in China, initiated by the Ministry of Civil Affairs, and using multiple regression models and a structural equation model, this study discusses whether and how having a rural household registration or being from a poor (*dibao*) family has an isolation effect on fear and inferiority in children's peer relationships. The research findings indicate that children with a rural household registration or those from a *dibao* family are at a disadvantage in peer interactions. Moreover, rural resident identity has an indirect effect on children's fear of peers and inferiority, mainly through psychological resilience, anxiety and depression, and mobile phone dependence. Being from a *dibao* family directly influences children's fear and inferiority in their peer relationships; it also indirectly influences fear of peers and inferiority through psychological resilience. This study suggests that more attention should be paid to fear of peers and inferiority in rural children or children from a *dibao* family.

Keywords: children; fear of peers and inferiority; urban–rural distinction; economic segmentation; China

Citation: Wang, S.; Liu, B.; Yang, Y.; Yang, L.; Zhen, M. Urban–Rural Distinction or Economic Segmentation: A Study on Fear and Inferiority in Poor Children's Peer Relationships. *Healthcare* **2022**, *10*, 2057. https://doi.org/10.3390/healthcare10102057

Academic Editors: Yasuhiro Kotera and Elaina Taylor

Received: 17 August 2022
Accepted: 14 October 2022
Published: 17 October 2022

1. Background

Previous studies have shown that peer relationships have a unique and irreplaceable role in the social and emotional development of children. These relationships impact the healthy development and social adaptation of children's social ability, cognition, emotion, self-conception, and personality [1,2]. Peer relationships and family environment are the two core systems of children's personality formation and socialization [3].

Peer relationships can be positive or negative. Early research primarily examined peer relationships from two perspectives: peer acceptance and friendship. Peer acceptance is a one-way structure of common opinions, reflecting the attitudes group members have toward individuals, such as likes or dislikes, acceptance, or exclusion. Friendship is an individual-oriented two-way structure, reflecting the emotional connection of individuals. With the development of research, scholars have gradually paid attention to difficulties in children's peer relationships [4], such as peer rejection [5] and peer victimization [6]. However, aside from peer exclusion and peer victimization, fear of peers and sense of inferiority are the individual subjective feelings of fear and inferiority in peer interactions, which are associated with social self-perception [7]. From a field perspective, peer exclusion and peer aggression are mainly concentrated in the middle and end of peer communication processes, while fear of peers and inferiority are mainly concentrated in the front end. Fear of peers and inferiority may be the major factors behind children's resistance to social participation and social integration. They might be detrimental to the production of prosocial behavior and could lead to aggressive behavior against others [8]. It is necessary to conduct an in-depth study on the antecedents of fear of peers and inferiority, which will increase the understanding of why some children fall into an adverse situation in peer relationships; such a study will also identify the effective measures needed to help children build and maintain good peer relationships.

Therefore, this study focused on fear of peers and inferiority in Chinese children. In reality, the reasons for children's fear of peers and inferiority may be multifaceted. However, in China, it is important to pay particular attention to two unique social backgrounds that are very important for children's growth. First, there is an urban–rural separation effect in children's development in China [9]. The differences in the geographic regions and the human environments between urban and rural China lead to a gap between urban and rural children's environmental adaptation and interpersonal communication skills [10]. Second, the rapid development of China's economy has produced serious economic divisions between people. Some vulnerable groups have fallen into poverty and have also suffered from social exclusion in terms of political participation and interpersonal communication [10]. Children from poor families, with low social skills and a lack of communication experience, are prone to psychological problems, such as feelings of inferiority [11]. In the study discussed in this paper, we aimed to determine whether a rural household registration or being from a poor family (*dibao* family) has an isolation effect on fear and inferiority in children's peer relationships and to investigate the specific mechanisms underlying this.

2. Research Questions and Analytic Frameworks

Although fear of peers and inferiority in children's peer interactions are influenced by socio-economic status (SES), some empirical studies have suggested that the effect of SES on children's peer relationships is very weak [12]. These controversies suggest that the effects of SES on children's fear of peers and inferiority may be influenced by mediating mechanisms. Therefore, the present study aimed to examine the association between SES and fear of peers and inferiority among children in China. Moreover, it explored the intermediary mechanisms of the SES stratification constituted by urban–rural distinction and economic segregation that affects fear of peers and inferiority from the perspective of social stratification.

2.1. Socio-Economic Stratification and Peer Fear and Inferiority

In China, urban–rural distinction and economic segmentation constitute two important aspects of socio-economic stratification. Urban–rural distinction has become an undeniable fact in China. Due to the long-term existence of the country's household registration system, there is a clear distinction between urban and rural areas. Moreover, this system has a comprehensive effect on urban–rural distinction [13], and a disadvantage in peer interactions has been confirmed, thus becoming a typical problem of children's development [9]. Studies have indicated that, due to urban–rural geographical and cultural differences, a certain degree of difference also exists among college students, in terms of environmental adaptation and interpersonal communication [14]. In comparison to rural students, urban students have better interpersonal communication skills [15]. Regarding group cooperation, rural children get along with others significantly better than children from villages and towns [16]. Therefore, this study proposes:

Hypothesis 1. *Urban–rural distinction has a negative effect on children's fear of peers and inferiority.*

In addition to urban–rural distinction, acute economic segmentation arises from China's rapid economic development. Long-term emphasis on economic efficiency and neglect of social justice have led to a significant gap between rich and poor, as well as a wide social class divide. Economically, some vulnerable groups of society not only fall into poverty, but also suffer all-around social exclusion with regard to political participation and interpersonal communication [13]. Among the associated disadvantages, social exclusion resulting from interpersonal communication is particularly harmful to the growth of poor children and students. Family poverty not only causes and aggravates students' psychological burden, it also negatively impacts their communication needs leading to a low level of social skills and a relative lack of contact experience. They have a high probability of confronting many psychological problems, such as an inferiority complex, impacting their interpersonal communication skills [11]. Students with family economic difficulties are a high-risk group for psychological poverty [17], and there is a gap between poor and non-poor undergraduates [18]. Therefore, this study proposes:

Hypothesis 2. *Economic segmentation has a negative effect on children's fear of peers and inferiority.*

2.2. Psychological Resilience

Resilience research began in the 1970s. Some scholars believe that resilience refers to an individual's ability to cope with changes and stressful events in a healthy way [19], while others emphasize that resilience is a process of reintegration. When children encounter serious sources of pressure, they can return to normal with the support of protective factors [20]. Studies have indicated that resilience helps diminish children's depressive symptoms and enables them to initiate peer relationships and cultivate more of them. Children with resilience are more popular among their peers. Consequently, they enjoy better interpersonal relations and social support networks [21] and have relatively more stable and effective social support resources [22].

However, other studies have suggested that SES indirectly reflects the abundance of resources that individuals can mobilize and utilize [23,24]. Individuals with lower SES may incur more health costs in maintaining psychological resilience and they may exhibit poorer mental health [25]. For example, children from rural areas typically have a lower SES, which in turn reduces their level of psychological resilience [26]; this may increase their fear and feelings of inferiority in their peer interactions. Therefore, this study proposes:

Hypothesis 3. *Psychological resilience partially mediates the relationship between urban–rural distinction and children's fear of peers and inferiority.*

Hypothesis 4. *Psychological resilience partially mediates the relationship between economic segmentation and children's fear of peers and inferiority.*

2.3. Anxiety and Depression

Anxiety and depression are the commonly diagnosed psychological disorders among children. Anxiety is a group of mental disorders characterized by anxiety and fear, often accompanied by severe depression or other personality disorders. There is a statistically significant correlation between anxiety and depression [27]. Studies have shown that adolescent anxiety and depression has a significant negative correlation with peer relationships. The higher the degree of anxiety and depression is, the worse the child's peer relationships [28]. Children with a higher level of anxiety and depression have poorer social functioning, less classmate support, and less social acceptance in social communication [29]. High anxiety and social insecurity will increase the risk of children's low-quality friendships and peer abuse, and a low level of social support and peer relationships will further deepen children's psychological distress, such as anxiety and depression [30].

Further research has shown that anxiety and depression are closely related to SES. Individuals with lower SES showed stronger anxious depression than individuals with higher economic and social status [31]. Therefore, this study proposes:

Hypothesis 5. *Anxious depression partially mediates the relationship between urban–rural distinction and children's fear of peers and inferiority.*

Hypothesis 6. *Anxious depression partially mediates the relationship between economic segmentation and children's fear of peers and inferiority.*

2.4. Mobile Phone Dependence

Attention overload theory considers that individual psychological resources are limited, and the maintenance of target information depends on the number of available psychological resources. The failure of sustained attention comes from limited psychological resources [32]. When individuals with high dependence on mobile phones input a large number of cognitive resources into those devices, they reduce the resources that should have been allocated to other personal activities. Consequently, excessive dependence on mobile phones will lead to children's cognitive failure in social communication as well as many adverse psychological characteristics, such as stress susceptibility and low self-evaluation [33]. Studies have found that, in a group of young people with mobile phone addiction, the negative factors impacting peer relationship quality are more significant than the positive factors. Furthermore, the higher the degree of mobile phone addiction, the more negative the impact is on the quality of peer relationships [34]. Social phobia is significantly associated with the risk of smartphone addiction in young people. Individuals with psychosocial problems, such as social phobia and loneliness, prefer mobile devices rather than face-to-face communication because online communication can reduce anxiety [35].

Research on mobile phone dependence has shown that it is closely related to economic and social status. Students from *dibao* families have higher levels of cell phone addiction than students from non-*dibao* families [36–38]. Therefore, this study proposes:

Hypothesis 7. *Mobile phone dependence partially mediates the relationship between urban–rural distinction and children's fear of peers and inferiority.*

Hypothesis 8. *Mobile phone dependence partially mediates the relationship between economic segmentation and children's fear of peers and inferiority.*

3. Methods

3.1. Participants

The data used in this study were collected by the Peking University Chinese Social Sciences Survey Center in 2018, extracted from a survey project called Chinese Social Policy Support System for Vulnerable Families (CSPSS). The Ministry of Civil Affairs of the People's Republic of China appointed and funded the Institute of Social Science Survey (ISSS) at Peking University to deliver the related project and organize a research team to write the report. It is a national large-scale sample survey project supported by the Chinese Ministry of Civil Affairs, aiming to be representative of the vulnerable Chinese families targeted by the government's social assistance program. Using stratified sampling methods, the project adopted the computer-assisted personal interviewing (CAPI) method to investigate more than 1800 villages in 29 provinces from July 2018 to September 2018. The project has compiled three questionnaire databases: disability, the elderly, and children. Among them, parents and their children were interviewed for the children questionnaire, which included detailed information of the demographic, socio-economic, health, learning, and psychological and social interactions of the respondent parents and their children (aged 8–16 years). The respondent parents and children had to complete separate questionnaires without communicating their opinions with each other. The children were required to answer the questionnaires about children's psychological health and school performance. If a child needed help during the procedure, an interviewer read and explained the questions. If the parents of the children (such as left-behind children in a rural area) were not at home when the interviewers were visiting, the questionnaires for parents could be also completed through a telephone survey. The database has 3342 samples, including 991 samples of urban poor families (*dibao* families), with 1032 urban non-*dibao* families; and 543 samples of rural *dibao* families, with 776 rural non-*dibao* samples. After deleting the missing and abnormal values in the database, 3334 observations were finally obtained.

3.2. Measurements

3.2.1. Dependent Variable

To evaluate the children's fear of peers and inferiority, we used the 10-item Fear of peers and Inferiority Scale (PFIS). The participants answered items (e.g., You feel afraid if you do something you have never done before in front of other students) on a 4-point rating scale ranging from 1 = completely disagree to 4 = completely agree. A mean score was computed to yield the composite score, and higher scores indicated higher fear of peers and inferiority. A cumulative score was created by adding the responses of all 10 indicators (ranging from 10 to 40). The higher the total score of the fear of peers and inferiority subscale, the higher the level of fear and inferiority in peer interactions, and the more negative the self-perception. In our study, the Cronbach's alpha coefficient for the PFIS was 0.8357, demonstrating good internal consistency.

3.2.2. Independent Variables

To evaluate the existence of economic segmentation and urban–rural distinction, we used two dummy variables: *dibao* family (0 = no; 1 = yes) and urban family (0 = rural family; 1 = urban family). It should be noted that, in China, families receiving *dibao* are often at the bottom of the economic status hierarchy, which can be considered to be the poorest group.

3.2.3. Mediating Variables

In this study, psychological resilience, anxiety and depression, and mobile phone addiction are the mediator variables.

We used the Child and Youth Resilience Measure Scale (CYRM-R) to measure children's psychological resilience. This scale was developed by Professor Michael Ungar et al. by integrating the results of 35 researchers from 11 countries and 14 regions on psychological resilience in 2009 [39]. The scale consists of 28 items, including three dimensions: individual level, relative level, and social and cultural level. They are evaluated on a five-point Likert scale, with a total score ranging from 28 to 140 points. A higher score indicates a better level of psychological resilience. In our study, the Cronbach's alpha coefficient for the CYRM-R (psychological resilience) was 0.9045, indicating good internal consistency.

We used the Revised Child Anxiety and Depression Scale (RCADS 25) to measure the respondents' depression tendencies. RCADS 25, which includes two dimensions (depression and anxiety), is a revised children's anxiety and depression scale tailored for children and adolescents ranging in age from 8 to 18. The RCADS 25 uses a four-point Likert scale, with 1 representing "never" and 4 representing "always" [40]. A cumulative score (ranging from 24 to 91) was obtained by adding the responses of all 25 indicators. A higher score indicates a higher degree of anxiety and depression. In our study, the Cronbach's alpha coefficient for the RCADS 25 was 0.8583, demonstrating good internal consistency.

To assess the tendency of mobile phone addiction, we used the Chinese version of the self-report 17-item Mobile Phone Addiction Index (MPAI), which was based on the English version of the MPAI. MPAI consists of four dimensions of mobile phone addiction: inability to control cravings, feeling anxious and lost, withdrawal/escape, and productivity loss. The participants answer items (e.g., You feel anxious if you have not checked for messages or switched on your mobile phone for some time) on a 5-point rating scale ranging from "1 = not at all" to "5 = always" [41]. A cumulative score (ranging from 14 to 83) was obtained by adding the responses of all 17 indicators. A higher score indicated a stronger tendency toward mobile phone addiction. In our study, the Cronbach's alpha coefficient for the MPAI was 0.8528, demonstrating good internal consistency.

3.2.4. Covariates

Based on the existing studies on the factors influencing peer relationships [42], the present study included three sets of covariates: school characteristics, family characteristics, and personal characteristics. School characteristics mainly included three variables: key school (*zhongdianxuexiao*), public school, and boarding school (*jisuxuexiao*); three of them are dummy variables (0 = no; 1 = yes). Family characteristics mainly included six variables: whether the parents are alive (1 = both; 0 = either or neither), whether the parents are divorced (0 = no; 1 = yes), if the parents quarrel (0 = never or rarely; 1 = occasionally or often), family gatherings (0 = never; 1 = several times a year; 2 = once a month; 3 = two or three times a month; 4 = several times a week: 5 = every day), parent-child communication (0 = never or occasionally; 1 = always or often), and parents' beating and scolding (0 = never or occasionally; 1 = always or often). Personal characteristics mainly included five variables: gender (0 = female; 1 = male), only child (0 = no; 1 = yes), health status (0 = bad; 1 = moderate or good), physical disability (0 = no; 1 = yes), and student leader (0 = no; 1 = yes).

3.3. *Analytical Strategies*

Stata 14.0 was used as the data analysis tool for this study. First, we used the t-test to check the differences in the characteristics of two groups of participants (urban vs. rural and *dibao* vs. non-*dibao*). Then, a multiple regression model was used to examine the impact of economic segmentation and urban–rural distinction on the respondents' fear of peers and inferiority. Finally, we used the structural equation model method of maximum likelihood with default values for model estimation.

4. Results

4.1. *Descriptive Analysis*

Table 1 shows the descriptive analysis results of the core dependent variables. The average score of fear of peers and inferiority of all children was 19.942. The average score was higher for rural children (20.746) than urban children (19.415). The average score was higher for children from *dibao* families (20.381) than children from non-*dibao* families (19.570).

4.2. *Analysis of the Multiple Regression Model*

To enhance the robustness of the statistical results of the independent variables, the independent variables and three sets of control variables were gradually put into a series of multiple regression models, as shown in Table 2. Model 1 reflects the regression results when only the independent variables are included. Model 2 reflects the regression results when the independent variables and school level control variables are included. Model 3 shows the regression results when the independent variables and the school and personal level control variables are included. Model 4 shows the regression results when the independent variables and the school, personal, and family level control variables are included. With the gradual inclusion of the school, personal, and family characteristic variables, the R^2 of the model gradually increased, indicating that the fitting degree of the model was increasingly higher. Moreover, in Model 4, the variance inflation factor (VIF) results were all lower than in Model 2 (specific results are not listed), indicating that there was no multicollinearity issue among the explanatory variables.

Table 1. Differences in the characteristics of the two group of participants.

Variable Type	Variable	All Samples		Urban vs. Rural					*Dibao* Family vs. Non-*dibao* Family				
				Urban Samples		Rural Samples		p-Values	*Dibao* Family Samples		Non-*dibao* Family Samples		p-Values
		Mean	Std. Dev.	Mean	Std. Dev.	Mean	Std. Dev.		Mean	Std. Dev.	Mean	Std. Dev.	
Dependent variable	Fear of peers and inferiority	19.942	5.723	19.417	5.803	20.746	5.504	0.000	20.381	5.800	19.570	5.632	0.000
Independent variables	Economic segmentation	0.459	0.498	0.489	0.500	0.412	0.492	0.000	—				
	Urban–rural distinction	0.605	0.489	—					0.646	0.479	0.571	0.495	0.000
Mediating variables	Psychological resilience	104.473	17.539	106.120	17.421	101.950	17.426	0.000	102.503	17.953	106.141	17.009	0.000
	Anxiety and depression	42.882	9.371	41.926	9.270	44.347	9.340	0.000	43.141	9.340	42.663	9.395	0.000
	Mobile phone addiction	34.611	10.311	34.048	10.340	35.528	10.204	0.000	34.832	10.401	34.433	10.238	0.160
Covariates	Key school	0.105	0.306	0.125	0.331	0.073	0.260	0.000	0.091	0.288	0.116	0.320	0.010
	Public school	0.916	0.278	0.926	0.262	0.900	0.300	0.005	0.928	0.258	0.905	0.293	0.008
	Boarding school	0.211	0.408	0.124	0.329	0.344	0.475	0.000	0.211	0.408	0.211	0.408	0.499
	Whether parents are alive	0.871	0.335	0.869	0.337	0.874	0.332	0.355	0.795	0.404	0.934	0.248	0.000
	Whether parents are divorced	0.166	0.372	0.188	0.390	0.132	0.339	0.000	0.224	0.417	0.117	0.321	0.000
	Parents' quarrel	0.281	0.450	0.278	0.448	0.287	0.453	0.278	0.270	0.444	0.291	0.454	0.089
	Family gathering	1.903	1.482	2.126	1.490	1.563	1.405	0.000	1.694	1.499	2.081	1.445	0.000
	Parent-child communication	0.785	0.411	0.804	0.397	0.758	0.429	0.000	0.742	0.437	0.822	0.383	0.000
	Parents' beating and scolding	0.092	0.289	0.091	0.288	0.093	0.291	0.420	0.082	0.275	0.100	0.300	0.039
	Gender	0.499	0.500	0.513	0.500	0.477	0.500	0.021	0.517	0.500	0.483	0.500	0.028
	Only child	0.364	0.481	0.480	0.500	0.187	0.390	0.000	0.387	0.487	0.345	0.476	0.007
	Health status	0.953	0.211	0.953	0.211	0.953	0.212	0.480	0.926	0.261	0.976	0.154	0.000
	Physical disability	0.022	0.147	0.021	0.143	0.024	0.152	0.303	0.028	0.166	0.017	0.128	0.012
	Student leader	0.423	0.494	0.459	0.498	0.368	0.482	0.000	0.378	0.485	0.461	0.499	0.000

Notes: t-tests were used for continuous variables, and proportion tests were used for variables in proportions.

Table 2. Analysis results of the multiple regression models.

Variable	Model 1	Model 2	Model 3	Model 4
Urban–rural distinction	−1.399 ***	−1.173 ***	−0.966 ***	−0.681 ***
	(0.199)	(0.210)	(0.212)	(0.215)
Economic segmentation	0.916 ***	0.897 ***	0.708 ***	0.573 ***
	(0.198)	(0.199)	(0.204)	(0.203)
Key school		−0.116	−0.102	−0.0102
		(0.333)	(0.329)	(0.331)
Public school		0.729 **	0.671 *	0.846 **
		(0.349)	(0.357)	(0.356)
Boarding school		1.074 ***	0.965 ***	0.934 ***
		(0.242)	(0.246)	(0.244)
Whether parents are alive			0.119	0.0383
			(0.312)	(0.310)
Whether parents are divorced			0.428	0.616 **
			(0.288)	(0.292)
Parents' quarrels			1.256 ***	1.162 ***
			(0.220)	(0.218)
Family gatherings			−0.286 ***	−0.261 ***
			(0.0707)	(0.0699)
Parent-child communication			−1.036 ***	−0.955 ***
			(0.255)	(0.255)
Parents' beating and scolding			1.630 ***	1.651 ***
			(0.368)	(0.364)
Gender				−0.867 ***
				(0.195)
Only child				−0.547 **
				(0.222)
Health status				−0.584
				(0.509)
Physical disability				0.944
				(0.687)
Student leader				−1.290 ***
				(0.199)
Constant	20.37 ***	19.36 ***	20.08 ***	21.48 ***
	(0.171)	(0.379)	(0.536)	(0.746)
Observed value	3334	3303	3240	3238
R2	0.019	0.025	0.058	0.078

Robustness standard errors are reported in parentheses. *** $p < 0.01$, ** $p < 0.05$, * $p < 0.1$.

In Model 1, both the independent variables—of whether the child is from a *dibao* family and whether the child holds an urban household registration—passed the significance test at the 1% level. The data show that the score of fear of peers and inferiority of urban children was 1.399 points lower than that of rural children. The score of fear of peers and inferiority of children from *dibao* families was 0.916 points higher than that of children from non-*dibao* families. The results demonstrate that peer interactions were impacted by urban–rural distinction and from barriers arising from basic living allowances. In comparison to urban children and those with better economic conditions from non-*dibao* families, rural children and children from *dibao* families faced more communication barriers and had stronger fear of peers and inferiority in peer interactions.

In Model 2, both independent variables again passed the significance test at the 1% level. The children's fear of peers and inferiority was still closely related to household registration and family economic conditions. Moreover, among the school characteristic variables, a public school or not and a boarding school or not were significantly associated with fear of peers and inferiority. The data show that the score of fear of peers and inferiority of children in public schools was 0.729 points higher than that of children in private schools. The score of fear of peers and inferiority of children in boarding schools was 1.074 points higher than that of children in the control group. The control variable of a key school or not was found to have no significant correlation with children's fear of peers and inferiority.

Like Model 1 and Model 2, in Model 3, both independent variables passed the significance test at the 1% level. The results demonstrate that children's fear of peers and inferiority in peer interactions were still closely related to household registration and family economic conditions. Consistent with Model 2, the control variables of a public school or not and a boarding school or not passed the significance test. Moreover, the family characteristic variables of parents' quarrels, family gatherings, parent-child communication, and parents' beating and scolding all passed the significance test.

In Model 4, both independent variables were significantly associated with fear of peers and inferiority, showing the same result as the other three models. Consistent with Model 2 and Model 3, the control variables of a public school or not, a boarding school or not, gender, parents' quarrels, family gatherings, parent-child communication, and parents' beating and scolding all showed the same significance. Personal characteristics, such as gender, only child or not, and a student leader or not, were significantly associated with fear of peers and inferiority.

Based on the descriptive analysis and multiple regression results, Hypothesis 1 and Hypothesis 2 were supported. Specifically, the score of fear of peers and inferiority was significantly higher for rural children than for urban children. The score of fear of peers and inferiority was significantly higher for children from *dibao* families than for children from non-*dibao* families. The results demonstrate that urban–rural distinction and economic segmentation impacted the children's peer interactions.

4.3. Results of the Structural Equation Model

The multiple regression models demonstrated that urban–rural distinction and acute economic segmentation had an impact on the children's peer interactions. Children with a rural household registration or those from *dibao* families suffered more from fear of peers and inferiority. However, the models cannot explain how the two factors led to a higher level of children's fear and inferiority in peer interactions. To identify the mechanisms, this study used the structural equation model method of maximum likelihood with default values for model estimation based on the literature review and the research hypotheses. In comparison to the multiple regression analyses based on OLS (ordinary least squares), the structural equation model enabled us to conduct a path analysis more efficiently. Table 3 shows the model estimation results based on unstandardized regression coefficients.

The three mediator variables all had a significant direct effect on the score of fear of peers and inferiority of children's peer interaction. Specifically, the score of fear of peers and inferiority decreased by 0.070 points for each point increase in children's psychological resilience. The score of fear of peers and inferiority increased by 0.284 points for each point increase in children's anxiety and depression. The score of fear of peers and inferiority increased by 0.048 points for each point increase in children's mobile phone dependence.

This study found that the independent variable of whether the child is from a *dibao* family had a significant direct effect on the score of fear of peers and inferiority of children's peer interaction; it also affected fear of peers and inferiority through the mediating mechanism of psychological resilience. The psychological resilience score was 2.180 points lower for children from a *dibao* family than for children from non-*dibao* families. However, whether the child was from a *dibao* family did not have a significant effect on the degree of anxiety and depression or on mobile phone dependence. Therefore, the influencing mechanism of a *dibao* family on children's fear and inferiority in peer interactions was the synthesis of the direct effect brought about by basic living allowances and the indirect effect brought about by the mediating variable of psychological resilience. Thus, Hypothesis 4 was supported, but Hypothesis 6 and Hypothesis 8 were not.

Table 3. Estimation results of the structural equation model.

	Psychological Resilience	Anxiety and Depression	Mobile Phone Addiction	Fear of Peers and Inferiority in Peer Interactions		
	Direct	Direct	Direct	Direct	Indirect	Total Effect
Urban–rural distinction	2.040 ***	−1.609 ***	−1.061 **	0.359	−0.652 ***	−0.689 **
	(0.630)	(0.403)	(0.432)	(0.164)	(0.126)	(0.214)
Economic segmentation	−2.180 ***	0.163	0.046	−0.037 **	0.202 *	0.561 ***
	(0.594)	(0.328)	(0.406)	(0.174)	(0.119)	(0.201)
Psychological resilience	-	-	-	−0.070 ***	-	−0.070 ***
				(0.005)		(0.005)
Anxiety and depression	-	-	-	0.284 ***	-	0.284 ***
				(0.010)		(0.010)
Mobile phone addiction	-	-	-	0.048 ***	-	0.048 ***
				(0.010)		(0.010)
Key school	3.143 ***	−0.115	−0.145	0.320	−0.261	0.059
	(0.929)	(0.513)	(0.625)	(0.256)	(0.186)	(0.315)
Public school	0.050	−0.260	0.228	0.922 ***	−0.066	0.855 **
	(1.041)	(0.575)	(0.704)	(0.286)	(0.207)	(0.353)
Boarding school	−0.525	0.754 *	1.999 ***	0.604 ***	0.348 **	0.952 ***
	(0.750)	(0.415)	(0.511)	(0.207)	(0.150)	(0.255)
Whether parents are alive	−0.644	−0.582	−0.807	0.147	−0.159	−0.012
	(0.879)	(0.485)	(0.603)	(0.242)	(0.175)	(0.298)
Whether parents are divorced	−0.860	1.857 ***	1.467 ***	−0.014	0.659 ***	0.645 **
	(0.788)	(0.435)	(0.548)	(0.218)	(0.158)	(0.267)
Parents' quarrel	−4.333 ***	2.581 ***	1.579 ***	0.052	1.115 ***	1.166 **
	(0.638)	(0.351)	(0.433)	(0.178)	(0.130)	(0.216)
Family gathering	1.460 ***	−0.502 ***	−0.323 **	−0.010	−0.261 ***	−0.272 ***
	(0.199)	(0.110)	(0.138)	(0.055)	(0.040)	(0.068)
Parent-child communication	1.460 ***	−1.086 ***	−0.183	−0.040	−0.858 ***	−0.898 ***
	(0.199)	(0.394)	(0.503)	(0.200)	(0.147)	(0.242)
Parents' beating and scolding	−5.602 ***	5.477 ***	3.631 ***	−0.470 *	2.126 ***	1.656 ***
	(0.990)	(0.545)	(0.684)	(0.277)	(0.203)	(0.335)
Gender	−0.951 *	−0.289	2.050 ***	−0.958 ***	0.084	−0.874 ***
	(0.568)	(0.314)	(0.387)	(0.158)	(0.115)	(0.193)
Only child	1.297 **	−1.196 ***	0.212	−0.161	−0.421 ***	−0.582 ***
	(0.629)	(0.347)	(0.427)	(0.174)	(0.126)	(0.213)
Health status	2.063	−1.496 **	0.886	0.036	−0.527 *	−0.491
	(1.358)	(0.750)	(0.985)	(0.375)	(0.272)	(0.461)
Physical disability	−2.957	−0.053	−2.299	0.830	0.082	0.912
	(1.948)	(1.075)	(1.404)	(0.537)	(0.389)	(0.661)
Student leader	5.174 ***	−0.222	−0.928 **	−0.838 ***	−0.472 ***	−1.311 ***
	(0.583)	(0.321)	(0.395)	(0.162)	(0.119)	(0.198)

Standard errors are reported in parentheses. *** $p < 0.01$, ** $p < 0.05$, * $p < 0.1$.

This study found that the independent variable of whether the child holds an urban household registration did not have a significant direct effect on children's fear of peers and inferiority in peer interactions. However, this variable had significant effects on children's psychological resilience, anxiety and depression, and mobile phone dependence (all passed the significant positive test). Therefore, this variable had an indirect effect on children's fear of peers and inferiority through the three mediating variables. Specifically, the psychological resilience score was 2.040 points higher for urban children than for rural children. Moreover the anxiety and depression score was 1.609 points lower for urban children than for rural children and the mobile phone dependence score was 1.061 lower for urban children than for rural children. These three mediating mechanisms jointly strengthened the urban–rural distinction in children's peer interactions. Thus, Hypothesis 3, Hypothesis 5, and Hypothesis 7 were supported.

Figure 1 shows the model path diagram based on standardized regression coefficients, which allows us to understand the influencing mechanisms of urban and rural areas and basic living allowances on children's fear of peers and inferiority more intuitively.

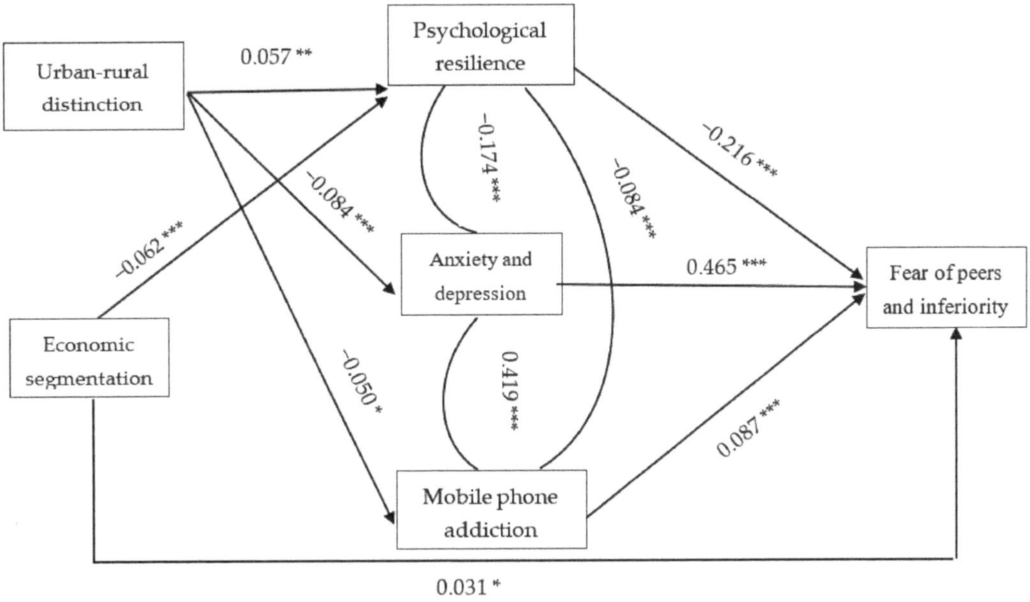

Figure 1. Model path diagram based on the standardized regression coefficients. *** $p < 0.01$, ** $p < 0.05$, * $p < 0.1$.

5. Conclusions and Policy Suggestions

5.1. Conclusions

Using data from the CSPSS national survey, this study found that urban–rural distinction and economic segmentation have an impact on children's fear of peers and inferiority in China. First, rural children suffer much more from fear of peers and inferiority than do urban children. Hypothesis 1 was supported. In China, compared to urban children, rural children have a higher probability of becoming left behind children and they are more likely to have the characteristics of imbalance, sensitive personality, psychological isolation, inferiority and discord. Second, children from *dibao* families are more vulnerable to fear of peers and inferiority than those from families without basic living allowances. Hypothesis 2 was also supported. Because of welfare stigma, children from *dibao* families are at a disadvantage in peer interactions.

This study also examined the mechanisms of the relationship between children's fear of peers and inferiority and urban–rural distinction and economic segmentation. The findings show that urban or rural household registration has no direct effect on children's fear of peers and inferiority, but a rural resident identity indirectly makes rural children suffer more from fear of peers and inferiority by affecting their psychological resilience, anxiety and depression, and mobile phone dependence. Hypotheses 3, 5, and 7 were supported. Similar to previous research [23–26,31,36–38], this study also found higher economic and social status have a significant impact on children's psychological resilience anxiety and depression, and mobile phone dependence. Being from *dibao* families has a direct effect on children's fear of peers and inferiority; it also indirectly leads to a higher level of fear of peers and inferiority by affecting their psychological resilience. Thus, Hypothesis 4 was supported, but Hypothesis 6 and Hypothesis 8 were not. Compared to previous research [31,36–38], this study further found that urban–rural distinction has a significant impact only on children's anxiety and depression, and mobile phone dependence.

Finally, this study found that, aside from urban or rural household registration and being from *dibao* families, children's personal characteristics, family environment, and school environment all affect their fear of peers and inferiority in peer interactions. Specifically, as the children grow older, their fear of peers and inferiority become more serious. Girls face more serious fear of peers and inferiority than boys. Children who are not an only child have a higher level of fear of peers and inferiority than those who are an only child. Children's health status is positively correlated with fear of peers and inferiority. Student leaders can help children diminish fear of peers and inferiority. Children from divorced families have more severe fear of peers and inferiority. Parental relationship, family relationship, and parent-child relationship are all negatively correlated with children's fear of peers and inferiority. Children in public schools have a higher level of fear of peers and inferiority than those in private schools. Children in boarding schools have a higher level of fear of peers and inferiority than those in day schools. Children's relationship with their teachers has a significantly negative correlation with fear of peers and inferiority.

5.2. Policy Suggestions

Positive peer interactions and developing and maintaining good peer relationships are conducive to children's healthy growth. The government, family, and school should pay active attention to the problem of children's fear of peers and inferiority, warranting timely interventions and help. They should encourage children to actively participate in peer interactions and create and maintain good peer relationships. Therefore, based on the research findings, this study provides the following suggestions.

First, policy makers should focus on the fear of peers and inferiority of rural children and children from *dibao* families. For rural children, it is necessary to concentrate on strengthening their psychological resilience, alleviating their anxiety and depression, and diminishing their dependence on mobile phones. For children from *dibao* families, it is essential to enhance their psychological resilience and prevent the *dibao* family from creating feelings of inferiority in their children's social interactions.

Second, policy makers should pay special attention to fear of peers and inferiority in exceptional children (*teshuertong*). Psychological changes of social cognition in older children are worth our attention, and it is necessary for us to solve the problem of fear of peers and inferiority. More support for and attention to girls and children who are not an only child is needed to enhance their peer interactions. Policy makers should work to improve children's health status to avoid the psychological problem of fear of peers and inferiority caused by health problems. Children from divorced families also deserve our attention, and we can help them by actively providing psychological support and encouraging them to engage in peer interactions to offset the negative psychological effects from divorce.

Third, families and schools should play important roles in solving the problem of fear of peers and inferiority. In a family, a harmonious conjugal relationship, supportive family atmosphere, and a close parent-child relationship are conducive to addressing children's problem of fear of peers and inferiority. Parents are advised to have fewer quarrels or avoid them. We suggest that parents should often organize family outings and regularly communicate with their children. For schools, children living on campus are the center of attention, and each school should support them in engaging in peer interactions. It is also necessary to build and maintain good relationships between teachers and students, and teachers should guide children and become role models. Teachers are advised to encourage children to actively participate in peer interactions. Children should be motivated to play an active role in class activities and to campaign for student leaders. In doing so, they can improve their ability to avoid fear of peers and inferiority.

Author Contributions: Conceptualization, S.W. and B.L.; methodology, S.W.; software, Y.Y.; validation, Y.Y., L.Y. and S.W.; formal analysis, B.L.; investigation, M.Z.; resources, B.L.; data curation,

L.Y.; writing—original draft preparation, B.L.; writing—review and editing, L.Y.; visualization, M.Z.; supervision, L.Y.; project administration, Y.Y. All authors have read and agreed to the published version of the manuscript.

Funding: This research received no external funding.

Institutional Review Board Statement: The study did not require ethical approval.

Informed Consent Statement: Informed consent was obtained from all the subjects involved in the study.

Data Availability Statement: The data that support the findings of this study are available from the corresponding author upon reasonable request.

Acknowledgments: The authors thank Keqing Han, a professor in the National Institute of Social Development, Chinese Academy of Social Sciences/School of Sociology and Ethnology, University of Chinese Academy of Social Science.

Conflicts of Interest: The authors declare no potential conflict of interest.

References

1. Wentzel, K.R.; Erdley, C.A. Strategies for Making Friends: Relations to Social Behavior and Peer Acceptance in Early Adolescence. *Dev. Psychol.* **1993**, *29*, 819–826. [CrossRef]
2. Tonks, J.; Yates, P.; Williams, W.H.; Frampton, I.; Slater, A. Peer-relationship difficulties in children with brain injuries: Comparisons with children in mental health services and healthy controls. *Neuropsychol. Rehabil.* **2010**, *20*, 922–935. [CrossRef]
3. Harris, J.R. Where is the child's environment? A group socialization theory of development. *Psychol. Rev.* **1995**, *102*, 458–489. [CrossRef]
4. Bukowski, W.M.; Newcomb, A.F. Stability and Determinants of Sociometric Status and Friendship Choice: A Longitudinal Perspective. *Dev. Psychol.* **1984**, *20*, 941–952. [CrossRef]
5. Laird, R.D.; Jordan, K.Y.; Dodge, K.A.; Pettit, G.S.; Bates, J.E. Peer rejection in childhood, involvement with antisocial peers in early adolescence, and the development of externalizing behavior problems. *Dev. Psychol.* **2001**, *13*, 337–354. [CrossRef] [PubMed]
6. Mynard, H.; Joseph, S. Development of the multidimensional peer-victimization scale. *Aggress. Behav.* **2000**, *26*, 169–178. [CrossRef]
7. Gao, C.; Zeng, X. Fear of peers and Inferiority, and Aggression: Moderating Effect of Peer Acceptance in Middle School Students. *Educ. Res. Monthly* **2018**, *2*, 81–86. [CrossRef]
8. Spanovic, M.; Lickel, B.; Denson, T.F.; Petrovic, N. Fear and anger as predictors of motivation for intergroup aggression: Evidence from Serbia and Republika Srpska. *Group Process. Intergroup Relat.* **2010**, *13*, 725–739. [CrossRef]
9. Zhang, W. The Self-esteem of Junior High School Students. *Psychol. Sci.* **1997**, 504–508. [CrossRef]
10. Gao, H.; Liu, K. Policy Construction of Inclusive Society Construction in China: From the Perspective of Social Quality Theory. *Study Pract.* **2011**, 98–103. [CrossRef]
11. Jia, K.; An, J.; Du, X.; Liu, S.; Wu, J.; Ma, J. A Restudy on the Poverty of College Students: Investigation on the Poor Students and Poverty Alleviation Work in Shanxi University of Finance and Economics. *Youth Stud.* **2001**, 36–42.
12. Li, J.; Wang, J.; Li, J.-Y.; Qian, S.; Jia, R.-X.; Wang, Y.-Q.; Liang, J.-H.; Xu, Y. How do socioeconomic status relate to social relationships among adolescents: A school-based study in East China. *BMC Pediatr.* **2020**, *20*, 271–281. [CrossRef] [PubMed]
13. Wu, Y. Does Social Links Affect the Probability of Chinese Adults' Remarriage? *J. Soc. Dev.* **2018**, *5*, 138–158.
14. Deng, L.; Wang, R.; Zheng, R. The Relationship Between Loneliness, Coping Strategies and Temperament in College Students. *Stud. Psychol. Behav.* **2007**, *2*, 120–126.
15. Zhu, D. A Comparative Study on Behavior Habits of Urban and Rural Primary and Middle School Students. *Youth Stud.* **1996**, *5*, 19–23.
16. Liu, W.; Han, J.; Zhang, L. A Study on Relationship Between Interpersonal Ability and Mental Health of College Students. *Chin. J. Spec. Educ.* **2008**, *3*, 71–80.
17. Yu, H. Analysis on "Psychological Poverty" of College Students with Family Economic Difficulties. *China Youth Study* **2010**, *1*, 114–116.
18. Cai, L. Social Capital of Economically Disadvantaged Undergraduates: Model, Accumulation and Its Educational Compensation. *Educ. Econ.* **2017**, *6*, 14–20.
19. Catalano, R.F.; Berglund, M.L.; Ryan, J.A.M.; Lonczak, H.S.; Hawkins, J.D. Positive Youth Development in the United States: Research Findings on Evaluations of Positive Youth Development Programs. *Ann. Am. Acad. Pol. Soc. Sci.* **2004**, *591*, 98–124. [CrossRef]
20. Richardson, G.E. The Metatheory of Resilience and Resiliency. *J. Clin. Psychol.* **2002**, *58*, 307–321. [CrossRef] [PubMed]
21. Losel, F.; Bliesener, T. Some High-risk Adolescents Do Not Develop Conduct Problems: A Study of Protective Factors. *Int. J. Behav. Dev.* **1994**, *17*, 753–777. [CrossRef]

22. Armstrong, M.I.; Birnie-Lefcovitch, S.; Ungar, M.T. Pathways Between Social Support, Family Well Being, Quality of Parenting, and Child Resilience: What We Know. *J. Child Fam. Stud.* **2005**, *14*, 269–281. [CrossRef]
23. Masten, A.S. Global perspectives on resilience in children and youth. *Child Dev.* **2014**, *85*, 6–20. [CrossRef] [PubMed]
24. Hittner, E.F.; Rim, K.L.; Haase, C.M. Socioeconomic status as a moderator of the link between reappraisal and anxiety: Laboratory-based and longitudinal evidence. *Emotion* **2019**, *19*, 1478–1489. [CrossRef]
25. Brody, G.H.; Yu, T.; Chen, E.; Miller, G.E.; Kogan, S.M.; Beach, S.R. Is resilience only skin deep? rural African Americans' socioeconomic status-related risk and competence in preadolescence and psychological adjustment and allostatic load at age 19. *Psychol. Sci.* **2013**, *24*, 1285–1293. [CrossRef] [PubMed]
26. Narayanan, A. Predictors of resilience among adolescents of low socio-economic status in India. *Int. Rev. Psychiatry* **2015**, *27*, 204–217. [CrossRef]
27. American Psychiatric Association. *Diagnostic and Statistical Manual of Mental Disorders*; Arlington, V.A., Ed.; American Psychiatric Association: Washington, DC, USA, 2013.
28. De Matos, M.G.; Barrett, P.; Dadds, M.; Shortt, A. Anxiety, Depression, and Peer Relationships During Adolescence: Results from the Portuguese National Health Behaviour in School-aged Children Survey. *Eur. J. Psychol. Educ.* **2003**, *18*, 3–14. [CrossRef]
29. La Greca, A.M.; Lopez, N. Social Anxiety Among Adolescents: Linkages with Peer Relations and Friendships. *J. Abnorm. Child Psychol.* **1998**, *26*, 83–94. [CrossRef] [PubMed]
30. Boivin, M.; Hymel, S.; Bukowski, W.M. The Roles of Social Withdrawal, Peer Rejection, and Victimization by Peers in Predicting Loneliness and Depressed Mood in Childhood. *Dev. Psychopathol.* **1995**, *7*, 765–785. [CrossRef]
31. Cohen, A.; Gilman, S.E.; Houck, P.R.; Szanto, K.; Reynolds, C.F., 3rd. Socioeconomic status and anxiety as predictors of antidepressant treatment response and suicidal ideation in older adults. *Soc. Psychiatry Psychiatr. Epidemiol.* **2009**, *44*, 272–2777. [CrossRef] [PubMed]
32. Head, J.; Helton, W.S. Sustained Attention Failures Are Primarily Dee To Sustained Cognitive Load Not Task Monotony. *Acta Psychol.* **2014**, *153*, 87–94. [CrossRef] [PubMed]
33. Hu, Y.; Huang, H.; Zhang, Y.; Zhou, C. The Mediating Effect of Negative Emotions Between Mobile Phone Dependence and Cognitive Failure. *Chin. J. Clin. Psychol.* **2017**, *25*, 1088–1092. [CrossRef]
34. Lee, J.S.; Myung, S.Y. A Study of Adolescents' Peer Relation and School Life Mal Adjustment According to Their Mobile Phone Use and Addiction. *J. Korea. Home Manag. Assoc.* **2007**, *25*, 67–86.
35. Enez Darcin, A.; Kose, S.; Noyan, C.O.; Nurmedov, S.; Yılmaz, O.; Dilbaz, N. Smartphone Addiction and Its Relationship with Social Anxiety and Loneliness. *Behav. Inf. Technol.* **2016**, *35*, 520–525. [CrossRef]
36. Sahin, S.; Ozdemir, K.; Unsal, A.; Temiz, N. Evaluation of mobile phone addiction level and sleep quality in university students. *Pak. J. Med. Sci.* **2013**, *29*, 913–918. [CrossRef]
37. Lin, Y.; Qin, L. Perceived subjective social status and smartphone addiction tendency among Chinese adolescents: A sequential mediation model. *Child. Youth Serv. Rev.* **2020**, *116*, 105222. [CrossRef]
38. Mazaheri, A.M.; Najarkolaei, F.R. Cell phone and internet addiction among students in Isfahan university of medical sciences-Iran. *J. Health Policy Sustain. Health* **2014**, *1*, 101–105.
39. Ungar, M.; Liebenberg, L. Assessing Resilience Across Cultures Using Mixed Methods: Construction of the Child and Youth Resilience Measure. *J. Mix. Methods Res.* **2011**, *5*, 126–149. [CrossRef]
40. Ebesutani, C.; Reise, S.P.; Chorpita, B.F.; Ale, C.; Regan, J.; Young, J.; Higa-McMillan, C.; Weisz, J.R. The Revised Child Anxiety and Depression Scale-Short Version: Scale Reduction via Exploratory Bifactor Modeling of the Broad Anxiety Factor. *Psychol. Assess.* **2012**, *24*, 833–845. [CrossRef]
41. Leung, L. Linking Psychological Attributes to Addiction and Improper Use of the Mobile Phone Among Adolescents in Hong Kong. *J. Child. Media* **2008**, *2*, 93–113. [CrossRef]
42. O'Hare, D.; Eapen, V.; Helmes, E.; McBain, K.; Reece, J.; Grove, R. Factors impacting the quality of peer relationships of youth with Tourette's syndrome. *BMC Psychol.* **2015**, *30*, 34. [CrossRef] [PubMed]

healthcare

Article

Obsessive-Compulsive Disorder with Psychotic Features: Is It a Clinical Entity?

Yasushi Okamura [1,2,*], Yuki Murahashi [1], Yuna Umeda [1], Toshihiro Misumi [3], Takeshi Asami [2], Masanari Itokawa [4], Hirohiko Harima [1], Masafumi Mizuno [1], Hisato Matsunaga [5] and Akitoyo Hishimoto [2]

1 Department of Psychiatry, Tokyo Metropolitan Matsuzawa Hospital, 2-1-1 Kamikitazawa, Setagaya-ku, Tokyo 156-0057, Japan
2 Department of Psychiatry, Yokohama City University Graduate School of Medicine, 3–9 Fukuura, Kanazawa-ku, Yokohama 236-0004, Japan
3 Department of Biostatistics, Yokohama City University School of Medicine, 3–9 Fukuura, Kanazawa-ku, Yokohama 236-0004, Japan
4 Schizophrenia Research Project, Tokyo Metropolitan Institute of Medical Science, 2-1-6 Kamikitazawa, Setagaya-ku, Tokyo 156-8506, Japan
5 Department of Neuropsychiatry, Hyogo Medical University, 1-1 Mukogawacho, Nishinomiya 663-8501, Japan
* Correspondence: oka221kaku@yahoo.co.jp

Abstract: (1) Background: Even though the comorbidity of obsessive-compulsive disorder (OCD) and a psychotic disorder (PD), such as schizophrenia, is being increasingly recognized, the impact of this comorbidity on the clinical presentation, including insight into obsessive-compulsive symptoms and the functioning of OCD, remains unclear. (2) Methods: To investigate clinical differences between OCD patients with and without PD, 86 Japanese outpatients who met the DSM-IV-TR criteria for OCD were recruited and divided into two groups: 28 OCD patients with PD, and 58 OCD patients without PD. The two groups were cross-sectionally compared in terms of their sociodemographic profiles and clinical characteristics, including the DSM-IV-TR insight specifier and the Global Assessment of Functioning (GAF). (3) Results: The results showed that OCD patients with PD scored lower on both the insight and GAF assessments. (4) Conclusions: The present study suggests that comorbid PD in OCD is a clinical entity.

Keywords: obsessive-compulsive disorder; psychotic disorder; schizophrenia; comorbidity; insight; functioning

Citation: Okamura, Y.; Murahashi, Y.; Umeda, Y.; Misumi, T.; Asami, T.; Itokawa, M.; Harima, H.; Mizuno, M.; Matsunaga, H.; Hishimoto, A. Obsessive-Compulsive Disorder with Psychotic Features: Is It a Clinical Entity? *Healthcare* **2022**, *10*, 1910. https://doi.org/10.3390/healthcare10101910

Academic Editor: Brandon Gaudiano

Received: 8 August 2022
Accepted: 24 September 2022
Published: 29 September 2022

1. Introduction

1.1. Background

The relationships between obsessive-compulsive disorder (OCD) and various psychotic disorders (PDs), such as schizophrenia, have long been noted [1–6]. OCD, schizophrenia, and their comorbidity interact with each other in a manner that may affect prognosis and treatment [7–10]. The comorbidity of OCD and schizophrenia is currently being gradually recognized [11], and a common biological basis may underlie the higher-than-expected comorbidity rate [12,13].

Most of previous studies have focused on patients with schizophrenia, with or without obsessive-compulsive symptoms or OCD [14], but not vice versa. It is clear from the assessments by a number of investigators over the last two decades that a subgroup of patients with schizophrenia holds co-occurring obsessions and compulsions, while early studies on psychotic symptoms in patients with primary OCD did not use standardized diagnostic criteria [15]. Prior to the DSM-III-R [16], the diagnosis of OCD was ruled out by the presence of schizophrenia, and obsessive-compulsive symptoms in patients with schizophrenia were interpreted as symptoms of schizophrenia, as is still the case in ICD-10 [13,17]. Therefore, research on the comorbidity and relationship between the two disorders has mainly emphasized broad variations in the psychopathological aspects of schizophrenia. DSM-5 [18] noted that the prevalence of OCD was higher in patients with schizophrenia than in that in a general population. Achim et al. (2011) showed that 12% of patients with schizophrenia had OCD. While obsessive-compulsive symptoms in schizophrenia have been proposed as a defense against psychotic deterioration [19] and, thus, are a predictor of a positive prognosis [20,21], many studies reported that the comorbidity of OCD among patients with schizophrenia had a negative impact on their prognosis [8,22–24]. Claims have been made [25–29] that the comorbidity of schizophrenia and OCD is a subtype of schizophrenia, with specific clinical features, but not without disagreement [30].

Less attention has been paid to OCD patients who are with or without PD. Comorbid PDs are frequently indicated as exclusion criteria in the majority of clinical research on OCD. Hence, limited information is currently available on the clinical characteristics of OCD patients with schizophrenia [13,31]. Numerous studies on psychotic traits comorbid with OCD have focused on schizotypal personality traits [32–35]. Epidemiological studies showed that 1–12.5% of patients with previously diagnosed OCD developed PD [36], and about one out of ten (12%) of patients with OCD met the diagnostic criteria for schizophrenia [25]. Recent meta-analyses found that individuals with OCD are more likely to have psychosis than the general population [37], and recent large-scale studies found that OCD increased the risk of developing schizophrenia after the onset of OCD [38–41], which is in contrast to the findings of previous studies showing that OCD was not associated with an increased risk of schizophrenia [42,43]. OCD with schizophrenia was more common in men [7], was associated with a lower score on the Global Assessment of Functioning (GAF) [9], had a more deteriorative course [7], was more resistant to conventional OCD treatments [44], and was susceptible to the exacerbation of psychotic symptoms when an anti-OCD agent was administered [10].

As used in Matsunaga et al.'s (2002) study, the GAF is internationally well known and widely used for scoring the severity of illness in psychiatry [45], and is recommended for routine clinical use [46]. The GAF summarizes the clinician's view of the patient's current degree of impairment in terms of psychosocial and occupational or educational function. Despite this, the GAF has been significantly less used in OCD studies.

Regarding insight into the obsessive-compulsive symptoms of pivotal clinical importance [42], Matsunaga et al. (2002) reported that a large percentage of patients with OCD had poor insight [9], and this may affect treatment and prognosis [47]. However, many studies discussing insight in patients with OCD excluded a psychotic comorbidity [35,48–51]. Studies on insight in patients with OCD with PD, especially schizophrenia, are rare [9,52]. It is still debatable whether comorbid PD is associated with a better [13] or worse [7,9,30] outcome of OCD.

A specifier of OCD with poor insight was first introduced in DSM-IV-TR [53]. The DSM-5 provides 3 insight specifiers: (1) good to fair, (2) poor, and (3) absent/delusional beliefs. Similarly, the ICD-11 has recently employed simpler dichotomous insight specifiers: (1) good to fair insight or (2) poor to absent insight [47]. As far as we know, there is one previous study on clinical characteristics and insight in patients with OCD using dichotomous insight specifiers [49]; this study, however, evaluated only OCD patients without PDs.

1.2. Objectives

The present study compared clinical characteristics between OCD patients with and without PD to identify what would differentiate the two groups, with particular focus on insight into OCD and the Global Assessment of Functioning. Our hypotheses were that OCD patients with PD would show poorer insight and lower GAF compared with OCD patients without PD.

2. Methods

2.1. Participants

Between April 2015 and April 2022, 86 outpatients at the Department of Psychiatry at Tokyo Metropolitan Matsuzawa Hospital were enrolled after submitting their written informed consent to participate in this study. Our hospital, the largest psychiatric center in Tokyo, receives approximately 7600 new patients annually, including 45 new OCD patients. After conducting surveys of outpatient medical records, all patients had been diagnosed with OCD (based on the DSM-IV-TR criteria) independently by an experienced psychiatrist (Y.O.) with more than five years of experience in the treatment of OCD, who was different from the attending psychiatrists. Inclusion criteria were patients with OCD based on the DSM-IV-TR criteria. No specific exclusion criteria were established. All patients were assessed using the Mini International Neuropsychiatric Interview, Japanese version 5.0.0 2003 (MINI), administered by one of the authors (Y.O.) or the attending psychiatrist. The MINI is a reliable and valid structured interview that may be administered by clinicians or trained non-clinicians to screen for 17 of the most common mental disorders listed in the ICD-10 and DSM-IV-TR [54–58]. The interview confirmed the diagnosis of OCD in all patients for the previous month, except for six who clearly had a history of OCD symptoms and irrational feelings about their obsessive-compulsive symptoms, but who were free from OCD symptoms in the month prior to the interview. Comorbidities not covered by the MINI were diagnosed according to the ICD-10. We also ensured that in OCD patients with PD, psychotic symptoms did not consist only of the delusional nature of insight into their obsessive-compulsive symptoms, but included other symptoms related to PD, including delusions other than those only related to poor insight, hallucinations, thought disorders, and negative symptoms. A total of 86 patients with OCD were divided into 2 groups: 28 patients with comorbid PD (26 with schizophrenia, one with schizotypal disorder, and one with schizoaffective disorder, depressive type, according to ICD-10) and 58 patients without PD.

2.2. Ethical Considerations

All of the procedures in the present study complied with the ethical standards of the relevant national and institutional committees on human experimentation and were conducted according to the guidelines of Declaration of Helsinki. The present study was approved by the Ethics Committee of Tokyo Metropolitan Matsuzawa Hospital. Detailed explanations of the study procedures were provided to each participant prior to their informed consent.

2.3. Clinical Evaluation

A detailed interview was conducted by one of the authors (Y.O.) covering the information on the patients' demographic profiles, clinical features, social background factors, and family and medical histories, as well as the clinical course and characteristics of their OCD. Insight levels were dichotomously assessed using the DSM-IV-TR insight specifier, which defines poor or absent/delusional insight as an individual's lack of awareness that his or her obsessions and compulsions are irrational during most episodes of OCD [53]. Patients other than those with poor to absent/delusional insight were assigned to the group characterized as having good to fair insight. If a patient had multiple obsessive-compulsive symptoms and no insight into at least one of these symptoms, the patient was defined as having poor to absent/delusional insight. Then, 20 out of 86 participants were randomly selected from the patients attending in January and February 2022 to examine the inter-rater reliability of the assessment of insight based on a joint interview by two of the authors (Y.O. and Y.M.), and the kappa coefficient [59] was 0.88.

Global severity and the prevalence of obsessive and compulsive symptoms was evaluated using the Yale–Brown Obsessive Compulsive Scale (Y-BOCS) [60]; general functioning was assessed using the Global Assessment of Functioning (GAF) [16,61,62]; disease severity was evaluated using the Clinical Global Impressions and Severity of Illness (CGI-S) [63];

and the severity of depression was rated using the Japanese version of the GRID-Hamilton Rating Scale for Depression (GRID-HAMD) [64,65], a structured interview incorporating the Hamilton Rating Scale for Depression (HAMD) [66], the international gold standard for assessing the severity of depression with high inter-rater reliability.

2.4. Statistical Analysis

Continuous variables were summarized with means (standard deviation, SD) regarding OCD with and without PD, and compared using a *t*-test between groups. Categorical variables were presented as frequencies and percentages by groups, and compared using the chi-square test or Fisher's exact test. The significance level was set at $p < 0.05$. A multivariate logistic regression analysis was used to extract the most influential clinical variables distinguishing OCD, with and without a psychotic comorbidity. The independent variables used were all seven variables that were significant in the bivariate analysis—the five clinical factors reported in previous studies as significant for distinguishing the two groups: gender, marital status, age at OCD onset, GAF, and CGI-S, plus the two new items significantly differing between the two groups in the present study, namely, insight and an involuntary initial visit to a healthcare provider for a consultation regarding obsessive-compulsive symptoms. Statistical analyses were performed using IBM SPSS Statistics (Version 23; IBM Corporation, Armonk, NY, USA, 1989, 2015) (SPSS Inc., Chicago, IL, USA).

3. Results

3.1. Sociodemographic and Clinical Characteristics

No significant differences were observed in age, housemates, education, employment, or a medical or family history of a psychiatric disease, or comorbidity of psychiatric disorders other than PD between the OCD cases with and without PD (Table 1). The gender ratio was different between the two groups: men were overrepresented in OCD patients with PD. None of the OCD patients were married among OCD patients with PD, while one quartier of the OCD patients without PD were married.

Table 1. Comparison between OCD with psychotic disorder (n = 28) and OCD without psychotic disorder (n = 58): sociodemographic profiles and clinical characteristics.

	OCD with PD (n = 28) (%)	OCD without PD (n = 58) (%)	p
Gender			
Men	20 (71.4)	26 (44.8)	0.02
Women	8 (28.6)	32 (55.2)	
Age, mean (SD)	29.7 (10.7)	31.2 (11.9)	0.579
Marital status			
Married	0 (0.0)	14 (24.1)	0.004
Unmarried	28 (100.0)	44 (75.9)	
Housemates			
Living alone	4 (14.3)	11 (19.0)	0.765
Cohabiting or living in an institution	24 (85.7)	47 (81.0)	
Educational level			
High school or higher, including current students	20 (71.4)	51 (87.9)	0.059
Junior high school, including current students	8 (28.6)	7 (12.1)	
Present employment status			
Employed, housewife, or current student	5 (17.9)	18 (31.0)	0.298
Unemployed or on a leave of absence from duty or school	23 (82.1)	40 (69.0)	

Table 1. *Cont.*

	OCD with PD (*n* = 28) (%)	OCD without PD (*n* = 58) (%)	*p*
Physical comorbidity			
Yes	10 (35.7)	18 (31.0)	0.584
No	17 (60.7)	40 (69.0)	
Family history of psychiatric illness			
Yes	18 (64.3)	29 (50.0)	0.212
No	10 (35.7)	29 (50.0)	
Psychiatric comorbidity			
Mood disorder	13 (46.4)	27 (46.5)	0.991
Anxiety disorder	10 (35.7)	16 (27.6)	0.462
Autism spectrum disorder	3 (10.7)	14 (24.1)	0.247
Self-harm attempt			
Yes	12 (42.9)	18 (31.0)	0.281
No	16 (57.1)	40 (69.0)	

The *t*-test or chi-square test were used to compare the groups, and Fisher's exact test was used if there were cells with expected frequencies of five or less.

3.2. OCD-Related Aspects

A total of 41 out of the 86 patients examined (47.7%) had poor to absent/delusional insight. Patients in the OCD with PD group were significantly more likely than those without PD to have poor to absent/delusional insight (23/28 vs. 18/58, $p < 0.001$) (Table 2).

The mean age of onset of OCD was lower in OCD patients with PD. The mean (SD) duration of untreated OCD was not different between the two groups. OCD patients with PD were more likely than those without PD to initially make an involuntary visit to a healthcare provider for a consultation regarding obsessive-compulsive symptoms (Table 2).

Table 2. Comparison between OCD with psychotic disorder (*n* = 28) and OCD without psychotic disorder (*n* = 58): clinical features and measures.

Variables	OCD with PD (*n* = 28)	OCD without PD (*n* = 58)	*p*
Insight into obsessive compulsive symptoms, *n* (%)			
Poor to absent/delusional	23 (82.1)	18 (31.0)	<0.001
Good to fair	5 (17.9)	40 (69.0)	
Age at OCD onset, mean (SD)	16.3 (6.2)	19.9 (9.4)	0.033
Duration of untreated OCD (yr), mean (SD)	7.4 (7.4)	5.6 (7.4)	0.295
Duration of OCD (yr), mean (SD)	14.0 (9.8)	11.3 (9.9)	0.232
First consultation with a health care provider for OCD			
Voluntary	9 (32.1)	35 (60.3)	0.014
Involuntary	19 (67.9)	23 (39.7)	
GRID-HAMD, mean (SD)	16.9 (8.6)	16.8 (8.9)	0.961
Y-BOCS, mean (SD)	27.5 (9.7)	26.2 (9.0)	0.555
Types of obsession, *n* (%)			
Aggression	11 (39.3)	24 (41.4)	0.853
Contamination	22 (78.6)	39 (67.2)	0.278
Sexual	1 (3.6)	2 (3.4)	1.000
Hoarding	2 (7.1)	11 (19.0)	0.207
Religious	1 (3.6)	1 (1.7)	0.548
Symmetry/exactness	4 (14.3)	16 (27.6)	0.276
Somatic	4 (14.3)	11 (19.0)	0.765
Miscellaneous	9 (32.1)	21 (36.2)	0.711

Table 2. *Cont.*

Variables	OCD with PD (*n* = 28)	OCD without PD (*n* = 58)	*p*
Types of compulsion, *n* (%)			
Cleaning/washing	22 (78.6)	38 (65.5)	0.217
Checking	12 (42.9)	34 (58.6)	0.170
Repeating	11 (39.3)	24 (41.4)	0.615
Counting	1 (3.6)	5 (8.6)	0.659
Ordering/arranging	2 (7.1)	6 (10.3)	1.000
Hoarding/collecting	2 (7.1)	7 (12.1)	0.712
Miscellaneous	5 (17.9)	8 (13.8)	0.749
GAF, mean (SD)	22.7 (13.8)	38.0 (15.9)	<0.001
CGI-S, mean (SD)	6.7 (0.48)	6.1 (1.1)	<0.001

GAF = Global Assessment of Functioning; CGI-S = Clinical Global Impressions of Severity scale; GRID-HAMD = GRID-Hamilton Rating Scale for Depression; Y-BOCS = Yale–Brown Obsessive-Compulsive Scale. The *t*-test or chi-square test was used to compare the groups, and Fisher's exact test was used if there were cells with expected frequencies of five or less.

OCD patients with and without PD were not different in terms of the mean Y-BOCS scores and the types of OCD symptoms, as well as the mean (SD) GRID-HAMD score (Table 2).

The GAF score was significantly lower, and the CGI-S was significantly higher in the OCD with PD group than in the OCD without PD group (Table 2).

3.3. Temporal Course of OCD and Schizophrenia

Among the 26 OCD patients with schizophrenia, the onset of OCD preceded schizophrenia in a majority of cases (22 patients); the onset of the disorders was simultaneous in two patients, and the onset of schizophrenia preceded OCD in two patients. In the 22 patients in whom the onset of OCD preceded schizophrenia, the mean (SD) age of OCD onset was 15.0 (5.5) years. The mean (SD) delay of schizophrenia onset after OCD was 9.7 (7.5) years, and the mean (SD) onset age of schizophrenia was 24.7 (10.4) years.

3.4. Multivariate Analysis

The multivariate logistic regression analysis was performed to identify the clinical variables that best distinguish OCD with PD from OCD without PD (Table 3). Two items significantly predicted group membership: poor to absent/delusional insight (odds ratio: 0.065; $p < 0.001$) and lower GAF (odds ratio: 0.927; $p = 0.012$).

Table 3. Multivariate logistic regression analysis of factors of OCD with psychotic disorder (*n* = 28) and OCD without psychotic disorder (*n* = 58): sociodemographic profiles, clinical features, and measures.

Variables	Level	Odds Ratio	95%CI	*p*
Insight into obsessive compulsive symptoms	Poor to absent/delusional vs. good to fair	0.065	0.013–0.318	0.001
Gender	Male vs. female	0.443	0.123–1.604	0.215
GAF		0.927	0.874–0.983	0.012
CGI-S		0.895	0.262–3.053	0.859
Age at OCD onset		0.969	0.881–1.065	0.513
First consultation for OCD	Voluntary vs. involuntary	0.513	0.113–2.328	0.387
Marital status	Unmarried vs. married	0.000	0.000	0.998

CI = confidence interval.

4. Discussion

The main results of this study were that OCD patients with PD scored lower in insight and GAF evaluations compared with OCD patients without PD.

4.1. Insight

In the present study, 41 patients (47.7%) were considered to have poor to absent/delusional insight, which is consistent with previous findings. Matsunaga et al. (2002) reported that a large percentage of patients with OCD had poor insight. Although a lack of insight is generally assumed to be rare in patients with OCD [7], recent studies demonstrated that the frequency of poor insight was 15–31% in patients with OCD without PD [48–51] and 9–36% in patients with OCD with PD [9,52]. The present results showed a slightly higher prevalence of OCD patients with poor insight, which may be attributed to the higher rate of psychotic comorbidity (32.6%) than in previous studies (20% [9] and 1.7% [52]). The prevalence rate of poor insight in patients with OCD may be influenced by the degree of treatment for OCD at the time of interview, since previous studies have reported that insight can be improved after treatment for OCD [9].

The results of the present study are consistent with previous findings showing that OCD with schizophrenia was correlated more strongly with poor insight than OCD without schizophrenia [9,30,67]. As for the mechanism of poor insight in OCD with schizophrenia, it is possible that the schizophrenic thought disorders may contribute to the poor insight into OCD. Further study considering the degree and effect of thought disorders due to schizophrenia on insight into OCD are required.

The assessment of insight or comorbid schizophrenia in OCD is important because it is relevant to treatment planning. Some previous studies reported that OCD with poor insight was closely associated with a poorer response to medication [50], and some reported that OCD patients with schizophrenia were often resistant to typical OCD treatments [44]. As for pharmacological treatment, for example, in the American Psychiatric Association practice guideline for the treatment of patients with OCD, the pharmacological treatment of OCD with poor insight or with comorbid schizophrenia has not been established, and psychiatrists must rely on clinical judgment in formulating a treatment plan, since no large, controlled trials have yet been conducted [68]. Some reports stated that the use of an augmenting atypical antipsychotic was effective in patients with poor insight and an early age of OCD onset [69], and olanzapine monotherapy has been beneficial for patients with co-occurring schizophrenia in two case series, while second-generation antipsychotics were reported to exacerbate obsessive-compulsive symptoms [68]. Therefore, since there is no consensus on pharmacological treatment of OCD patients with poor insight or with the comorbidity of schizophrenia, those groups of patients may be resistant to conventional OCD treatment, and further research is needed to improve pharmacological therapeutic strategies.

4.2. Sociodemographic Profiles and Clinical Characteristics

In the present study, the OCD group with PD was significantly more likely than the OCD group without PD to exhibit the following clinical characteristics: unmarried, male, with poor prognostic factors, including a lower GAF and a higher CGI-S score in the univariate analysis, which is consistent with previous findings [7,9,30]. Among these factors, the multivariate logistic regression analysis revealed that the GAF score was a significant predictor of the OCD group with PD, suggesting the negative clinical impact of the comorbidity of PD and OCD. Since GAF measures the degree of mental illness by rating psychological, social, and occupational functioning [45], further research is needed to evaluate this issue using more specific rating scales that evaluate each function separately.

In the present study, the mean untreated duration of OCD was 7.1 years, which closely corresponded to the 7 years reported by previous studies [70]; however, the period may be as long as 17 years [71]. The prevalence of OCD in the general population is reportedly as high as 1.1–1.8% [18] or 2.5% [25]. Many patients with OCD hesitate to seek medical care and exhibit a low rate of hospital visits. Comorbidities may complicate the disease and make it enduring, and a chronic course is one of the poorest prognostic factors [72]. Therefore, early interventions, such as sharing knowledge about OCD and the importance of early consultation as part of medical care, with the general public and family physicians and enabling early access to treatment by OCD experts, is important for the prevention of chronicity and severity in OCD.

4.3. Temporal Course of OCD and Schizophrenia

There are four possibilities regarding the pathogenesis of the comorbidity of OCD and schizophrenia. First, the two disorders may coexist by chance, without affecting each other. Second, OCD may develop first as a prodromal symptom of schizophrenia, or schizophrenia with OCD may represent a subtype of schizophrenia [28]. Previous studies on OCD in individuals at ultra-high risk of psychosis reported various conclusions [73,74]. Third, coexistent OCD and schizophrenia may have a mutually exacerbating or amplifying effect [29,75], and persistent OCD may predispose a patient to an increased risk of developing schizophrenia [40,41,73]. OCD may also become schizophrenia [76] or another form of psychosis [42]. Finally, patients with the comorbidity of OCD and schizophrenia may have common risk factors or a neurobiological basis [12,77,78]. In the present study, the onset of OCD preceded that of schizophrenia in most of the cases with PD, which is in line with previous findings [39,79,80]. In comorbidity cases, average ages at the onsets of OCD and schizophrenia were 15.0 years (slightly lower than the DSM-5 figure of 19.5 years, which was similar to those reported in previous studies [7,13]) and 24.7 years, respectively. Among the above four possibilities, the second, third, and fourth possibilities are the most likely, and the coexistence of the two disorders suggests more than a chance occurrence, because OCD with PD may be distinct from OCD without PD, based on poor to absent/delusional insight and the lower GAF score found in the present study. In addition, previous reports may configure such OCD with PD groups characterized by male predominance, younger age at OCD onset, and poor prognosis. To explore these possibilities, increasing the sample size and conducting cluster analysis or latent class analysis in a heterogeneous group of OCD patients with various comorbidities may also be useful in elucidating the pathophysiology of OCD. Such studies would provide a more detailed picture of the relationship between the two disorders, or of the cases of comorbidity of OCD and schizophrenia, especially in patients characterized by poor insight, low functioning, male predominance, younger age at OCD onset, and poor prognosis.

4.4. Strengths and Limitations

The present study has three strengths. First, it is one of the few studies to focus on the clinical characteristics of OCD with and without PD, in contrast to the large number of previous studies on schizophrenia with and without OCD. Recognizing the difference between OCD patients with and without PD may improve current treatments. For example, clinicians may provide adequate psychoeducation to patients with a comorbidity of OCD and psychosis in order to improve their insight, or consider a sufficient number of treatment options, such as the addition of antipsychotics if patients are resistant to conventional OCD treatment, or introduce the use of measures to prevent the development of schizophrenia.

Second, the present study contains rather common and practical information on clinical judgement in daily psychiatric settings. Psychiatrists need to make treatment plans for patients with OCD with various comorbidities, including PD, and to seek multidimensional information about the patients based on the results of the GAF; with these processes, we believe, better treatment of the patients with OCD can be achieved.

Third, the present study was conducted at a general psychiatric center in Japan, with no outpatient clinic specializing in OCD treatment. In Japan, there are not sufficient OCD specialists, and many psychiatric institutions do not have outpatient clinics specializing in OCD. Patients with OCD hesitate to disclose their obsessive-compulsive symptoms, resulting in a long period of time before they receive specialized treatment. Furthermore, clinicians may overlook obsessive-compulsive symptoms because OCD patients with poor insight may present to general psychiatric outpatient clinics for other prominent psychiatric symptoms, such as depression and psychomotor agitation. While this study was conducted in a more general psychiatric institution with no outpatient clinic specializing in OCD, it is important to note that our hospital is a medical institution that treats many patients with schizophrenia, particularly the most severe forms of the disorder.

The present study is not without limitations. It contained a selection bias. The study center is the largest general psychiatric center in Tokyo, but has no specialized OCD treatment clinic. We included one patient with schizotypal disorder and one with schizoaffective disorder under rubric of PD. This may have biased the results. However, we obtained virtually the same results (Supplementary Tables S1-1, S1-2, and S1-3) when we repeated the analysis using 26 patients with schizophrenia and 58 OCD patients without PD (thus excluding the above 2 patients). Our OCD patients with PD included four patients with OCD onset that did not precede the onset of PD. This may have biased the results. However, we obtained virtually the same results (Supplementary Tables S2-1, S2-2, and S2-3) when we repeated the analysis using 24 patients whose OCD onset preceded that of PD and 58 OCD patients without PD (thus excluding the above 4 patients). Furthermore, the present study was exploratory; hence, the results of the multivariate logistic regression were not conclusive. Moreover, this study did not use a more detailed OCD insight scale. Notwithstanding these limitations, we believe this study provides clinically essential results, with practical applications.

5. Conclusions

The present study demonstrated that patients with a comorbidity of OCD and psychotic disorders were more likely to have poor insight and a lower GAF score than patients with OCD without psychotic disorders, and suggested that a psychotic disorder is one of the important clinical factors in assessing patients with OCD. The present results indicate that patients with this comorbidity may have a clinically different phenotype from that of patients with OCD without psychotic disorders. Future studies that enroll a larger cohort and employ more quantitative and qualitative assessment scales or multicenter research, including other institutions specializing in OCD clinics, are needed to generalize the results.

Supplementary Materials: The following supporting information can be downloaded at: https://www.mdpi.com/article/10.3390/healthcare10101910/s1, Table S1-1: Comparison between OCD with schizophrenia (n = 26) and OCD without psychotic disorder (n = 58): sociodemographic profiles and clinical characteristics; Table S1-2: Comparison between OCD with schizophrenia (n = 26) and OCD without psychotic disorder (n = 58): clinical features and measures; Table S1-3: Multivariate logistic regression analysis of factors of OCD with schizophrenia (n = 26) and OCD without psychotic disorder (n = 58): sociodemographic profiles, clinical features, and measures; Table S2-1: Comparison between OCD with the onset of OCD preceding the onset of psychotic disorder (n = 24) and OCD without psychotic disorder (n = 58): sociodemographic profiles and clinical characteristics; Table S2-2: Comparison between OCD with the onset of OCD preceding the onset of psychotic disorder (n = 24) and OCD without psychotic disorder (n = 58): clinical features and measures; Table S2-3: Multivariate logistic regression analysis of factors of OCD with the onset of OCD precede the onset of psychotic disorder (n = 24) and OCD without psychotic disorder (n = 58): sociodemographic profiles, clinical features, and measures.

Author Contributions: All authors discussed the research idea and were involved in the data collection and analysis; Y.O. wrote the first draft of the manuscript; Y.M., Y.U., T.M., T.A., M.I., H.H., M.M., H.M. and A.H. reviewed the manuscript and added comments. All authors have read and agreed to the published version of the manuscript.

Funding: This research was funded by the Tokyo Metropolitan Government, grant numbers H27080303, H29080301, R01080303, R03080302 and R04080302.

Institutional Review Board Statement: The study was conducted according to the guidelines of the Declaration of Helsinki and was approved by the Ethics Committee of the Tokyo Metropolitan Matsuzawa Hospital, IRB code numbers 2015-1, 2016-1, 2017-2, 2018-22, and 2021-19.

Informed Consent Statement: Written informed consent was obtained from all subjects involved in the study.

Data Availability Statement: All data were generated at the Tokyo Metropolitan Matsuzawa Hospital, Japan. The derived data supporting the findings of this study are available from the corresponding author on request.

Acknowledgments: The authors would like to express their deepest gratitude to Toshinori Kitamura, Director of the Kitamura Institute of Mental Health Tokyo, who provided significant guidance in this research.

Conflicts of Interest: The authors declare no conflict of interest.

References

1. Göppert, H. *Zeitschrift für Psycho-Somatische Medizin*; Vandenhoeck & Ruprecht: Göttingen, Germany, 1966; Volume 12, pp. 111–118.
2. Oulis, P.; Konstantakopoulos, G.; Lykouras, L.; Michalopoulou, P.G. Differential diagnosis of obsessive-compulsive symptoms from delusions in schizophrenia: A phenomenological approach. *World J. Psychiatry* **2013**, *3*, 50–56. [CrossRef]
3. Bleuler, E. *Dementia Praecox oder Gruppe der Schizophrenien*; Franz Deuticke: Leipzig und Wien, Austria, 1911.
4. Schneider, K. *Klinische Psychopathologie*; Georg Thime Verlag: Stuttgart, Germany, 2007.
5. Berrios, G.E. Obsessional disorders during the nineteenth century: Terminological and classificatory issues. In *The Anatomy of Madness Volume I. People and Ideas*, 1st ed.; Bynum, W.F., Porter, R., Shepherd, M., Eds.; Tavistock Publications: London, UK; New York, NY, USA, 1985.
6. Berrios, G.E. Obsessive-Compulsive Disorder. In *A History of Clinical Psychiatry*; Berrios, G.E., Porter, R., Eds.; The Athlone Press: London, UK, 1995.
7. Eisen, J.L.; Rasmussen, S.A. Obsessive compulsive disorder with psychotic features. *J. Clin. Psychiatry* **1993**, *54*, 373–379.
8. Berman, I.; Kalinowski, A.; Berman, S.M.; Lengua, J.; Green, A.I. Obsessive and compulsive symptoms in chronic schizophrenia. *Compr. Psychiatry* **1995**, *36*, 6–10. [CrossRef]
9. Matsunaga, H.; Kiriike, N.; Matsui, T.; Oya, K.; Iwasaki, Y.; Koshimune, K.; Miyata, A.; Stein, D.J. Obsessive-compulsive disorder with poor insight. *Compr. Psychiatry* **2002**, *43*, 150–157. [CrossRef] [PubMed]
10. Hadi, E.; Greenberg, Y.; Sirota, P. Obsessive-compulsive symptoms in schizophrenia: Prevalence, clinical features and treatment. A literature review. *World J. Biol. Psychiatry* **2012**, *13*, 2–13. [CrossRef]
11. Nechmad, A.; Ratzoni, G.; Poyurovsky, M.; Meged, S.; Avidan, G.; Fuchs, C.; Bloch, Y.; Weizman, R. Obsessive-compulsive disorder in adolescent schizophrenia patients. *Am. J. Psychiatry* **2003**, *160*, 1002–1004. [CrossRef]
12. Tibbo, P.; Warneke, L. Obsessive-compulsive disorder in schizophrenia: Epidemiologic and biologic overlap. *J. Psychiatry Neurosci.* **1999**, *24*, 15–24.
13. Ozdemir, O.; Tukel, R.; Turksoy, N.; Ucok, A. Clinical characteristics in obsessive-compulsive disorder with schizophrenia. *Compr. Psychiatry* **2003**, *44*, 311–316. [CrossRef]
14. Bottas, A.; Cooke, R.G.; Richter, M.A. Comorbidity and pathophysiology of obsessive-compulsive disorder in schizophrenia: Is there evidence for a schizo-obsessive subtype of schizophrenia? *J. Psychiatry Neurosci.* **2005**, *30*, 187–193.
15. Eisen, J.E.; Yip, A.G.; Mancebo, M.C.; Pinto, A.; Rasmussen, S.A. Phenomenology of Obsessive-Compulsive Disorder. In *Textbook of Anxiety Disorders*, 2nd ed.; Stein, D.J., Hollander, E., Rothbaum, B.O., Eds.; American Psychiatric Publishing: Washington, DC, USA, 2010.
16. American Psychiatric Association. *Diagnostic and Statistical Manual of Mental Disorders*, 3rd ed.; Text Revision; American Psychiatric Association: Washington, DC, USA, 1987.
17. World Health Organization. *International Statistical Classification of Diseases and Related Health Problems*, 10th ed.; World Health Organization: Geneva, Switzerland, 1990.
18. American Psychiatric Association. *Diagnostic and Statistical Manual of Mental Disorders*, 5th ed.; American Psychiatric Association: Washington, DC, USA, 2013.
19. Stengel, E. A Study on Some Clinical Aspects of the Relationship between Obsessional Neurosis and Psychotic Reaction Types. *J. Ment. Sci.* **1945**, *91*, 166–187. [CrossRef]
20. Poyurovsky, M.; Fuchs, C.; Weizman, A. Obsessive-compulsive disorder in patients with first-episode schizophrenia. *Am. J. Psychiatry* **1999**, *156*, 1998–2000. [CrossRef]
21. Poyurovsky, M.; Dorfman-Etrog, P.; Hermesh, H.; Munitz, H.; Tollefson, G.D.; Weizman, A. Beneficial effect of olanzapine in schizophrenic patients with obsessive-compulsive symptoms. *Int. Clin. Psychopharmacol.* **2000**, *15*, 169–173. [CrossRef]
22. Fenton, W.S.; McGlashan, T.H. The prognostic significance of obsessive-compulsive symptoms in schizophrenia. *Am. J. Psychiatry* **1986**, *143*, 437–441. [CrossRef]
23. Lysaker, P.H.; Whitney, K.A. Obsessive-compulsive symptoms in schizophrenia: Prevalence, correlates and treatment. *Expert Rev. Neurother.* **2009**, *9*, 99–107. [CrossRef]
24. de Haan, L.; Hoogenboom, B.; Beuk, N.; van Amelsvoort, T.; Linszen, D. Obsessive-compulsive symptoms and positive, negative, and depressive symptoms in patients with recent-onset schizophrenic disorders. *Can. J. Psychiatry* **2005**, *50*, 519–524. [CrossRef]
25. Karno, M.; Golding, J.M.; Sorenson, S.B.; Burnam, M.A. The epidemiology of obsessive-compulsive disorder in five US communities. *Arch. Gen. Psychiatry* **1988**, *45*, 1094–1099. [CrossRef]

26. Reznik, I.; Mester, R.; Kotler, M.; Weizman, A. Obsessive-compulsive schizophrenia: A new diagnostic entity? *J. Neuropsychiatry Clin. Neurosci.* **2001**, *13*, 115–116. [CrossRef]
27. Poyurovsky, M.; Zohar, J.; Glick, I.; Koran, L.M.; Weizman, R.; Tandon, R.; Weizman, A. Obsessive-compulsive symptoms in schizophrenia: Implications for future psychiatric classifications. *Compr. Psychiatry* **2012**, *53*, 480–483. [CrossRef] [PubMed]
28. Berman, I.; Merson, A.; Viegner, B.; Losonczy, M.F.; Pappas, D.; Green, A.I. Obsessions and compulsions as a distinct cluster of symptoms in schizophrenia: A neuropsychological study. *J. Nerv. Ment. Dis.* **1998**, *186*, 150–156. [CrossRef] [PubMed]
29. Rajkumar, R.P.; Reddy, Y.C.; Kandavel, T. Clinical profile of "schizo-obsessive" disorder: A comparative study. *Compr. Psychiatry* **2008**, *49*, 262–268. [CrossRef] [PubMed]
30. Frias, A.; Palma, C.; Farriols, N.; Salvador, A.; Bonet, J.; Bernaldez, I. Psychopathology and quality of life among patients with comorbidity between schizophrenia spectrum disorder and obsessive-compulsive disorder: No evidence for a "schizo-obsessive" subtype. *Compr. Psychiatry* **2014**, *55*, 1165–1173. [CrossRef] [PubMed]
31. de Haan, L.; Dudek-Hodge, C.; Verhoeven, Y.; Denys, D. Prevalence of psychotic disorders in patients with obsessive-compulsive disorder. *CNS Spectr.* **2009**, *14*, 415–417. [CrossRef] [PubMed]
32. Jenike, M.A.; Baer, L.; Minichiello, W.E.; Schwartz, C.E.; Carey, R.J., Jr. Concomitant obsessive-compulsive disorder and schizotypal personality disorder. *Am. J. Psychiatry* **1986**, *143*, 530–532. [CrossRef] [PubMed]
33. Stanley, M.A.; Turner, S.M.; Borden, J.W. Schizotypal features in obsessive-compulsive disorder. *Compr. Psychiatry* **1990**, *31*, 511–518. [CrossRef]
34. Sobin, C.; Blundell, M.L.; Weiller, F.; Gavigan, C.; Haiman, C.; Karayiorgou, M. Evidence of a schizotypy subtype in OCD. *J. Psychiatr. Res.* **2000**, *34*, 15–24. [CrossRef]
35. Bellino, S.; Patria, L.; Ziero, S.; Bogetto, F. Clinical picture of obsessive-compulsive disorder with poor insight: A regression model. *Psychiatry Res.* **2005**, *136*, 223–231. [CrossRef]
36. Dowling, F.G.; Pato, M.T.; Pato, C.N. Comorbidity of obsessive-compulsive and psychotic symptoms: A review. *Harv. Rev. Psychiatry* **1995**, *3*, 75–83. [CrossRef]
37. Rowe, C.; Deledalle, A.; Boudoukha, A.H. Psychiatric comorbidities of obsessive-compulsive disorder: A series of systematic reviews and meta-analyses. *J. Clin. Psychol.* **2022**, *78*, 469–484. [CrossRef]
38. Tibbo, P.; Kroetsch, M.; Chue, P.; Warneke, L. Obsessive-compulsive disorder in schizophrenia. *J. Psychiatr. Res.* **2000**, *34*, 139–146. [CrossRef]
39. Tien, A.Y.; Eaton, W.W. Psychopathologic precursors and sociodemographic risk factors for the schizophrenia syndrome. *Arch. Gen. Psychiatry* **1992**, *49*, 37–46. [CrossRef]
40. Meier, S.M.; Petersen, L.; Pedersen, M.G.; Arendt, M.C.; Nielsen, P.R.; Mattheisen, M.; Mors, O.; Mortensen, P.B. Obsessive-compulsive disorder as a risk factor for schizophrenia: A nationwide study. *JAMA Psychiatry* **2014**, *71*, 1215–1221. [CrossRef] [PubMed]
41. Cederlof, M.; Lichtenstein, P.; Larsson, H.; Boman, M.; Ruck, C.; Landen, M.; Mataix-Cols, D. Obsessive-Compulsive Disorder, Psychosis, and Bipolarity: A Longitudinal Cohort and Multigenerational Family Study. *Schizophr. Bull.* **2015**, *41*, 1076–1083. [CrossRef] [PubMed]
42. Insel, T.R.; Akiskal, H.S. Obsessive-compulsive disorder with psychotic features: A phenomenologic analysis. *Am. J. Psychiatry* **1986**, *143*, 1527–1533. [CrossRef] [PubMed]
43. Goodwin, D.W.; Guze, S.B.; Robins, E. Follow-up studies in obsessional neurosis. *Arch. Gen. Psychiatry* **1969**, *20*, 182–187. [CrossRef]
44. Hwang, M.Y.; Yum, S.Y.; Losonczy, M.F.; Mitchell, G.; Kwon, J.S. Schizophrenia with obsessive compulsive features. *Psychiatry* **2006**, *3*, 34–41. [PubMed]
45. Aas, I.H. Global Assessment of Functioning (GAF): Properties and frontier of current knowledge. *Ann. Gen. Psychiatry* **2010**, *9*, 20. [CrossRef]
46. Salvi, G.; Leese, M.; Slade, M. Routine use of mental health outcome assessments: Choosing the measure. *Br. J. Psychiatry* **2005**, *186*, 146–152. [CrossRef]
47. Stein, D.J.; Kogan, C.S.; Atmaca, M.; Fineberg, N.A.; Fontenelle, L.F.; Grant, J.E.; Matsunaga, H.; Reddy, Y.C.J.; Simpson, H.B.; Thomsen, P.H.; et al. The classification of Obsessive-Compulsive and Related Disorders in the ICD-11. *J. Affect. Disord.* **2016**, *190*, 663–674. [CrossRef]
48. Marazziti, D.; Dell'Osso, L.; Di Nasso, E.; Pfanner, C.; Presta, S.; Mungai, F.; Cassano, G.B. Insight in obsessive-compulsive disorder: A study of an Italian sample. *Eur. Psychiatry* **2002**, *17*, 407–410. [CrossRef]
49. Turksoy, N.; Tukel, R.; Ozdemir, O.; Karali, A. Comparison of clinical characteristics in good and poor insight obsessive-compulsive disorder. *J. Anxiety Disord.* **2002**, *16*, 413–423. [CrossRef]
50. Ravi Kishore, V.; Samar, R.; Janardhan Reddy, Y.C.; Chandrasekhar, C.R.; Thennarasu, K. Clinical characteristics and treatment response in poor and good insight obsessive-compulsive disorder. *Eur. Psychiatry* **2004**, *19*, 202–208. [CrossRef] [PubMed]
51. De Berardis, D.; Campanella, D.; Gambi, F.; Sepede, G.; Salini, G.; Carano, A.; La Rovere, R.; Pelusi, L.; Penna, L.; Cicconetti, A.; et al. Insight and alexithymia in adult outpatients with obsessive-compulsive disorder. *Eur. Arch. Psychiatry Clin. Neurosci.* **2005**, *255*, 350–358. [CrossRef] [PubMed]
52. Cherian, A.V.; Narayanaswamy, J.C.; Srinivasaraju, R.; Viswanath, B.; Math, S.B.; Kandavel, T.; Reddy, Y.C. Does insight have specific correlation with symptom dimensions in OCD? *J. Affect. Disord.* **2012**, *138*, 352–359. [CrossRef]

53. American Psychiatric Association. *Diagnostic and Statistical Manual of Mental Disorders*, 4th ed.; Text Revision; American Psychiatric Association: Washington, DC, USA, 2000.
54. Otsubo, T.; Tanaka, K.; Koda, R.; Shinoda, J.; Sano, N.; Tanaka, S.; Aoyama, H.; Mimura, M.; Kamijima, K. Reliability and validity of Japanese version of the Mini-International Neuropsychiatric Interview. *Psychiatry Clin. Neurosci.* **2005**, *59*, 517–526. [CrossRef]
55. Lecrubier, Y.; Sheehan, D.V.; Weiller, E.; Amorim, P.; Bonora, I.; Sheehan, E.H.; Janavs, J.; Dunbar, G.C. The Mini International Neuropsychiatric Interview (MINI). A short diagnostic structured interview: Reliability and validity according to the CIDI. *Eur. Psychiatry* **1997**, *12*, 224–231. [CrossRef]
56. Sheehan, D.V.; Lecrubier, Y.; Sheehan, K.H.; Amorim, P.; Janavs, J.; Weiller, E.; Hergueta, T.; Baker, R.; Dunbar, G.C. The Mini-International Neuropsychiatric Interview (M.I.N.I.): The development and validation of a structured diagnostic psychiatric interview for DSM-IV and ICD-10. *J. Clin. Psychiatry* **1998**, *59* (Suppl. S20), 22–33. [PubMed]
57. Sheehan, D.V.; Lecrubier, Y.; Sheehan, K.H.; Janavs, J.; Weiller, E.; Keskiner, A.; Schinka, J.; Knapp, E.; Sheehan, M.F.; Dunbar, G.C. The validity of the Mini International Neuropsychiatric Interview (MINI) according to the SCID-P and its reliability. *Eur. Psychiatry* **1997**, *12*, 232–241. [CrossRef]
58. Amorim, P.; Lecrubier, Y.; Weiller, E.; Hergueta, T.; Sheehan, D. DSM-IH-R Psychotic Disorders: Procedural validity of the Mini International Neuropsychiatric Interview (MINI). Concordance and causes for discordance with the CIDI. *Eur. Psychiatry* **1998**, *13*, 26–34. [CrossRef]
59. Cohen, J. A Coefficient of Agreement for Nominal Scales. *Educ. Psychol. Meas.* **1960**, *20*, 37–46. [CrossRef]
60. Goodman, W.K.; Price, L.H.; Rasmussen, S.A.; Mazure, C.; Fleischmann, R.L.; Hill, C.L.; Heninger, G.R.; Charney, D.S. The Yale-Brown Obsessive Compulsive Scale. I. Development, use, and reliability. *Arch. Gen. Psychiatry* **1989**, *46*, 1006–1011. [CrossRef]
61. Luborsky, L. Clinician's judgments of mental health. *Arch. Gen. Psychiatry* **1962**, *7*, 407–417. [CrossRef] [PubMed]
62. Endicott, J.; Spitzer, R.L.; Fleiss, J.L.; Cohen, J. The global assessment scale. A procedure for measuring overall severity of psychiatric disturbance. *Arch. Gen. Psychiatry* **1976**, *33*, 766–771. [CrossRef]
63. Guy, W. *ECDEU Assessment Manual for Psychopharmacology*; U.S. Dept. of Health, Education, and Welfare, Public Health Service, Alcohol, Drug Abuse, and Mental Health Administration, National Institute of Mental Health, Psychopharmacology Research Branch, Division of Extramural Research Programs: Rockville, MD, USA, 1976; pp. 602–603.
64. Williams, J.B.; Kobak, K.A.; Bech, P.; Engelhardt, N.; Evans, K.; Lipsitz, J.; Olin, J.; Pearson, J.; Kalali, A. The GRID-HAMD: Standardization of the Hamilton Depression Rating Scale. *Int. Clin. Psychopharmacol.* **2008**, *23*, 120–129. [CrossRef] [PubMed]
65. Tabuse, H.; Kalali, A.; Azuma, H.; Ozaki, N.; Iwata, N.; Naitoh, H.; Higuchi, T.; Kanba, S.; Shioe, K.; Akechi, T.; et al. The new GRID Hamilton Rating Scale for Depression demonstrates excellent inter-rater reliability for inexperienced and experienced raters before and after training. *Psychiatry Res.* **2007**, *153*, 61–67. [CrossRef] [PubMed]
66. Hamilton, M. A rating scale for depression. *J. Neurol. Neurosurg. Psychiatry* **1960**, *23*, 56–62. [CrossRef] [PubMed]
67. Rasmussen, S.A.; Eisen, J.L. The epidemiology and clinical features of obsessive compulsive disorder. *Psychiatr. Clin. N. Am.* **1992**, *15*, 743–758. [CrossRef]
68. American Psychiatric Association. Practice Guidline for the Treatment of Patients With Obsessive-Compulsive Disorder. 2007. Available online: psychiatryonline.org (accessed on 5 August 2022).
69. Sumitani, S.; Ueno, S.; Ishimoto, Y.; Taniguchi, T.; Tomotake, M.; Motoki, I.; Yamauchi, K.; Ohmori, T. Clinical features of patients with obsessive-compulsive disorder showing different pharmacological responses. *Seishin Shinkeigaku Zasshi* **2006**, *108*, 1282–1292.
70. Matsumoto, Y.; Nakamae, T.; Abe, Y.; Watanabe, A.; Narumoto, J. Duration of untreated illness of patients with obsessive-compulsive disorder in Japan. *Early Interv. Psychiatry* **2021**, *15*, 1644–1649. [CrossRef]
71. Hollander, E.; Greenwald, S.; Neville, D.; Johnson, J.; Hornig, C.D.; Weissman, M.M. Uncomplicated and comorbid obsessive-compulsive disorder in an epidemiologic sample. *Depress. Anxiety* **1996**, *4*, 111–119. [CrossRef]
72. Skoog, G.; Skoog, I. A 40-year follow-up of patients with obsessive-compulsive disorder [see commetns]. *Arch. Gen. Psychiatry* **1999**, *56*, 121–127. [CrossRef]
73. Fontenelle, L.F.; Lin, A.; Pantelis, C.; Wood, S.J.; Nelson, B.; Yung, A.R. A longitudinal study of obsessive-compulsive disorder in individuals at ultra-high risk for psychosis. *J. Psychiatr. Res.* **2011**, *45*, 1140–1145. [CrossRef] [PubMed]
74. Niendam, T.A.; Berzak, J.; Cannon, T.D.; Bearden, C.E. Obsessive compulsive symptoms in the psychosis prodrome: Correlates of clinical and functional outcome. *Schizophr. Res.* **2009**, *108*, 170–175. [CrossRef]
75. Ongur, D.; Goff, D.C. Obsessive-compulsive symptoms in schizophrenia: Associated clinical features, cognitive function and medication status. *Schizophr. Res.* **2005**, *75*, 349–362. [CrossRef]
76. Lo, W.H. A follow-up study of obsessional neurotics in Hong Kong Chinese. *Br. J. Psychiatry* **1967**, *113*, 823–832. [CrossRef] [PubMed]
77. Frias-Ibanez, A.; Palma-Sevillano, C.; Farriols-Hernando, N. Comorbidity between obsessive-compulsive disorder and schizophrenia: Prevalence, explanatory theories, and nosological status. *Actas Esp. Psiquiatr.* **2014**, *42*, 28–38. [PubMed]
78. Frias, A.; Palma, C.; Farriols, N.; Gonzalez, L. Comorbidity between obsessive-compulsive disorder and body dysmorphic disorder: Prevalence, explanatory theories, and clinical characterization. *Neuropsychiatr. Dis. Treat.* **2015**, *11*, 2233–2244. [CrossRef] [PubMed]

79. Poyurovsky, M.; Faragian, S.; Shabeta, A.; Kosov, A. Comparison of clinical characteristics, co-morbidity and pharmacotherapy in adolescent schizophrenia patients with and without obsessive-compulsive disorder. *Psychiatry Res.* **2008**, *159*, 133–139. [CrossRef] [PubMed]
80. Rosen, I. The clinical significance of obsessions in schizophrenia. *J. Ment. Sci.* **1957**, *103*, 773–785. [CrossRef]

healthcare

MDPI

Article

Sharing Visual Narratives of Diabetes on Social Media and Its Effects on Mental Health

Syed Ali Hussain

Walter Cronkite School of Journalism and Mass Communication, Arizona State University, Phoenix, AZ 85004, USA; sahussa8@asu.edu

Abstract: Diabetes is a chronic illness affecting over six percent of the global population. Visual social media sites such as Tumblr provide a unique opportunity to understand visual illness narratives of type 1 and type 2 diabetes and its effects on mental health. We qualitatively analyze 259 Tumblr images with a "diabetes" hashtag. The results provide rich insights into the lives of diabetes patients, including personal and social life interactions, visual narratives portraying one's acceptance and maintenance of diabetes, difficulty in social interactions, and how patients' identity and beliefs are shaped by the daily struggles and failures of living with diabetes. We discuss the findings in the context of the chaos, quest, and restitution narratives of illness proposed by Arthur Frank. The results have implications for the visual representation of chronic diseases on social media and for improving patient–provider interactions and treatment of type 1 and type 2 diabetes patients.

Keywords: visual narratives; social media; diabetes; mental health; chronic illnesses; Tumblr

Citation: Hussain, S.A. Sharing Visual Narratives of Diabetes on Social Media and Its Effects on Mental Health. *Healthcare* **2022**, *10*, 1748. https://doi.org/10.3390/healthcare10091748

Academic Editors: Yasuhiro Kotera and Alyx Taylor

Received: 1 August 2022
Accepted: 8 September 2022
Published: 12 September 2022

1. Introduction

Diabetes is a chronic illness [1] and affects over six percent of the world's population [2] Similar to other chronic illnesses, living with diabetes is a daily battle of self-care and decision making [3,4], patient education and psychosocial support, anxiety stress management [5], and better treatment adherence [6].

Diabetes patients have to follow a strict regimen in terms of diet and exercise. The regimens often bring communication breakdown between patients vs. caregivers, often referred to as "diabetes police" [7], or significant others (e.g., wife of a patient) overtly monitoring food consumption of diabetes patients. In such situations, many patients switch to social media for social support [8]. Social media enables them to share new forms of narratives that are quick, symbolic, and ad hoc [9], providing instant means for self-expression and identity formation [10]. In particular, visual narratives, such as photographs, help them to express in forms meanings that are otherwise not easily conveyed through words. Such visual forms of self-expression produce illness narratives that help patients with diabetes to make sense of life events, such as career, family, marriage, sex, and intimacy, among others.

Sharing narratives of personal experiences through social media is found useful in improved self-management behavior [11]. Illness narratives seem to have therapeutic effects for the writer and strongly influence the readers to understand the writer's context [12]. Diaries, blogs, and other illness narrative expressions have also shown to produce therapeutic effects on social relationships and self-help [13].

Specifically, this study examines the visual illness narratives of diabetes on the Tumblr social media site. The findings have implications for employing visual narratives to improve chronic illness care for diabetes patients. Studying visual illness narratives can also help healthcare providers, researchers, and caregivers to understand the challenges that patients experience and can have design implications for better disease treatment.

1.1. Visual Research Methodology

Visual research methods have a long history in qualitative health research. For example, Kaley, Hatten, and Milligan [14] conducted a visual ethnographic study to explore how individuals with intellectual disabilities engaged in green care. Harvey and Brooks [15] explored the role of stock images in forming perceptions of patients with dementia. Cheezum, Rosse, Niewolak, and Cobb [16] conducted a PhotoVoice study to assess the determinants of health and wellness in Detroit's homeless community. Additionally, researchers have used photo-videography in data collection and in the analysis of daily computing environments [17] and visual methods to explore human–computer interactions [18], including design projects and visual ethnography [19]. Other visual research methodologies include visual ethnography [20], ethnomethodology [21], documentary photography [22], photo elicitation [20], and photovoice [23]. Considering the abundant use of visuals on social media, visual research methodologies are also experiencing a transition from traditional objectivist forms of data collection and analysis to more subjective ways of looking into culture and social life as evidenced by more recent work in sociology and anthropology using photographs, illustrations, and videos [24,25].

1.2. Illness Narratives

Studies of illness narratives describe two distinct languages of discourse around health. The first is objective language, which includes medical information provided by physicians and other health providers. The second is subjective language, which talks about the intimate, non-verifiable experiences around having an illness. Frank [26] discussed the critical role of such subjective language in illness narratives as "ill people learn by hearing themselves tell their stories, absorbing others' reactions and experiencing their stories being shared." Illness narrative is shaped by a mix of interactions a patient has with other patients similar to himself, family, and providers. These could be in-person interactions and interactions taking place in mediated environments, such as websites or mass media (e.g., radio, movies, TV).

Researchers found that giving the opportunity to tell personal stories about their illness helps the participants to actively participate in their health improvement process [27]. In addition, by hearing stories from similar others they do not feel alone in experiencing the illness [26]. The conventional method to achieve this is through maintaining patient diaries. For instance, patients have produced narratives of their illness experiences and hospital stays in the ICU [28]. Healthcare providers have also recognized the importance of illness narratives. They use illness narratives to learn about the patient's life after disease [29]. Intensive-care nurses use diaries to record their experience, which is often co-authored by other healthcare providers and even patients' family members [30]. For example, a study examined the effects of digital photo-sharing on the social connectedness of patients with spinal cord injury and elderly living in a nursing home [31]. The results showed that the sharing of photos served as "food for the talk" and improved the bonding with family members.

1.3. Visual Narratives

With the presence of new media and communication technologies, people have found more elaborate ways to express their daily life stories. Among these, people use photographs to communicate through signs and symbols about concepts that are generally hard to convey in words [27]. The method has the potential to reveal hidden meanings as well as new opportunities to understand and improve disease management. Patients' lived experiences might not always be possible to verbally describe. Narratives presented visually provide a unique window into the subjective experiences of the narrator [27]. For instance, a picture of swollen feet and the person's painful gesture provides a window to see beyond physical symptoms and feel the emotional side of the patient's struggle with the illness. Such visuals also show subtle details that are often not shared during medical visits. Visual narratives take several forms, including but not limited to photography [32] and graffiti [33].

The ubiquity of visual social media sites has led to a significant increase in the visual sharing of illness narratives, thereby increasing social interactions around it. This has been studied in many health contexts such as HIV AIDS [23], chemotherapy [19], and post-hospital recovery [32]. Visual narratives have also been used to study embodiment [34], body image, and how it shapes the personal–social identity [35]. To that extent, images provide a unique avenue to express and understand illness narratives with more richness.

1.4. Tumblr: A Place for Sharing Visual Narratives

Tumblr is a visual social media site that is popular among internet users, with its fast sign-up procedure and micro-text, images, and photographs sharing abilities. The website hosts over 518 million blogs [36]. It publishes over 12 million blog posts daily and has over 327 million unique visitors around the globe. The posts are categorized as photo, video, audio, text, link, chat, and answer (Figure 1). Several of these posts are re-blogs, with roughly 10% original posts [37]. In terms of posts' sentiment, image posts express more intense emotions and positive valence compared with text-only posts [38].

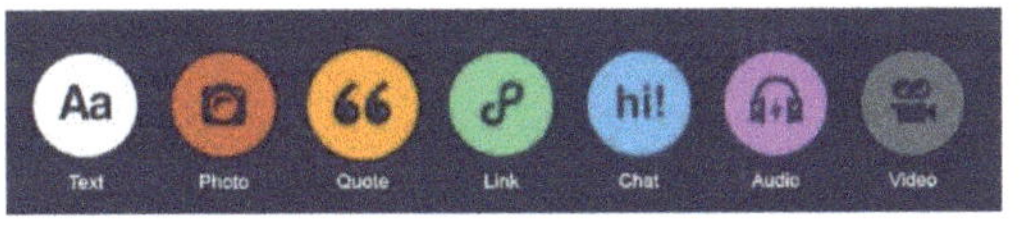

Figure 1. Tumblr dashboard.

Even though Tumblr appears similar to other visual social media sites, some differences still exist. For example, Pinterest lets users curate images. Flickr lets people view and like others' photos but not share or re-blog. On Flickr, tagging behavior occurs significantly among friends. Accordingly, popular photos do not spread widely and quickly on the network. Over 50% of information exchange is between online friends but with a significant delay at each connection [39]. However, Tumblr makes possible the quick dissemination of information through the re-blogging feature, similar to Twitter.

Several researchers have used Tumblr blogs to study health topics. These include visual representation of anorexia and found blogs containing emotionally intense and triggering images [40]. Tumblr has also been used to examine visuals of domestic violence through a feminist participatory action approach [41].

Most research on illness narrative is based on textual data generated from written diaries, surveys, focus discussion groups, and in-depth interviews. However, qualitative data exist in other non-textual forms, such as signs, symbols, non-verbal gestures, paintings, and other visual artifacts. Visual communication offers significant opportunities to explore lived experience and the subjectivity of patients' illness experience [27]. This is particularly relevant for patients with diabetes that puts them in awkward social situations such as having to inject a needle before a dinner date and having to worry about telling a blind date that one has a disease that is not curable. As patients tell their story, they untangle their illness experience and begin to accommodate to life as the person they have become.

Based on this discussion, this article presents a qualitative analysis of diabetes-related images on Tumblr and explores how patients use Tumblr in the context of expressing their illness. Our study informs how visual social media sites can be developed to effectively enhance patients' illness experiences and help stakeholders such as providers and caregivers to better understand the patient's perspective, leading to improved patient care.

2. Methods

2.1. Data Collection

First, the author searched the term "Diabetes" on Tumblr.com, which resulted in posts associated with the tag "diabetes". Tumblr shows 15 posts on the first page and shows more posts as users scroll down. The search was repeated multiple times for four months. Each time new images were found. Those images were recorded in a dataset. Because we were interested in understanding how patients expressed their illness experiences in visual narratives, we collected only those posts that contained images. If the post contained images showing only text embedded in an image file (e.g., banner), the image was still included as our data. We captured the images and the text for each post. Posts that only included text or videos, but not images, were not captured. As a result, we collected 295 posts comprising photos, illustrations, cartoons, drawings, and images with embedded text. All posts were de-identified during the collection. We did not record the poster information and other meta-data, including re-blog and like counts, due to the scope of this study being exploratory. We focused our data collection and analysis on the images as units of analysis.

2.2. Data Analysis

We analyzed the images from Tumblr using open coding analysis [42]. We chose this method to identify emerging themes and patterns of an unknown phenomenon. The codes were iteratively refined and grouped using an affinity diagram [43] to make linkages among themes, categories, and sub-categories. The method allowed us to understand the social construction of illness experiences expressed by the posters. Additionally, the analysis provided an alternative understanding of patients' beliefs and actions that may vary from those expressed in a clinical setting [43]. We asked the following questions during the analysis:

1. What is portrayed in these images?
2. What diabetes-related problems are highlighted in the images?
3. How does the poster cope with the problem, if any?
4. What are the distinctive characteristics across the images?
5. What constrains or facilitates the process of disease management?

With these questions in mind, the author began the descriptive coding, translating the visual images into a textual description of what each image contained using Atlas.Ti qualitative software. The textual data were also coded using open coding schema. Next, the first author moved on to interpretive coding, identifying possible causes, implications, and connections among the images by iteratively comparing images and codes produced from the descriptive codes. The author then conducted axial coding, attempting to find emerging themes of clusters and connections among the codes.

For eight months, the author regularly discussed and debated the concepts and themes arising from data with a team of researchers once a month at research meetings. The meetings comprised graduate students and a faculty member with expertise in health and risk communication, communicative science and disorders, games, and human–computer interactions. The author solicited feedback from the group on the interpretation of the images, gaining consensus on balancing interpretation with a more objective view of what each image conveyed relative to the poster's illness experiences. Comments and questions generated from this process provided further avenues for reflection during the analysis. The quotes from the textual images appear in italics with double quotation marks throughout the article.

3. Results

3.1. Finding 1: The Beginning

Tumblr users posted metaphors, body images, illustrations, cartoons, selfies, photoshopped images, memes, and photographs of their daily experiences. These images expressed their frustration of accepting diabetes as part of their lives, dealing with fluctuating emotional and physical challenges, along with the determination to manage their illness.

3.2. Acceptance of Diabetes as a Chronic Illness

Tumblr users posted revelations about how long it took to embrace both the good and the worse parts of living with diabetes. Seeing other Tumblr users go through similar situations helped them to move on to accepting the disease. One post stated: *"It is never really been easy to talk about [diabetes] to other people, but I have come to realize over time that diabetes has become a big part of me. It's been the greatest part (I literally get to eat skittles … daily), and the worst part (I … get to take insulin shots … daily). But it's something that I can't change."* Here, the quote illustrates the poster's coming to agreement with "the greatest" and "the worst" parts about having diabetes as a major life transformation.

Another poster, after spending 15 years with diabetes, accepted "the fact that I'm a diabetic" and stated how "Finding this blog has helped so much, and knowing others struggle the same as I do [was] reassuring." Similarly, another poster stated: "I have never met an immature diabetic. We have all had to grow up and face life's reality." (Figure 2a). One poster talked about learning from the past, when injecting insulin was a big deal, which has now become a "peace of cake." (Figure 2b). The images of these posts showed peoples' perceptions about the self and the illness transforming overtime, finally leading to a lifelong commitment to the illness.

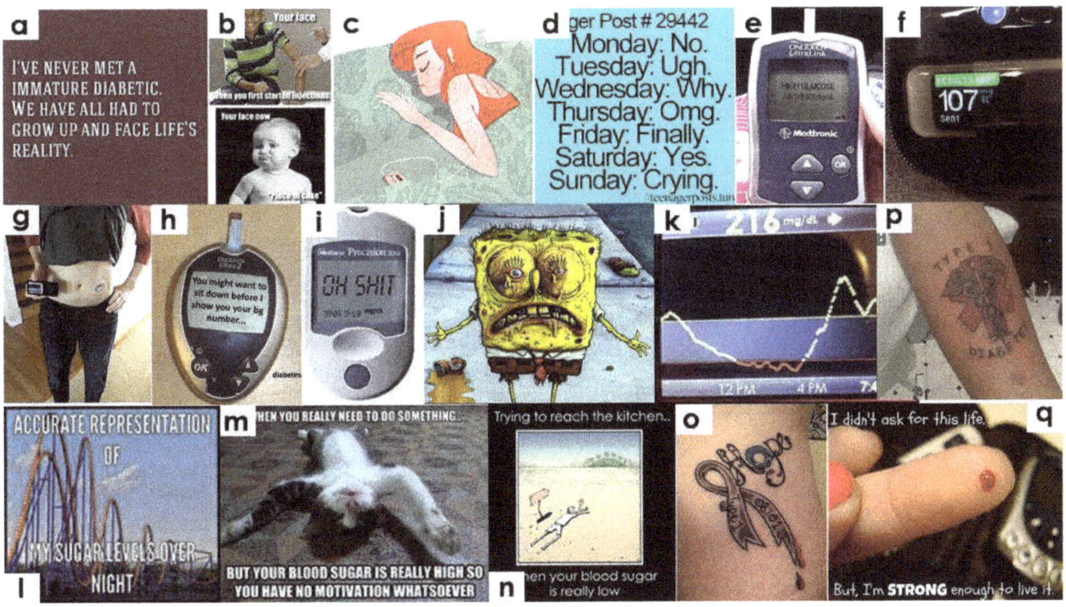

Figure 2. Images related to the beginning of diabetes and related illness narratives on Tumblr.

3.3. Emotional and Physical Fluctuations

Tumblr images depicted the emotional fluctuations and physical symptoms that people with diabetes pass through on a daily basis. For example, one post expressed how people with diabetes find it hard to have a good peaceful sleep, showing an image of a woman sleeping peacefully (Figure 2c). Another post, tagged as "anger post," described their weekly routine as: *"no, ugh, why, omg, finally, yes and ends at crying"* (Figure 2d), showing the constant challenges they face with diabetes throughout the week. Another poster expressed: *"I hate my life and need to roll over and die"* (Figure 2e). Another image showed that diabetes *"loves to complicate"* things in their life. Other posts showed similar feelings of frustration when they have important work to do or meet deadlines. These images showed the emotional trauma and despair that diabetes patients pass through daily as they struggle to manage blood glucose.

Measuring blood sugar takes a large part of diabetes patients' everyday life. This was evident through the substantial number of images about glucometers and insulin pumps in our data (Figure 2f). Posters expressed several emotions surrounding the use of a glucometer, for example, anxiously waiting for the glucometer and the excitement upon finally receiving it. Reading the instructions and using the glucometer for the first time held particular importance (Figure 2g). In addition, images conveyed the emotional state of waiting for the blood sugar number to appear on the glucometer and expressed feelings of extreme disappointment upon receiving a high blood sugar number (Figure 2h). For instance, one image showed a glucometer screen displaying *"Oh Shit!"* instead of the actual glucometer reading (Figure 2i). The images showed their constant fear of uncontrolled diabetes. This image was related to another image showing the cartoon character Sponge Bob Square Pants (Figure 2j) depicting panic attacks and depression. The findings consist of a mix of both excitement and despair, depending upon the test results. We realized that it is not only the physical feeling but also the emotional state of sadness upon receiving a low glucometer reading that adds up to their stress.

Feeling low (blood sugar) or high (blood sugar) are critical indicators for tracking how well blood sugar is controlled. Feeling low happens when patients' medications are overworked compared with the level of sugar in their blood. Patients need to take sugar as a remedy not to fall into shock. Diabetes patients can feel low at one point and high the next. Many posts expressed such fluctuating physical symptoms as contributing to fluctuating emotional states. For instance, a screenshot image of a fluctuating blood sugar monitor (Figure 2k) was coupled with the text: *"a long day."* Another image expressed the low sugar levels at night as similar to a frantic roller coaster (Figure 2l). When patients are low, they want to keep sleeping, sedated with sugar, tired of the struggle, and scared of the consequences (Figure 2m). Some posts described the state of feeling low and attempting to take sugar as crawling on a desert: *"Trying to reach the kitchen ... when your blood sugar is very low"* (Figure 2n). We found strong indications of emotional burnout and fatigue in managing the illness throughout the day. Additionally, there is no particular time of day when they could be feeling worse or better. Instead, images related to feeling high and low appeared across the board at all times of the day.

3.4. Determination to Fight against Diabetes

Tattoos emerged as a dominant theme in our data (Figure 2o), mainly as an expression of determination and hope. These images included a tattooed text, shape, or design that captured their health condition. The designs expressed emotions such as hope, the cycle of life, dependency on disease, and expressions of staying calm while bravely surviving the illness. One post read *"Tatted up finally. Type 2 Diabetes for five years now."* (Figure 2p) Despite all difficulties, posts shared encouraging messages and showed determination to keep fighting with the disease until its cure is found (Figure 2q).

Posts other than tattoo images also conveyed determination. For instance, a post showed comic hero characters such as "insu-man" and captain glucose to keep them driven. Many posts compared improvements made over time and the associated pride with these accomplishments.

In sum, the images portrayed their beginning journey as a diabetes patient, starting with struggles and acceptance, while continuously being challenged with emotional and physical fluctuations, but being determined to stay healthy. Although posters often used symbolic metaphors that only insiders would understand, the posters attempted to talk to the broader audience about how challenging it is to go through this initial phase of having diabetes.

3.5. Finding 2: Living with Diabetes

Posters also expressed the need for others to understand the daily struggles of a person with diabetes. We found several posts indicating that people with diabetes strongly wish others could feel their emotional and physical pain. Many images were shared about the misunderstandings and stereotypes associated with diabetes. One post read: *"Don't tell me I am a bad diabetic. You don't have diabetes. You don't know how hard it is. I try my best to manage what I didn't ask for"* (Figure 3s).

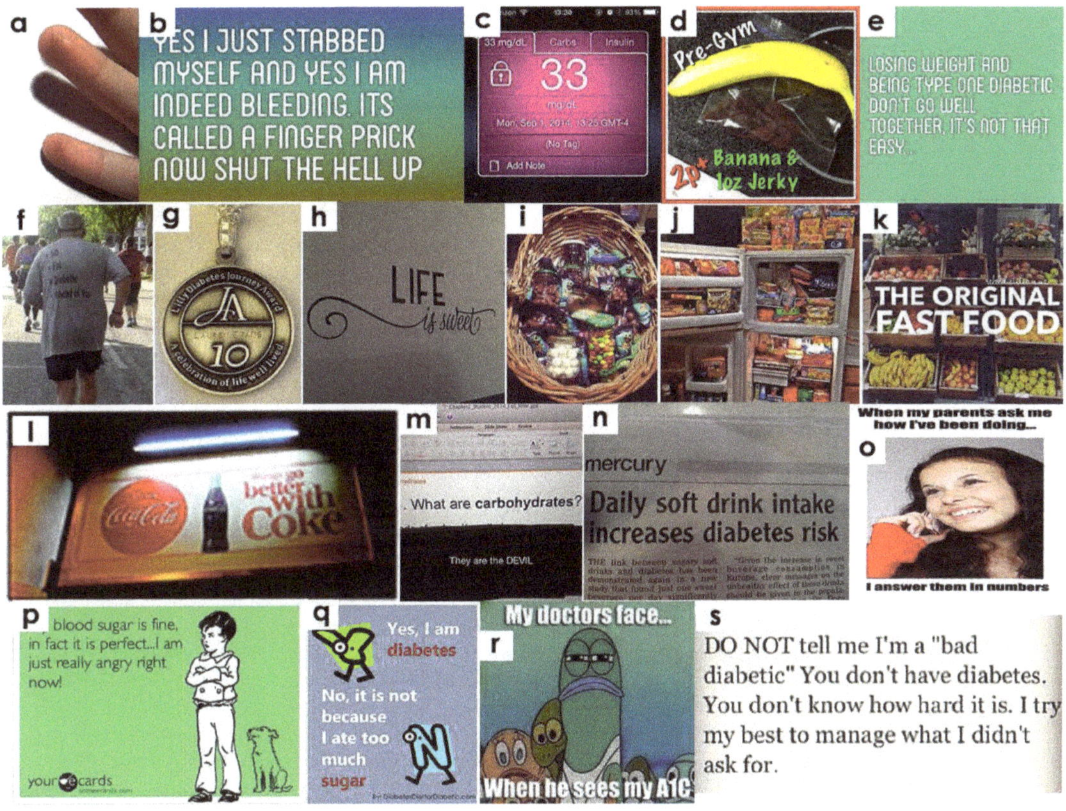

Figure 3. Images related to the maintenance of diabetes posted by Tumblr users.

Several other posts expressed a similar level of annoyance regarding when someone says too much sugar relates to diabetes or when people mix up type 1 with type 2 diabetes. Many posters tried to raise awareness regarding type 1 diabetes, which is not limited to children only and it is not always caused by eating sugar: *"I have type-1 diabetes because I ate too much sugar"* (Figure 3q). A similar post showing what people with diabetes do not agree with stated: *"you didn't care for yourself."* On the whole, we found posters trying to clarify that there are no good or bad diabetes patients and that each patient is different in terms of acquiring and experiencing the illness.

Speaking about the daily social interactions, one post mentioned: *"people with diabetes notice the knee-jerk responses by others when they order low-calorie food in a public space."* Another post mentioned that people with diabetes feel bad and hurt when they are asked to check their blood sugar and strongly want other people to stop judging them. Overall, we found that people with diabetes react with hostility towards social control and pressurizing behavior of others.

With regards to diabetes, information spread through mass media was found to be both helping and misleading. For example, posters mentioned that mass media spreads the misperception that blood sugar is correlated with anger. One post mentioned a person with diabetes hearing from others: *"You are not fat, you don't look fat."* Few posts touched upon the misperceptions that exist within the diabetes community, such as the proper usage of flex pens vs. syringe and the right way to inject insulin. Mass media have also helped raise awareness about diabetes; for example, one poster thanked news reporters for clarifying that: *"daily soft drink intake increases diabetes risk"* (Figure 3n). On the whole, we found a considerable portion of images posted on Tumblr about mass-media-based messages on diabetes. We noticed that Tumblr users aggressively responded to such misperceptions and showed a strong urge to clarify these misperceptions as much as they could.

3.6. Interactions with Friends, Family, and Providers

Tumblr images also included different ways a person with diabetes interacts with family, friends, and healthcare providers. These posts revealed the close monitoring of their parents on the posters' diabetes status, which the posters took as sometimes excessive and sometimes supported. For example, one post mentioned: *"When parents ask, 'how you are doing?' I reply in numbers. My friends don't understand what I said, but my parents do."* (Figure 3o). Another image showed parents with an insulin pump painted on their stomach as a gesture of support to their child who has diabetes. Other posts showed images of adolescent diabetes patients being panicked upon seeing high glucose readings but trying to hide this from parents. A similar post stated: *"Sometimes when I act stupid, my blood sugar is just fine ... so don't ask me to check it."* Another post read: *"My blood sugar is fine, in fact, it is perfect ... I am just really angry right now!"* (Figure 3p). All these images expressed frustration and distress about the comments they receive from others who do not know about diabetes. We found that posters strongly protested about being judged by others for having diabetes or not managing their illness.

Tumblr users portrayed their friends as being unsupportive or uninformed about their diabetes status. One image showed how posters illustrated their friends being the diabetes police—meaning that the posters felt that the friends judged the posters' behavior. One quote mentioned above illustrated that friends did not understand the diabetes conversation they have with their parents. We also found images in which posters strongly opposed the misperception that a person gets diabetes because their parents did not take care of them (Figure 3q). Few posters showed their gratitude towards parents and appreciated their help, albeit over-protective, in correcting their diet and exercise.

We also found posts related to interactions with healthcare providers. For example, some posts mentioned to avoid self-diagnosis and consult a doctor if you think you have diabetes. One image mainly talked about the perceived facial expressions of the doctor when he knows the A1C score (Figure 3r). Overall, we found a tendency of hiding diabetes-related worries from both friends and family. At the same time, we noticed a silent want of support from diabetes patients. They perceived their larger social circle as being unsupportive and ignorant to their real condition. Overall, we found reluctance towards both friends and family members as the groups to open up to about their situation.

3.7. Finding 3: Self-Reflection

This section describes findings around Tumblr images portraying the posters' reflection about their illness. This section describes the core value that illness narratives offer—the beliefs that patients have and how those beliefs drive their behavior; rich, historical stories of failure, recovery, and restoration; and how those experiences built up who the patients are.

3.8. Life: Past, Present, and Future

Our analysis of Tumblr images found that people with diabetes expressed a profound view of both past and future about the illness. In terms of looking to the past, we found images about how their health had improved or worsened with time. In many images, patients expressed that they hate to think about past events that they do not want to remember. We also found forward-looking images in which people with diabetes expressed a mix of hope and distress. One such post mentioned that distress is associated with how many more years one would have to live with diabetes, especially in the absence of a permanent cure. Another image showed the Garden of Eden in heaven with a tree's hanging fruits being insulin bottles (Figure 4a). Overall, we found a mix of bittersweet experience with regards to diabetes management. It is not surprising considering the chronic nature of the disease; however; we think that feelings differ depending on the time spent with the illness. Although we could not assess the age of the person posting on Tumblr, we found indications that older patients seemed to have negotiated with their illness as they made an effort to maintain health and wait for better cures.

Figure 4. Images related to the self-reflection on diabetes.

We also found a strong sense of helplessness expressed in the images posted on Tumblr, such as expressing feelings of depression because of the restrictions imposed by the illness (Figure 4b). One such image showed that diabetes had defined their territory, which is limited to the glucose test result. Explicitly, the younger patients expressed added frustration of not being able to enjoy *"junk"* food, which they loved so much. One post expressed the same feeling of reliance and helplessness by relating diabetes with a voodoo doll controlled by strings in the hands of the disease (Figure 4d) and thinking of oneself as an insulin junkie for life (Figure 4c). Some posts indicated that people with diabetes find it hard to believe that their whole life depends on the medical supplies in a pouch that continuously dictates their daily choices and actions (Figure 4e). Some posts talked about the added complexity of diabetes when coupled with other illnesses such as the common cold. Overall, we found that living on the terms and conditions defined by diabetes is particularly frustrating for diabetes patients. We noticed that diabetes patients longed for the past when they had the freedom and liberty to enjoy food.

We also found images in which posters talked about accepting the disease as a permanent part of their life, again with an understanding of there being no escape from its complexities. They showed to console themselves by thinking that diabetes is a disease with a long lifespan, and they can survive it even with low-functioning organs. We found images in which people appreciated the invention of insulin but with the recognition that it is not a replacement for a fully functional pancreas. Images depicted the reality of being used to insulin shots, daily monitoring, and pricking with the realization that it is not getting any easier. With regards to reactions towards glucometer readings, we found posters expressing optimism to keep the morale high and hope for a better score in the coming months. On the whole, we found that people with diabetes are fully aware of the fact that their illness is without cure and that it may become worse if they do not put in concerted effort to control it.

With regards to diabetes management, we found images related to routine day-to-day activities and their impact on the illness. For example, one poster mentioned that working from home is particularly helpful for a person with diabetes. A post about medication intake mentioned that unlike other illnesses, diabetes patients do not have a fixed time for medication because it depends on their glucometer reading and daily fluctuations in the sugar level. One post was about the extra space needed to carry diabetes supplies such as medicines and the glucometer. Few posts talked about the challenge of having to inject insulin in a moving vehicle (Figure 4f). Another image showed insulin injection as an old, long-time friend (Figure 4g). With regards to educating newly diagnosed patients, one post explained the difference between CGM and an insulin pump by clarifying that CGM measures blood sugar levels and not insulin.

3.9. Self-Perceptions about Body Shape and Organs

The images posted on Tumblr also showed how people with diabetes might view their physical body and internal body organs, especially stomach and pancreas, in the context of the illness. One image showed that people with diabetes tend to treat their body as a very sacred place, such as a temple, and critically reject the views of treating it as trash where one can just throw anything (Figure 4h). Other posts showed that people with diabetes idealize a very smart and slim body (Figure 4i). One post talked about the color and size of the insulin pump because this determines its placement on the body and also the effect on the choice of clothing. In the case of the stomach, images showed insulin pumps pasted with tape (Figure 4j). One post showed a patient filling out a driver's license form and not checking the field for organ donation of the pancreas (Figure 4k). Other posts showed a healthy body but a dead pancreas. Such posts expressed emotions associated with sadness because the pancreas does not perform as well as other body parts. Another image showed a cartoon character talking with his pancreas and saying: *"Diabetes, you will not compromise recovery!"* From these posts, we found that people with diabetes are very conscious of their physical changes and how the illness affects their perception about their physique and perceptions of smartness. It is not surprising that the pancreas and insulin production remain a dominant thought on the minds of diabetes patients. Thus, anything related to pancreas either on mass media, print advertisements, or news immediately catches their attention. Additionally, diabetes patients stay on the lookout for more ways to paste the insulin pump on their body. The stomach is the most used place for this because it hides the pump under the clothing.

4. Discussion

The ultimate goal of this study is to examine how visual illness narratives can be used to help various stakeholders, such as peer patients, health professionals, researchers, and caregivers, to better provide support for patients. We learned that micro-visual illness narratives provide a rich medium for Tumblr users to express struggles, share help, and reflect in depth. The images triggered productive conversations about posters' illness experiences that we might have not otherwise captured elsewhere. Below, we discuss how visual narrative illness extracted from Tumblr images provide utility and implications around patient care.

4.1. Chaos, Quest, and Restitution Narratives of Diabetes

In his book *The wounded storyteller: Body, Illness, and Ethics*, Frank classified three broad categories of illness narratives, i.e., restitution, chaos, and quest narratives (Frank, 1995). Based on these classifications, we elaborate on the illness narratives found on Tumblr. First, *chaos narratives* are disjointed and without temporal sequence. Tumblr images showed posters' progress in their diabetes management to fluctuate between sick vs. worse, no matter how hard they tried to maintain it. As a result, posters showed signs of despair and loss of hope. Chaos narratives describe experiencing illnesses with no cure and/or unreliable treatments. Patients suffering from chronic illness find listening to chaos narrative helpful because they no longer feel alone.

Patients with experiences associated with the narrative are often not in a position to fully articulate their condition. Accordingly, chaos narratives are difficult to retrieve from patients. Close family members and friends mostly witness such narratives. However, through Tumblr images, the public could witness chaos narratives around posters' earlier struggles and acceptance of the illness. The narrative depicted anxiety, frustration, and distress from illness. It revealed vulnerability and weaknesses, which most people hide, thus helping us understand the more in-depth, subjective side of the illness experience.

Quest narratives are about patients' fighting back. These illness narratives consist of stories about patients' strong will and making a concerted effort to fight the disease. Accordingly, the narrative includes signs of improvement. Quest narratives also depict alternative ways of being well by focusing on how to live longer and healthier. We found quest narratives to be most commonly appearing after chaos narratives on Tumblr. Patients' quest narratives consisted of images showing efforts to manage the diet, perform regular physical activity, and showing a positive attitude towards life, even though pricking emerged as one of the most painful aspects of patients' life, even when posters showed signs of resilience and commitment to get better.

Restitution narratives point toward the belief that health is restorable. These narratives are stories of the recovery and restoration to better health. For example, "in the past, I was healthy, today I have an illness, but I will be healthy again tomorrow." The restitution narrative originally denotes recovery. The connection between the restitution narrative and our data was the hopefulness and positive attitude of posters. For instance, images showed recovery and improvement in maintaining the glucose level, performing physical activity, controlling the diet, battling temptations to eat sugary food, and managing stress with a sense of optimism and hopefulness for the future self. These images provided insights about the everyday matters of people with diabetes and subtle realities that they may not openly disclose in a clinical setting. Tumblr also provides users with an opportunity to express aspects of their illness that may be considered taboo to openly share. Such narratives can open up conversations about sensitive topics that many want to talk about but never find the opportunity to do so.

4.2. Implications for Stakeholders and Chronically Ill Patients

Tumblr images with the diabetes hashtag showed rich expressions of diabetes illness narratives. The images communicated the anxiety, frustration, and vulnerability of posters who might have found it difficult or not appropriate to easily express such emotions in an offline social setting. We also found stories of hope and struggle in which posters were seen facing the illness with a strong will to get better. Such narratives consisted of posters who made a concerted effort to fight the disease and showed enormous signs of improvement. These images also showed alternatives ways of being well by focusing on how to live longer and healthier under the existing health conditions.

Social media are increasingly becoming mostly visual media, and enormous possibilities exist to utilize these data for pro-social benefits. Our analysis contributes unique insights about how diabetes patients self-disclose their illness narratives and experience on Tumblr. Patients and researchers can use the data to understand patients' experience, follow one's trajectory of illness, and emerging trends by predicting possible outcomes from a patients' current physical and emotional state. For instance, we can use crowd-sourced tags, meta-data of images, and automated visual analytic technologies to group images showing struggles, success, tips, and self-reflection. These groups can be used to provide tailored help to peer patients going through struggles. Posters can be matched as recommended friends to provide peer help if they post about similar emotional distress status.

Researchers can use these images to understand emerging trends in a cross-cultural context. Currently, Tumblr posters are predominantly western-oriented. However, with time, Tumblr membership may grow to include other nationalities, thus providing even richer insights for diabetes management in an international public health setting. Research organizations such as the Center for Disease Control and the National Cancer Institute can benefit from these findings to formulate funding announcements employing visual illness narratives on social media to better understand and serve the diabetes community.

During the analysis, a strong sense of disapproval was found towards the social stereotypes and misunderstandings about diabetes in the Tumblr community. We suggest that this topic be brought into discussion by healthcare providers during hospital visits and counseling sessions to provide patients with an opportunity to express and build strong patient–provider relationships. In addition, diabetes support groups and community health educators can use this information for community engagement and for supporting existing health education interventions. For instance, patients with speaking disabilities or limited English proficiency may find it hard to verbalize their illness experience. Such patients can be shown visuals to use as a medium for expressing their concerns and questions.

Additionally, the findings of this study can contribute to the design and development of glucometers, insulin pumps, and future wearable devices for diabetes care. We found specific emotional and physical behaviors associated with the use of these mobile devices that have implications in patients' choice of clothing and preference for mobile devices. For example, children with diabetes may show specific resistance to wearing an insulin pump. It is both uncomfortable and insecure for them to wear during school and play. The findings can help design future devices which are more convenient and more comfortable to wear for all age groups.

The study findings also have implications for health marketing and advertisement. We found that people with diabetes have a critical view of mass media advertisement, specifically about food products. A strong sense of skepticism and lack of trust was seen towards soda and sugary food manufacturers. Health marketers can use social media images to gain insights and design useful print and mass media advertisement and public service messages.

Our study is not without limitations. We do not know detailed information about the posters in terms of demographics and whether they were diagnosed patients. Our findings are instead based on the analysis of the images that the posters shared on Tumblr. Interviewing the Tumblr posters of these would help us to further understand how our interpretation differed and matched with the posters' intention. Our next step is to involve healthcare providers, caregivers, and patients to extract their interpretations of these images. This process could unveil hidden communication breakdowns and hints for improving patient–provider relationships.

Funding: This research received no external funding.

Institutional Review Board Statement: The study was approved by the institutional review board at Michigan State University, IRB # i047709.

Informed Consent Statement: The study only analyzed publicly available images and no interaction with human subjects was involved.

Data Availability Statement: The images used in the analysis are embedded within the manuscript. All data used in the study is publicly available on Tumblr.

Conflicts of Interest: The authors declare no conflict of interest.

References

1. Wild, S.; Roglic, G.; Green, A.; Sicree, R.; King, H. Global prevalence of diabetes: Estimates for the year 2000 and projections for 2030. *Diabetes Care* **2004**, *27*, 1047–1053. [CrossRef] [PubMed]
2. Owen, J.; Reisin, E. Non-communicable disease: A welcome and long needed addition to the WHO's 2012, world health statistics. *Curr. Hypertens. Rep.* **2012**, *14*, 475–477. [CrossRef] [PubMed]
3. Chiu, C.; Wray, L.A. Healthy lifestyle and psychological well-being buffer diabetes-related cognitive decline: Findings from the taiwan longitudinal study on aging. *Gerontologist* **2011**, *51*, 518.

4. Hernandez-Tejada, M.A.; Lynch, C.P.; Strom, J.L.; Egede, L.E. Effect of perceived control on quality of life in indigent adults with type 2 diabetes. *Diabetes Educ.* **2012**, *38*, 256–262. [CrossRef]
5. Griffith, L.S.; Field, B.J.; Lustman, P.J. Life stress and social support in diabetes: Association with glycemic control. *Int. J. Psychiatry Med.* **1990**, *20*, 365–372. [CrossRef]
6. Glasgow, R.E.; Toobert, D.J. Social environment and regimen adherence among type II diabetic patients. *Diabetes Care* **1988**, *11*, 377–386. [CrossRef]
7. Davidson, N.K.; Moreland, P. Dealing with the Diabetes 'Police'. 2012. Available online: http://www.mayoclinic.org/diseases-conditions/diabetes/expert-blog/diabetes-police/bgp-20056530 (accessed on 25 August 2022).
8. Ravert, R.D.; Hancock, M.D.; Ingersoll, G.M. Online forum messages posted by adolescents with type 1 diabetes. *Diabetes Educ.* **2004**, *30*, 827–834. [CrossRef]
9. Xiaomeng, H.; Yue, C. Psychology and micro-blogging: Self-presentation, social interaction and social culture. In Proceedings of the 2010 IEEE 2nd Symposium on Web Society, Beijing, China, 16–17 August 2010; IEEE: New York, NY, USA, 2010; pp. 643–647.
10. Jong, S.T.; Drummond, M.J. Shaping adolescent girls' body image perceptions: The effect of social media on Australian adolescent girls. *Res. Sci. Committee* **2013**, *2013*, 74.
11. Skinner, H.; Biscope, S.; Poland, B.; Goldberg, E. How adolescents use technology for health information: Implications for health professionals from focus group studies. *J. Med. Internet Res.* **2003**, *5*, e32. [CrossRef]
12. Hiles, D.; Cermák, I. Narrative psychology. In *The Sage Handbook of Qualitative Research in Psychology*; Sage Publications: London, UK, 2008; pp. 147–164.
13. White, M.; White, M.K.; Wijaya, M.; Epston, D. *Narrative Means to Therapeutic Ends*; WW Norton & Company: New York, NY, USA, 1990.
14. Kaley, A.; Hatton, C.; Milligan, C. More Than Words: The Use of Video in Ethnographic Research with People with Intellectual Disabilities. *Qual. Health Res.* **2018**, *29*, 1049732318811704. [CrossRef] [PubMed]
15. Harvey, K.; Brookes, G. Looking through dementia: What do commercial stock images tell us about aging and cognitive decline? *Qual. Health Res.* **2018**, *29*, 1049732318814542. [CrossRef] [PubMed]
16. Cheezum, R.R.; Rosso, M.T.; Niewolak, N.; Cobb, T. Using PhotoVoice to Understand Health Determinants of Formerly Homeless Individuals Living in Permanent Housing in Detroit. *Qual. Health Res.* **2018**, *29*, 1049732318816670. [CrossRef] [PubMed]
17. Dourish, P. The Foundations of Embodied Interaction. *MIT. Semiot. Compassion Value-Cent. Des.* **2001**, *13*, 133–187.
18. Fleron, B.R.; Pedersen, C. Exploring the roles in a photo elicitation dialogue. In Proceedings of the 11th Biennial Participatory Design Conference, Sydney, Australia, 29 November–3 December 2010; ACM: New York, NY, USA, 2010; pp. 175–178.
19. Frith, H.; Harcourt, D. Using photographs to capture women's experiences of chemotherapy: Reflecting on the method. *Qual. Health Res.* **2007**, *17*, 1340–1350. [CrossRef]
20. Banks, M.; Zeitlyn, D. *Visual Methods in Social Research*; Sage Publishing: Newbury Park, CA, USA, 2015.
21. Mondada, L. Working with video: How surgeons produce video records of their actions. *Vis. Stud.* **2003**, *18*, 58–73. [CrossRef]
22. Brown, R. Photography as process, documentary photographing as discourse. In *Visual Research Methods in the Social Sciences: Awakening Visions*; Routledge: Oxfordshire, UK, 2011; pp. 199–224.
23. Mitchell, C.; DeLange, N.; Moletsane, R.; Stuart, J.; Buthelezi, T. Giving a face to HIV and AIDS: On the uses of photo-voice by teachers and community health care workers working with youth in rural South Africa. *Qual. Res. Psychol.* **2005**, *2*, 257–270. [CrossRef]
24. Pink, S. *The Future of Visual Anthropology: Engaging the Senses*; Routledge: Oxfordshire, UK, 2006.
25. Pink, S. *Situating Everyday Life: Practices and Places*; Sage Publications: Newbury Park, CA, USA, 2012.
26. Frank, A.W. The Wounded Storyteller: Body, Illness, and Ethics. University of Chicago Press: Chicago, IL, USA, 2013.
27. Reavey, P.; Johnson, K. Visual approaches: Using and interpreting images. In *The SAGE Handbook of Qualitative Research in Psychology*; Sage Publications: Newbury Park, CA, USA, 2008; pp. 296–314.
28. Rier, D. The missing voice of the critically ill: A medical sociologist's first-person account. *Sociol. Health Illn.* **2000**, *22*, 68–93. [CrossRef]
29. Mattingly, C. Healing Dramas and Clinical Plots: The Narrative Structure of Experience. Cambridge University Press: Cambridge, UK, 1998; Volume 7.
30. Roulin, M.J.; Hurst, S.; Spirig, R. Diaries written for ICU patients. *Qual. Health Res.* **2007**, *17*, 893–901. [CrossRef]
31. van Dijk, B.; Dadlani, P.; van Halteren, A.; Biemans, M. Life changes, connection stays: Photo sharing and social connectedness for people with special needs. In Proceedings of the 28th Annual European Conference on Cognitive Ergonomics, Delft, The Netherlands, 25–27 August 2010; ACM: New York, NY, USA; pp. 135–142.
32. Radley, A.; Hodgetts, D.; Cullen, A. Visualizing homelessness: A study in photography and estrangement. *J. Community Appl. Soc. Psychol.* **2005**, *15*, 273–295. [CrossRef]
33. Lynn, N.; Lea, S.J. 'Racist'graffiti: Text, context and social comment. *Visual Commun.* **2005**, *4*, 39–63. [CrossRef]
34. Gillies, V.; Harden, A.; Johnson, K.; Reavey, P.; Strange, V.; Willig, C. Painting pictures of embodied experience: The use of nonverbal data production for the study of embodiment. *Qual. Res. Psychol.* **2005**, *2*, 199–212. [CrossRef]
35. Gleeson, K.; Frith, H. (De) constructing body image. *J. Health Psychol.* **2006**, *11*, 79–90. [CrossRef]
36. Eira, A. Number of Tumblr Blogs in 2022/2023: User Demographics, Growth, and Revenue. 2021. Available online: https://financesonline.com/number-of-tumblr-blogs/ (accessed on 25 August 2022).

37. Xu, J.; Compton, R.; Lu, T.C.; Allen, D. Rolling through tumblr: Characterizing behavioral patterns of the microblogging platform. In Proceedings of the 2014 ACM conference on Web science, Bloomington, IN, USA, 23–26 June 2014; ACM: New York, NY, USA, 2014; pp. 13–22.
38. Bourlai, E.; Herring, S.C. Multimodal communication on tumblr: I have so many feels! In Proceedings of the 2014 ACM Conference on Web Science, Bloomington, IN, USA, 23–26 June 2014; ACM: New York, NY, USA, 2014; pp. 171–175.
39. Anagnostopoulos, A.; Kumar, R.; Mahdian, M. Influence and correlation in social networks. In Proceedings of the 14th ACM SIGKDD International Conference on Knowledge Discovery and Data Mining, Las Vegas, NE, USA 24–27 August 2008; ACM: New York, NY, USA, 2008; pp. 7–15.
40. De Choudhury, M. Anorexia on tumblr: A characterization study. In Proceedings of the 5th International Conference on Digital Health, Florence, Italy, 18–20 May 2015; ACM: New York, NY, USA; pp. 43–50.
41. Clarke, R.; Wright, P.; Balaam, M.; McCarthy, J. Digital portraits: Photo-sharing after domestic violence. In Proceedings of the SIGCHI Conference on Human factors in Computing Systems, Paris, France, 27 April–2 May 2013; ACM: New York, NY, USA, 2013; pp. 2517–2526.
42. Strauss, A.; Corbin, J. *Basics of Qualitative Research*; Sage Publications: Newbury Park, CA, USA, 1990.
43. Beyer, H.; Holtzblatt, K.J. Contextual Design. *Interactions* **1999**, *6*, 32–42. [CrossRef]

healthcare
MDPI

Article

Association between Locomotive Syndrome and Physical Activity in Long-Term Inpatients of Psychiatric Care Wards in Japan: A Preliminary Study

Yusuke Ishibashi [1], Muneyoshi Nishida [1], Motoharu Hirai [1], Sae Uezono [2], Sosuke Kitakaze [3], Munetsugu Kota [4], Yukihide Nishimura [5], Fumihiro Tajima [6] and Hideki Arakawa [7,*]

1 Akitsu Kounoike Hospital, Nara 639-2273, Japan
2 Hirakawa Hospital, Tokyo 192-0152, Japan
3 Maplehill Hospital, Hiroshima 739-0651, Japan
4 Faculty of Health Science, Department of Rehabilitation, Hiroshima Cosmopolitan University, Hiroshima 731-3166, Japan
5 Department of Rehabilitation Medicine, Iwate Medical University, Shiwa-gun 028-3694, Japan
6 Department of Rehabilitation Medicine, Wakayama Medical University, Wakayama 641-8509, Japan
7 Rehabilitation Center, Faculty of Medicine, University of Miyazaki, Miyazaki 889-1692, Japan
* Correspondence: hideki_arakawa@med.miyazaki-u.ac.jp; Tel.: +81-985-85-1510

Abstract: The aim of this cross-sectional study was to determine the status of locomotive syndrome (LS) and the level of physical activity (PA) in long-term inpatients in a psychiatric care ward and to investigate the association between the severity of LS and the level of PA. The study participants consisted of 25 patients aged 55 years or older who had been admitted to a psychiatric care ward for more than one year. The participants' LS stage was determined and their level of PA was measured using an accelerometer. We also analyzed the correlations between the LS stage test results, level of PA, and values for each assessment item. The LS stage test showed that 84.0% of the participants were at stages 3. The participants' mean step count was 3089.8 ± 2346.5 steps. The participants' mean sedentary time was 349.7 ± 68.9 min, which is more than 70% of the total measuring time. Overall, the results indicate that LS stage was significantly correlated to age, ADL, and level of PA. Patients who stay in a psychiatric care ward experience declining motor functioning and lack PA. Deterioration of motor functioning is associated with lack of PA, suggesting the need for physical intervention.

Keywords: motor function; physical activity; psychiatric inpatients; physical therapist

Citation: Ishibashi, Y.; Nishida, M.; Hirai, M.; Uezono, S.; Kitakaze, S.; Kota, M.; Nishimura, Y.; Tajima, F.; Arakawa, H. Association between Locomotive Syndrome and Physical Activity in Long-Term Inpatients of Psychiatric Care Wards in Japan: A Preliminary Study. *Healthcare* **2022**, *10*, 1741. https://doi.org/10.3390/healthcare10091741

Academic Editors: Yasuhiro Kotera and Elaina Taylor

Received: 13 August 2022
Accepted: 9 September 2022
Published: 11 September 2022

Publisher's Note: MDPI stays neutral with regard to jurisdictional claims in published maps and institutional affiliations.

1. Introduction

In Japan, aging and long-term hospitalization pose serious problems for psychiatric care inpatients. In 2017, 16.2%, 23.8%, and 38.1% of psychiatric care inpatients were 55–64, 65–74, and >75 years of age, respectively [1]. In Japan, the average length of stay (LOS) for the treatment of psychiatric and behavioral impairments is 277.1 days [1]. Moreover, long periods of hospitalization are common in psychiatric care wards; 51.1% and 82.2% of inpatients have LOSs that exceed 5 and 1 year, respectively [2]. Older and longer-term psychiatric inpatients are likely in poor physical health and they tend to develop more serious complications, including physical disability and falling incidents.

Over 80 years ago, the BMJ reported that mental illness was associated with poor physical health [3]. Subsequent studies have shown that the risk for premature mortality of people with severe mental illness is 2–3 times higher than that of the general population [4], with cardiovascular disease causing most of the early mortalities [5]. In the general population, there is ample evidence demonstrating that low cardiorespiratory fitness is a strong and independent predictor of cardiovascular disease and all-cause mortality [6]. A meta-analysis of published works indicates that cardiorespiratory fitness in people with severe mental illness is severely reduced compared to age- and sex-matched controls [7].

Thus, physical fitness is a major factor determining the health of patients with mental illness. The following factors associated with low physical fitness have been found in patients with schizophrenia: lack of physical activity (PA), illness duration, smoking, metabolic syndrome, and the presence of more severe negative, depressive, and cognitive symptoms [8]; and patients with bipolar disorder are often associated with older age, lack of PA, positive and negative affect, and consumption of antipsychotic medication [9].

Only a few studies have investigated the decline in physical fitness and the factors contributing to such decline in elderly, long-term hospitalized psychiatric patients. In our previous study [10], we investigated the motor functioning of long-term psychiatric Japanese inpatients who were suffering from the locomotive syndrome (LS), taking into account the characteristics associated with aging. The Japanese Orthopaedic Association (JOA) defines LS as a condition in which the mobility function of the patient is impaired due to musculoskeletal disorders, and as it progresses, patients tend to require more nursing care [11]. Our previous study found that an overwhelming majority of long-stay psychiatric care inpatients suffer from declining motor function due to aging and long-term hospitalization, consequently affecting their activities of daily living (ADL). Psychiatric care inpatients often have fixed schedules for treatment and daily living. However, hospital life tends to involve low levels of PA, which may be associated with a decline in physical fitness. Therefore, PA is an important consideration for psychiatric care inpatients; however, in our previous studies, we have not objectively measured PA.

There is growing evidence that patients with mental illness undergo a decrease in PA [12–16], which is a serious problem because it may be associated with cardiovascular disease, premature mortality, and a decline in physical fitness.

Therefore, the purpose of this study was to determine the status of LS and the level of PA in long-term inpatients in a psychiatric care ward and to examine the association between the severity of LS and the level of PA.

2. Materials and Methods

2.1. Study Participants

This is a single-center study. The subjects consisted of 49 patients aged 55 years or older who had been admitted to psychiatric care wards for more than one year. Sixteen patients were unable to walk and seven did not give their informed consent and were excluded from the study. Finally, 25 patients participated in this study.

2.2. Basic Information

We obtained the following information from the participants' medical records: age, sex, F code in the International Classification of Diseases 10th edition (ICD-10), LOS, body mass index (BMI), and chlorpromazine-equivalent dose of antipsychotics (CP dose) [17]. Physical therapists evaluated each of the participant's independence level in ADL using the Functional Independence Measure (FIM) [18]; mental function was assessed with the Global Assessment of Functioning (GAF) [19]; cognitive function was assessed with the Japanese version of the Montreal Cognitive Assessment (MoCA-J) [20].

2.3. Physical Activity and Sedentary Behavior

All participants were fitted with a triaxial accelerometer (Panasonic; Acti-marker EW4800, Kadoma, Japan), which they wore for 2 days from 9:00 to 17:00. PA was measured in steps taken and expressed as metabolic equivalents (MET; in kcal/h/kg), an indicator of daily energy expenditure [21]. Based on the model proposed by Pate et al. [22] for classifying the MET intensity of PAs, PA was classified as sedentary, 1.0–1.5 METs; light (LPA), 1.6–3.0 METs; moderate-to-vigorous (MVPA), >3.0 METs. Daily average times spent in sedentary, LPA, and MVPA were calculated.

2.4. Locomotive Syndrome Stage Tests

To assess the LS risk, we administered a stand-up test, a two-step test, and the 25-item Geriatric Locomotive Function Scale (GLFS-25) test, as described below. Based on clinical levels developed by the JOA [23], the participants were classified as having no LS risk or having LS risk levels of 1 (LS stage 1), 2 (LS stage 2), or 3 (LS stage 3). Descriptions of these stages are given below.

In the stand-up test, physical therapists measured the participants' ability to adopt a single-leg or double-leg stance from stools that were 40, 30, 20, and 10 cm high. The result was reported as the minimum stool height from which the participant could stand.

In the two-step test, the physical therapists measured the two-stride length of each participant, starting from a starting line to the final position of the tips of the toes. The score was calculated by dividing the maximal length of two steps by the participant's height.

The GLFS-25 is a self-reported, comprehensive survey limited to information relevant to the preceding month and included 4 items on pain, 16 items on ADL, 3 items on social functions, and 2 items on mental health status. Each item was graded from no impairment (0 points) to severe impairment (4 points), and thus total scores ranged from 0 (no symptoms) to 100 (most severe symptoms) points.

The criteria for LS stage 1 were any of the following: (1) difficulty rising on 1 leg (either leg) from a 40-cm-high stool in the stand-up test; (2) a two-step test score of <1.3; and (3) a GLFS-25 score of ≥ 7. Any of these conditions resulted in a diagnosis of LS stage 1, indicating the beginning of a decline in mobility. The criteria for LS stage 2 were as follows: (1) difficulty rising on both legs from a 20-cm-high stool in the stand-up test; (2) a two-step test score of <1.1; and (3) a GLFS-25 score of Th16. Any participant who met these conditions was diagnosed with LS stage 2, indicating progression toward decline in mobility that increases the risk of losing the ability to live independently. The criteria for LS stage 3 were as follows: (1) difficulty rising on both legs from a 30-cm-high stool in the stand-up test; (2) a two-step test score of <0.9; and (3) a GLFS-25 score of ≥ 24. Any participant who met these conditions was diagnosed with LS stage 3, indicating a progressive decline in mobility that interferes with social participation.

2.5. Statistical Analysis

Basic statistics for each survey item and correlation analysis were performed using statistical analysis software R, version 4.0.2 (The R Foundation, Boston, MA, USA). Spearman's rank correlations were analyzed between the LS stage tests (stand-up test, two-step test, and GLFS-25) and age, LOS, BMI, CP dose, FIM score, GAF score, MoCA-J score, and PA (step counts, LPA, MVPA, sedentary time). Correlations were considered significant at a p-value of 5%.

3. Results

3.1. Characteristics of Study Participants

Table 1 lists the mean values of age, LOS, BMI, and CP dose of the 25 participants included in the final analysis. The participants' mean age was 70.1 ± 9.3 years, with ages ranging from 55 to 92 years; 80.0% of the participants were aged 65 years or older. The participants' mean LOS was 16.2 ± 14.1 years. Over half of the participants had stayed in the psychiatric ward for >10 years. The participants' mean CP dose was 646.9 ± 741.6 mg. We defined a high CP dose regimen as a prescription of >1000 chlorpromazine-equivalent mg of antipsychotics per day, a regimen that 16.0% of the participants received. Based on ICD-10 F code sorting, 92.0% of all participants were diagnosed as either F2 or F3 (Table 2). Table 3 lists the mean FIM and GAF scores, as well as the means of other measured variables. The participants' mean scores for FIM motor and cognitive items were 82.0 ± 9.1 and 27.4 ± 5.5 points, respectively. More than 90% of the participants' GAF scores were below 40 (GAF scores below 40 had a higher probability of readmission to the hospital [19]). The participants' mean MoCA-J score was 13.9 ± 6.8 points, with all scoring below 25 points, the cut-off value for mild cognitive impairment.

Table 1. Characteristics of the study participants.

Variables	Overall (n = 25)	Male (n = 5)	Female (n = 20)
Age, y	70.1 ± 9.3	67.0 ± 7.6	70.9 ± 9.7
≤59 y	5 (20.0%)	1 (20.0%)	4 (20.0%)
60–69 y	7 (28.0%)	2 (40.0%)	5 (25.0%)
70–79 y	9 (36.0%)	2 (40.0%)	7 (35.0%)
≥80 y	4 (16.0%)	0 (0.0%)	4 (20.0%)
LOS, y	16.2 ± 14.1	17.5 ± 16.9	15.8 ± 13.8
≥1y, <5y	6 (24.0%)	2 (40.0%)	4 (20.0%)
≥5y, <10y	6 (24.0%)	0 (0.0%)	6 (30.0%)
≥10y	13 (52.0%)	3 (60.0%)	10 (50.0%)
BMI (kg/m^2)	22.5 ± 3.9	20.9 ± 4.1	22.9 ± 3.9
CP dose (mg/day)	646.9 ± 741.6	745.9 ± 611.3	621.1 ± 763.0

Continuous variables are expressed as mean ± SD. Categorical variables are expressed as n (%). LOS, length of stay; BMI, body mass index; CP dose, chlorpromazine equivalent dose of antipsychotics.

Table 2. Diagnoses (F-codes) of all participants based on the ICD-10 classification of mental and behavioral disorders.

Categories	Overall (n = 25)	Male (n = 5)	Female (n = 20)
F0	1 (4.0%)	0 (0.0%)	1 (5.0%)
F2	18 (72.0%)	3 (60.0%)	15 (75.0%)
F3	5 (20.0%)	2 (40.0%)	3 (15.0%)
F7	1 (4.0%)	0 (0.0%)	1 (5.0%)

Categorical variables are expressed as n (%). F0, Organic, including symptomatic, mental disorders; F2, Schizophrenia, schizotypal and delusional disorders; F3, Mood [affective] disorders; F7, Mental retardation; ICD-10, the International Classification of Diseases 10th edition.

Table 3. Independence level in ADL, mental function, and cognitive function of the participants.

Variables		Overall (n = 25)	Male (n = 5)	Female (n = 20)
FIM-motor		82.0 ± 9.1	83.8 ± 10.5	81.6 ± 8.9
FIM-cognitive		27.4 ± 5.5	27.6 ± 7.5	27.4 ± 5.2
GAF	11–20	1 (4.0%)	1 (20.0%)	0 (0.0%)
	21–30	16 (64.0%)	3 (60.0%)	13 (65.0%)
	31–40	6 (24.0%)	1 (20.0%)	5 (25.0%)
	41–50	1 (4.0%)	0 (0.0%)	1 (5.0%)
	51–60	1 (4.0%)	0 (0.0%)	1 (5.0%)
MoCA-J		13.9 ± 6.8	15.0 ± 9.2	13.6 ± 6.6

Continuous variables are expressed as mean ± SD. Categorical variables are expressed as n (%). FIM, Functional Independence Measure; GAF, Global Assessment of Functioning; MoCA-J, Japanese version of The Montreal Cognitive Assessment.

3.2. Physical Activity Levels of the Participants

Table 4 lists the mean values of step counts, sedentary time, LPA time, and MVPA time. The participants' mean step count was 3089.8 ± 2346.5 steps, with all the participants scoring below 8000 steps per day, the recommended value for physical health [24–27]. The participants' mean sedentary time was 349.7 ± 68.9 min, which is more than 70% of the time between 9:00 to 17:00.

Table 4. Physical activity levels of the participants measured from 09:00 to 17:00.

Variables	Overall (n = 25)	Male (n = 5)	Female (n = 20)
Step counts (steps)	3089.8 ± 2346.5	4243.3 ± 1685.3	2955.5 ± 2450.1
Sedentary time (min)	349.7 ± 68.6	366.3 ± 34.7	342.1 ± 74.7
LPA time (min)	122.4 ± 66.9	107.0 ± 32.1	129.8 ± 72.8
MVPA time (min)	9.2 ± 7.7	8.0± 9.2	9.3 ± 7.9

Continuous variables are expressed as mean ± SD. LPA, Light-intensity physical activity; MVPA, Moderate-to-vigorous physical activity.

3.3. Distribution of Risk Levels for Locomotive Syndrome

Figure 1 shows the distribution of LS risk levels based on the results of the stand-up, two-step, and GLFS-25 risk assessment tests and the total assessment for all participants. In the total assessment, all of them were determined to be LS (stages 1, 2, and 3).

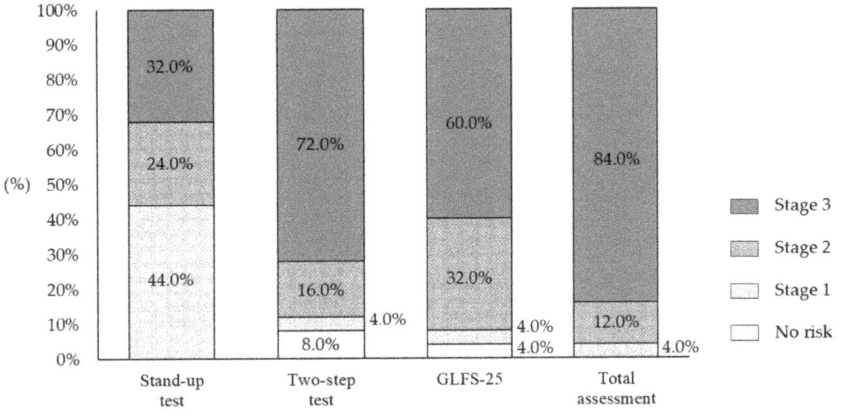

Figure 1. Distribution of risk levels for locomotive syndrome based on the results of the stand-up test, two-step test, and GLFS-25.

3.4. Association between Locomotive Syndrome and Measured Variables

Table 5 shows the correlations between the scores of the stand-up, two-step, and GLFS-25 tests and the measured variables. The stand-up test scores were significantly negatively correlated with age (r = −0.65, p = 0.02), and significantly positively correlated with the FIM motor items score (r = 0.78, $p < 0.01$) and step counts (r = 0.68, $p < 0.01$). The two-step test scores were significantly positively correlated with the FIM motor items score (r = 0.73, $p < 0.01$) and step counts (r = 0.27, p = 0.05). The GLFS-25 scores were significantly positively correlated with age (r = 0.65, $p < 0.01$) and significantly negatively correlated with the FIM motor items score (r = −0.68, $p < 0.01$).

Table 5. Correlations between the stand-up test, two-step test, and GLFS-25 scores and measured variables in participants.

Variables	Stand-Up Test	Two-Step Test	GLFS-25
Age	−0.65 *	−0.59	0.65 **
LOS	0.02	−0.12	0.27
BMI	0.42	0.30	−0.33
CP dose	0.18	0.42	−0.21
FIM-motor	0.78 **	0.73 **	−0.68 **
FIM-cognitive	0.44	0.49	−0.32
GAF	0.31	0.25	−0.19
MoCA-J	0.21	−0.06	−0.15
Step counts	0.68 **	0.27 *	−0.41
Sedentary time	−0.21	0.11	0.28
LPA	0.22	−0.06	−0.30
MVPA	0.35	0.11	−0.36

GLFS-25, 25-Geriatric Locomotive Function Scale; LOS, Length of stay; BMI, Body Mass Index; CP dose, Chlorpromazine equivalent dose of antipsychotics; FIM, Functional Independence Measure; GAF, Global Assessment of Functioning; MoCA-J, Japanese version of The Montreal Cognitive Assessment; LPA, Light-intensity physical activity; MVPA, Moderate-to-vigorous physical activity. Data are Spearman's rank correlation coefficients (r). * $p < 0.05$, ** $p < 0.01$.

4. Discussion

The purpose of this study was to determine the status of LS and PA in long-term inpatients in a psychiatric care ward and to investigate the association between the severity of LS and the level of PA. Our results show that most of the long-term inpatients in the psychiatric ward were at LS stages 2 or 3, and that the levels of PA in these patients were also declining. The severity of LS was related not only to age and ADL, but also to the level of PA, suggesting the need for physical intervention. The novelty of this study is that it focused on long-term hospitalization and aged patients, which characterize psychiatric care patients in Japan. We evaluated the motor functioning and measured the level of PA in elderly patients who had been admitted to psychiatric care wards for long periods of time. In addition, previous studies have often measured the level of PA subjectively, such as by self-reporting by questionnaires; however, compared to objective measurements, this method underestimates sedentary behavior while overestimating MVPA [12–16]. Therefore, we believe that the use of an accelerometer in this study was appropriate and provided an objective measure of PA.

The motor functioning of patients with mental illness is often impaired. In addition, the risk for premature mortality in people with severe mental illness is 2–3 times higher than that of the general population [4]. For example, schizophrenia is associated with accelerated aging syndrome, in which the physiological changes in body structures and functions that are associated with normal aging occur approximately 25 years earlier [28]. In the present study, 84.0% of the participants had the most severe form of LS (stage 3), confirming that motor functioning declines in long-term inpatients in psychiatric care wards. The prevalence of LS risk increases with age, as does the prevalence of age-related decline in motor function. Yoshimura et al. [29] estimated the prevalence of LS risk by age, based on the results of LS grade tests administered to the general population of a community. The prevalence of LS stage 2 was 28.2% and 39.0% in men and women in their 70s, respectively; and 62.1% and 76.0% in men and women in their 80s and over, respectively. Although the average age of the participants in the present study was 70 years old, all were diagnosed with LS, with 84.0% and 12.0% of them having LS stage 3 and 2, respectively. This indicates that inpatients in psychiatric care wards are at a higher risk of LS than members of the same age in the general population.

In a previous study, Aoyagi et al. reported that elderly members of a community performed an average of 6500 steps per day and an average of 17 min of moderate- to high-intensity activities [24]. In the present study, the participants per day and an av steps taken per day was 3089, and their average MVPA duration was 9 min, suggesting a lack of PA, although it is necessary to take into account that the measurement period of the present study was from 9:00 to 17:00. Aoyagi et al. established that physical health requires at least 8000 steps of activity per day [25–27]. None of the participants in the present study exceeded 8000 steps, thus, they did not reach the level of activity needed to maintain physical health. Moreover, the average duration of sedentary behavior in this study was 350 min, which represents more than 70% of the total measuring time. Several meta-analyses on the level of PA in mentally ill patients agree that compared to the general population, mentally ill patients spend significantly more time sedentary [12–16], averaging 660 min per day [14]. This corresponds to 60–70% of awake time, a level comparable to the results of the present study. These results indicate that patients admitted to psychiatric care wards spend long periods sedentary.

The current study showed that step counts, as a primary physical activity index, were correlated with the scores in stand-up and two-step tests but not with GLFS-25. The participants who had been admitted to a psychiatric care ward for >1 year showed low GAF and MOCA-J scores (e.g., negative symptoms, depressive symptoms, and cognitive function), which may have affected the results of the GLFS-25, a self-reported questionnaire. The correlation between PA and LS was only observed in step counts but not in sedentary time, LPA, and MVPA. The participants could walk of their own free will in the psychiatric care wards; however, most participants led a low-stimulation, low-activity hospital life due to the fixed daily treatment and living schedule. This suggests that the environment may have affected the MET intensity of PAs. In the present study, we found an association between the LS stage test results and PA, which is consistent with many previous studies that have found an association between motor functioning and PA in patients with mental illness [8,9,30]. Psychiatric inpatients are deficient in PA, and the fact that LS risk is associated with PA suggests that efforts to increase PA may prevent the development of metabolic syndrome and decreased health-related quality of life (QOL) [31–33]. Previous studies have also reported that PA is beneficial for mental health and QOL, and it helps alleviate psychiatric symptoms in psychiatric patients [34–36]. Factors contributing to inactivity in psychiatric inpatients include their daily fixed schedule that lacks time PA, their limited living space, the effects of antipsychotic medication, and psychiatric symptoms (such as negative, depressive, and cognitive symptoms) that hinder engaging in PA. Physical therapists can contribute to the solution of these problems by increasing their understanding of the characteristics of schizophrenia and other mental and behavioral disorders getting involved in the daily activities of patients, improving physical mobility, and expanding the living space. Moreover, other interventions can also be considered, such as establishing mobility patterns and living environments that allow patients to move safely and comfortably, conducting group exercise programs developed by psychiatric occupational therapists, and cooperating with ward staff in creating care plans.

This study has a number of limitations. First, we conducted a single-center study involving a small number of subjects. It will be necessary to conduct a multi-center study involving a larger number of subjects to determine whether our results are consistent across other psychiatric wards. Second, this was a cross-sectional study that only showed the association between LS risk and PA. Therefore, longitudinal studies that clarify causal relationships are needed, along with studies that examine the effects of efforts aiming to prevent the decline of motor functioning, including LS, in psychiatric inpatients. Finally, Kota et al. [37] reported that among patients admitted to a psychiatric ward, LS risk as judged by the two-step test is associated with the decision of discharge to the community. We believe that it is necessary to investigate whether improving motor functioning can facilitate patient discharge. Our results indicate that a decline in motor functioning of psychiatric inpatients is associated with a lack of PA, suggesting an urgent need for physical rehabilitation. We recommend that physical therapists play a larger role in inpatient intervention in psychiatric care wards.

5. Conclusions

Patients who stayed in a psychiatric care ward experience declining motor function and lack PA. Outcomes of the present study revealed that 84.0% of the participants had the most severe form of LS (stage 3), which confirmed a long-term decline in motor functioning among inpatients in psychiatric care wards. The average number of steps taken by the participants per day was 3089, whereas their average MVPA duration was 9 min, thereby suggesting a lack of PA. Deterioration of motor function was associated with a lack of PA, and indicated the need for physical intervention. We recommend that physical therapists should have a larger role in the intervention of psychiatric care ward inpatients.

Author Contributions: Conceptualization, Y.I., S.U., S.K., M.K. and H.A.; Formal analysis, Y.I.; Funding acquisition, H.A. and F.T.; Investigation, Y.I.; Methodology, Y.I. and H.A.; Project administration, M.N.; Resources, M.H.; Supervision, H.A.; Writing—original draft, Y.I.; Writing—review and editing, M.N., M.H., S.U., S.K., M.K., Y.N., F.T. and H.A. All authors have read and agreed to the published version of the manuscript.

Funding: This research was supported by the Grants-in-Aid for Scientific Research of Japan Society for the Promotion of Science, Grant Number JP20K19379 and MHLW Comprehensive Research on Disability Health and Welfare program, Grant Number 201817032B.

Institutional Review Board Statement: The study was conducted in accordance with the Declaration of Helsinki, and approved by the Ethics Committee of the Akitsu Kounoike Hospital (approval number: Kou18-009).

Informed Consent Statement: Written informed consent has been obtained from the participants to publish this paper.

Data Availability Statement: Not applicable.

Acknowledgments: The authors thank the rehabilitation staff at Akitsu Kounoike Hospital for assistance with the assessment of the participants.

Conflicts of Interest: The authors declare no conflict of interest.

References

1. Ministry of Health, Labour and Welfare. Patient Survey 2017. 2017. Available online: https://www.mhlw.go.jp/toukei/saikin/hw/kanja/17/index.html (accessed on 22 November 2021). (In Japanese).
2. Ministry of Health, Labour and Welfare. Central Social Insurance Medical Council; 40th Result Verification Committee for Medical Fee Revision. 2014. Available online: https://www.mhlw.go.jp/stf/shingi/0000049078.html (accessed on 22 November 2021). (In Japanese)
3. Robert, J.; Phillips, M.B. Physical disorder in 164 consecutive admissions to a mental hospital: The incidence and significance. *BMJ* **1937**, *2*, 363–366.
4. Reininghaus, U.; Dutta, R.; Dazzan, P.; Doody, G.A.; Fearon, P.; Lappin, J.; Heslin, M.; Onyejiaka, A.; Donoghue, K.; Lomas, B.; et al. Mortality in schizophrenia and other psychoses: A 10-year follow-up of the SOP first-episode cohort. *Schizophr. Bull.* **2015**, *41*, 664–673. [CrossRef] [PubMed]
5. Walker, E.R.; McGee, R.E.; Druss, B.G. Mortality in mental disorders and global disease burden implications: A systematic review and meta-analysis. *JAMA Psychiatry* **2015**, *72*, 334–341. [CrossRef] [PubMed]
6. Kodama, S.; Saito, K.; Tanaka, S.; Maki, M.; Yachi, Y.; Asumi, M.; Sugawara, A.; Totsuka, K.; Shimano, H.; Ohashi, Y.; et al. Cardiovascular fitness as a quantitative predictor of all-cause mortality and cardiovascular events in healthy men and women: A meta-analysis. *JAMA* **2009**, *301*, 2024–2035. [CrossRef] [PubMed]
7. Vancampfort, D.; Rosenbaum, S.; Schuch, F.; Ward, P.B.; Richards, J.; Mugisha, J.; Probst, B. Cardiorespiratory fitness in severe mental illness: A systematic review and meta-analysis. *Sports Med.* **2017**, *47*, 343–352. [CrossRef]
8. Vancampfort, D.; Probst, M.; Scheewe, T.; De Herdt, A.; Sweers, K.; Knapen, J.; van Winkel, R.; De Hert, M. Relationships between physical fitness, physical activity, smoking and metabolic and mental health parameters in people with schizophrenia. *Psychiatry Res.* **2013**, *207*, 25–32. [CrossRef]
9. Vancampfort, D.; Sienaert, P.; Wyckaert, S.; Probst, M.; De Herbt, A.; De Hert, M.; Stubbs, B.; Buys, R. Cardiorespiratory fitness in outpatients with bipolar disorder versus matched controls: An exploratory study. *J. Affect. Disord.* **2016**, *199*, 1–5. [CrossRef]
10. Ishibashi, Y.; Arakawa, H.; Uezono, S.; Kitakaze, S.; Kota, M.; Daikuya, S.; Hirakawa, J.; Nakamura, T.; Chosa, E. Association between long-term hospitalization for mental illness and locomotive syndrome. *J. Orthop. Sci.* **2022**, *27*, 473–477. [CrossRef]
11. Nakamura, K. A "super-aged" society and the "locomotive syndrome". *J. Orthop. Sci.* **2008**, *13*, 1–2. [CrossRef]
12. Vancampfort, D.; Firth, J.; Schuch, F.; Rosenbaum, S.; De Hert, M.; Mugisha, J.; Probst, M.; Stubbs, B. Physical activity and sedentary behavior in people with bipolar disorder: A systematic review and meta-analysis. *J. Affect. Disord.* **2016**, *201*, 145–152. [CrossRef]
13. Schuch, F.; Vancampfort, D.; Firth, J.; Rosenbaum, S.; Ward, P.; Reichert, T.; Bagatini, N.C.; Bgeginski, R.; Stubbs, B. Physical activity and sedentary behavior in people with major depressive disorder: A systematic review and meta-analysis. *J. Affect. Disord.* **2017**, *210*, 139–150. [CrossRef] [PubMed]
14. Stubbs, B.; Williams, J.; Gaughran, F.; Craig, T. How sedentary are people with psychosis? A systematic review and meta-analysis. *Schizophr. Res.* **2016**, *171*, 103–109. [CrossRef] [PubMed]
15. Stubbs, B.; Firth, J.; Berry, A.; Schuch, F.B.; Rosenbaum, S.; Gaughran, F.; Veronesse, N.; Williams, J.; Craig, T.; Yung, A.R.; et al. How much physical activity do people with schizophrenia engage in? A systematic review, comparative meta-analysis and meta-regression. *Schizophr. Res.* **2016**, *176*, 431–440. [CrossRef] [PubMed]
16. Vancampfort, D.; Firth, J.; Schuch, F.B.; Rosenbaum, S.; Mugisha, J.; Hallgren, M.; Probst, M.; Ward, P.B.; Gaughran, F.; De Hert, M.; et al. Sedentary behavior and physical activity levels in people with schizophrenia, bipolar disorder and major depressive disorder: A global systematic review and meta-analysis. *World Psychiatry* **2017**, *16*, 308–315. [CrossRef] [PubMed]
17. Inada, T.; Inagaki, A. Psychoropic dose equivalence in Japan. *Psychiatry Clin. Neurosci.* **2015**, *69*, 440–447. [CrossRef] [PubMed]
18. Linacre, J.M.; Heinemann, A.W.; Wright, B.D.; Granger, C.V.; Hamilton, B.B. The structure and stability of the Functional Independence Measure. *Arch. Phys. Med. Rehabil.* **1994**, *75*, 127–132. [CrossRef]
19. Endicott, J.; Spitzer, R.L.; Fleiss, J.L.; Cohen, J. The Global Assessment Scale: A procedure for measuring overall severity of psychiatric disturbance. *Arch. Gen. Psychiatry* **1976**, *33*, 766–771. [CrossRef] [PubMed]

20. Fujiwara, Y.; Suzuki, H.; Yasunaga, M.; Sugiyama, M.; Ijuin, M.; Sakuma, N.; Inagaki, H.; Iwasa, H.; Ura, C.; Yatomi, N.; et al. Brief screening tool for mild cognitive impairment in older Japanese: Validation of the Japanese version of the Montreal Cognitive Assessment. *Geriatr. Gerontol. Int.* **2010**, *10*, 225–232. [CrossRef] [PubMed]
21. Yamada, Y.; Yokoyama, K.; Noriyasu, R.; Osaki, T.; Adachi, T.; Itoi, A.; Naito, Y.; Morimoto, T.; Kimura, M.; Oda, S. Light-intensity activities are important for estimating physical activity energy expenditure using uniaxial and triaxial accelerometers. *Eur. J. Appl. Physiol.* **2009**, *105*, 141–152. [CrossRef]
22. Pate, R.R.; Pratt, M.; Blair, S.N.; Haskell, W.L.; Macera, C.A.; Bouchard, C.; Buchner, D.; Ettinger, W.; Heath, G.W.; King, A.C.; et al. Physical activity and public health. A recommendation from the Centers for Disease Control and Prevention and the American College of Sports Medicine. *JAMA* **1995**, *273*, 402–407. [CrossRef]
23. Japan Locomo Challenge Promotion Conference, The Japanese Orthopaedic Association 2018. 2018. Available online: https://locomo-joa.jp/check/judge (accessed on 22 November 2021). (In Japanese).
24. Aoyagi, Y.; Park, H.; Watanabe, E.; Park, S.; Shephard, R.J. Habitual physical activity and physical fitness in older Japanese adults: The Nakanojo Study. *Gerontology* **2009**, *55*, 523–531. [CrossRef] [PubMed]
25. Park, H.; Park, S.; Shephard, R.J.; Aoyagi, Y. Yearlong physical activity and sarcopenia in older aduls: The Nakanojo Study. *Eur. J. Appl. Physiol.* **2010**, *109*, 953–961. [CrossRef] [PubMed]
26. Park, H.; Togo, F.; Watanabe, E.; Yasunaga, A.; Park, S.; Shephard, R.J.; Aoyagi, Y. Relationship of bone health to yearlong physical activity in older Japanese adults: Cross-sectional data from the Nakanojo Study. *Osteoporos. Int.* **2007**, *18*, 285–293. [CrossRef]
27. Aoyagi, Y.; Park, H.; Kakiyama, T.; Park, S.; Yoshiuchi, K.; Shephard, R.J. Yearlong physical activity and regional stiffness of arteries in older adults: The Nakanojo Study. *Eur. J. Appl. Physiol.* **2010**, *109*, 455–464. [CrossRef] [PubMed]
28. Kirkpatrick, B.; Messias, E.; Harvey, P.D.; Fernandez-Egea, E.; Bowie, C.R. Is schizophrenia a syndrome of accelerated aging? *Shizophr. Bull.* **2008**, *34*, 1024–1032. [CrossRef] [PubMed]
29. Yoshimura, N.; Muraki, S.; Nakamura, K.; Tanaka, S. Epidemiology of the locomotive syndrome: The research on osteoarthritis/osteoporosis against disability study 2005–2015. *Mod. Rheumatol.* **2017**, *27*, 1–7. [CrossRef]
30. Vancampfort, D.; Stubbs, B.; Sienaert, P.; Wyckaert, S.; De Hert, M.; Richards, J.; Kinyanda, E.; Probst, M. Depressive symptoms and muscular fitness contribute independently to the ability to perform daily life activities in people with bipolar disorder. *Nord. J. Psychiatry* **2016**, *70*, 477–482. [CrossRef]
31. Vancampfort, D.; Sweers, K.; Probst, M.; Maurissen, K.; Knapen, J.; Minguet, P.; De Hert, M. Association of the metabolic syndrome with physical activity performance in patients with schizophrenia. *Diabetes Metab.* **2011**, *37*, 318–323. [CrossRef]
32. Vancampfort, D.; Probst, M.; Scheewe, T.; Maurissen, K.; Sweers, K.; Knapen, J.; De Hert, M. Lack of physical activity during leisure time contribute to an impaired health related quality of life in patients with schizophrenia. *Schizophr. Res.* **2011**, *129*, 122–127. [CrossRef]
33. Deenik, J.; Kruisdijk, F.; Tenback, D.; Braakman-Jansen, A.; Taal, E.; Hopman-Rock, M.; Beekman, A.; Tak, E.; Hendriksen, I.; van Harten, P. Physical activity and quality of life in long-term hospitalized patients with severe mental illness: A cross-sectional study. *BMC Psychiatry* **2017**, *17*, 298–307. [CrossRef]
34. Zschucke, E.; Gaudlitz, K.; Ströhle, A. Exercise and physical activity in mental disorders: Clinical and experimental evidence. *J. Prev. Med. Public Health* **2013**, *46*, S12–S21. [CrossRef] [PubMed]
35. Vera-Garcia, E.; Mayoral-Cleries, F.; Vancampfort, D.; Stubbs, B.; Cuesta-Vargas, A. A systematic review of the benefits of physical therapy within a multidisciplinary care approach for people with schizophrenia: An update. *Psychiatry Res.* **2015**, *229*, 828–839. [CrossRef] [PubMed]
36. Silveira, H.; Moraes, H.; Oliveira, N.; Coutinho, E.; Laks, J.; Delandes, A. Physical exercise and clinically depressed patients: A systematic review and meta-analysis. *Nueropsychobiology* **2013**, *67*, 61–68. [CrossRef] [PubMed]
37. Kota, M.; Uezono, S.; Ishibashi, Y.; Kitakaze, S.; Arakawa, H. Relationship between whether the planned discharge destination is decided and locomotive syndrome for admitted patients in psychiatric long-term care wards. *Phys. Ther. Res.* **2020**, *23*, 180–187. [CrossRef] [PubMed]

healthcare

MDPI

Article

Caregiver Representations of Therapeutic Patient Education Programmes for People with Schizophrenia: A Qualitative Study

Corinne Rat [1], Nicolas Meunier-Beillard [2,3], Samuel Moulard [1,2] and Frédéric Denis [4,5,6,*]

1 Clinical Research Unit of la Chartreuse Hospital, 21000 Dijon, France
2 Clinical Investigation Center—EC Inserm 1432, 21000 Dijon, France
3 Delegation for Clinical Research and Innovation, Dijon Bourgogne University Hospital, 21000 Dijon, France
4 Department of Odontology, Tours University Hospital Center, 37000 Tours, France
5 Faculty of Dentistry, Nantes University, 44035 Nantes, France
6 EA 75-05 Education, Ethics, Health, Faculty of Medicine, François-Rabelais University, 37000 Tours, France
* Correspondence: frederic.denis@chu-tours.fr; Tel.: +33-6-77-15-69-68

Abstract: Background: In France, there are two main types of psychosocial educational therapies for people with mental disorders: (1) therapeutic patient education (TPE) or "training", and (2) psychoeducation. Both types of educational therapy aim to improve disease morbidity, treatment compliance and patient quality of life, but they have very different modes of application. The aim of this study was to interview mental health professionals in order to explore and identify the determinants (barriers and enablers) underlying their acceptance of therapeutic patient education (TPE) in order to facilitate the implementation of these programmes among people with severe mental illness such as schizophrenia. **Methods:** In this multicentre cross-sectional study, we opted for a qualitative approach based on individual semi-structured interviews with 21 mental health professionals trained in TPE, regardless of whether they had practiced it before or not. In accordance with the "Jardé" law (Decree no 2016-1537 dated 16 November 2016 published on 17 November 2016 in the Official Journal of the French Republic). No regulatory approval was required for this study. **Results:** The major themes that emerged from the analysis were grouped into the following conceptual framework: (1) mental health professionals (MHPs) highlight important organizational and institutional challenges that they feel are beyond their scope; (2) MHPs mention in parallel their own perceptions and representations of TPE in the context of mental health care; and (3) MHPs' representations could hide a lack of knowledge or awareness that would prevent them from appropriating TPE programmes. For each major theme, the sub-themes identified are presented. **Conclusions:** Although TPE is of interest in the process of patient empowerment, we found that caregivers were reluctant to appropriate this approach to care. Efforts must be made in the initial and ongoing training of MHPs to move from a paternalistic model to a patient partnership model, which is made possible by TPE. These efforts must also be firmly supported by health care facilities, and proactive governance is required for the successful implementation of TPE.

Keywords: schizophrenia; therapeutic patient education programme; mental healthcare; patient partnership

Citation: Rat, C.; Meunier-Beillard, N.; Moulard, S.; Denis, F. Caregiver Representations of Therapeutic Patient Education Programmes for People with Schizophrenia: A Qualitative Study. *Healthcare* **2022**, *10*, 1644. https://doi.org/10.3390/healthcare10091644

Academic Editors: Yasuhiro Kotera and Elaina Taylor

Received: 17 July 2022
Accepted: 24 August 2022
Published: 29 August 2022

1. Background

Schizophrenia is a severe and persistent mental disorder that affects 0.7 to 1% of the world population [1] and 600,000 people in France [2]. Schizophrenic patients are exposed to higher mortality and to numerous comorbidities, and their life expectancy is 10 to 15 years lower than the general population (excluding suicide) [3]. The gap in life expectancy between the general population and schizophrenic patients is not acceptable, and all potential means of reducing it must be explored.

In France, there are two main types of psychosocial educational therapies for people with mental disorders: (1) therapeutic patient education (TPE) or "training" and (2) psychoeducation. Both types of educational therapy aim to improve disease morbidity, treatment compliance and patient quality of life, but they have very different modes of application [4].

Psychoeducation aims, usually through group workshops, to inform patients (or their relatives) about their psychiatric disorder and to promote coping skills by providing structured information about the illness and its treatments [5]. Psychoeducation promotes the resolution of emotional, psychological, behavioural and cognitive problems [5]. It has been shown to have a positive impact on patients' adherence to treatment, autonomy and recovery, as well as being useful for family and friends [6,7]. It is recommended in particular for bipolar disorders and schizophrenia [8,9]. It is considered to be particularly suitable for psychiatry, and, in contrast to TPE, its implementation does not require a legislative framework or compulsory training [8,9]. The evaluation procedures for the programmes developed are under the sole authority of the practitioners and the institutions that offer them. Specific institutional funding for this activity is infrequent and it is mainly part of a project for a single unit or for the psychiatric department in general, or as part of an annual operating grant. It is often carried out on an ongoing basis, through shared staff or redeployment of staff [8,9] (Table 1).

Table 1. A comparison between therapeutic patient education or "training" and psychoeducation in France.

Educational Therapies	Therapeutic Patient Education or "Training"	Psychoeducation
Aims	Aims to encourage a process of empowerment in order to make the patient more autonomous and active in his/her behavioural changes throughout the care project.	Promotes the resolution of emotional, psychological, behavioural and cognitive problems
Disease type	For individuals with chronic diseases	For individuals with mental disorders
Legislative framework	Framed by the "Hospital, Patients, Health and Territory" law (HPST) of 21 July 2009 [10].	Does not require a legislative framework or compulsory training
Assessment	The "Haute Autorité de Santé" (HAS) has set recommendations for the annual and four-year self-assessment required for the renewal of programmes [11].	Evaluation procedures for the programmes developed are under the sole authority of the practitioners and the institutions that manage them.
Funding	Conditional funding by the "Agence Régionale de Santé (ARS)	No specific institutional funding

According to the World Health Organisation (WHO), TPE programmes aim to help patients acquire or maintain the skills they need to manage their lives with a chronic dis-ease [12]. The "Hôpital Patients Santé et Territoire" (HPST) law dated 21 July 2009 [10] gave a legislative framework to TPE in France and recognised it as a right for patients, while the decree no 2010-906 dated 2 August 2010 specifies the skills required to deliver TPE programmes as well as the specifications for the authorisation of these programmes by the regional health agencies (ARS) [13]. They provide for a formalised "shared educational assessment", the definition of a personalised care programme with the patient, the planning of workshops and sessions, and the evaluation of their acquisition and effectiveness [14]. The French National Authority for Health (HAS) has provided a guide to facilitate the annual self-assessment and four-year assessment that are required to continue running the programmes [14]. Table 1 should be moved to this position

Despite the mentioned differences between psychoeducation and TPE, these educational programmes also have similarities in that they help the person to understand their psychological disorders and to acquire self-care skills to manage them in their daily lives.

They also provide information in a personalised way, helping to rebuild their identity, develop coping skills and explore the emotions generated by the disorder [8–10]. Most TPE programmes are currently built on the basis of the caregivers' knowledge of the disease. A caregiver in mental health care refers to any person who is involved in the treatment or prevention of the illness or its complications. The doctor, nurse and care assistant can be considered caregivers. Some TPE programmes are delivered in collaboration with a multidisciplinary team and aim to help the patient gain mastery and skills in order to increase their sense of self-efficacy and help them to be more active in their management [15,16]. New approaches even take into account the patient's experience of the disease and the experiential knowledge gained during their illness [17]. This new role attributed to patients has led to a redefinition of the relationship between caregivers and patients. Thus, the process of empowerment [18] and otherness relationship patterns [19] have been conceptualised to describe an educational position that "allows the subject to exist in his or her health choices and apprehensions in the face of disease", and also to redefine the caregiver–patient relationship. In these models, patients seem to be at least as well placed as carers to recognize their needs. Thus, TPE works towards changing the practices and attitudes of caregivers towards patients in general. In France, however, TPE is having difficulty gaining momentum in the field of mental health. Only 2.3% of the 1800 TPE programmes listed in France in 2011 were programmes for patients with psychiatric disorders [20]. Although, in 2016, the number of TPEs in psychiatry had doubled since 2011 [21], a search of the ARS websites shows that in 2021 there was a great disparity in the deployment of TPEs in psychiatry. For example, 13 programmes were listed in the Auvergne-Rhône Alpe region, 6 in Bourgogne Franche-Comté and only 4 in the Grand Est region and 2 in Normandy. The care of psychiatric disorders is of course very different from that of other systemic illnesses. Firstly, because health care alone is not enough, and above all, because it alters relationships with others and is a source of stigmatization [22,23]. Moreover, the relationship between the carer and the patient has until now been oriented towards a paternalistic, benevolent and protective position. In this context, the patient's attempts to decide for him or herself can be perplexing for caregivers because of their representations of how to care for a patient with a mental health disorder [24]. Moreover, beyond the relational dimension of care in psychiatry, the educational approach raises ethical questions and forces caregivers to assess their personal biases about providing TPE to people with psychiatric disorders [22].

Indeed, recent studies on the representations of the main providers of TPE in psychiatry have highlighted that some mental health professionals (MHPs) have a skewed image of TPE, wrongly considering that it is limited to improving treatment compliance [24,25]. The few studies that have looked at the representations of those providing TPE to patients suffering from psychiatric disorders highlight a feeling of lack of exchange and infantilization by some caregivers [25].

Insofar as the effectiveness of a TPE programme is based on the synergy of the patient–caregiver relationship, individual MHP representations must not hinder the support provided to patients in these programmes. It is therefore important to identify the gaps between the expected and actual roles of caregivers, the mechanisms of resistance employed by the individuals involved, and/or the structural and contextual blockages complicating the implementation of health education for schizophrenic patients.

Qualitative research is a social-sciences approach that is based on observation and listening to bring out new non-quantitative data [26]. It seeks to answer the questions of "why?", "how?" and "what?", taking subjectivity into account, rather than validating pre-established hypotheses. It may highlight the ambivalence often found in open questions involving the interviewees' convictions and feelings. Grounded theory in particular is a systematic set of techniques and procedures that enable researchers to identify concepts and build theories or conceptual frameworks from qualitative data. More specifically, grounded theory is focused on psycho-social processes of behaviour and seeks to identify and explain how and why people behave in certain ways, in similar and different contexts [27]. Data generation is aimed at explaining how changes in action–interaction come about in response to different conditions by capturing data in a way that is amenable to identifying and explaining these processes and phenomena. This leads to a better interpretation of the results and to more complex and relevant hypotheses [28].

2. Aims

Our aim was to interview MHPs in order to explore and identify the underlying determinants (barriers and enablers) of their acceptance of TPE, in order to facilitate the implementation of these programmes among people with severe mental illness such as schizophrenia, and how changes can be introduced and sustained.

3. Methods

In this multicentre cross-sectional study, we opted for a qualitative approach based on individual semi-structured interviews with MHPs trained in TPE, whether they had practiced it or not. The interviewer was a male sociologist trained in qualitative health research who was not involved in the follow-up of patients. A purposive sample was constituted with MHPs from six French psychiatric care centres.

3.1. Participant Selection

Interviewees were paramedical and medical caregivers (nurses, nursing staff, nurses' aides, social workers, pharmacists, and physicians) practicing in psychiatry and trained in TPE. They may or may not have provided a TPE program for patients with schizophrenic disorders in the last 6 months.

The interviews were conducted until the phenomenon of saturation was reached [29]. Theoretical saturation occurs when questioning additional participants does not bring out new information or themes.

3.2. Participants

Interviews with 21 participants came from 7 different organisations were conducted between August 2019 and December 2020 (Table 2); 13 (62%) participants had already practiced TPE and 8 (38%) had not. All participants had completed the mandatory 40 h training to conduct a TPE program [14]. Before a TPE program is started, the patient and caregiver choose together the themes to be worked on following an educational diagnosis. They then set goals to be achieved. Individual or group TPE sessions are then carried out or the patient is involved in his or her education. A specific session aims to evaluate the skills acquired and the changes implemented by the patient in his daily life. For the success of this program, the coordination of health professionals involved in the management of chronic disease around and with the patient is essential [11,14]. The characteristics of the 21 participants in the study are presented in Table 2.

3.3. Data Analysis

3.3.1. Inclusion Process

In each participating centre, the study investigator identified eligible MHPs. The investigator explained the objectives of the research to each potential participant and mailed them the written information. If the MHP accepted the interview, their email address was given to the sociologist to schedule an appointment for a face-to-face interview.

3.3.2. Interview Guide

The interview guide was based on: (1) the data available in the literature in order to identify the area of focus [21,27]; and (2) the exploratory semi-directive interviews conducted by a health sociologist with a nurse, a nurse manager, a care assistant and a physician corresponding to the study's inclusion criteria. The semi-structured interview is a qualitative data collection strategy in which the researcher asks informants a series of predetermined but open-ended questions [27].

Table 2. Population characteristics.

Demographics	n	(%)	Minimum	Maximum
Age (years) (SD) 46.3 (8.2)	21	100%	34.2	59.6
Gender				
Male	4	19%		
Female	17	81%		
Profession			Cumulated frequency	Cumulative percentage
Nurse	8	38.1%	8	38.1%
Nurse Manager	4	19%	12	57.1%
Nursing assistant	2	9.5%	14	66.7%
Social worker	4	19%	18	85.7%
Pharmacist	1	4.8%	19	90.5%
Physician (1 doctor and 1 psychiatrist)	2	9.5%	21	100%
Experience in psychiatry				
0 to 5 years	3	14.3%	3	14.3%
6 to 10 years	5	28.2%	8	38.1%
11 to 15 years	2	9.5%	10	47.6%
>15 years	11	52.4%	21	100%

SD: Standard deviation.

The following themes were thus addressed: experience, training and professional background, representations of mental health care, representations of and experience with therapeutic education, and needs and expectations for the development of therapeutic education.

3.3.3. Procedure

Due to the COVID-19 pandemic, some interviews were conducted by videoconference (n = 2) or by phone (n = 5), while the others were conducted in person in the mental health centres. Participants met up with the interviewer in a quiet place for 30 to 60 min. The interviewer first reviewed the principles of a qualitative study and reminded the participant that the interview would be audio recorded (anonymously), which was specified in the study information documents provided beforehand. Active listening with interrogative, reiterative or interpretative techniques was used to encourage the professional to explain and clarify their answers and feelings, but without influencing them. A pause was allowed between questions so that participants could recall events and feelings with precision. Notes on non-verbal communication were included in the transcription of interviews to enrich the data.

Interview records were transcribed in their entirety in a text format for later analysis. Data were encoded to guarantee the anonymity of the participants. Starting from the realities of the field and the participants' discourse, the inductive approach was favoured, in accordance with the requirements of grounded theory [30]. This involves identifying the themes addressed, grouped into major thematic categories (divided into sub-categories), and then proceeding, with a higher level of inference, to a conceptual interpretation of their interactions. The analysis of the interviews proceeded in six distinct main steps:

1. Open coding of the transcribed interviews in order to bring out as many themes as possible from the initial corpus.
2. Categorisation of the codified elements: a careful rereading of the entire corpus to ensure that each category is clearly defined, its properties identified, and the different forms and conditions of appearance of the phenomena specified.
3. Relating the categories: writing more detailed memos and designing explanatory diagrams.
4. Integration of the previous steps in order to identify the essence of the phenomenon.

5. Modelling: the phenomenon, in addition to being described, defined and explained, will then be examined and conceptualised in terms of its dynamics. The structural and functional relationships of each of its constituents were then highlighted.
6. Theorising: careful and exhaustive construction of the "multi-dimensionality" and "multi-causality" of the phenomenon of the relationships between the needs, expectations and representations of the different actors (physicians, nurses, nursing assistants, social workers, and pharmacists).

In order to reduce analysis bias and ensure cross-validation of the data, the interview data were analysed and interpreted first by the sociologist involved in the field. Then, this initial coding framework or set of codes was discussed in an interdisciplinary meeting with the study's steering committee who could redefine the boundaries of the themes that emerged from the data.

3.4. Ethics

After participants were provided with information about the study, their oral consent was collected and they were included in the study. Their contact details were sent to the sociologist in charge of conducting the interviews so that an appointment could be made in accordance with the "Jardé" law (Decree no 2016-1537 dated 16 November 2016 published on 17 November 2016 in the Official Journal of the French Republic). No regulatory approval was required for this study.

4. Results

4.1. Findings

The major themes that emerged from the analysis were grouped into the following conceptual framework: (1) MHPs highlight important organizational and institutional challenges that they feel are beyond their scope; (2) MHPs mention in parallel their own perceptions and representations of TPE in the context of mental health care; and (3) MHPs' representations could hide a lack of knowledge or awareness that would prevent them from appropriating TPE programmes. For each major theme, the sub-themes identified are presented.

4.2. MHPs Highlight Important Organizational and Institutional Challenges That They Feel Are beyond Their Scope

4.2.1. The TPE Would Be Too Time Consuming

The professionals described a lack of institutional will to provide TPE in their facility due to the lack of time dedicated to this additional activity.

> "I think that in fact the difficulty of TPE comes from the fact that it has not been supported by anyone specific..." (Doctors (D)) and (Nursing Manager (NM)).
>
> "They don't have much time. They are really busy with all the protocols. They have more and more things to do..."…/… "They don't have much time…/…" (NM).
>
> "And then there's so much work too, so much nursing and care assistant work, that it's not easy to detach yourself" (NM).
>
> "We could do it but at the expense of something else." [Nursing assistant (NA)].
>
> "In the evening, they have finished their day, they want to go home, I understand them, I really don't blame them, but they want to go home. They have their children too, they have their mother to visit, they have their life. And to commit oneself like that to something very regular is clearly a constraint." [NA].

In France, psychiatry suffers from a significant shortage of caregivers, which has a major impact on the implementation of new activities.

4.2.2. High Turnover of Professional Staff within the Units

The respondents also presented the issue of high staff turnover because of the difficulty of the work as a factor that was detrimental to the sustainability of the programmes, whether this was due to personal or institutional causes.

> "I was running a workshop with a health executive who left the hospital. And so I found myself... a bit on my own with this..."…/… [NA].

> "Another obstacle that everyone in the hospital knows, and this was the case for the unit I was talking about, is that in fact all the care workers change departments every 5 years..."…/… [NM and Psychiatrist (P)].

> "This is really a big obstacle. Because if there isn't a team of carers who can come and re-mobilise, re-explain the origin and re-initiate work on these programmes, on these tools, people find it difficult to re-appropriate them and they have to run a programme that was created by others..." [NM and P].

4.2.3. Institutional Communication Needs to Be Improved

The lack of centralisation and institutional organisation also seemed to be lacking within the institutions and accentuated the feeling of having to manage alone and of lacking visibility in the running of the programmes.

> "There must have been four or five professionals who really wanted to get involved. We were lucky enough to have the support of the managers, who made it possible for us to make up for the time taken up personally to structure our workshops." [P and NM].

> "When we asked for a bit of material, we had it all, right away. No, we were lucky in that respect, yes." [NM].

On the other hand, "At the beginning, it was really word of mouth, there would have been no communication if the doctor who was at the initiative of this project had not done a bit of pushing, creating meetings, inviting people, inviting the different partners, we made flyers, brochures, we tried to talk to a lot of people. And it's true that little by little our colleagues began to understand the interest in better perceiving the indications and why we could propose that, why we could also think like that." [NM].

4.3. MHPs Mention in Parallel Their Own Perceptions and Representations of TPE in the Context of Mental Health Care

4.3.1. TPE and the Emergence of a Conflict of Values

In its approach to care and the new type of relationship with the patient, TPE is contrary to the culture of psychiatric care that carers are used to. The culture around medical care seems to be still strongly anchored in a paternalistic model where the doctor is the decision maker on what the quality and philosophy of life of the patient should be.

> "I think that doctors, or even nurses, like to control, to know everything, and not to leave too much autonomy." [NA and Social Worker [SW)].

Generally speaking, the idea emerged that the initial training of MHPs leaves too little room for the personalisation of work, group work and the development of social skills, which hinders the development of cooperation.

> "We are all used to receiving educational content in a rather formatted, vertical way . . . ", "Really one-sided, yes. And so that's totally the opposite of what TPE can be, totally . . . " [D].

> "And, in psychiatry, perhaps we are a little behind in this respect, a little behind, because... well yes, clearly diabetes and pneumology have made a lot of progress in all this. And I think that for years, psychiatry has been watching all this progress, that's it." [P].

The participants recognize that the organisation of practices is rigid and difficult to change because it calls into question the meaning of "care" in psychiatry.

> "When I told colleagues from other departments that were doing therapeutic education in psychiatry, they looked at me wide-eyed when in fact... it's still shocking, that's all." [P].

Even though the general management of the participants in the study had approved the training for TPE, contradictory orders in the field made it difficult to implement these practices and created a conflict of values and a loss of meaning for them.

> "Behind all this, it is the importance of the meaning we give to all this!" [P].

4.3.2. TPE Seen as a Fad

Some carers, in the context of the current strain on psychiatry units, considered that it was necessary to distance themselves from TPE by considering it to be an ephemeral

> "trend" within the context of the current issues in psychiatry. [NA].

> "There is a new thing, it's a trend, it will go as it came, that's it..." [NA].

4.4. MHPs' Representations Could Hide a Lack of Knowledge or Awareness That Would Prevent Them from Appropriating TPE Programmes

4.4.1. TPE Is Not Well Known

Some MHPs justified the opposition to TPE in psychiatry by a lack of knowledge of TPE among psychiatric carers. Others acknowledged that they did not have sufficient knowledge of TPE and had difficulty distinguishing it from psychoeducation, which is more widely practiced.

> "There was nothing to stop me from advocating TPE: I just didn't know about it!" ... / ... ."For me, it wasn't part of the care." [P].

Psychoeducation programmes have been developed for people living with psychiatric disorders. They are defined as systematic didactic and psychotherapeutic interventions that aim to inform patients and their relatives about the psychiatric disorder and promote coping skills. Beyond the transmission of information, psychoeducation is a pedagogical method with the aims of identity clarification and empowerment [10]. The aim of TPE is to help patients acquire or maintain the skills they need to manage their lives with a chronic illness to the best of their ability, with the aim of acquiring and maintaining self-care skills and mobilising or acquiring coping skills based on the patient's previous experience. The distinction between psychoeducation and TPE remains blurred.

> "I talk about TPE, and in fact it's systematic: every time, when people manage to understand a little bit what TPE is, they say, 'But in fact, we do it every day!' Well, yes! Except that the representation that people have of TPE is not the right one! It's not just about giving information to patients, to make them overly responsible and then let them manage." [P].

> "The training in TPE proved to me that people were like me, ignorant. Because in fact, we do a workshop... without a goal, without an objective, without a care project with psychoeducation! Whereas in TPE, there is a care project. We know what it should lead to and we know where the patient is at." [NA].

Our study revealed a paradoxical situation in that although MHPs have been trained to provide TPE, they have difficulty distinguishing it from psychoeducation. It is possible that the quality of the training received needs to be reviewed.

4.4.2. The Highly Formatted Framework of TPE Is Not Adapted to the Problems of Psychiatry

In France, TPE is considered as part of the patient care pathway. It aims to make patients more autonomous by facilitating their adherence to prescribed treatments and improving their quality of life. These programmes must be authorised by the Regional Health Agency (ARS). The conditions of authorisation provide a strict framework for the composition of the educational teams, the skills required to provide TPE and the teams that take the courses.

> "We had our ARS file turned down many times; it was never right. The doctor persevered, we never got a positive response, we ended up giving up, and so each year we tried to modify the programme, to adapt it according to." [NM].

> "So, yes, honestly, it was all Greek to me. For me, it was really constraints that served no purpose; I didn't see the point of submitting a file to the ARS." [NA].

> "Because it was totally beyond me, I didn't really care" ... / ... "there are so many criteria... " ... / ... Doctor XXX, I think she must have made 4 or 5 reminders, files, and at the last one we said, is it really useful to continue, and we said no. So, we are not called a 'Therapeutic Education Group', we are called a 'Psychoeducation Group', we are not recognised, we are not registered in the official booklet of existing groups that we will see on the ARS." [NA].

The conditions required for authorisation are poorly adapted to the functioning of psychiatric professionals, and have discouraged a number of initiatives. Although mental disorders are chronic illnesses, the TPE framework seems more suited to the management of a physical disorder because it was originally designed on a regulatory basis within that framework. Adaptations seem necessary.

4.4.3. Mistaken Representations of TPE

Some caregivers reported that there was a form of refusal of the principle of TPE for some MHPs. This negative outlook was sometimes linked to a generational effect, explaining divergent representations of TPE.

> "I think that doctors, even perhaps doctors or even nurses, like to control, to know everything, and not to allow too much autonomy, but, well..." [NA and SW].

> "There was one major obstacle: a general practitioner who was totally opposed to all this.../.... Radically. So, it was impossible, even for things that are very common, namely, insulin injections, self-injections of insulin, it was impossible! You couldn't... and that's how it was!" [NA].

> "Afterwards, it is perhaps also a question of generation but also undoubtedly of personality..." [P and Pharmacist].

4.4.4. Lack of TPE Training in Initial/Continuing Education

Insufficient training in TPE, whether in initial and/or continuing education, has been identified as a major barrier to the dissemination of TPE knowledge and culture among psychiatric workers.

> "So no, it was not at all a transversal thing in my training...".../..."Yes, it was something optional, but it was optional, or rather over-optional." [NA].

> "It changed for the nurses, it was a portfolio to fill in with boxes, crosses and things, there were no more notes, so I was in my second year, so we saw, vaguely, therapeutic education in class, it must have been on the syllabus but really minor, eh: 2 h." [NA].

5. Discussion

While the analysis of the various interviews we conducted with MHPs revealed many barriers, it also highlighted the means of improving the implementation of TPE for people with severe mental illnesses such as schizophrenia.

The conflict of values that we have brought to light shows that paternalism seems to be more widespread in psychiatry than what is generally assumed. There is indeed a long tradition of paternalism in psychiatry, even though patient rights have been considerably more in focus in recent decades [31]. From the point of view of the interviewees, patients with psychiatric disorders simply do not seem to be able to develop a sufficient level of self-control. It is on the basis of these arguments that the paternalistic approach would be justified in psychiatry. In this context, psychiatric staff members should be aware of their responsibility and not exploit dependency by making offers that would coerce or manipulate the patient's free will [32,33]. One of the reasons cited for not moving towards more patient autonomy is the continuing reduction in the number of psychiatric beds [34]. This means that inpatients tend to be in more serious condition than in the past, and the proportion of those who are treated by restraint has increased.

One way to move towards a less paternalistic model could be to see psychiatry as a value-based practice [35] that would aim to (1) always start from the patient's perspective but also seek a balance between legitimately different perspectives and (2) ensure that communication skills play an important role in clinical decision making. Although it is not always possible to move towards this mode of decision making in psychiatry, staff should at all times try to establish an open dialogue in order to reach a compromise that is acceptable to the patient and sufficiently appropriate from a professional point of view. Only after such a strategy has failed should staff consider paternalistic decision making [36]. In this sense, TPEs could play an interesting role in the paradigm shifts.

There also appears to be a lack of knowledge about TPE. In psychiatry, the educational dimension has been promoted for some years in effective, internationally recognised experiments, which are essentially based on the notion of empowerment and recovery [37]. The American Psychological Association states that sustainable recovery (from mental illness) requires treatment that is comprehensive, coordinated, consistent, competent, empathetic and person-centred [38]. Therapeutic education offered in an inappropriate manner may be ineffective, if not counterproductive [39]. Despite the obstacles, psychiatric carers trained in therapeutic education validate the fact that it is efficient in care and beneficial to their practice; psychiatrists are thus increasingly taking an educational approach [4].

TPE has shown its value in developing positive attitudes towards mental illness and mentally ill people and has led to improvements in self-reported health [39]. However, in France, Cadiot et al. [4] showed that very few structured programs were underway, and that most dealt with bipolar disorders and schizophrenia, i.e., disorders for which the effectiveness of patient education has been demonstrated. The French National Institute for Prevention and Health Education (INPES) estimates that only 50% of initial training facilities for health professionals offer specific teaching in TPE [40]. Limited access to training due to a lack of specific financial resources and overcrowded curricula is also a reason given by professionals [41], even though the legislation provides, for example, in the decree of 31 July 2009, for 150 h of training on preventive care for students of the State Nursing Diploma [42].

It would seem that TPE care is viewed as creating additional work for caregivers, causing a fear of insufficient return on investment. However, TPE should be a priority in the care plan since an early return home has become one of the key aims for supporting recovery from mental illness [43]. Mental health professionals have a scientific, ethical and moral responsibility to guide the social, political and health care organisations involved in the process of meeting mental health needs.

6. Recommendations and Limitations

This study has some limitations inherent to any qualitative study, such as possible self-selection bias or potential response bias. Due to the COVID-19 pandemic, some interviews were conducted by video conference (n = 2) or by telephone (n = 5). Telephone calls are impersonal compared to a physical interview and do not allow for reading, for example, body language or facial expressions. Although video conferencing is an interesting alternative to the telephone, it requires a certain amount of time to adapt to the tool so that the interviewer feels at ease, as does the interviewee.

Another limitation is that the views of those involved in psycho-education were not formally explored. In this context, it is difficult to say whether the concerns of the participants in the study are specific to TPE or related to general frustrations with the development of educational programmes in their institutions. This grey area of this study will be explored in the future.

We can nevertheless highlight some of the elements identified herein that could foster the ability of mental health professionals to deliver TPE to people with psychiatric disorders.

1. Continue the fundamental work on the recognition of the rights of mental health users.
2. Continue efforts to destigmatise mental health, including among health professionals, and leave more place for the patient's voice.
3. Institutions must fully commit to accompanying the implementation of TPEs to achieve this, and it would be useful to organise a shared activity between professionals in order to create a dynamic for the implementation of TPE in psychiatry.
4. Organisational or even financial support could be provided through shared resources between institutions, which could be organised through a periodical magazine, events, conferences or associations.
5. Training for the provision of TPE should be fully integrated into initial training curricula and included in continuing education.
6. Mental health nurses have a key role to play in TPE insofar as it is part of their skill set. In many European countries, important steps have been taken in recent years to recognise and develop the role of nurses in health care teams, particularly with regard to prevention, clinical monitoring and the education of chronic patients.
7. The focus should be on the good practices needed to ensure equal access to care and on how to provide treatment at costs that are sustainable for health systems. In this case, ad hoc payments would promote the use of TPE.
8. The framework for building TPE programmes and reporting to the ARS should be simplified so they can be easily recorded in the establishment's care project.

7. Conclusions

Although TPE is of interest in the process of patient empowerment, we found that caregivers were reluctant to appropriate this approach to care. Efforts must be made in the initial and ongoing training of MHPs to move from a paternalistic model to a patient partnership model, which is made possible by TPE. These efforts must also be firmly supported by health care facilities, and proactive governance is required for the successful implementation of TPE.

Author Contributions: Conceptualization, C.R. and N.M.-B. methodology, N.M.-B. and F.D.; software, N.M.-B.; validation, C.R., N.M.-B. and F.D., formal analysis, C.R. and N.M.-B., investigation, C.R., N.M.-B., S.M. and F.D., resources, F.D. and C.R.; data curation, C.R. and N.M.-B. writing—original draft preparation, C.R., N.M.-B. and F.D.; writing—review and editing, C.R., N.M.-B., S.M. and F.D., visualization, F.D.; supervision, F.D.; project administration, F.D.; funding acquisition, C.R. All authors have read and agreed to the published version of the manuscript.

Funding: This research was funded by the French Ministry of Solidarity and Health, through a call for projects. The funders had no role in the study design, data collection analysis, decision to publish, or preparation of the manuscript.

Institutional Review Board Statement: After participants were provided with information about the study, their oral consent was collected and they were included in the study in accordance with the "Jardé" law (Decree no 2016-1537 dated 16 November 2016 published on 17 November 2016 in the Official Journal of the French Republic). No regulatory approval was required for this study.

Informed Consent Statement: Not applicable.

Data Availability Statement: Data are fully available and will be shared upon request to C.R.

Acknowledgments: The authors would like to thank the staff from the La Chartreuse Hospital Centre of Dijon for administrative and technical support. We thank all the participants in the study. We also thank Jessica Massenot and Suzanne Rankin for proofreading the text.

Conflicts of Interest: The authors declare no conflict of interest.

Abbreviations

World Health Organization (WHO), Therapeutic Patient Education (TPE), Haute Autorité de Santé (HAS), Mental Health Professionals (MHPs), Hôpital Patients Santé et Territoire" (HPST).

References

1. Jablensky, A. Epidemiology of schizophrenia: The global burden of disease and disability. *Eur. Arch. Psychiatry Clin. Neurosci.* **2000**, *250*, 274–285. [CrossRef] [PubMed]
2. Épidémiologie des Troubles Schizophréniques. | Documentary Base | BDSP [Internet]. Available online: http://www.bdsp.ehesp.fr/Base/348611/ (accessed on 30 March 2022).
3. Hautecouverture, S.; Limosin, F.; Rouillon, F. Epidémiologie des troubles schizophréniques. *Presse Med.* **2006**, *35*, 461–468. (In French) [CrossRef]
4. Cadiot, F.; Verdoux, H. Pratiques d'éducation thérapeutique en psychiatrie. Enquête auprès des psychiatres hospitaliers d'Aquitaine [Patient therapeutic education in psychiatry. A survey of practice in Aquitaine public hospitals]. *Encephale* **2013**, *39*, 205–211. (In French) [CrossRef] [PubMed]
5. Bonsack, C.; Rexhaj, S.; Favrod, J. Psychoeducation: Definition, history, interest and limits. *Ann. Médico-Psychol.* **2015**, *173*, 79–84. [CrossRef]
6. Lucksted, A.; McFarlane, W.; Downing, D.; Dixon, L. Recent developments in family psychoeducation as an evidence-based practice. *J. Marital Fam. Ther.* **2012**, *38*, 101–121. [CrossRef]
7. Magliano, L.; Fiorillo, A.; Fadden, G.; Gair, F.; Economou, M.; Kallert, T.; Schellong, J.; Xavier, M.; Pereira, M.G.; Gonzales, F.T.; et al. Effectiveness of a psychoeducational intervention for families of patients with schizophrenia: Preliminary results of a study funded by the European Commission. *World Psychiatry* **2005**, *4*, 45–49.
8. Maurel, M.; Drai, D.; Kaladjian, A.; Pauly, V.; Azorin, J.-M. Étude contrôlée randomisée des effets de la pharmaco-éducation sur une population française de patients souffrant de troubles schizophréniques et schizo-affectifs. *Can. J. Psychiatry* **2010**, *55*, 329–337. [CrossRef]
9. Haute Autorité de Santé (HAS). Guide—Affection de Longue Durée. *Schizophrénies*. 2007. Available online: www.hassante.fr/portail/upload/docs/application/pdf/guide_ald23 (accessed on 17 October 2018).
10. Loi Hôpital Patients Santé et Territoire du 21 Juillet 2009. Available online: http://solidaritessante.gouv.fr/IMG/pdf/vademecum_loi_HPST.pdf (accessed on 10 June 2022).
11. Foucaud Bury, J.A.; Balcou-Debussche, M.; Eymard, C. *Éducation Thérapeutique du Patient. Modèles, Pratiques et Evaluation*; Inpes, coll. Santé en action: Saint-Denis, France, 2010.
12. Decree N° 2010-2906 du 02 August 2010. Available online: https://www.legifrance.gouv.fr/eli/decret/2010/8/2 (accessed on 15 June 2022).
13. Ministère de la santé et des Sports. *Décret n° 2010-904 du 2 août 2010 Relatif aux Conditions D'autorisation des Programmes d'éducation Thérapeutique du Patient*; Journal Officiel n°0178 du 04 août 2010, Paris; Ministère de la santé et des Sports: Paris, France, 2010.
14. WHO-EUROPE Report, Published in 1996, Therapeutic Patient Education—Continuing Education Programme for Health Care Providers in the Field of Chronic Disease. Available online: https://www.euro.who.int/__data/assets/pdf_file/0007/145294/E63674.pdf (accessed on 15 June 2022).
15. Hodé, Y. Psychoeducation of patients and their family members during episode psychosis. *Encephale* **2013**, *39* (Suppl. S2), S110–S114. (In French) [CrossRef]
16. Programme d'Education Thérapeutique G.A.P.E. Available online: http://www.schizo.oui.com/non.classe/programme-deducation-therapeutique-g-a-p-e (accessed on 10 June 2022).
17. Rennick-Egglestone, S.; Ramsay, A.; McGranahan, R.; Llewellyn-Beardsley, J.; Hui, A.; Pollock, K.; Repper, J.; Yeo, C.; Ng, F.; Roe, J.; et al. The impact of mental health recovery narratives on recipients experiencing mental health problems: Qualitative analysis and change model. *PLoS ONE* **2019**, *14*, e0226201. [CrossRef]
18. Saleh, M.O.; Eshah, N.F.; Rayan, A.H. Empowerment Predicting Nurses' Work Motivation and Occupational Mental Health. *SAGE Open Nurs.* **2022**, *23*, 23779608221076811. [CrossRef]
19. Maurel, M. Education and recurrence prevention. *Encephale* **2009**, *35* (Suppl. S1), S20–S23. (In French) [CrossRef]
20. Instruction Relative aux Programmes D'éducation Thérapeutique du Patient. Instruction_DGS-ETP.pdf. Available online: https://solidarites-sante.gouv.fr/IMG/pdf/Rapport_Education_therapeutique_du_patient.pdf (accessed on 16 July 2022).

21. Lang, J.P.; Jurado, N.; Herdt, C.; Sauvanaud, F.; Lalanne Tongio, L. Education care in patients with psychiatric disorders in France: Psychoeducation or therapeutic patient education? *Rev. Epidemiol. Sante Publique* **2019**, *67*, 59–64. [CrossRef] [PubMed]
22. Hempler, N.F.; Pals, R.A.S.; Pedersbæk, L.; DeCosta, P. Barriers and facilitators of effective health education targeting people with mental illness: A theory-based ethnographic study. *BMC Psychiatry* **2018**, *18*, 353. [CrossRef] [PubMed]
23. Joly, A. The contribution of therapeutic education in mental health. *Rev. Infirm.* **2021**, *70*, 28–29. (In French) [CrossRef] [PubMed]
24. Viard, D.; Netillard, C.; Cheraitia, E.; Barthod, V.; Choffel, J.M.; Tartary, D.; Sauze, M.; Nouara, A.; Chalmendrier, C.; Tissot, E. Patient education in psychiatry: Representations of caregivers, patients and their families. *Encephale* **2016**, *42*, 4–13. (In French) [CrossRef] [PubMed]
25. Kaggwa, M.M.; Acai, A.; Rukundo, G.Z.; Harms, S.; Ashaba, S. Patients' perspectives on the experience of absconding from a psychiatric hospital: A qualitative study. *BMC Psychiatry.* **2021**, *21*, 371. [CrossRef]
26. Tong, A.; Sainsbury, P.; Craig, J. Consolidated criteria for reporting qualitative research (COREQ): A 32-item checklist for interviews and focus groups. *Int. J. Qual. Health Care* **2007**, *19*, 349–357. [CrossRef]
27. Ritchie, J.; Lewis, J. *Qualitative Research Practice: A Guide for Social Science Students and Researchers*; Sage Publications: London, UK, 2003.
28. Foley, G.; Timonen, V. Using Grounded Theory Method to Capture and Analyze Health Care Experiences. *Health Serv. Res.* **2015**, *50*, 1195–1210. [CrossRef]
29. Saunders, B.; Sim, J.; Kingstone, T.; Baker, S.; Waterfield, J.; Bartlam, B.; Burroughs, H.; Jinks, C. Saturation in qualitative research: Exploring its conceptualization and operationalization. *Qual. Quant.* **2018**, *52*, 1893–1907. [CrossRef]
30. Chun Tie, Y.; Birks, M.; Francis, K. Grounded theory research: A design framework for novice researchers. *SAGE Open Med.* **2019**, *7*, 2050312118822927. [CrossRef]
31. Pelto-Piri, V.; Engström, K.; Engström, I. Paternalism, autonomy and reciprocity: Ethical perspectives in encounters with patients in psychiatric in-patient care. *BMC Med Ethics* **2013**, *14*, 49. [CrossRef] [PubMed]
32. Kopelman, L.M. Moral Problems in Psychiatry. In *Medical Ethics, Jones & Bartlett, Sudbury/Mississauga/London 2*; Veatch, R., Ed.; Jones and Bartlett Publishers International: London, UK, 1997; pp. 275–320.
33. Matthews, E. Autonomy and the Psychiatric Patient. *J. Appl. Philos.* **2000**, *17*, 59–70. [CrossRef] [PubMed]
34. *The National Board of Health and Welfare: Innehållet i den Psykiatriska Tvångsvården, Article no 2009-126-81*; Socialstyrelsen: Stockholm, Sweden, 2009. (In Swedish)
35. Fulford, K.W.M.; Thornton, T.; Graham, G. *Oxford Textbook of Philosophy Andpsychiatry*; Oxford University Press: Oxford, UK, 2007.
36. Sandman, L.; Munthe, C. Shared decision-making and patient autonomy. *Theor. Med. Bioeth.* **2009**, *30*, 289–310. [CrossRef]
37. Greacen, T. User Empowerment: Implications for Training the Mental Health Workforce. In *Briefing Paper for the WHO Regional Office for Europe Meeting Empowerment in Mental Health—Working Together for Leadership*; WHO Regional Office for Europe: Leuven, Belgium; Copenhagen, Demark, 2010.
38. Kopelowicz, A.; Liberman, R.P. Integration of care: Integrating treatment with rehabilitation for persons with major mental illnesses. *Psychiatr. Serv.* **2003**, *54*, 1491–1498. [CrossRef] [PubMed]
39. Le Rhun, A.; Gagnayre, R.; Moret, L.; Lombrail, P. Analyse des tensions perçues par les soignants hospitaliers dans la pratique de l'éducation thérapeutique: Implications pour leur supervision. *Glob. Health Promot.* **2013**, *20*, 43–47. [CrossRef] [PubMed]
40. L'Inpes, Acteur et Soutien du Développement de L'éducation Thérapeutique du Patient. Saint-Denis, INPES 2009. Available online: http://inpes.santepubliquefrance.fr/70000/dp/09/dp090130.pdf (accessed on 17 October 2018).
41. Tomaras, V.D.; Ginieri-Coccossis, M.; Vassiliadou, M.; Malliori, M.; Ferentinos, S.; Soldatos, C.R.; Tylee, A. Education in mental health promotion and its impact on the participants' attitudes and perceived mental health. *Ann. Gen Psychiatry* **2011**, *10*, 33. [CrossRef] [PubMed]
42. Académie Nationale de Médecine. Résumé du Rapport: L'éducation Thérapeutique du Patient (ETP), une Pièce Maîtresse pour Répondre aux Nouveaux Besoins de la Médecine. 2013. Available online: https://www.academie-medecine.fr/leducation-therapeutique-du-patient-etp-une-piece-maitresse-pour-repondre-aux-nouveaux-besoins-de-la-medecine/ (accessed on 15 June 2022).
43. Wood, L.; Williams, C.; Billings, J.; Johnson, S. The therapeutic needs of psychiatric in-patients with psychosis: A qualitative exploration of patient and staff perspectives. *BJPsych Open* **2019**, *5*, e45. [CrossRef]

healthcare

MDPI

Article

Anxiety/Depression Predominance in Liaison-Psychiatry Users of a South-East Mexico Tertiary Hospital

Lizzette Gómez-de-Regil [1,*], Damaris F. Estrella-Castillo [2] and Miguel Cicero-Ancona [1]

1 Hospital Regional de Alta Especialidad de la Península de Yucatán, Mérida 97130, Mexico; hraeyucatan@gmail.com

2 Faculty of Medicine, School of Rehabilitation, Universidad Autónoma de Yucatán, Mérida 97000, Mexico; ecastill@correo.uady.mx

* Correspondence: gomezderegil@gmail.com

Abstract: Patients at tertiary hospitals may find themselves in need of mental health support due to the distress associated with the illness that may or not lead to a psychiatric condition. Here is an overview of the clinical cases treated by the liaison psychiatry service of a public tertiary hospital from Southeast Mexico during its first years of operation (2008–2018), with the purpose of gathering information about the status and needs of this population. A sample of 304 clinical records of patients treated for the first time by the psychiatry service was reviewed, and the distribution by demographic characteristics, diagnosis of mental illness and medical area of reference was analyzed. Anxiety and depression symptoms were the most frequent. Most patients were women, lived in Merida and returned after the first appointment. The neurology service referred most patients, yet most attended directly. General tertiary hospitals should prioritize integrating ad hoc mental and physical health care. Adult women with a profile of anxiety and/or depression would be the first target group. Some areas of opportunity for further research and improvement of mental health services are: preventive services for anxiety and depression, follow-up of patients, attention to relatives of patients at intensive care units, implementation of telehealth alternatives, training on mental health screening and inter- and intra-institutional collaboration.

Keywords: anxiety; depression; psychiatric department; hospital; Mexico

Citation: Gómez-de-Regil, L.; Estrella-Castillo, D.F.; Cicero-Ancona, M. Anxiety/Depression Predominance in Liaison-Psychiatry Users of a South-East Mexico Tertiary Hospital. *Healthcare* **2022**, *10*, 1162. https://doi.org/10.3390/healthcare10071162

Academic Editors: Yasuhiro Kotera and Elaina Taylor

Received: 12 May 2022
Accepted: 2 June 2022
Published: 22 June 2022

1. Introduction

Tertiary hospitals provide services to patients whose health conditions are severe and/or chronic and, thus, require highly specialized care. These patients may find themselves in need of mental health support due to the distress associated with the illness that may or not lead to a psychiatric disorder. Although the strong link between stress and illness is widely acknowledged, the underlying mechanisms are complex [1]. Chronic stress may increase vulnerability to illness [2,3] by suppressing the immune system, releasing histamine and/or altering insulin needs [4,5]. Conversely, when facing serious illness, the patients [6,7], and even their relatives [8–10], may feel stressed not only about the medical prognosis per se but due to their emotional representations and perceptions of possible adverse consequences in various life dimensions (e.g., work activity, household economy, independence for functioning and long-term treatment) [11].

Improvements in medical diagnoses and treatments have increased the survival rates from illnesses that, in the past, were terminal, yet the upcoming years may come with chronic conditions that require ongoing medical resources and/or limit daily functioning. The current definition of health goes beyond merely the absence of disease or infirmity but comprises a state of complete physical, mental and social well-being [12]. Given that "there is no health without mental health" [13], in the last decades, the relevance of mental health care has increased, being well-recognized and promoted by the World Health Organization (WHO).

Caring for mental health must be present at all levels, and particularly in tertiary hospitals, where patients often arrive already emotionally exhausted and overwhelmed by illnesses that demand time, energy and economic resources. In these circumstances, patients may experience psychological distress and require mental health care for preventing or ameliorating symptoms of prolonged stress, depression and/or anxiety.

In 2012, at the 65th World Health Assembly, the WHO adopted the Mental Health Action Plan 2013–2020 (last updated at the 74th Assembly in 2021 for 2013–2030), aiming at "promoting mental well-being, preventing mental disorders, providing care, improving recovery, promoting human rights and reducing mortality, morbidity and disability of people with mental disorders" [13]. Among its objectives, the Action Plan calls for strengthening information systems, scientific data and research on mental health. Governments have a responsibility for the health of their peoples, which can be fulfilled only by the provision of adequate health and social measures, planned accordingly to the evidenced needs of their populations. Responding to this demand, this work presents an overview of the clinical cases treated by the liaison psychiatry service of a public tertiary hospital from Southeast Mexico during its first years of operation. The authors had the goal of gaining evidence of the profiles and needs of the users in order to be taken into consideration for the optimal design, implementation and evaluation of mental health services.

2. Materials and Methods

This exploratory and retrospective study was performed at the HRAEPY (Regional Hospital of High Specialty of the Yucatan Peninsula), a public tertiary hospital in Southeast Mexico. Both the Research and the Research Ethics Committees of the HRAEPY approved the protocol. The HRAEPY started functioning in June 2008, and even though it does not have a mental health department, during its first 11 years, up to three active staff psychiatrists were available to provide liaison services for a total of 50,863 new users.

The study targeted patients that had ever attended a psychiatry appointment at the HRAEPY. First, from the Statistics Department, the authors obtained the list of patients (recorded by name) that had received care between 2008 (the hospital's start year) and 2018. The department in charge of the management of clinical files corroborated this list and filtered those cases that were inexistent, duplicated or had passed away. Additionally, it provided the total number of new clinical files opened each year, data to help estimate the rate of users who have ever attended an appointment at the HRAEPY's psychiatric service. Following this, 10 files were randomly selected to corroborate that the required information would be available and, if necessary, make adjustments to the data capture sheet. The required sample size by year was estimated (Table 1), and files were randomly selected. Up to the date of the data collection, medical records were not electronically archived; thus, the clinical files were manually reviewed to gather information regarding: age, sex, dwelling location, date of first and last appointments, mental diagnosis and service referring the patient. Additionally, the total number of appointments by patient from his/her first interview to the end of 2019 was recorded. In order to prevent bias, no recorded information from 2020 and on was considered, given that the year was still ongoing during data collection and the COVID-19 pandemic demanded substantial changes in the regular services provided by the hospital. The data was analyzed with descriptive statistics.

A stratified random sampling with a population size of 1450 (original files of living patients identified by name, file number and whose names are not duplicated); a confidence level of 95% and a 5% margin error required a sample of 304 cases (Figure 1) in a range from 2 to 42 cases by year (Table 1). A total of 374 files were randomly selected in order to complete the required sample size, given that, of the initially selected files, 61 were not found in the archives, and 9 were discarded because they did not have any notes from a psychiatrist.

Table 1. Estimation of the sample size.

	Total New Files (*n*)	Potential Cases * (*n*/%)	Distribution (%)	Required Sample (*n*)
2008	455	11 (2.418)	0.758	2
2009	4253	81 (1.905)	5.582	17
2010	5487	152 (2.770)	10.476	32
2011	6401	199 (3.109)	13.784	42
2012	5619	170 (3.025)	11.716	36
2013	3964	136 (3.431)	9.373	29
2014	6400	190 (2.969)	13.094	40
2015	5209	126 (2.419)	8.684	26
2016	4686	149 (3.180)	10.269	31
2017	4432	130 (2.933)	8.959	27
2018	3957	106 (2.679)	7.305	22
Total	50,863	1450 (2.851)	100	304

* Original files of patients identified by name, file number and whose names are not duplicated and are not registered as deceased. Percentage by total new cases each year.

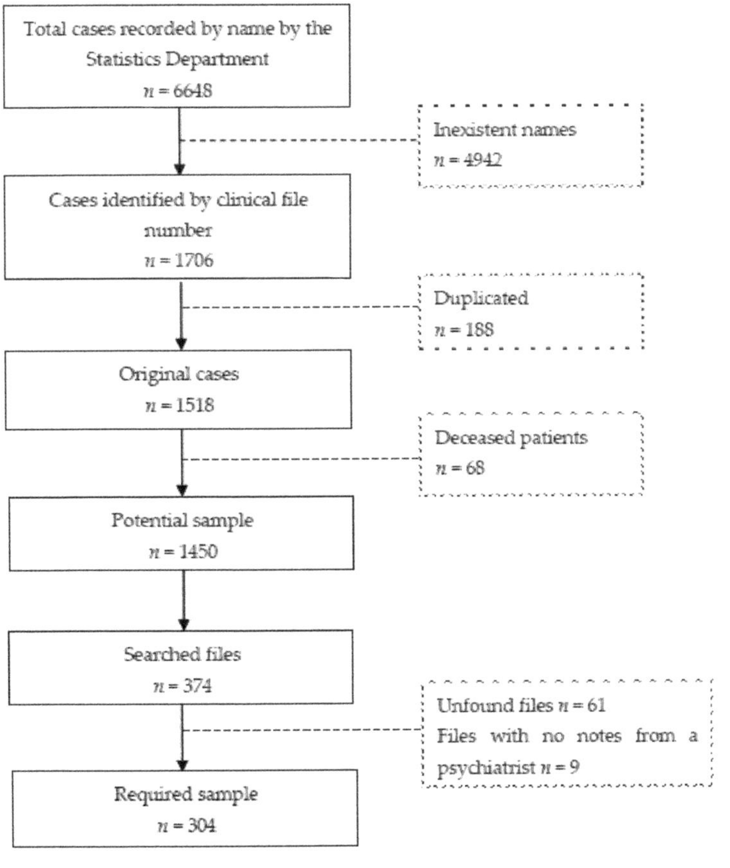

Figure 1. Sample flow diagram.

3. Results

Table 2 presents the descriptive statistics of the samples. A mean of 2.9 (1.9–3.4%) of the new cases each year attended the psychiatry service at least once. It must be remembered that not all 50,863 files ever opened at the HRAEPY were available during data collection; some might have been lost and some other currently found in the dead archives or being used by a service at the moment of request.

Table 2. Descriptive statistics.

Distribution by Year											
Year of file	2008	2009	2010	2011	2012	2013	2014	2015	2016	2017	2018
New files (*n* = 50,863) *n*	455	4253	5487	6401	5619	3964	6400	5209	4686	4432	3957
Potential cases * (*n* = 1450) *n*	11	81	152	199	170	136	190	126	149	130	106
% over total yearly new cases	(2.4)	(1.9)	(2.8)	(3.1)	(3.0)	(3.4)	(3.0)	(2.4)	(3.2)	(2.9)	(2.7)
Final sample (*n* = 304) *n*	2	17	32	42	36	29	40	26	31	27	22
Time (months) from first to last contact (*n* = 304)											
Minimum	33.0	0.0	0.0	0.0	0.0	0.0	0.0	0.0	0.0	0.0	0.0
Maximum	93.1	120.4	116.3	108.1	73.9	79.7	67.5	53.1	41.5	35.3	19.7
Mean	63.1	47.5	31.6	27.6	18.7	15	11.7	17.4	12	11.1	4.9
Standard Deviation	42.5	41.0	35.4	36.2	20.6	22.5	19.3	16.2	13.3	10.7	6.2
Total of appointments (*n* = 304)											
Minimum	21	1	1	1	1	1	1	1	1	1	1
Maximum	36	55	36	50	25	33	25	32	25	17	8
Mean	28.5	12.2	9.8	9.2	6.9	6.4	5.1	8.2	6.3	5.5	3.6
Standard Deviation	10.6	13.2	10.2	12.2	6	8.1	6.3	7.4	6.9	4.3	2.4

Characteristics of sample (*n* = 304)					
		Diagnosis (*n*/%)			
Mixedanxiety/depression	81 (26.6)	Surgery protocol	25 (8.2)	Psychosis	10 (3.3)
Anxiety	79 (26.0)	No diagnosis/symptoms	16 (5.3)	Substance abuse	6 (2.0)
Depression	55 (18.1)	Dementia	12 (3.9)	Other	20 (6.6)
		Referring service (*n*/%)			
Psychiatry	90 (29.6)	Internal medicine	30 (9.9)	Transplant protocol	10 (3.3)
Neurology	64 (21.1)	Emergency admission	20 (6.6)	Gastroenterology	10 (3.3)
Surgery	56 (18.4)	Bariatric surgery	17 (5.6)	Other	7 (2.3)
Sex (*n*/%)		Age (years at the time of first interview)		Dwelling location (*n*/%)	
Female	219 (72.0)	Minimum	17.3	Yucatan	268 (88.2)
Male	85 (28.0)	Maximum	92.1	Quintana Roo Campeche	19 (6.3)
		Mean	48	Other	12 (3.9)
		Standard Deviation	16.8		5 (1.6)

* Original files of patients identified by name, file number and whose names are not duplicated and are not registered as deceased. Percentage by total new cases each year.

Out of the final sample (n = 304), most patients (70.7%) reported symptoms related to depression and/or anxiety disorders. In the case of the people interviewed as part of a surgery protocol, 84.0% were declared suitable. For the remaining cases, 16 files did not have any reference to a psychiatric symptom or disorder. Most of the cases (29.6%) attended the psychiatry service as their first appointment, whereas most medical referrals came from the neurology service (21.1%). As part of their protocols, the bariatric and the transplant surgery services referred 6.6% and 5.6% cases, respectively.

Most patients were females, and the mean age at the time of their first appointment was 48.0 years, with a range from 17.3 to 92.1 and no significant differences by sex (p = 0.25). Most patients lived in the State of Yucatan (88.2%), either in the city of Merida (40.5%), where the hospital is located, or in adjacent locations (11.5%).

Regarding the time from first to last appointment with the psychiatry service, the average was 19.4 (SD = 26.7) months. When excluding the cases that attended only once (21.1%), the average was 24.6 (SD = 27.9) months, ranging from 2 days to 120 months. During their first year, 36.5% of the patients returned but dropped out, 13.8% during the second year and 11.5% during the third year, and only 17.1% returned after that time. Yet, it could be the case that a patient interrupted treatment for some time and later retuned. Most patients (78.9%) returned after their first appointment, with 38.8% having between 2 and 5 visits, 17.8% between 6 and 10 visits and 22.8% between 11 and up to 55 visits. The average number of visits was 7.4 (SD = 8.6), ranging from 1 to 55. Here, it must be considered that the older the file, the more chances the case added follow-up appointments. Moreover, it is very likely that the physical files of those patients that had not returned for some time had already been removed from the clinical archive at some depuration point.

4. Discussion

An important number of people with long-term physical health conditions also have mental health problems; this comorbidity can severely impact health outcomes and reduce the quality of life [1,14]. For instance, depression is twice as common in people with diabetes relative to the general population, and in the other direction, comorbid depression in diabetes has an additive negative effect on patients' mental quality of life and is associated with an increased risk of debilitating complications, further increasing the disease burden [15]. It is estimated that up to 18% of the amount expended on long-term conditions is in relation to deteriorated mental health. By interacting with and exacerbating physical illnesses, comorbid mental health problems may raise the total health care costs by at least 45%. Innovative forms of liaison psychiatry demonstrate that providing better support for comorbid mental health needs can reduce physical health care costs [1].

At tertiary hospitals, most patients arrive by medical reference and dealt with their illnesses for some time. It is not unusual that they experience burden and despair that may lead to the manifestation of incipient or full-blown mental disorders. Learning about the main features and needs of patients is a mandatory step towards the design, implementation and validation of strategic health programs.

At the HRAEPY, most patients attending its psychiatric service are women in their late-middle adulthood years. This may reflect that females are not only more prone to present mental/emotional symptoms [16–18] but also to seek treatment earlier [17–19]. It is important to provide early intervention in mental health to these users, not only to prevent a mental disorder and the associated economic costs of a comorbidity but also considering that, in Mexican culture, women, at home full-time or not, are often the family's primary caregivers, so their emotional well-being would have a significant effect on their family. Lara and colleagues, considering social and gender factors, developed psycho-educational interventions for women with depressive symptoms [20] to prevent women from developing postpartum depression [21] and the Help for Depression (HDep) program [22,23], one of the few unguided web-based interventions available in Latin America, all with promising results. Regarding age, it is important to notice that the HRAEPY does not provide pediatric care (except for the Pediatric Cardiac Intensive Care

Unit), so all patients attending the psychiatric service are adults. However, the HRAEPY must consider that every Pediatric Intensive Care Unit should be able to address not only the physical but also the psychosocial, emotional and spiritual needs of patients with life-threatening conditions and their immediate families [24].

Most patients of the psychiatric service presented symptoms of anxiety and/or depression. These mental disorders are not only the most prevalent in patients with chronic illnesses [6,25–27] but also in the general population [16,28]. It must be underlined that presence of "symptoms of" does not equal the "clinical diagnoses of". Patients who experience mild symptoms of anxiety and/or depression (an expected response when facing a severe, chronic and/or disabling physical illness) may better benefit from preventive services (e.g., counseling, group therapy, psycho-education and support groups) to help them manage their cognitions and emotions and increase adherence to treatment [2,29,30]. This highlights the importance of expanding efforts to enable society to identify anxiety and depressive symptoms as obstacles to well-being and productivity and to accept them as challenges that can and must be overcome with the decisive and organized participation of everyone.

Most patients returned after their first appointment; yet, it must be considered that the ones that had not returned for a long time might have already been taken out of the archive. The patient might have dropped off treatment due to the amelioration of symptoms or receiving care somewhere else. However, on the contrary, it might be the case that the patient did not continue under treatment due to limitations in accessing care (e.g., economic and geographical), a lack of insight regarding mental illness or even death. Research on health trajectories [31–33] comes forward as an ideal model to monitor the routes of care in mental health and chronic diseases, learn about the likely limitations for access and improve health care provisions.

With most patients living in the hospital's city, it would be feasible to provide in-person individual and/or psychoeducational group interventions to develop self-management of health. These interventions must be designed accordingly to the target group, taking into consideration not only clinical features but also psychological and social factors that may intervene [34]. In Mexican culture, there is a marked influence of a male-dominated society and the stigmatized concept of mental disorders. Within the indigenous Mayan population (mostly located at the Yucatan Peninsula), the traditional healers provide a remedy for a wide range of mental health problems, so it is important to reach a compromise that include the individual's cosmovision (worldview) [35]. In addition to this, Yucatan is positioned in the first places in alcoholism nationwide, where the rates of suicides and alcohol intoxications are among the leading causes of mortality [36]. Patients may not report the use of alcohol and substances, either from shame and/or as it is normalized in the community and not even seen as a risk for physical and mental health.

Patients with limitations to attend in-person appointments (e.g., due to distanced dwelling, mobility impairments and/or limited schedule) may benefit from alternative provision of services (not yet available at the HRAEPY). Telehealth allows the delivery of universal health coverage through technology in order to provide quality and cost-effective health services and eliminate the need for either the patient or practitioner to travel to appointments, offering potential help in reducing the 'mental health gap' in low- and middle-income countries [37,38]. Although remote therapy offers a number of advantages, it brings about a variety of challenges that are unique to this modality. For example, pre-recorded video modules are less interactive with the patient, and teleconferences in real time require trained staff for technological support. Partial or total contents of online interventions in mental health may be accessed 24/7, and research has shown favorable results in the clinical [37,39] and economic [40] outcomes. Yet, this option is not always suitable. From the age, education and socioeconomic levels of most people attended to at the HRAEPY, it can be assumed that they are unlikely to be computer literate and own an electronic device with reliable internet access. Moreover, an important segment of users only speak (but not write) Maya, the local indigenous language rarely known by health professionals. The use of telehealth is expected to increase as the upcoming generations of patients and health professionals are more skilled in the use of technology and internet access spreads, so further research and training will be needed to provide an effective and more widely distributed low-cost mental health approach, a task that the HRAEPY must adopt.

At the HRAEPY, the psychiatric service does not offer hospitalization and functions mainly as a support to patients treated by other medical specialities; thus, intra-institutional referrals are predominant. Despite the benefits, referral to consultation–liaison psychiatry remains low; a thoughtful analysis would help clarify whether this is due to low rates of psychiatric symptoms or to non-detection and/or nonreferrals from other specialists. Education should be provided to hospital doctors but also to nurses, residents and social workers to better recognize mental illness so referrals can see an increase [41,42]. Collaborative screening and accurate referrals would require the consolidation of a multi- and interdisciplinary mental health service [43–45].

From clinical records, it was observed that most of the patients directly sought the psychiatry services; it would be worth exploring the source of this informal referral, how patients acknowledged their need for mental health care, how they reached for treatment at HRAEPY and whether or not they had previously received psychiatric treatment. Among the medical departments at the HRAEPY, the one with the most percentage of patient referrals to psychiatry was the neurology department. Although this study does not count with the specific diagnosis, it does concur with previous findings [43,46,47], where the neurology team stands out as the highest number of referrals to psychiatry, mainly in relation to brain organic disorders, mood and dissociative disorders.

At the HRAEPY, a psychosocial evaluation is mandatory for all patients before transplant or bariatric or surgery; yet, it seems to be limited to meeting the requirement, as no follow-up appointments are routinely scheduled. The psychological care for transplant recipients and donors should continue throughout the postoperative period to guarantee treatment adherence and a successful process [48]. In the case of bariatric patients, an evaluation is recommended to identify potential contraindications to surgical intervention or barriers for lifestyle changes [49]. The published recommendations point out that the optimal frequency for follow-up visits for bariatric patients are seven or more visits per year [50].

The data collection was limited by the COVID-19 pandemic, where all recorded information (if any) regarding the psychiatric service from 2020 and on was not included to prevent bias. Populations were exposed to prolonged uncertainty and lockdown, and this might have had a significant effect on mental health, with an increment in rates of symptoms of depression and/or anxiety and an exacerbation of illness in psychiatric patients. In England, it was estimated that up to 30% of people in the general population experienced trajectories with symptoms in the clinical range during the lockdown; yet, the trajectory patterns were diverse [33]. Depressive and anxiety disorders were the leading causes of the global health-related burden in the years prior to the pandemic, but the social restrictions created an environment where psychological distress and mental illness were likely exacerbated. The COVID-19 pandemic has created an increased urgency to strengthen mental health systems, demanding the early detection and management of mental and emotional health issues for managing the global prevalence and burden of depressive and anxiety disorders in the upcoming populations of patients [51].

This study follows a simple design; however, the value of this research relies, above all, on the utility of the analysis and the feasibility of replicating the method in other hospitals from the same hospital network. At the moment, across Mexico, there are five HRAEs (Regional Hospitals of High Specialization). Apart from being tertiary hospitals dependent on the Federal Government, they also have in common the task of performing up-to-date research in benefit of their populations, and this study would be easy to replicate across institutions alike, in Mexico and/or abroad.

Given the inner conditions for research at the site, the study followed a descriptive/observational design. The results came from information available in nonelectronic files, which was not only scarce but, in many cases, missing. This natural nonmanipulated condition, evidencing the usual functioning of the services, implies that a deeper analysis could not proceed without questioning the validity and reliability of data for a generalization of the findings. Further projects must consider the standardized inclusion of other relevant variables for more precise characterizations of users. For instance, coded diagnoses (CIE and/or DSM) would help establish whether symptoms are present at the subclinical or clinical level and/or in comorbidity with a physical illness, which increases the economic [52,53] and psychological costs [1,14]. Additionally, they would reflect more accurately the evolution of the patient's health status.

Lastly, the implementation of electronic files [54–56] would standardize the recording of relevant information for follow-up (e.g., date, updated diagnosis and prescription) from all services each patient attends and would also reduce discrepancies between Statistics and Archive departments' information. Given that the descriptive statistics reported in this study were based on only 304 samples, all findings and implications are not conclusive.

5. Conclusions

General tertiary hospitals should prioritize integrating ad hoc mental and physical health care more closely as a fundamental strategy for improving the quality and functionality of their provision of health care. Adult women with a profile of anxiety and/or depression would be the first target group to attend. Some areas of opportunity for further research and improvement of mental health services are: preventive services for anxiety and depression, follow-up of patients, attention to relatives of patients at intensive care units, the implementation of telehealth alternatives, training in mental health screening and inter- and intra-institutional collaborations.

Author Contributions: All authors contributed to the study conception and design, data collection, analysis and interpretation of results and manuscript preparation. All authors have read and agreed to the published version of the manuscript.

Funding: The Hospital Regional de Alta Especialidad de la Península de Yucatán supported this work by covering the Article Processing Charges.

Institutional Review Board Statement: The study was conducted in accordance with the Declaration of Helsinki and approved by the Institutional Review Board and the Research Ethics Committee of the Hospital Regional de Alta Especialidad de la Península de Yucatán (protocol code: 2018–073, date of approval: 15 August 2019).

Informed Consent Statement: Not applicable.

Data Availability Statement: The dataset generated and/or analyzed during the current study is available from the corresponding author upon reasonable request.

Conflicts of Interest: The authors declare no conflict of interest.

References

1. Naylor, C.; Parsonage, M.; McDaid, D.; Knapp, M.; Fossey, M.; Galea, A. *Long Term Conditions and Mental Health: The Cost of Comorbidities*; The King's Fund and Centre for Mental Health: London, UK, 2012.
2. Karademas, E.C.; Tsalikou, C.; Tallarou, M.-C. The impact of emotion regulation and illness-focused coping strategies on the relation of illness-related negative emotions to subjective health. *J. Health Psychol.* **2011**, *16*, 510–519. [CrossRef] [PubMed]
3. Li, A.; Rosella, L.C.; Kurdyak, P.; Wodchis, W.P. Depression as a risk factor for physical illness and multimorbidity in a cohort with no prior comorbidity. *Can. J. Psychiatry* **2021**, *66*, 726–736. [CrossRef] [PubMed]
4. Salleh, M.R. Life event, stress and illness. *Malays. J. Med. Sci.* **2008**, *15*, 9–18. [PubMed]
5. McLachlan, K.J.J.; Gale, C.R. The effects of psychological distress and its interaction with socioeconomic position on risk of developing four chronic diseases. *J. Psychosom. Res.* **2018**, *109*, 79–85. [CrossRef]
6. Brady, A.M.; Deighton, J.; Stansfeld, S. Psychiatric outcomes associated with chronic illness in adolescence: A systematic review. *J. Adolesc.* **2017**, *59*, 112–123. [CrossRef]

7. Cal, S.F.; Ribeiro de Sá, L.; Glustak, M.E.; Santiago, M.B. Resilience in chronic diseases: A systematic review. *Cogent. Psychol.* **2015**, *2*, 1024928. [CrossRef]
8. Hickman, R.L.; Douglas, S.L. Impact of chronic critical illness on the psychological outcomes of family members. *AACN Adv. Crit. Care* **2010**, *21*, 80–91.
9. Knecht, C.; Hellmers, C.; Metzing, S. The perspective of siblings of children with chronic illness. *J. Pediatr. Nurs.* **2015**, *30*, 102–116. [CrossRef]
10. Pitceathly, C.; Maguire, P. The psychological impact of cancer on patients' partners and other key relatives: A review. *Eur. J. Cancer* **2003**, *39*, 1517–1524. [CrossRef]
11. Dempster, M.; Howell, D.; McCorry, N.K. Illness perceptions and coping in physical health conditions: A meta-analysis. *J. Psychosom. Res.* **2015**, *79*, 506–513. [CrossRef]
12. World Health Organization. World Health Organization Constitution. 1948. Available online: https://www.who.int/about/governance/constitution (accessed on 2 March 2022).
13. World Health Organization. Mental Health: Strengthening Our Response. 2013. Available online: https://www.who.int/news-room/fact-sheets/detail/mental-health-strengthening-our-response (accessed on 2 March 2022).
14. Mujica-Mota, R.E.; Roberts, M.; Abel, G.; Elliott, M.; Lyratzopoulos, G.; Roland, M.; Campbell, J. Common patterns of morbidity and multi-morbidity and their impact on health-related quality of life: Evidence from a national survey. *Qual. Life Res.* **2015**, *24*, 909–918. [CrossRef] [PubMed]
15. Moussavi, S.; Chatterji, S.; Verdes, E.; Tandon, A.; Patel, V.; Ustun, B. Depression, chronic diseases, and decrements in health: Results from the world health surveys. *Lancet* **2007**, *370*, 851–858. [CrossRef]
16. World Health Organization. Depression and other common mental disorders. In *Global Health Estimates*; World Health Organization: Geneva, Switzerland, 2017.
17. Simo, B.; Bamvita, J.M.; Caron, J.; Fleury, M.J. Patterns of health care service utilization by individuals with mental health problems: A predictive cluster analysis. *Psychiatr. Q.* **2018**, *89*, 675–690. [CrossRef] [PubMed]
18. Harris, M.G.; Baxter, A.J.; Reavley, N.; Diminic, S.; Pirkis, J.; Whiteford, H.A. Gender-related patterns and determinants of recent help-seeking for past-year affective, anxiety and substance use disorders: Findings from a national epidemiological survey. *Epidemiol. Psychiatr. Sci.* **2016**, *25*, 548–561. [CrossRef]
19. Matheson, F.I.; Smith, K.L.W.; Fazli, G.S.; Moineddin, R.; Dunn, J.R.; Glazier, R.H. Physical health and gender as risk factors for usage of services for mental illness. *J. Epidemiol. Community Health.* **2014**, *68*, 971–978. [CrossRef]
20. Lara, M.A.; Navarro, C.; Navarrete, L.; Liliana Mondragón, R.N. Seguimiento a dos años de una intervención psicoeducativa para mujeres con síntomas de depresión, en servicios de salud para población abierta. *Salud. Ment.* **2003**, *26*, 27–36.
21. Lara, M.A.; Navarro, C.; Navarrete, L. Outcome results of a psycho-educational intervention in pregnancy to prevent PPD: A randomized control trial. *J. Affect Disord.* **2010**, *122*, 109–117. [CrossRef]
22. Lara, M.A.; Tiburcio, M.; Aguilar Abrego, A.; Sánchez-Solís, A. A four-year experience with a web-based self-help intervention for depressive symptoms in Mexico. *Rev. Panam. Salud. Publica* **2014**, *35*, 399–406.
23. Lara, M.A.; Patiño, P.; Tiburcio, M.; Navarrete, L. Satisfaction and acceptability ratings of a web-based self-help intervention for depression: Retrospective cross-sectional study from a resource-limited country. *JMIR Form. Res.* **2022**, *6*, e29566. [CrossRef]
24. Rosenberg, D.I.; Moss, M.M. Guidelines and levels of care for pediatric intensive care units. *Crit Care Med* **2004**, *32*, 2117–2127. [CrossRef]
25. Chapman, D.P.; Perry, G.S.; Strine, T.W. The vital link between chronic disease and depressive disorders. *Prev. Chronic. Dis.* **2005**, *2*, A14. [PubMed]
26. Voinov, B.; Richie, W.D.; Bailey, R.K. Depression and chronic diseases: It is time for a synergistic mental health and primary care approach. *Prim. Care Companion CNS Disord.* **2013**, *15*, 26226. [CrossRef] [PubMed]
27. DeJean, D.; Giacomini, M.; Vanstone, M.; Brundisini, F. Patient experiences of depression and anxiety with chronic disease: A systematic review and qualitative meta-synthesis. *Ont. Health Technol. Assess Ser.* **2013**, *13*, 1–33. [PubMed]
28. Medina-Mora, M.E.; Borges, G.; Lara, M.C.; Benjet, C.; Blanco, J.J.; Fleiz, B.C.; Villatoro, V.J.; Rojas, G.E.; Zambrano, R.J.; Casanova, R.L.; et al. Prevalencia de trastornos mentales y uso de servicios: Resultados de la encuesta nacional de epidemiología psiquiátrica en México. *Salud. Ment.* **2003**, *26*, 1–16.
29. Wierenga, K.L.; Lehto, R.H.; Given, B. Emotion regulation in chronic disease populations: An integrative review. *Res. Theory Nurs. Pract.* **2017**, *31*, 247–271. [CrossRef]
30. Wierenga, K.L.; Fresco, D.M.; Alder, M.; Sattar, A.; Moore, S.M. Preliminary efficacy of an emotion regulation intervention on physical activity and depressive and anxious symptoms in individuals in cardiac rehabilitation. *J. Cardiovasc. Nurs.* **2022**, *37*, 296–305. [CrossRef]
31. Pinaire, J.; Azé, J.; Bringay, S.; Landais, P. Patient healthcare trajectory. An essential monitoring tool: A systematic review. *Health Inf. Sci. Syst.* **2017**, *5*, 1. [CrossRef]
32. Henly, S.J.; Wyman, J.F.; Findorff, M.J. Health and illness over time: The trajectory perspective in nursing science. *Nurs. Res.* **2011**, *60*, S5–S14. [CrossRef]
33. Saunders, R.; Buckman, J.E.J.; Fonagy, P.; Fancourt, D. Understanding different trajectories of mental health across the general population during the COVID-19 pandemic. *Psychol. Med.* **2021**, 1–9. [CrossRef]

34. Lara, M.A.; Navarro, C.; Navarrete, L. Influencia de los sucesos vitales y el apoyo social en una intervención psicoeducativa para mujeres con depresión. *Salud Publica Mex.* **2004**, *46*, 378–387. [CrossRef]
35. Torres López, T.M.; Munguía Cortés, J.A.; Salazar Estrada, J.G.; Aranda Beltrán, C. Concepciones culturales sobre la salud mental de personas que asisten a procesos terapéuticos psicológicos. *Actual Psicol.* **2015**, *29*, 35–46. [CrossRef]
36. Moreno-Cabrera, S.A. Entre "saber y no saber tomar": Representaciones y prácticas de varones y mujeres sobre el consumo de alcohol en Yucatán. *Salud Colect.* **2020**, *16*, e2533. [CrossRef] [PubMed]
37. Arjadi, R.; Nauta, M.H.; Chowdhary, N.; Bockting, C.L.H. A systematic review of online interventions for mental health in low and middle income countries: A neglected field. *Glob. Ment. Health* **2015**, *2*, e12. [CrossRef] [PubMed]
38. World Health Organization. Global diffusion of eHealth: Making universal health coverage achievable. In *Report of the Third Global Survey on eHealth*; World Health Organization: Geneva, Switzerland, 2016.
39. van der Feltz-Cornelis, C.; Allen, S.F.; Holt, R.I.G.; Roberts, R.; Nouwen, A.; Sartorius, N. Treatment for comorbid depressive disorder or subthreshold depression in diabetes mellitus: Systematic review and meta-analysis. *Brain Behav.* **2021**, *11*, e01981. [CrossRef]
40. Mitchell, L.M.; Joshi, U.; Patel, V.; Lu, C.; Naslund, J.A. Economic evaluations of internet-based psychological interventions for anxiety disorders and depression: A systematic review. *J. Affect Disord.* **2021**, *284*, 157–182. [CrossRef] [PubMed]
41. Scantlebury, A.; Parker, A.; Booth, A.; McDaid, C.; Mitchell, N. Implementing mental health training programmes for non-mental health trained professionals: A qualitative synthesis. *PLoS ONE* **2018**, *13*, e0199746. [CrossRef]
42. Cornwall, P.L. Doctors in all specialties need more training in mental health. *BMJ* **2014**, *349*, g7527. [CrossRef]
43. Lipowski, Z.J.; Wolston, E.J. Liaison psychiatry: Referral patterns and their stability over time. *Am. J. Psychiatry* **1981**, *138*, 1608–1611.
44. Chen, K.Y.; Evans, R.; Larkins, S. Why are hospital doctors not referring to consultation-liaison psychiatry?—A systemic review. *BMC Psychiatry* **2016**, *16*, 390. [CrossRef]
45. Onnis, L.A.; Kinchin, I.; Pryce, J.; Ennals, P.; Petrucci, J.; Tsey, K. Evaluating the implementation of a mental health referral service "Connect to Wellbeing": A quality improvement approach. *Front. Public Health* **2020**, *8*, 585933. [CrossRef]
46. Dawood, S.; Poole, N.; Fung, R.; Agrawal, N. Neurologists' detection and recognition of mental disorder in a tertiary in-patient neurological unit. *BJPsych Bull.* **2018**, *42*, 19–23. [CrossRef] [PubMed]
47. Gurram, S.; Macharapu, R.; Kumar Mv Reddy, P.K. A study of psychiatric referrals from other specialties at a tertiary care center. *Arch. Ment. Health* **2018**, *19*, 129–135.
48. Rodríguez-Rodríguez, T.; Fonseca-Fernández, M. Guide for psychological care of patients and families in the organ donation and transplantation process. *Medisur* **2015**, *13*, 560–568.
49. Consejo de Salubridad General. *Tratamiento Quirúrgico del Paciente con Obesidad Mórbida*; Consejo de Salubridad General: México City, México, 2009.
50. Mechanick, J.I.; Youdim, A.; Jones, D.B.; Garvey, W.T.; Hurley, D.; McMahon, M.M.; Heinberg, L.J.; Kushner, R.; Adams, T.D.; Shikora, S.; et al. Clinical practice guidelines for the perioperative nutrition, metabolic, and nonsurgical support of patients undergoing bariatric procedures—2019 update. *Surg. Obes. Relat. Dis.* **2013**, *9*, 159–191. [CrossRef] [PubMed]
51. Santomauro, D.F.; Herrera, A.M.M.; Shadid, J.; Zheng, P.; Ashbaugh, C.; Pigott, D.M.; Cristiana, A.; Christopher, A.; Joanne, O.A.; Aleksandr, Y.A.; et al. Global prevalence and burden of depressive and anxiety disorders in 204 countries and territories in 2020 due to the COVID-19 pandemic. *Lancet* **2021**, *398*, 1700–1712. [CrossRef]
52. Sartorius, N. Comorbidity of mental and physical disorders: A key problem for medicine in the 21st century. *Acta Psychiatr. Scand.* **2018**, *137*, 369–370. [CrossRef] [PubMed]
53. Merikangas, K.R.; Ames, M.; Cui, L.; Stang, P.E.; Ustun, T.B.; Von Korff, M.; Kessler, R.C. The impact of comorbidity of mental and physical conditions on role disability in the US adult household population. *Arch. Gen. Psychiatry* **2007**, *64*, 1180–1188. [CrossRef]
54. McGinn, C.A.; Grenier, S.; Duplantie, J.; Shaw, N.; Sicotte, C.; Mathieu, L.; Leduc, Y.; Légaré, F.; Gagnon, M.-P. Comparison of user groups' perspectives of barriers and facilitators to implementing electronic health records: A systematic review. *BMC Med.* **2011**, *9*, 46. [CrossRef]
55. Nguyen, L.; Bellucci, E.; Nguyen, L.T. Electronic health records implementation: An evaluation of information system impact and contingency factors. *Int. J. Med. Inform.* **2014**, *83*, 779–796. [CrossRef]
56. Janssen, A.; Donnelly, C.; Elder, E.; Pathmanathan, N.; Shaw, T. Electronic medical record implementation in tertiary care: Factors influencing adoption of an electronic medical record in a cancer centre. *BMC Health Serv. Res.* **2021**, *21*, 23. [CrossRef]

healthcare

MDPI

Article

Differential Methylation Analysis of Suicidal Ideation Severity in Schizophrenia with the Illumina MethylationEPIC Array

Kevin Z. Wang, Zanib Chaudhary, Jessica Qian, Christopher Adanty, Ariel Graff-Guerrero, Philip Gerretsen, Clement C. Zai and Vincenzo De Luca *

Department of Psychiatry, University of Toronto, Toronto, ON M5T 1R8, Canada; kevinz.wang@mail.utoronto.ca (K.Z.W.); zn857242@dal.ca (Z.C.); jessica.qian@camh.ca (J.Q.); christopher.adanty@camh.ca (C.A.); ariel.graff@camh.ca (A.G.-G.); philip.gerretsen@camh.ca (P.G.); clement.zai@camh.ca (C.C.Z.)

* Correspondence: vincenzo.delucal@camh.ca

Abstract: There is a multitude of factors that makes difficult to identify those at risk for suicide, especially among schizophrenia patients. Suicide cannot be explained by genetics alone, therefore epigenetic mechanisms including DNA methylation are thought to play a role. DNA methylation could be a valuable tool in helping predict those at-risk individuals. This cross-sectional study comprised 112 subjects diagnosed with schizophrenia spectrum disorders, and were grouped according to the current suicidal ideation severity. DNA methylation across the genome was measured with the Infinium® MethylationEPIC BeadChip. We utilized the dmpFinder and bumphunter functions within the Bioconductor minfi package to identify differentially methylated positions (DMPs) and differentially methylated regions (DMRs), respectively. Following quality control, we removed one sample from the analysis and reported the most significant DMPs and DMRs associated with suicidal ideation severity. All positions and regions identified in this analysis were only found to have suggestive levels of significance at the genome-wide level. The present study was one of the first to investigate genome-wide methylation and suicidal ideation severity. While there were many strengths of our study, including investigating both differentially methylated positions and regions, further larger-scale studies are necessary to replicate, support, and validate our findings presented here.

Keywords: DNA methylation; schizophrenia; suicide; Illumina EPIC; C-SSRS

Citation: Wang, K.Z.; Chaudhary, Z.; Qian, J.; Adanty, C.; Graff-Guerrero, A.; Gerretsen, P.; Zai, C.C.; De Luca, V. Differential Methylation Analysis of Suicidal Ideation Severity in Schizophrenia with the Illumina MethylationEPIC Array. *Healthcare* **2022**, *10*, 809. https://doi.org/10.3390/healthcare10050809

Academic Editors: Yasuhiro Kotera and Elaina Taylor

Received: 5 April 2022
Accepted: 22 April 2022
Published: 27 April 2022

1. Introduction

Suicidal behavior has consistently been ranked as one of the leading causes of injury and death worldwide. However, patients with schizophrenia have a tenfold increase in suicide risk compared to that of the general population, with an estimated 25–50% of these individuals making at least one suicide attempt during the course of their lifetime [1,2]. Suicidal ideation greatly increases the risk for later suicide attempts and can serve as an important early warning sign for family, caregivers, and clinicians alike [3,4].

The symptoms of schizophrenia can often complicate the ability to accurately predict patients at high risk for suicide. For instance, positive symptoms such as auditory hallucinations and delusions are known to increase the risk for suicide, yet these same symptoms frequently hinder proper and adequate communication with the patient [5]. Depression, also strongly associated with suicide, can easily be confused with the negative symptoms of schizophrenia, such as apathy and reduced expression, or can even be attributed to the side effects of prescribed antipsychotic regimens [1,6]. In light of these considerations, it becomes necessary to consider other factors to better predict and prevent suicide. The National Alliance for Suicide Prevention proposed the identification of peripheral biomarkers for screening and risk assessment [7]. We, in turn, adopted this approach to predict suicidal ideation specifically in the schizophrenia population.

Family studies have long supported the notion that suicidal behaviors, including both suicide attempts and completion, have a strong genetic component [8]. However, genome-wide association studies (GWAS) have been unable to consistently and reliably identify risk associated with DNA sequence variants alone. As such, in recent years, epigenetic mechanisms have been suggested to play a role in "quantifying the missing heritability" of suicide in schizophrenia [9]. DNA methylation and suicide have been well-studied in relation to the dysfunction of the adaptive stress response by the hypothalamic-pituitary-adrenal (HPA) axis, particularly in the NR3C1 and SKA2 genes [10–14]. Therefore, the evidence suggests that these epigenetic markers can prove valuable in investigating suicide attempts and completed suicides. However, predicting suicidal ideation may very well yield different results from those studies utilizing suicidal behaviors as the outcome variable [15]. As such, it is necessary to replicate these studies in the context of suicidal ideation. The present study will investigate genome-wide methylation status in association with suicidal ideation severity. Based on previous studies investigating DNA methylation and suicide [10–14], we further predict that suicide ideation will be associated with genome-wide methylation.

2. Materials and Methods

2.1. Participants

The present study included 112 individuals with a diagnosis of schizophrenia spectrum disorders, recruited from the Centre for Addiction and Mental Health (CAMH) in Toronto, Canada. Subjects were between the ages of 18 and 75 years and did not have a past history of head trauma with loss of consciousness, nor diagnosis of an intellectual disability, major neurological disorder, or substance-induced psychosis. All participants were prescribed antipsychotics. The study was approved by the CAMH Research Ethics Board. Each subject provided consent to participate. The Columbia- Suicide Severity Rating Scale (C-SSRS) [16] was administered to determine whether subjects were experiencing suicidal ideation at the time of the assessment, and if so, the severity of ideation.

2.2. Sample Collection and DNA Methylation

Venous blood was collected from participants after administering the C-SSRS. Genomic DNA was extracted using the QIAamp® DNA Blood Maxi Kit (QIAGEN Inc., Hilden, Germany). Samples were then sent to The Centre for Applied Genomics (TCAG) at the Hospital for Sick Children for further processing. There, 500 ng of DNA was treated with sodium bisulfite using the EZ DNA Methylation Kit (Zymo Research, Irvine, CA, USA). Genome-wide DNA methylation was quantified using the Infinium® MethylationEPIC BeadChip array (Illumina) to interrogate over 850,000 CpG loci at single-nucleotide resolution. The confocal laser scanning iScan® (Illumina) system was utilized to output signal intensities of the bisulfite-converted DNA methylation as IDAT files [17,18].

2.3. Identification of Differentially Methylated Positions and Regions

All analyses were conducted using the minfi Bioconductor package run in the R- 3.5.1 (64-bit) statistical analysis environment on the CAMH Specialized Computing Cluster. The methylation intensity data files (IDAT) and suicidal ideation severity for each corresponding subject were first analyzed. For quality control purposes, the preprocessRaw function was implemented to convert IDAT data into methylation β-values without normalization. We generated a probe intensity scatterplot and β-value density plot; samples with poor quality were excluded from further analyses [19,20]. For further downstream analyses, the preprocessFunnorm function was utilized for functional normalization to remove biological and technical variation [21]. To identify differentially methylated positions (DMPs), the dmpFinder function was used to test individual CpG sites for associations between methylation level and the suicidal ideation severity phenotype. The dmpFinder function performed univariate linear regressions of the C-SSRS score by each of the CpG positions. Considering multiple testing, a significance threshold of $p < 5.8 \times 10^{-8}$ was used to determine CpG sites that are significantly associated with the C-SSRS score.

Differentially methylated regions (DMRs) were identified through the implementation of the bumphunter function, with a methylation differential cut-off of 0.2 corresponding to a 20% difference in *p*-values [21]. The utility of bump hunting allows for the consideration of methylation levels between nearby CpG sites, and hence the identification of regions that are differentially methylated [17,22]. For DMR, the *p*-value was used to determine the significance of the association between candidate bumps and the C-SSRS score. For all analyses, only autosomal positions and regions were included and no other probe filtering were applied [23]. Cell composition was estimated using the Horvath algorithm (https://horvath.genetics.ucla.edu/html/dnamage (accessed on 31 March 2022)).

3. Results

3.1. Demographic and Clinical Characteristics

Among our cohort of 112 participants, we found that 19 subjects, or approximately 16.8%, currently reported experiencing suicidal ideation, while the remaining 94 subjects did not report suicidal ideation. The demographic and clinical variables from these participants are summarized in Table 1. Consistent with previous findings, we identified significant group differences between ideators and non-ideators with respect to psychosis severity, depression, hopelessness, and perceived stress [1,5,24,25].

Table 1. Demographic and clinical variables in the study cohort presented as mean ± standard deviation. Variables were also tested for group differences between subjects with and without current suicidal ideation with the Mann-Whitney U test for continuous variables and the Chi-square test for categorical variables.

Total (N = 112)	Ideator (n = 19)	Non-Ideator (n = 93)	p-Value
Sex (male/female)	11/8	59/34	0.377
Age (years)	44.7 ± 9.3	44.9 ± 12.8	0.961
Age-of-Onset (years)	21.2 ± 6.1	22.9 ± 6.8	0.345
Duration-of-Illness (years)	23.1 ± 9.9	21.5 ± 13.3	0.503
BPRS (Brief Psychiatric Rating Scale)	33.2 ± 7.8	28.3 ± 6.6	0.013
CDSS (Calgary Depression Scale for Schizophrenia)	6.7 ± 5.3	2.8 ± 3.2	0.031
BHS (Beck Hopelessness Scale)	7.6 ± 6.8	4.1 ± 4.3	0.043
PSS (Perceived Stress Scale)	31.2 ± 6.6	24.1 ± 6.8	0.012
SAI (Schedule for Assessment of Insight)	11.1 ± 2.6	11.6 ± 3.1	0.707

3.2. Quality Control Assessments

The probe intensity scatterplot graphically represents the logarithm of the median intensity of the methylated signal against the logarithm of the median intensity of the unmethylated signal (Figure 1A). Typically, high-quality samples cluster together with high median methylated and unmethylated intensities, whereas low-quality samples are located separately from the main cluster at lower medians [26]. Our results indicated that one sample was considered to be an outlier based on the above definitions. Following the conversion of raw IDAT data into methylation levels, we then generated a β-value density plot (Figure 1B), a visual representation that allows for the identification of sample outliers with poor quality. We observed that there was an overall level of consistency in the density plot, though one sample showed relatively poor quality. This was confirmed to be the same subject as identified with the probe intensity scatterplot and was removed from further analyses. The QC analysis confirms the findings illustrated in Figure 1A,B.

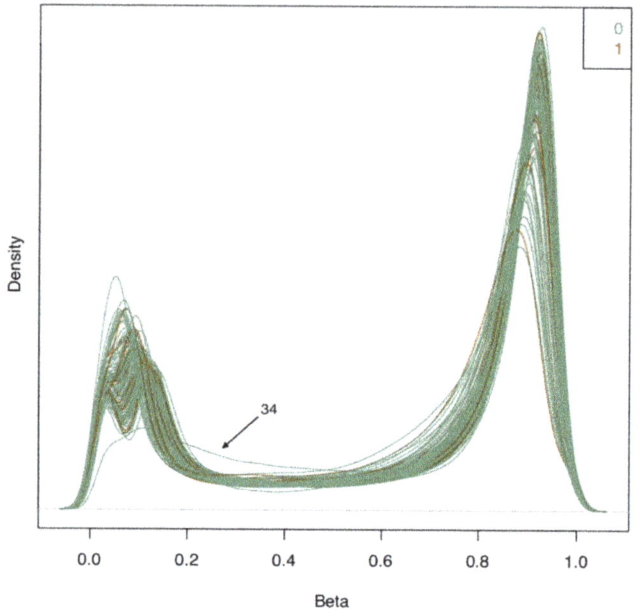

(A)

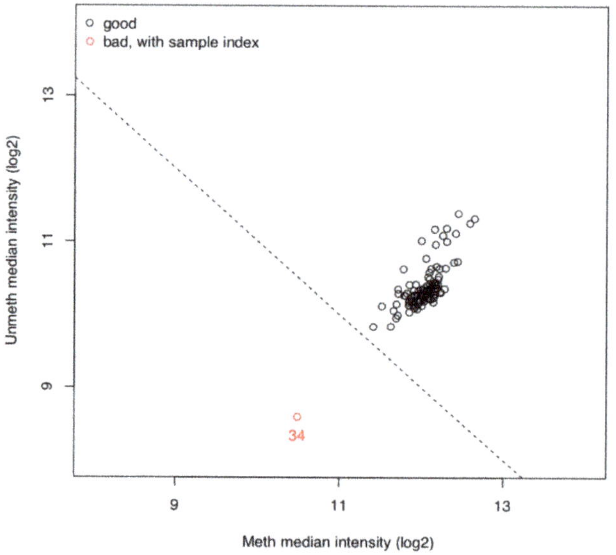

(B)

Figure 1. (**A**): β-value density plot for quality control purposes. Each line represents the density distribution of methylation levels for each CpG site throughout the genome. The number 0 and associated green lines represent samples from subjects not reporting suicidal ideation, whereas the number 1 and corresponding red lines represent samples from subjects reporting current suicidal ideation. The two peaks in the distribution curve are seen near 0.0 and 1.0 and refer to the theoretical states of CpG sites being completely methylated or unmethylated. (**B**): Probe intensity scatterplot for quality control purposes. The logarithm of median methylated and unmethylated signal intensities was plotted. High quality appears to cluster with high signal intensities (black), whereas low-quality samples are located separately from the main cluster with lower signal intensities (red). One sample with bad quality was identified and indicated in red, along with the sample index number. The sample index number allows for the identification of a given sample for removal in subsequent analyses.

3.3. *Differentially Methylated Positions (DMPs)*

The dmpFinder function was utilized to identify DMPs between subjects with and without suicidal ideation, considering their suicidal ideation severity scores. As defined by the C-SSRS, this continuous phenotype ranged from a score of zero to five, with increasing severity. The ten most significant CpG sites that were identified using this approach are reported in Table 2. Using LIMMA, we further analyzed the top ten CpG sites using a regression model that incorporated age, sex, ethnicity, and plate and cell composition (granulocytes, lymphocytes, and monocytes) as covariates. Only two of the top ten CpG sites remained significant after the addition of covariates (Table 2). The association *p*-values for all CpG sites across the genome before correction are graphically represented with volcano and Manhattan plots (Figure 2). In the original analysis, we determined that cg27077219, located within the LINC01356 gene on chromosome 1, was hypomethylated in subjects experiencing current suicidal ideation. However, after covariate correction, this CpG site was no longer significant.

Table 2. List of top ten differentially methylated positions (DMPs) associated with current suicidal ideation severity. Methylation at these CpG sites were identified to be associated with suicidal ideation severity, as a continuous phenotype, with a linear regression-based algorithm.

Chr	Position	CpG Site	Gene	*p*-Value	β Coefficient	*p*-Value after Correction *
1	113392580	cg27077219	*LINC01356*	7.85×10^{-8}	−0.600	0.247351
6	97285662	cg14723344	*GPR63*	3.85×10^{-7}	−0.478	0.015739
1	153044071	cg13950674	*SPRR2B*	4.08×10^{-6}	−0.851	9.9181×10^{-7}
17	33772796	cg00888402	*SLFN13*	4.54×10^{-6}	−0.766	0.000008
6	168629778	cg01801443	Intergenic	5.47×10^{-6}	−0.306	0.000003
16	89299756	cg27334271	Intergenic	6.39×10^{-6}	0.574	0.000029
10	30692613	cg02903852	Intergenic	7.65×10^{-6}	0.302	0.001251
6	5951562	cg12116564	Intergenic	1.92×10^{-5}	−0.257	6.2654×10^{-7}
2	88355002	cg06459916	*KRCC1*	2.20×10^{-5}	−0.305	0.178508
8	55380008	cg17993900	Intergenic	2.37×10^{-5}	−0.188	0.005229

Chr = chromosome number; Position = base-pair coordinate of the CpG site; *p*-Value = significance of the differentially methylated position associated with current suicidal ideation severity; β Coefficient = regression coefficient. * Results corrected for age, sex, ethnicity, batch, monocyte count, lymphocyte count and granulocyte count.

3.4. *Differentially Methylated Regions (DMRs)*

In the investigation of DMRs with *bumphunter*, we identified a total of 575 regions that were differentially methylated and nominally associated with current suicidal ideation. A list of the ten most significant DMRs are shown in Table 3. The methylation difference value represents the percent difference in methylation at a particular DMR between subjects with and without suicidal ideation. Positive methylation differences were indicative of a particular region being hypermethylated, and negative differences indicated hypomethylation of the DMR in subjects with current suicidal ideation. We report that a DMR located in chromosome 10 with a start position at 79655482 was hypermethylated in subjects with current suicidal ideation (top hit).

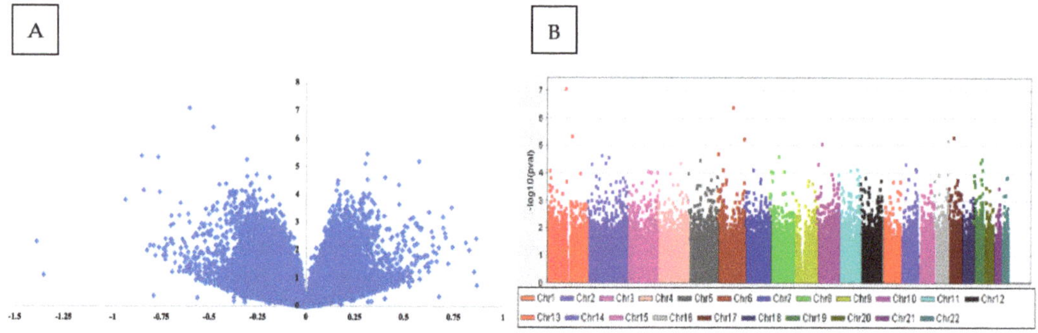

Figure 2. (**A**): Volcano plot of DMPs associated with suicidal ideation severity, a scatterplot representation of the association *p*-values for all CpG sites; the x-axis indicates the beta coefficient. The y-axis represents the effect size. (**B**): Manhattan plot of DMPs associated with suicidal ideation severity. Scatterplot representation of the association *p*-values for all CpG sites across the genome, arranged in order based on chromosome and position. The *y*-axis represents the negative logarithm of *p*-values.

Table 3. List of top ten differentially methylated regions (DMRs) associated with current suicidal ideation. These regions were identified to be the most differentially methylated between subjects with and without current suicidal ideation.

Chr	Position	Gene	Methylation Difference (%)	*p*-Value
10	79655482	*DLG5*	27.78	1.04×10^{-3}
1	2100232	*CACNB4*	27.65	1.14×10^{-3}
12	49074303	*KANSL2*	−26.56	1.66×10^{-3}
22	43168851	Intergenic	−23.77	4.89×10^{-3}
11	118022607	*SCN4B*	−23.54	5.41×10^{-3}
13	103423502	*TEX30*	−23.53	5.41×10^{-3}
9	135937572	*CEL*	−22.90	7.59×10^{-3}
1	152572665	*LCE3C*	−15.32	7.80×10^{-3}
12	123757860	*CDK2AP1*	21.95	9.88×10^{-3}
1	152586240	*LCE3B*	−21.63	1.10×10^{-2}

Chr = chromosome number; Position = base-pair coordinate of the beginning of the DMR; Methylation Difference = difference in the methylation levels (%) between subjects with and without suicidal ideation; *p*-Value = significance of the differentially methylated region associated with current suicidal ideation. Nominal *p*-value presented.

4. Discussion

In the present study, we assessed differential DNA methylation across the genome at the level of individual positions (DMPs) and regions (DMRs). To the best of our knowledge, this study is the first to investigate genome-wide methylation in relation to suicidal ideation severity in schizophrenia. We identified several DMPs and DMRs associated with suicidal ideation severity, although there was no consistency between DMR and DMP analyses and we were unable to conclude that these findings were significant at the genome-wide level. When applying a Bonferroni correction for 850,000 CpG sites, as in the DMP test, the genome-wide significance would require a $p < 5.8 \times 10^{-8}$. For the DMRs, the threshold would be less stringent. Considering that the human genome has approximately 30,000 CpG islands [27], the genome-wide significance would require a $p < 1.6 \times 10^{-6}$. Our top DMPs and DMRs were therefore only found to have suggestive levels of significance at the genome-wide level since they were not corrected for multiple tests.

Previous studies on DNA methylation and suicidal behavior have specifically pointed to hypomethylation of the NR3C1 gene [10]. However, we were unable to find similar results in our analysis of DNA methylation and current suicide ideation severity.

Moreover, the CpG sites highlighted in our study also did not coincide with a previously published genome-wide methylation study on suicide attempts [9]. Additionally, a recent genome-wide methylation analysis showed hypomethylation of the PSORS1C3 in suicide victims, although we were unable to find similar results [28].

Despite these results, our study had several strengths. From a technical standpoint, we utilized the latest Infinium® MethylationEPIC BeadChip array with the most comprehensive coverage of over 850,000 CpG sites, compared to previous generations of methylation arrays which covered either only 450,000 or 27,000 sites. The MethylationEPIC array removed approximately 10% of CpG sites found in the 450K chip due to poor performance, and among others, added 333,265 CpG sites located on intergenic and gene enhancer regions [29,30]. Prior to the identification of DMPs and DMRs, we conducted an array of quality control assessments to identify and remove samples of poor quality, as well as extra steps to process and normalize methylation measures.

Furthermore, our study design considered both differentially methylated positions (DMPs) and regions (DMRs) in an annotation-free approach. We were not the first to utilize these methods, with another study identifying both DMPs and DMRs associated with psychotic experiences [31]. Nonetheless, it is a strength of this study. In fact, it was even suggested that both approaches be run in tandem, since individual DMPs are not necessarily evenly spaced across the genome, and in many cases are not located within 1 kbp of a neighboring site [32]. While the identification of DMPs is certainly of interest, the region analysis is generally considered more robust than individual probes. Bump hunting for regions that are differentially methylated is more likely to identify differentially expressed genes than probing for individual CpG sites within the genome [19]. The identified genes would then have the potential to lead to the identification of downstream associated pathways and a greater understanding of etiological factors contributing to suicidal ideation.

Several limitations are present in this study. While our study had a reasonable size, a larger sample would have been indicated for improved interpretation of genome-wide results [33]. Furthermore, in the investigation of psychotic experiences mentioned above, the authors utilized a sample size of 845 participants [31]. Further studies with increased sample sizes are thus required. Additionally, the present study employs a cross-sectional design to investigate DNA methylation in relation to the severity of suicidal ideation; however, the causal impact of DNA methylation should be investigated longitudinally in the future so that additional factors such as financial, health, and family stress can be analyzed. Probes that are directly affected by SNPs, such as those associated with suicidality or schizophrenia, should be included in further investigations. Finally, future studies should examine methylation change in post-mortem neural tissue, especially considering the fact that genetic liability of schizophrenia has long been emphasized, with gray matter reductions in the anterior cingulate being reported as a marker of genetic liability for psychosis [34]. Although previous literature demonstrates that there is a limited, albeit significant degree of correspondence between peripheral blood methylation and brain methylation, causal effects should be interpreted as tentative since CpG sites may differ by tissue type.

5. Conclusions

We investigated individual sites and regions across the genome that were differentially methylated in association with current suicidal ideation severity. While the present hypothesis-free study did not determine any positions or regions differentially methylated that were significant at the genome-wide level, our findings suggest trends toward significance. While epigenome-wide association studies are still in their infancy, further work is required to replicate, support, and validate our findings presented here. The present analysis can be further expanded to determine the clinical utility of using genome-wide methylation markers to determine current suicide ideation severity.

In conclusion, despite the lack of statistical power to discover genome-wide changes for biomarker identification, this study might stimulate further investigation addressing the biological relevance of genome-wide methylation in psychiatric disorders.

Author Contributions: Conceptualization, V.D.L.; Data curation, C.A.; Formal analysis, K.Z.W.; Writing—original draft, K.Z.W.; Writing—review & editing, A.G.-G., P.G., C.C.Z., Z.C. and J.Q. All authors have read and agreed to the published version of the manuscript.

Funding: This work was supported by a Miner's Lamp award from the Department of Psychiatry at the University of Toronto and an American Foundation for Suicide Prevention standard research grant.

Institutional Review Board Statement: The study was approved by the CAMH Research Ethics Board.

Informed Consent Statement: Each subject provided consent to participate.

Data Availability Statement: Data available on request.

Acknowledgments: V.D.L. is supported by a Scholar Award from the Department of Psychiatry at the University of Toronto.

Conflicts of Interest: The authors declare no conflict of interest.

References

1. Carlborg, A.; Winnerbäck, K.; Jönsson, E.G.; Jokinen, J.; Nordström, P. Sucide in schizophrenia. *Expert Rev. Neurother.* **2010**, *10*, 1153–1164. [CrossRef]
2. Meltzer, H.Y. Treatment of suicidality in schizophrenia. *Ann. N. Y. Acad. Sci.* **2001**, *932*, 44–60. [CrossRef] [PubMed]
3. Hocaoglu, C.; Babuc, Z.T. Suicidal Ideation in Patients with Schizophrenia. *Isr. J. Psychiatry Relat. Sci.* **2009**, *46*, 195–203.
4. Simon, G.E.; Shortreed, S.M.; Johnson, E.; Beck, A.; Coleman, K.J.; Rossom, R.C.; Penfold, R.B. Between-visit changes in suicidal ideation and risk of subsequent suicide attempt. *Depress. Anxiety* **2016**, *34*, 794–800. [CrossRef] [PubMed]
5. Hor, K.; Taylor, M. Suicide and schizophrenia: A systematic review of rates and risk factors. *J. Psychopharmacol.* **2010**, *24* (Suppl. 4), 81–90. [CrossRef]
6. Klaus, F.; Kaiser, S.; Kirschner, M. Negative Symptoms in Schizophrenia- an Overview. *Ther. Umsch.* **2018**, *75*, 51–56. [CrossRef]
7. National Institute of Mental Health [NIMH]; The Research Prioritization Task Force. *A prioritized Research Agenda for Suicide Prevention: An Action Plan to Save Lives*; National Action Alliance for Suicide Prevention: Rockville, MD, USA, 2014.
8. Zai, C.C.; De Luca, V.; Strauss, J.; Tong, R.P.; Sakinofsky, I.; Kennedy, J.L. Genetic Factors and Suicidal Behavior. In *The Neurobiological Basis of Suicide*; Dwivedi, Y., Ed.; Taylor & Francis: Boca Raton, FL, USA, 2012; pp. 213–254.
9. Bani-Fatemi, A.; Jeremian, R.; Wang, K.Z.; Silveira, J.; Zai, C.C.; Kolla, N.J.; De Luca, V. Epigenome-wide association study of suicide attempt in schizophrenia. *J. Psychiatr. Res.* **2018**, *104*, 192–197. [CrossRef]
10. Guintivano, J.; Brown, T.; Newcomer, A.; Jones, M.; Cox, O.; Maher, B.S.; Kaminsky, Z.A. Identification and Replication of a Combined Epigenetic and Genetic Biomarker Predicting Suicide and Suicidal Behaviors. *Am. J. Psychiatry* **2014**, *171*, 1287–1296. [CrossRef]
11. Labonté, B.; Yerko, V.; Gross, J.; Mechawar, N.; Meaney, M.J.; Szyf, M.; Turecki, G. Differential Glucocorticoid Receptor Exon 1B, 1C, and 1H Expression and Methylation in Suicide Completers with a History of Childhood Abuse. *Biol. Psychiatry* **2012**, *72*, 41–48. [CrossRef] [PubMed]
12. McGowan, P.O.; Sasaki, A.; D'Alessio, A.C.; Dymov, S.; Labonté, B.; Szyf, M.; Meaney, M.J. Epigenetic regulation of the glucocorticoid receptor in human brain associates with childhood abuse. *Nat. Neurosci.* **2009**, *12*, 342–348. [CrossRef]
13. Pandey, G.N.; Rizavi, H.S.; Zhang, H.; Bhaumik, R.; Ren, X. The Expression of the Suicide-Associated Gene SKA2 Is Decreased in the Prefrontal Cortex of Suicide Victims but Not of Nonsuicidal Patients. *Int. J. Neuropsychopharmacol.* **2016**, *19*, pyw015. [CrossRef] [PubMed]
14. Sadeh, N.; Wolf, E.J.; Logue, M.W.; Hayes, J.P.; Stone, A.; Griffin, L.M.; Miller, M.W. Epigenetic Variation at *SKA2* Predicts Suicide Phenotypes and Internalizing Psychopathology. *Depress. Anxiety* **2016**, *33*, 308–315. [CrossRef] [PubMed]
15. Le-Niculescu, H.; Levey, D.F.; Ayalew, M.; Palmer, L.; Gavrin, L.M.; Jain, N.; Niculescu, A.B. III. Discovery and validation of blood biomarkers for suicidality. *Mol. Psychiatry* **2013**, *18*, 1249–1264. [CrossRef] [PubMed]
16. Posner, K.; Brown, G.K.; Stanley, B.; Brent, D.A.; Yershova, K.V.; Oquendo, M.A.; Mann, J.J. The Columbia- Suicide Severity Rating Scale: Initial validity and internal consistency findings from three multisite studies with adolescents and adults. *Am. J. Psychiatry* **2011**, *168*, 1266–1277. [CrossRef] [PubMed]
17. Li, D.; Xie, Z.; Le Pape, M.; Dye, T. An evaluation of statistical methods for DNA methylation microarray data analysis. *BMC Bioinform.* **2015**, *16*, 217. [CrossRef] [PubMed]
18. Smith, M.L.; Baggerly, K.A.; Bengtsson, H.; Ritchie, M.E.; Hansen, K.D. *illuminaio*: An open source IDAT parsing tool for Illumina microarrays. *F1000Research* **2013**, *2*, 264. [CrossRef] [PubMed]

19. Aryee, M.J.; Jaffe, A.E.; Corrada-Bravo, H.; Ladd-Acosta, C.; Feinberg, A.P.; Hansen, K.D.; Irizarry, R.A. Minfi: A flexible and comprehensive Bioconductor package for the analysis of Infinium DNA methylation microarrays. *Bioinformatics* **2014**, *30*, 1363–1369. [CrossRef]
20. Hansen, K.D.; Aryee, M. *The Minfi User's Guide: Analyszing Illumina 450K Methylation Arrays, Version 1.4.0*; 2014; pp. 1–21. Available online: https://bioconductor.riken.jp/packages/2.14/bioc/vignettes/minfi/inst/doc/minfi.pdf (accessed on 31 March 2022).
21. Fortin, J.P.; Labbe, A.; Lemire, M.; Zanke, B.W.; Hudson, T.J.; Fertig, E.J.; Hansen, K.D. Functional normalization of 450k methylation array data improves replication in large cancer studies. *Genome Biol.* **2014**, *15*, 503. [CrossRef]
22. Jaffe, A.E.; Murakami, P.; Lee, H.; Leek, J.T.; Fallin, M.D.; Feinberg, A.P.; Irizarry, R.A. Bump hunting to identify differentially methylated regions in epigenetic epidemiology studies. *Int. J. Epidemiol.* **2012**, *41*, 200–209. [CrossRef]
23. Yousefi, P.; Huen, K.; Davé, V.; Barcellos, L.; Eskenazi, B.; Holland, N. Sex differences in DNA methylation assessed by 450 K BeadChip in newborns. *BMC Genom.* **2015**, *16*, 911. [CrossRef]
24. Pompili, M.; Amador, X.F.; Girardi, P.; Harkavy-Friedman, J.; Harrow, M.; Kaplan, K.; Tatarelli, R. Suicide risk in schizophrenia: Learning from the past to change the future. *Ann. Gen. Psychiatry* **2007**, *6*, 10. [CrossRef]
25. Young, A.S.; Neuchterlein, K.H.; Mintz, J.; Ventura, J.; Gitlin, M.; Liberman, R.P. Suicidal Ideation and Suicide Attempts in Recent-Onset Schizophrenia. *Schizophr. Bull.* **1998**, *24*, 629–634. [CrossRef] [PubMed]
26. Fortin, J.P.; Hansen, K.D. Analysis of 450k Data Using Minfi. 2015. Available online: https://www.bioconductor.org/help/course-materials/2015/BioC2015/methylation450k.html (accessed on 29 October 2018).
27. Jeziorska, D.M.; Murray, R.J.S.; De Gobbi, M.; Gaentzsch, R.; Garric, D.; Ayyub, H.; Tufarelli, C. DNA methylation of intragenic CpG islands depends on their transcriptional activity during differentiation and disease. *Proc. Natl. Acad. Sci. USA* **2017**, *114*, E7526–E7535. [CrossRef]
28. Murphy, T.; Crawford, B.; Dempster, E.; Hannon, E.; Burrage, J.; TureckI, G.; Kaminsky, Z.; Mill, J. Methylomic profiling of cortex samples from completed suicide cases implicates a role for PSORS1C3 in major depression and suicide. *Transl. Psychiatry* **2017**, *7*, e989. [CrossRef] [PubMed]
29. Solomon, O.; Maclsaac, J.; Quach, H.; Tindula, G.; Kobor, M.S.; Huen, K.; Holland, N. Comparison of DNA methylation measured by Illumina 450K and EPIC BeadChips in blood of newborns and 14-year old children. *Epigenetics* **2018**, *13*, 655–664. [CrossRef] [PubMed]
30. Zaimi, I.; Pei, D.; Koestler, D.C.; Marsit, C.J.; De Vivo, I.; Tworoger, S.S.; Michaud, D.S. Variation in DNA methylation of human blood over a 1-year period using the Illumina MethylationEPIC array. *Epigenetics* **2018**, *13*, 1056–1071. [CrossRef]
31. Roberts, S.; Suderman, M.; Zammit, S.; Watkins, S.H.; Hannon, E.; Mill, J.; Fisher, H.L. Longitudinal investigation of DNA methylation changes preceding adolescent psychotic experiences. *Transl. Psychiatry* **2019**, *9*, 69. [CrossRef] [PubMed]
32. Wright, M.L.; Dozmorov, M.G.; Wolen, A.R.; Jackson-Cook, C.; Starkweather, A.R.; Lyon, D.E.; York, T.P. Establishing an analytic pipeline for genome-wide DNA methylation. *Clin. Epigenetics* **2016**, *8*, 45. [CrossRef] [PubMed]
33. Tsai, P.C.; Bell, J.T. Power and sample size estimation for epigenome-wide association scans to detect differential DNA methylation. *Int. J. Epidemiol.* **2015**, *44*, 1429–1441. [CrossRef]
34. Fusar-Poli, P.; Smieskova, R.; Serafini, G.; Politi, P.; Borgwardt, S. Neuroanatomical markers of genetic liability to psychosis and first episode psychosis: A voxelwise meta-analytical comparison. *World J. Biol. Psychiatry* **2014**, *15*, 219–228. [CrossRef]

healthcare MDPI

Article

Assessing Psychological Impact of COVID-19 among Parents of Children Returning to K-12 Schools: A U.S. Based Cross-Sectional Survey

Kavita Batra [1,*], Jennifer R. Pharr [2], Emylia Terry [2] and Brian Labus [3]

1 Department of Medical Education/Trauma and Critical Care, Kirk Kerkorian School of Medicine, University of Nevada, Las Vegas, NV 89102, USA
2 Department of Environmental and Occupational Health, School of Public Health, University of Nevada, Las Vegas, NV 89119, USA; jennifer.pharr@unlv.edu (J.R.P.); emylia.terry@unlv.edu (E.T.)
3 Department of Epidemiology and Biostatistics, School of Public Health, University of Nevada, Las Vegas, NV 89119, USA; brian.labus@unlv.edu
* Correspondence: kavita.batra@unlv.edu

Abstract: Background and Purpose: While impacts of the pandemic on family well-being have been documented in the literature, little is known about the psychological challenges faced by children and their parents as schools reopen after mandated closures. Therefore, the purpose of this study was to determine if sending children back to in-person school impacts the mental health of parents and the perceived mental health of their children. Methods: This cross-sectional descriptive study recruited a nationally representative, non-probability sample of parents or guardians (n = 2100) of children attending grades K-12 in the United States (U.S.) through a 58-item web-based survey. The univariate, bivariate, and multivariate statistical tests were used to analyze the data. Results: The mean scores of parental Coronavirus anxiety and Coronavirus obsession were significantly different between race/ethnic groups of parents. Parents with children going to private schools had significantly higher mean scores for Coronavirus anxiety and obsession compared to parents whose children are attending public schools. Nearly 55% of parental Coronavirus anxiety was explained by the generalized anxiety, separation anxiety, child's vulnerability to infection, and school type of the child. Similarly, 52% of parental Coronavirus obsession was explained by the generalized anxiety, separation anxiety, child's vulnerability to infection, and social phobia of the children. Conclusions: The COVID-19 pandemic has a substantial impact on psychological well-being of parents and their school-going children. Findings of this study will inform policy makers in developing targeted interventions to address unique needs of families with school-going children.

Keywords: COVID-19; K-12; coronavirus anxiety; coronavirus obsession; psychosocial impact

Citation: Batra, K.; Pharr, J.R.; Terry, E.; Labus, B. Assessing Psychological Impact of COVID-19 among Parents of Children Returning to K-12 Schools: A U.S. Based Cross-Sectional Survey. *Healthcare* **2022**, *10*, 775. https://doi.org/10.3390/healthcare10050775

Academic Editors: Yasuhiro Kotera and Elaina Taylor

Received: 12 March 2022
Accepted: 19 April 2022
Published: 22 April 2022

1. Introduction

COVID-19 has negatively impacted mental health outcomes around the world. The pandemic has been associated with an increase in mental health disorders, elevated anxiety, and an overall disruption to the delivery of mental health services in most countries [1–3]. According to UNICEF (2021), one in five of individuals ages 15–24 reported feeling depressed and experiencing anhedonia (the inability to feel pleasure) during the pandemic [4]. Within the United States (U.S.), an increase in mental health disorders and symptoms have been attributed to COVID-19; communities of color, women, immigrants, and parents with school going children present higher rates of psychological symptoms [5–7].

While American adults were eight times more likely to suffer severe mental distress during the pandemic, adult caregivers and households with children reported even worse outcomes, including increased anxiety and depression-related symptoms, higher rates of substance use, and elevated suicidal ideation [5,8,9]. Almost half of parents reported higher levels of stress during the pandemic, and this rate increased to over 60% for parents with children attending remote learning environments [10]. In addition, nearly 50% of mothers whose children attended remote learning environments reported a decline in mental health, compared to 30% of fathers. Parents were more likely to access mental health services during the pandemic than non-parents with 75% of parents indicating a need for increased emotional support [10]. Parents were also more likely than non-parents to have been diagnosed with a mental health disorder and exhibited high rates of anxiety, depression, and burnout during the pandemic [10,11]. Job loss, health issues, and death of loved ones were attributed to lower levels of pandemic-related resilience and exacerbated mental health issues among parents [12]. The previous research highlighted a need for emotional, financial, and caregiving support to ease burdens on parents [11–13].

While 54% of parents indicated that their children could benefit from mental health services, considerably less is known about the needs and wellbeing of school going children during COVID-19 [10]. However, studies do show that COVID-19 reduced family functioning and children's psychological wellbeing. He et al. found that caregiver stress during the pandemic was felt by many children, who internalized household discord and felt responsible for causing it [14]. Female students and older students have exhibited higher levels of anxiety during the pandemic [15]. In Canada, a quarter of children returning to in-person school had stress level above the critical threshold [16]. Further, nearly 25% of high school students in schools with closures reported elevated levels of pandemic-related worry, with students of color, low-income students, students in higher grades, and female students exhibiting the highest rates of COVID-19-related concerns [17]. As people of color and people with low income experienced higher rates of COVID-19, the elevated levels of pandemic-related worry among students of color and low-income students are additional pandemic-related disparities [17].

While pandemic impacts on family wellbeing have been documented in the literature [10–15], little is known about the mental health challenges faced by children and their parents as schools across the country reopen after mandated closures. Therefore, the purpose of this study was to determine if sending children back to in-person school impacts the mental health of parents and the perceived mental health of their children as they return to school. Due to the established relationship between child–parent anxiety and phobias, we wanted to understand this relationship within the context of COVID-19 and children's return to school [18]. Here, we explore the association between parental COVID-19 anxiety, obsession, and anxiety (generalized and separation) and social phobia in their children. As schools reopened across the country and children transitioned away from home schooling, we hypothesized that their generalized anxiety, separation anxiety, and social phobia would be associated with COVID-19 anxiety and obsession in their parents.

2. Materials and Methods

2.1. Study Design and Eligibility Criteria

This cross-sectional descriptive study recruited a nationally representative sample of parents or guardians of children who were enrolled in grades K-12 (kindergarten, elementary, middle, high school) in the U.S. In addition, only participants who could comprehend English and were capable of providing voluntary consent were included in this study.

2.2. Data Collection and Sampling Procedure

Data were collected between November and December 2021. This study utilized a commercial service offered through the Qualtrics Research Marketing Team to manage the data collection. A contractual agreement was established between the study's investigators and Qualtrics to recruit a quota sample that would mirror the U.S. Census with regard to gender, race, and region of the country of participating parents. Qualtrics recruited a high-quality sample through multiple avenues, including apps, games, social media platforms, and their dashboard-type system. A detailed sampling strategy used by the Qualtrics can be found at https://www.qualtrics.com/panels-project/ (accessed on 18 January 2022) [19]. A few screening questions related to the study's inclusion criteria were asked at the start of the survey to determine eligibility of the participants and to prevent response bias. If participants did not self-identify as a parent or guardian of a child going to school in grades K-12, the survey was programmed to automatically terminate. Given the use of multiple sources for the data collection, calculation of the response rate was not possible. Eligible participants who completed the survey were given incentives per terms and conditions set forth by Qualtrics and its data collection partners.

2.3. Ethical Considerations

This study (protocol ID: UNLV-2021-223 dated 2 December 2021) was granted an exempt status from the Institutional Review Board at the University of Nevada, Las Vegas. This study was exempted under category 2, as it includes protocols survey procedures, interviews, observation of public behavior, etc. Participants were provided detailed information about the study objectives, and participation was voluntary. Personal identifiers of the participants were not collected to abide by the ethical guidelines.

2.4. Quality Assurance of Data

The Qualtrics Research Marketing Team provided a complete dataset with only complete survey responses as a part of the contractual agreement. Qualtrics first collected only 10 percent of the total sample size as a part of a "soft launch" to identify any potential discrepancies before the full data set was collected [19]. Once the data quality was assessed by the study's investigators, the rest (90%) of the sample was collected. Algorithms such as digital fingerprinting and "prevent ballot-box stuffing" were used to ensure unique responses (only one response per participant). Once the data collection effort was completed, investigators were given an additional 7 days to review the quality of the data.

2.5. Survey Instrument

The 58-item questionnaire included 5 items related to parental Coronavirus anxiety, (Appendix A.1), 4 items related to parental Coronavirus obsession (Appendix A.2), and 22 items to measure separate anxiety, social phobia, and generalized anxiety among children using the Revised Child Anxiety Disorder Scale-Parent version (RCADS-P) [20–23]. As described by the previous studies, Coronavirus anxiety is an unhealthy state of mind, which stems from behavioral and psychological reactions following exacerbation of fear and worries related to coronavirus and its stimuli [24]. Coronavirus, on the other hand, is repetitive and maladaptive thinking about COVID-19 most of the time (Lee, 2020a; Lee 2020b; Chen et al., 2021). The remaining items were related to school safety measures and demographics of parents and their youngest child. Both Coronavirus anxiety and obsession were measured on a 5-point Likert scale ranging from "not at all" (0) to "nearly every day over the last 2 weeks" (4) [20,21]. Separation anxiety, social phobia, and generalized anxiety were measured on a 4-point Likert scale ranging from "never" (0), "sometimes" (1), "often" (2), and "always" (3) [19]. The RCADS-P has a good internal consistency, test–retest reliability, and construct validity [22,23].

2.6. Sample Size Justification

Prior to running analyses, the minimum sample size required was predetermined using the formula $n = (z)2\ p\ (1 - p)/d^2$ with a 95% confidence interval (alpha = 0.05, Z = 1.96), and a margin of error d = 5% [25]. This sample size calculation used the normal approximation to the binomial distribution [21]. The proportion (p) of worst mental health symptoms among parents of school going children was 27% in June 2020 based on the data reported by a study performed by Patrick et al. [26]. The estimated sample size after accounting for 10% non-response was 334. We also predetermined our sample size separately for t, chi-square, ANOVA, and multiple regression tests by using "small" Cohen's effect size conventions (effect size = 0.1 for t-tests, chi square, one-way ANOVA; 0.02 for linear regression with 10 predictors) [27]. The total sample size estimated with a power of 0.95 for was 1289, 1979, 1865, and 1229 for t test, chi square, one-way ANOVA, and linear regression test, respectively. The sample size with the greatest value (n = 1.979) was considered appropriate since it satisfies the minimum requirement of all the statistical tests used.

2.7. Data Analysis

SPSS software v.26 (IBM Corp.: Armonk, NY, USA) was used to analyze the data. Univariate statistics were used to describe the sample population. Independent-samples-t-test/Welch test or one-way ANOVA was used to draw group comparisons for continuous outcomes. Homogeneity of variance assumption was assessed by the Levene's test. Tukey or Games–Howell analysis was conducted to derive multiple group comparisons (wherever appropriate). The Pearson test was conducted to assess bivariate associations. Two models of multiple linear regression were fit to predict parental Coronavirus anxiety and obsession by parent demography, child demography, school characteristics, and the child's generalized anxiety, social phobia, and separation anxiety. All independent variables, including the polytomous variables, were dummy coded to aid in the accurate computations of regression coefficients and slopes. A Checklist for Statistical Assessment of Medical Papers (CHAMP statement) was used for our reporting [28]. The significance level was set at 5%. A detailed methodology can be seen in Figure 1.

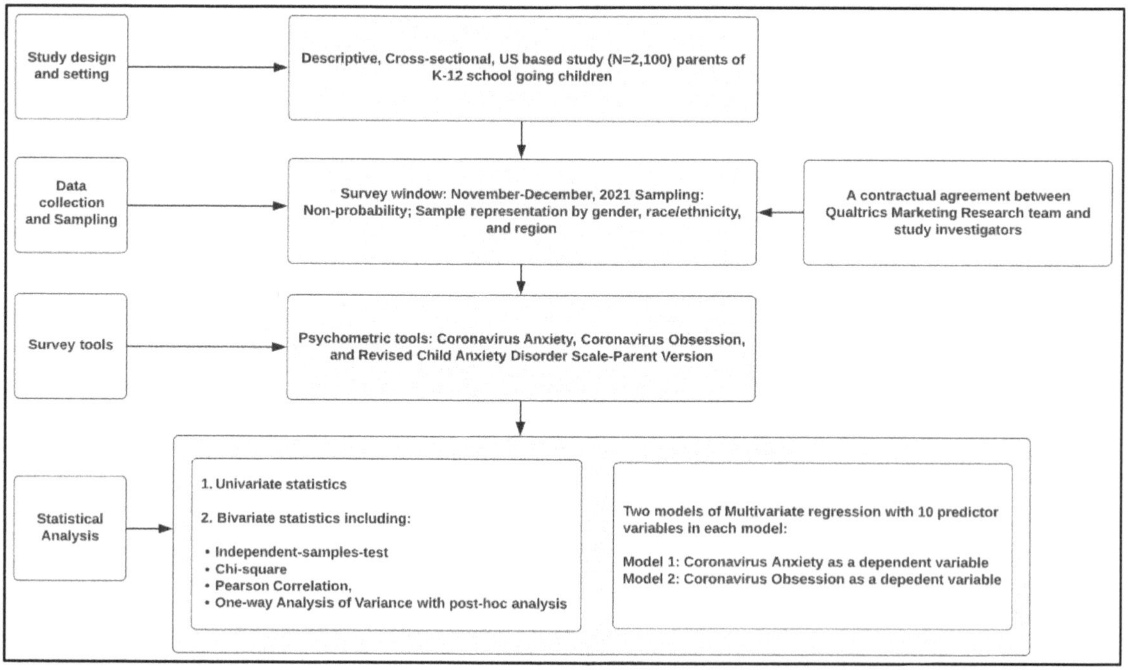

Figure 1. Flowchart detailing the study methodology.

3. Results

3.1. Univariate Statistics

A total of 2100 parents completed the survey. The mean age of the participants was 49.9 ± 11.2 years. The gender, race/ethnicity, and regional distribution was comparable to the U.S. Census parameters. Over thirty percent of parents had a 4-year college degree followed by nearly 27% reporting they had attended some college (Table 1). Over sixty percent of parents were married and had full-time paid jobs (Table 1). As reported by the parents, nearly 51% of children were male as compared to 47% being female (Table 2). The mean age of the youngest child reported was 9.10 ± 4.11 years and over 40% of children were attending elementary school (Table 2). A larger proportion (83.4%) of parents reported having their children attending public schools. Over 35% of parents reported having a child vulnerable to infections or were living with a vulnerable family member. Nearly 57% of parents relied on local health departments for COVID-19 related information.

Table 1. Demographic characteristics of the respondents (n = 2100).

Variable	Categories	n (%)	95% CI (LCL, UCL)
Age (Mean ± SD)	-	49.9 ± 11.2	49.4, 50.4
Gender	Female	1065 (50.7)	48.5, 52.9
	Male	999 (47.6)	45.4, 49.7
	Other, including non-binary, Transgender	29 (1.4)	0.93, 2.0
Race/ethnicity	Non-Hispanic White	1197 (57.0)	54.9, 59.1
	Non-Hispanic Black	273 (13.0)	11.6, 14.5
	Hispanic	378 (18.0)	16.4, 19.7
	Asian or Pacific Islanders	126 (6.0)	5.0, 7.1
	Other	126 (6.0)	5.0, 7.1
Education	Some high school	54 (2.6)	1.9, 3.3
	High school diploma or GED	417 (19.9)	18.2, 21.6
	Some College	560 (26.7)	24.8, 28.6
	4-year college degree	656 (31.2)	29.3, 33.3
	Graduate level degree	383 (18.2)	16.6, 19.9
	Other	30 (1.4)	1.0, 2.0
Marital status	Married	1294 (61.6)	59.5, 63.7
	Single, never married	364 (17.3)	15.7, 19.0
	Divorced or Separated	192 (9.1)	7.9, 10.5
	Widowed	42 (2.0)	1.4, 2.7
	A member of an unmarried couple	208 (9.9)	8.6, 11.3
Employment status	Full-time paid job	1390 (66.2)	64.1, 68.2
	Part-time paid job	175 (8.3)	7.2, 9.6
	Self-employed	146 (7.0)	5.9, 8.1
	Unable to work	103 (4.9)	4.0, 5.9
	Unemployed and looking for work	104 (5.0)	4.1, 5.9
	Unemployed but not looking for work (e.g., retired, homemaker, student, etc.)	182 (8.7)	7.5, 10.0
Annual Gross Income	0 to $10,000	153 (7.3)	6.2, 8.5
	$10,001 to $25,000	202 (9.6)	8.4, 11.0
	$25,001 to $50,000	473 (22.5)	20.8, 24.4
	$50,001 to $100,000	606 (28.9)	26.9, 30.9
	$100,001 to $250,000	528 (25.1)	23.3, 27.1
	>$250,001	111 (5.3)	4.4, 6.3
Region	Midwest	448 (21.3)	19.6, 23.2
	Northeast	358 (17.0)	15.5, 18.7
	South	819 (39.0)	36.9, 41.1
	West	475 (22.6)	20.8, 24.5

SD: Standard Deviation; CI: Confidence interval; LCL: Lower Confidence Level; UCL: Upper Confidence Level. The percentages may not add up to 100% for some variables (e.g., gender and income), as a few participants preferred not to say.

Table 2. Characteristics of family and youngest school going child as reported by parent respondents (n = 2100).

Variable	Categories	n (%)	95% CI (LCL, UCL)
Age of the youngest child (Mean ± SD)	-	9.10 ± 4.11	8.93, 9.27
Gender of the youngest child	Female	983 (46.8)	44.7, 49.0
	Male	1077 (51.3)	49.1, 53.4
	Other, including non-binary, transgender	26 (1.2)	0.8, 1.8
School type of the youngest child	Private	349 (16.6)	15.1, 18.3
	Public	1751 (83.4)	81.7, 85.0
School level of the youngest child	Kindergarten	536 (25.5)	23.7, 27.5
	Elementary	852 (40.6)	38.5, 42.7
	Middle school	357 (17.0)	15.4, 18.7
	High school	355 (16.9)	15.3, 18.6
Vulnerability of the youngest child to infections	Yes	741 (35.3)	33.2, 37.4
	No	1094 (52.1)	50.0, 54.3
	Not sure	265 (12.6)	11.2, 14.1
Living with vulnerable family member	Yes	755 (36.0)	33.9, 38.1
	No	1345 (64.0)	62.0, 66.1
Friends/family tested positive for COVID-19	Yes	955 (45.5)	43.3, 47.6
	No	1129 (53.8)	51.6, 55.9
COVID-19 information source	Local health department	1194 (56.9)	54.7, 58.9
	School administration	367 (17.5)	15.9, 19.2
	Parent association	205 (9.8)	8.5, 11.1
	Superintendent or official district communications	166 (7.9)	6.8, 9.1

The percentages may not add up to 100% for some variables (e.g., gender and family member tested positive for COVID-19), as a few participants preferred not to say. CI: Confidence interval; LCL: Lower Confidence Level; UCL: Upper Confidence Level.

3.2. Bivariate Statistics

A Welch-t-test was run to determine if there were differences in mean scores of Coronavirus anxiety, Coronavirus obsession, children separation anxiety, and generalized anxiety by parental gender. There were no outliers in the data, as assessed by inspection of a boxplot. All scores for each level of gender were normally distributed, as assessed by Shapiro–Wilk's test ($p > 0.05$), and there was no homogeneity of variances as assessed by Levene's test for equality of variances ($p < 0.05$). For social phobia, the assumption of homogeneity of variance was not met ($p > 0.564$), for which an independent-samples-t-test was used to compare the means. The differences in the means scores of parental Coronavirus anxiety and Coronavirus obsession, and children separation anxiety, social phobia, and generalized anxiety by parental gender, were not significant (Table 3). A one-way Welch ANOVA was conducted to compare mean scores among different race/ethnic groups of parents. There were no outliers, and the data were normally distributed for each group, as assessed by boxplot and Shapiro–Wilk test ($p > 0.05$), respectively. Homogeneity of variances was violated, as assessed by Levene's Test of Homogeneity of Variance ($p < 0.001$). Coronavirus anxiety, Welch' F (3, 2096) = 14.829, $p < 0.001$, $\omega2 = 0.02$, and Coronavirus obsession, F (3, 2096) = 8.536, $p < 0.001$, $\omega2 = 0.011$, scores were significantly different between race/ethnic groups of parents. Coronavirus anxiety score increased from the other group (M = 3.30, SD = 4.37) to the Black (M = 4.20, SD = 4.96), White (M = 4.60, SD = 5.24), and Hispanic (M = 6.00, SD = 5.32) groups, in that order. Games-Howell post hoc analysis revealed that the mean score increases from Black to Hispanic (1.77, 95% CI [0.72, 2.82]), White to Hispanic (1.36, 95% CI [0.55, 2.17]), as well as White to other (1.30, 95% CI [0.49, 2.11]) were significant. Similarly, the mean scores of separation anxiety, social phobia, and generalized anxiety of children (reported by parents) were significantly different, with Hispanic parents reporting the highest scores (Table 3). The assumption of homogeneity of variance was met for groups of parents living in different regions ($p = 0.06$). By regional location of parents, only Coronavirus obsession mean scores were significantly different. F (3, 2096) = 3.455, $p = 0.02$, $\omega 2 = 0.003$. Games-Howell post hoc analysis revealed that the mean increases from Midwest to Northeast region (0.93, 95% CI [0.14, 1.72]), as well as Midwest to Southern region (0.67, 95% CI [0.017, 1.33]) were significant ($p < 0.05$). There was no homogeneity of variances in, as assessed by Levene's test for equality of variances ($p < 0.05$) by school type. Parents with children going to private schools had significantly higher mean scores for Coronavirus anxiety and obsession compared with parents whose children are attending public schools (Table 3). In addition, parents of children going to private schools reported that their children had higher separation anxiety, social phobia, and generalized anxiety as compared to those going to public schools (Table 3). Children's separation anxiety (as reported by the parents), F (3, 2096) = 15 139, $p < 0.001$, $\omega 2 = 0.020$, scores were significantly different by school level of the children. Separation anxiety score decreased from the Kindergarten (M = 7.05, SD = 5.19) to the Elementary (M = 6.17, SD = 5.00), Middle (M = 5.81, SD = 5.78), and High (M = 4.67, SD = 5.29) school level, in that order. The mean scores of all the constructs were significantly different by the gender of the child with parents having male children and other gender identities reporting higher scores for coronavirus anxiety and obsession (Table 3).

Table 3. Group-wise comparisons for parental and children anxiety as reported by parents ($n = 2100$).

Variable	Groups	Coronavirus Anxiety		Coronavirus Obsession		Separation Anxiety		Social Phobia		Generalized Anxiety	
		M ± SD	*p* Value	M ± SD	*p* Value	M ± SD	*p* Value	M ± SD	*p* Value	M ± SD	*p* Value
Gender of the parent	Male	4.64 ± 5.42	0.8	4.60 ± 4.66	0.3	6.21 ± 5.61	0.2	8.93 ± 6.68	0.6	5.22 ± 4.73	0.8
	Female	4.60 ± 4.93		4.39 ± 4.23		5.93 ± 4.98		9.07 ± 6.55		5.19 ± 4.43	
Race/ethnicity of the parent	White	4.60 ± 5.24	**<0.001**	4.37 ± 4.48	**<0.001**	6.12 ± 5.36	**<0.001**	9.18 ± 6.46	**<0.001**	5.33 ± 4.54	**<0.001**
	Black	4.20 ± 4.96		4.38 ± 4.39		5.62 ± 5.26		8.34 ± 7.03		4.45 ± 4.31	
	Hispanic	6.00 ± 5.32		5.51 ± 4.51		7.07 ± 5.31		10.09 ± 6.89		6.27 ± 4.87	
	Other	3.30 ± 4.37		3.89 ± 4.01		4.91 ± 4.68		7.46 ± 6.07		4.06 ± 4.26	
Region	Northeast	5.05 ± 5.30	0.05 *	4.93 ± 4.43	**0.02**	6.37 ± 5.62	0.08	9.37 ± 7.08	0.3	5.53 ± 4.94	0.09
	South	4.74 ± 5.15		4.67 ± 4.52		6.12 ± 5.16		8.89 ± 6.48		5.13 ± 4.48	
	West	4.66 ± 5.29		4.44 ± 4.48		6.31 ± 5.50		9.33 ± 6.90		5.53 ± 4.84	
	Midwest	4.07 ± 4.94		3.99 ± 4.24		5.54 ± 4.99		8.66 ± 6.11		4.88 ± 4.16	
School type	Public	4.40 ± 5.06	**<0.001**	4.38 ± 4.39	**0.002**	5.85 ± 5.19	**<0.001**	8.84 ± 6.50	**0.005**	5.10 ± 4.46	**0.006**
	Private	5.81 ± 5.53		5.21 ± 4.63		7.25 ± 5.64		9.98 ± 7.04		5.91 ± 5.12	
School level	Kindergarten	5.05 ± 5.25	0.2	4.75 ± 4.45	0.4	7.05 ± 5.19	**<0.001**	8.48 ± 6.66	0.09	4.90 ± 4.59	0.2
	Elementary	4.53 ± 5.00		4.47 ± 4.27		6.17 ± 5.00		9.14 ± 6.25		5.28 ± 4.33	
	Middle school	4.63 ± 5.30		4.55 ± 4.61		5.81 ± 5.78		9.59 ± 6.97		5.56 ± 4.89	
	High school	4.25 ± 5.28		4.24 ± 4.57		4.67 ± 5.29		9.00 ± 6.95		5.30 ± 4.83	
Gender of the youngest child	Male	4.77 ± 5.20	**<0.001**	4.58 ± 4.50	**0.03**	6.27 ± 5.44	**0.04**	9.20 ± 6.65	**0.003**	5.45 ± 4.64	**<0.001**
	Female	4.37 ± 5.07		4.38 ± 4.33		5.82 ± 5.11		8.74 ± 6.55		4.91 ± 4.47	
	Other	8.35 ± 6.01		6.62 ± 4.64		7.73 ± 5.30		12.96 ± 5.62		9.23 ± 4.65	

Note: For two groups comparisons, independent-sample-*t*-test was conducted. For more than two groups, one-way ANOVA was conducted. Bolded *p* values are significant <0.05 level. * $p < 0.05$.

Table 4 shows the Pearson correlation coefficient matrix of all continuous variables. Parental Coronavirus anxiety was directly and strongly correlated with Coronavirus obsession (r = 0.83, $p < 0.01$), separation anxiety of the child (r = 0.71, $p < 0.01$), and had direct moderate correlation with social phobia (r = 0.62, $p < 0.01$), and generalized anxiety of the child (r = 0.66, $p < 0.01$). Parental age was negatively or indirectly correlated with parental anxiety and obsession as well as separation anxiety, social phobia, and generalized anxiety of the child.

Table 4. Pearson correlations, and reliability estimates for study variables in the sample (n = 2100).

Variables	1	2	3	4	5	6
1. Coronavirus anxiety of parents	1	0.83 **	0.71 **	0.62 **	0.66 **	−0.14 **
2. Coronavirus obsession of parents	0.83 **	1	0.70 **	0.62 **	0.64 **	−0.13 **
3. Separation anxiety of child	0.71 **	0.70	1	0.75 **	0.76 **	−0.19 **
4. Social phobia of child	0.62 **	0.62 **	0.75 **	1	0.82 **	−0.12 **
5. Generalized anxiety of child	0.66 **	0.64 **	0.76 **	0.82 **	1	−0.11 **
6. Parental age	−0.14 **	−0.14 **	−0.19 **	−0.12 **	−0.11 **	1
Cronbach's Alpha	0.915	0.895	0.901	0.925	0.918	-

** $p < 0.01$.

3.3. Multiple Linear Regression

For two models with parental Coronavirus anxiety and Coronavirus obsession as dependent variables, there was independence of residuals, as assessed by a Durbin–Watson statistic nearly 2. There was homoscedasticity, as assessed by visual inspection of a plot of studentized residuals versus unstandardized predicted values. There was no evidence of multicollinearity, as assessed by tolerance values greater than 0.1. There were no studentized deleted residuals greater than ±3 standard deviations, no leverage values greater than 0.2, and values for Cook's distance above 1. The assumption of normality was met, as assessed by a P-P Plot (Appendix B.1. and Appendix B.2). The multiple regression model was significant for parental coronavirus anxiety, F (15, 2048) = 169.051, $p < 0.001$, adj. $R^2 = 0.55$. Only generalized anxiety of the child, separation anxiety of the child, child's vulnerability to infection, and school type variables added significantly to the prediction, $p < 0.05$. Regression coefficients and standard errors can be found in Table 5. The multiple regression model was significant for parental coronavirus obsession, F (15, 2048) = 151.182, $p < 0.001$, adj. $R^2 = 0.522$. Only generalized anxiety, separation anxiety, child's vulnerability to infection, and social phobia variables added significantly to the prediction, $p < 0.05$. Regression coefficients and standard errors can be found in Table 6.

Table 5. Multiple regression results for Coronavirus anxiety among parents.

Variables	B	95% CI for B		SE B	β	R^2	ΔR^2
		LL	UL				
Model	-					0.553	0.550 **
Constant	−0.020	−0.826	0.785	0.411	-		
Parental gender (ref: female)	−0.157	−0.494	0.179	0.172	−0.015		
Parental age	−0.007	−0.023	0.010	0.009	−0.012		
Race/ethnicity of the parent, white (ref: black)	0.016	−0.454	0.486	0.240	0.002		
Race/ethnicity of the parent, Hispanic (ref: black)	0.540	−0.016	1.096	0.283	0.040		
Race/ethnicity of the parent, other (ref: black)	−0.379	−0.985	0.226	0.309	−0.024		
Child gender, male (ref: female)	0.070	−0.257	0.397	0.167	0.007		
Child gender, other (ref: female)	1.870	−0.044	3.784	0.976	0.029		
Child's vulnerability to infections (ref: No)	0.768	0.433	0.999	0.171	0.071 **		
School type (ref: public)	0.460	0.051	0.869	0.209	0.028 *		
School level, elementary (ref: kindergarten)	−0.128	−0.517	0.260	0.198	−0.012		
School level, middle (ref: kindergarten)	0.052	−0.436	0.541	0.249	0.004		
School level, high (ref: kindergarten)	0.312	−0.194	0.819	0.258	0.023		
Social phobia of child	0.042 *	0.002	−0.084	0.022	0.053		
Generalized anxiety of child	0.25 **	0.19	0.32	0.032	0.22 **		
Separation anxiety of the child	0.467 **	0.417	0.52	0.025	0.48 **		

Note. Model = "Enter" method in SPSS Statistics; B = Unstandardized regression coefficient; CI = confidence interval; LL = lower limit; UL = upper limit; SE B = Standard error of the coefficient; β = standardized coefficient; R^2 = coefficient of determination; ΔR^2 = adjusted R; * $p < 0.05$; ** $p < 0.01$.

Table 6. Multiple regression results for Coronavirus obsession among parents.

Variables	B	95% CI for B		SE B	β	R^2	ΔR^2
		LL	UL				
Model	-					0.525	0.522 **
Constant	0.731	0.018	1.444	0.364			
Parental gender (ref: female)	0.166	−0.132	0.464	0.152	0.019		
Parental age	−0.008	−0.023	0.007	0.008	−0.018		
Race/ethnicity of the parent, white (ref: black)	−0.325	−0.741	0.091	0.212	−0.036		
Race/ethnicity of the parent, Hispanic (ref: black)	0.139	−0.353	0.631	0.251	0.012		
Race/ethnicity of the parent, other (ref: black)	0.013	−0.523	0.549	0.273	0.001		
Child gender, male (ref: female)	−0.174	−0.464	0.116	0.148	−0.020		
Child gender, other (ref: female)	−0.145	−1.839	1.550	0.864	−0.003		
Child's vulnerability to infections (Ref: No)	0.653 **	0.357	0.950	0.151	0.070 **		
School type (ref: public)	0.053	−0.310	0.415	0.185	0.004		
School level, elementary (ref: kindergarten)	0.009	−0.335	0.353	0.175	0.001		
School level, middle (ref: kindergarten)	0.126	−0.306	0.559	0.221	0.011		
School level, high (ref: kindergarten)	0.357	−0.092	0.805	0.229	0.030		
Social phobia of child	0.086 **	0.048	0.123	0.019	0.128 **		
Generalized anxiety of child	0.176 **	0.120	0.232	0.029	0.181 **		
Separation anxiety of the child	0.369 **	0.326	0.413	0.022	0.440 **		

Note. Model = "Enter" method in SPSS Statistics; B = Unstandardized regression coefficient; CI = confidence interval; LL = lower limit; UL = upper limit; SE B = Standard error of the coefficient; β = standardized coefficient; R^2 = coefficient of determination; ΔR^2 = adjusted R; ** $p < 0.01$.

4. Discussion

This study yielded several interesting findings. First, there was a variation in parental COVID-19 anxiety and obsession as well as child separation anxiety, social phobia, and generalized anxiety based on the parent's race/ethnicity, with Hispanic parents reporting the highest scores for all variables. Hispanic and Latino populations in the U.S. have been disproportionately impacted by COVID-19 [29]. Compared with White people, Hispanic/Latino people have had 1.5 times more cases, 2.4 times more hospitalizations, and 1.9 times more deaths [25]. These disparities in COVID-19 cases, hospitalizations, and deaths may have influenced Hispanic/Latino parents' COVID-19 anxiety and obsession and their children's anxiety and social phobia. Another study found that Hispanic parents were less likely to support schools reopening full-time/in person and were more likely to support homeschooling until a vaccine was available and mask mandates in school than White parents [29,30]. Additionally, one study of Latino parenting styles revealed that the majority of parents had a protective/warm parenting style, while another found that Latino parents are more likely to have a family focused parenting style with high scores for involvement, monitoring, agency, and familismo (the cultural orientation and sense of obligation to family) [31–34]. These might explain higher scores for COVID-19 anxiety and obsession among this parent group compared with the other groups.

COVID-19 obsession among parents varied by region, with parents living in the Northeast having the highest scores and parents living in the Midwest having the lowest scores. This may have been impacted by the rates of disease or the political response such as mask or vaccine mandate variations within the regions. The Northeast, which includes New York, was hit hard by the pandemic in the early months. This may have influenced people's attitudes about the seriousness of COVID-19 in the Northeast region [35]. Recent hotspot trends in COVID-19 cases show relatively low cases in the Midwest region of the country compared with other regions [35]. This may be a result of population density, as many of the Midwest states have large rural areas. People in the Midwest might not feel as threatened by COVID-19 as people living in the more densely packed Northeast part of the country. Interestingly, the Northeast was reported to have the highest rate of mask wearing than other regions in the U.S. in June of 2020 [36]. Additionally, states that comprise the Northeast regions have the highest rates of vaccination according to the Johns

Hopkins COVID Resource Center [37]. This higher COVID-19 vigilance may have led to more persistent thoughts about COVID-19 among people in the Northeast. Lastly, the survey took a few weeks to conduct, and COVID rates were in a constant state of flux in November and December, 2021. These differences could have affected responses. For example, news coverage of a pediatric death in the community could have driven people to be more worried that day

Parents whose children attended private school had higher scores for all constructs than parents whose children attended public schools. A large survey of school attendance in the U.S. during November–December of 2020 found that children attending private school were much more likely to attend school fully in-person and face to face (60%) compared with students in public school with virtual attendance (25%) [38]. Another study found that attending a private school is associated with almost a 40% increase in attending school fully in-person and almost a 30% decrease in attending school fully remotely when compared with public school attendance. These differences in full in-person attendance may have heightened parents' anxiety and obsession with COVID-19 as well have been associated with children's anxiety and social phobia. On the other hand, some parents who were anxious about COVID-19 and how prevention strategies were being handled in the public-school system sought out private school for their children [38]. Additionally, parents of children who have social phobia or generalized anxiety may be more likely to send their children to private school because of smaller class sizes and more resources for children with special needs, which may explain some of the difference in the reported children's anxiety and social phobia scores [39].

Parents' COVID-19 anxiety was significantly associated with their child's reported generalized anxiety and separation anxiety while parents' COVID-19 obsession was significantly associated with their child's reported generalized anxiety, separation anxiety, and social phobia. Due to the cross-sectional nature of this study, we cannot say if parents' COVID-19 anxiety and obsession resulted in higher anxieties and social phobia among children or if children's higher anxieties and social phobia created higher COVID-19 anxiety and obsession among parents. It could be either/or, or both. During the COVID-19 pandemic, parents have been more likely to have higher rates of anxiety and depression than non-parents [10,11]. Research has shown that parental anxiety increases the risk of anxiety and other mood disorders among their children [40,41]. Parental stress has been associated with distress in their children during the COVID-19 pandemic [14]. Additionally, when their child experiences distress, anxious parents tend to reciprocate their child's distress rather than help them regulate it [42]. Among parents who reported having high reactivity to imagining their child experiencing fear, greater parental anxiety was associated with higher reactivity in their children. During the distress associated with COVID-19, parents' and children's reactivity to the distress may have elevated anxiety in each [42].

Strengths and Limitations

As with any study, there are limitations to this study. Due to the cross-sectional nature of this study, causation could not be assessed. Additionally, we did not measure pre-COVID-19 measures of anxiety in parents or children and are not able to determine if anxiety increased during the pandemic. Parents who are more anxious may have been more likely to complete the survey which may have resulted in anxiety bias. Parents self-reported their COVID-19 anxiety and obsession which subjects this study to a self-report bias. Parents may have over or under reported their concerns. Additionally, we did not directly measure children's general and separation anxiety or social phobia, but rather had parents reported these data. Parents may have over or under reported their child's anxieties and social phobia based on their own perceptions or concerns. Despite these limitations, there are strengths to this study. We had a nationally representative sample of parents mirroring the census representation by gender, race/ethnicity, and region distribution, although generalizability may be somewhat compromised by the lack of representation by type of child's school. Still, the findings of this study provide baseline data to design targeted interventions.

5. Conclusions

The purpose of this study was to determine if sending children back to in-person school impacts the mental health of parents and the perceived mental health of their children. The mean scores of parental Coronavirus anxiety and Coronavirus obsession were significantly different between race/ethnic groups of parents and among parents with children going to private versus public schools. Over half of parental Coronavirus anxiety was explained by the generalized anxiety, separation anxiety, and school type of the child, while parental Coronavirus obsession was explained by the generalized anxiety, separation anxiety, and social phobia of the children. The COVID-19 pandemic has a substantial impact on psychological well-being of parents and their school-going children. Future studies could directly measure children's anxiety, separation anxiety, and social phobia rather than gleaning this data from parents. A future area for research may include children prior to returning to school, during the return, and after the return to school to draw comparisons. Additionally, research in this area should continue as the COVID-19 pandemic and precautions pertaining to the pandemic change over time. Findings of this study can inform policy makers to develop targeted interventions to address unique needs of families with school-going children. For example, parents of children in private school who were more likely to be in-person, face to face, may need more information about the precautions being implemented to keep their children safe while in school (e.g., social distancing, mask requirements, protective barriers). This information needs to be culturally tailored toward different races and ethnicities with the understanding that Hispanic/Latino parents may have elevated levels of concern. Due to the association between parental Coronavirus anxiety/obsession and children's anxiety (general and separation) and social phobia, interventions to increase parents' coping skills (e.g., mindfulness, mediation) or stress reduction (emotional, financial, or caregiving support) may help to reduce anxiety for both parents and children.

Author Contributions: Conceptualization, K.B., J.R.P.; methodology, K.B., J.R.P.; software, K.B., J.R.P.; validation, K.B., J.R.P.; formal analysis, K.B.; investigation, K.B., J.R.P., E.T.; resources, B.L., J.R.P.; data curation, K.B., J.R.P.; writing, K.B., J.R.P., E.T.; writing—review and editing, K.B., J.R.P., E.T., B.L.; visualization, K.B.; supervision, J.R.P., B.L.; project administration, K.B.; All authors have read and agreed to the published version of the manuscript.

Funding: This project was partially supported by the US Centers for Disease Control and Prevention as part of an Epidemiology and Laboratory Capacity subaward from the Nevada Division of Public and Behavioral Health (6 NU50CK000560-01-05).

Institutional Review Board Statement: This study was conducted according to the guidelines of the Declaration of Helsinki and approved by the Institutional Review Board (or Ethics Committee) of the University of Nevada, Las Vegas (UNLV-2021-223 dated 2 December 2021).

Informed Consent Statement: Informed consent was obtained from all subjects involved in the study.

Data Availability Statement: The data presented in this study are available on request from the corresponding author. The data are not publicly available due to ethical reasons.

Conflicts of Interest: The authors declare no conflict of interest.

Appendix A

Appendix A.1. Coronavirus Anxiety Scale

Coronavirus anxiety (Adapted)

How often have you experienced the following activities over the past 2 weeks?

Q1: I felt dizzy, lightheaded, or faint, when I read or listened to news about the coronavirus among school children.

Not at all (0), Rare or less than a day or two (1), Several days (2), More than 7 days (3), Nearly every day over the last 2 weeks (4)

Q2: I had trouble falling or staying asleep because I was thinking about the coronavirus transmission in school.

Not at all (0), Rare or less than a day or two (1), Several days (2), More than 7 days (3), Nearly every day over the last 2 weeks (4)

Q3: I felt paralyzed or frozen when I thought about or was exposed to information about the coronavirus transmission in school.

Not at all (0), Rare or less than a day or two (1), Several days (2), More than 7 days (3), Nearly every day over the last 2 weeks (4)

Q4: I lost interest in eating when I thought about or was exposed to information about the coronavirus transmission in school.

Not at all (0), Rare or less than a day or two (1), Several days (2), More than 7 days (3), Nearly every day over the last 2 weeks (4)

Q5: I felt nauseous or had stomach problems when I thought about or was exposed to information about the coronavirus transmission in school.

Not at all (0), Rare or less than a day or two (1), Several days (2), More than 7 days (3), Nearly every day over the last 2 weeks (4)

Appendix A.2. Coronavirus Obsession Scale

Coronavirus Obsession (Adapted): This will record the persistent and disturbed thinking about COVID-19.

How often have you experienced the following activities over the last 2 weeks?

Q1: I had disturbing thoughts that my child may have caught the coronavirus.

Not at all (0), Rare or less than a day or two (1), Several days (2), More than 7 days (3), Nearly every day over the last 2 weeks (4)

Q2: I had disturbing thoughts that certain people I saw around my child in school may have the coronavirus.

Not at all (0), Rare or less than a day or two (1), Several days (2), More than 7 days (3), Nearly every day over the last 2 weeks (4)

Q3: I could not stop thinking about the coronavirus.

Not at all (0), Rare or less than a day or two (1), Several days (2), More than 7 days (3), Nearly every day over the last 2 weeks (4)

Q4: I dreamed about the coronavirus.

Not at all (0), Rare or less than a day or two (1), Several days (2), More than 7 days (3), Nearly every day over the last 2 weeks (4)

Appendix B

Appendix B.1

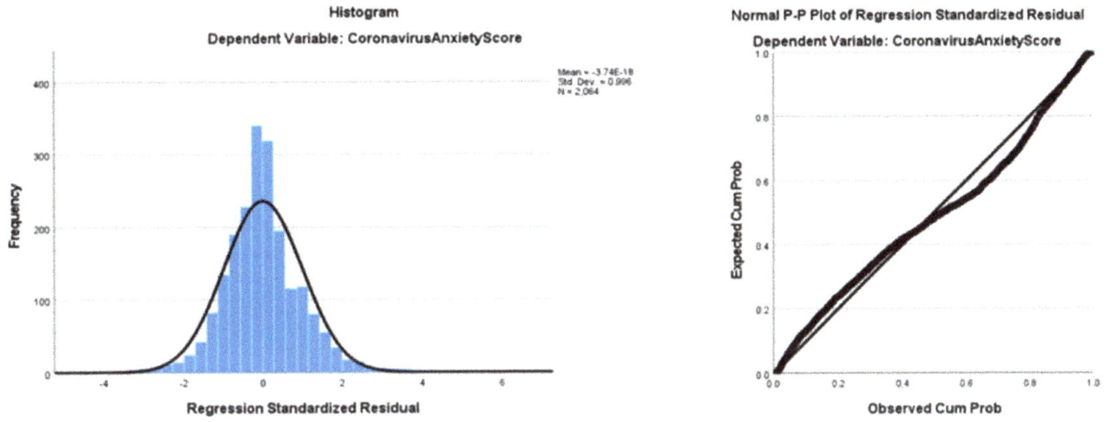

Figure A1. Figures showing residual analysis with Coronavirus Anxiety as dependent variable.

Appendix B.2

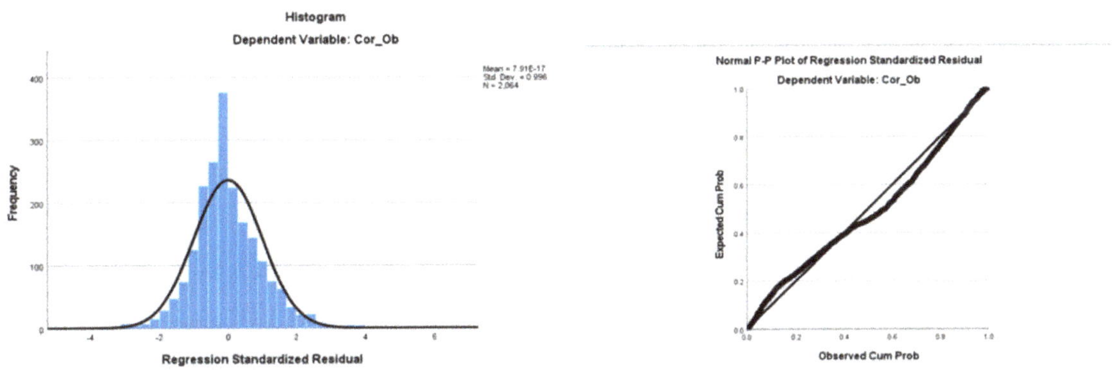

Figure A2. Figures showing residual analysis with Coronavirus Obsession as dependent variable.

References

1. Salari, N.; Hosseinian-Far, A.; Jalali, R.; Vaisi-Raygani, A.; Rasoulpoor, S.; Mohammadi, M.; Rasoulpoor, S.; Khaledi-Paveh, B. Prevalence of stress, anxiety, depression among the general population during the COVID-19 pandemic: A systematic review and meta-analysis. *Glob. Health* **2020**, *16*, 57. [CrossRef] [PubMed]
2. World Health Organization. COVID-19 Disrupting Mental Health Services in Most Countries, WHO Survey. 2020. Available online: https://www.who.int/news/item/05-10-2020-covid-19-disrupting-mental-health-services-in-most-countries-who-survey (accessed on 3 February 2022).
3. World Health Organization. (n.d.). Mental Health. Available online: https://www.who.int/health-topics/mental-health#tab=tab_1 (accessed on 3 February 2022).
4. UNICEF. Impact of COVID-19 on Poor Mental Health in Children and Young People 'Tip of the Iceberg'. [Press Release]. 4 October 2021. Available online: https://www.unicef.org/press-releases/impact-covid-19-poor-mental-health-children-and-young-people-tip-iceberg#:~{}:text=According%20to%20early%20findings%20from,little%20interest%20in%20doing%20things (accessed on 3 February 2022).
5. Fitzpatrick, K.M.; Harris, C.; Drawve, G. Fear of COVID-19 and the mental health consequences in America. *Psychol. Trauma* **2020**, *12*, S17–S21. [CrossRef] [PubMed]
6. Liu, C.H.; Zhang, E.; Wong, G.T.F.; Hyun, S.; Hahm, H.C. Factors associated with depression, anxiety, and PTSD symptomatology during the COVID-19 pandemic: Clinical implications for U.S. young adult mental health. *Psychiatry Res.* **2020**, *290*, 113172. [CrossRef] [PubMed]

7. Kamal, R.; Panchal, N.; Cox, C.; Garfield, R. The Implications of COVID-19 for Mental Health and Substance Use. KFF. 20 July 2021. Available online: https://www.kff.org/coronavirus-covid-19/issue-brief/the-implications-of-covid-19-for-mental-health-and-substance-use/ (accessed on 2 February 2022).
8. Czeisler, M.É.; Wiley, J.F.; Facer-Childs, E.R.; Robbins, R.; Weaver, M.D.; Barger, L.K.; Czeisler, C.A.; Howard, M.E.; Rajaratnam, S.M.W. Mental health, substance use, and suicidal ideation during a prolonged COVID-19-related lockdown in a region with low SARS-CoV-2 prevalence. *J. Psychiatr. Res.* **2021**, *140*, 533–544. [CrossRef]
9. Twenge, J.M.; Joiner, T.E. Mental distress among U.S. adults during the COVID-19 pandemic. *J. Clin. Psychol.* **2020**, *76*, 2170–2182. [CrossRef]
10. American Psychological Association. Mothers-and Fathers-report Mental, Physical Health Declines. American Psychological Association. 11 March 2021. Available online: https://www.apa.org/news/press/releases/stress/2021/one-year-pandemic-stress-parents (accessed on 13 April 2022).
11. Kerr, M.L.; Rasmussen, H.F.; Fanning, K.A.; Braaten, S.M. Parenting During COVID-19: A Study of Parents' Experiences Across Gender and Income Levels. *Fam. Relat.* **2021**, *70*, 1327–1342. [CrossRef]
12. Valero-Moreno, S.; Lacomba-Trejo, L.; Tamarit, A.; Pérez-Marín, M.; Montoya-Castilla, I. Psycho-emotional adjustment in parents of adolescents: A cross-sectional and longitudinal analysis of the impact of the COVID pandemic. *J. Pediatr. Nurs.* **2021**, *59*, e44–e51. [CrossRef]
13. University of Oregon RAPID-EC Project. How Long Can the Levee Hold? 10 September 2020. Available online: https://medium.com/rapid-ec-project/how-long-can-the-levee-hold-2a2cd0779914 (accessed on 7 February 2022).
14. He, Y.; Ortiz, R.; Kishton, R.; Wood, J.; Fingerman, M.; Jacobs, L.; Sinko, L. In their own words: Child and adolescent perceptions of caregiver stress during early COVID-19. *Child Abuse Negl.* **2022**, *124*, 105452. [CrossRef]
15. Chawla, N.; Tom, A.; Sen, M.S.; Sagar, R. Psychological impact of COVID-19 on children and adolescents: A systematic review. *Indian J. Psychol. Med.* **2021**, *43*, 294–299. [CrossRef]
16. Schwartz, K.D.; Exner-Cortens, D.; McMorris, C.A.; Makarenko, E.; Arnold, P.; Van Bavel, M.; Williams, S.; Canfield, R. COVID-19 and student well-being: Stress and mental health during return-to-school. *Can. J. Sch. Psychol.* **2021**, *36*, 166–185. [CrossRef]
17. Gazmararian, J.; Weingart, R.; Campbell, K.; Cronin, T.; Ashta, J. Impact of COVID-19 Pandemic on the Mental Health of Students From 2 Semi-Rural High Schools in Georgia. *J. Sch. Health* **2021**, *91*, 356–369. [CrossRef] [PubMed]
18. Murray, L.; Creswell, C.; Cooper, P.J. The development of anxiety disorders in childhood: An integrative review. *Psychol. Med.* **2009**, *39*, 1413–1423. [CrossRef] [PubMed]
19. Qualtrics Panels Project. Available online: https://www.qualtrics.com/panels-project/ (accessed on 31 December 2021).
20. Lee, S.A. Coronavirus Anxiety Scale: A brief mental health screener for COVID-19 related anxiety. *Death Stud.* **2020**, *44*, 393–401. [CrossRef] [PubMed]
21. Lee, S.A. How much "Thinking" about COVID-19 is clinically dysfunctional? *Brain Behav. Immun.* **2020**, *87*, 97–98. [CrossRef]
22. Chorpita, B.F.; Yim, L.; Moffitt, C.; Umemoto, L.A.; Francis, S.E. Assessment of symptoms of DSM-IV anxiety and depression in children: A revised child anxiety and depression scale. *Behav. Res. Ther.* **2000**, *38*, 835–855. [CrossRef]
23. Ebesutani, C.; Bernstein, A.; Nakamura, B.J.; Chorpita, B.F.; Weisz, J.R.; Research Network on Youth Mental Health. A psychometric analysis of the revised child anxiety and depression scale—Parent version in a clinical sample. *J. Abnorm. Child Psychol.* **2010**, *38*, 249–260. [CrossRef]
24. Chen, J.H.; Tong, K.K.; Su, X.; Yu, E.W.; Wu, A.M.S. Measuring COVID-19 related anxiety and obsession: Validation of the Coronavirus Anxiety Scale and the Obsession with COVID-19 Scale in a probability Chinese sample. *J. Affect. Disord.* **2021**, *295*, 1131–1137. [CrossRef]
25. Daniel, W.W. *Biostatistics: A Foundation for Analysis in the Health Sciences*, 7th ed.; John Wiley & Sons: New York, NY, USA, 1999.
26. Patrick, S.W.; Henkhaus, L.E.; Zickafoose, J.S.; Lovell, K.; Halvorson, A.; Loch, S.; Letterie, M.; Davis, M.M. Well-being of Parents and Children During the COVID-19 Pandemic: A National Survey. *Pediatrics* **2020**, *146*, e2020016824. [CrossRef]
27. Cohen, J. *Statistical Power Analysis for the Behavioral Sciences*, 2nd ed.; Department of Psychology, New York University: New York, NY, USA, 1988. Available online: http://utstat.toronto.edu/~{}brunner/oldclass/378f16/readings/CohenPower.pdf (accessed on 18 January 2022).
28. Mansournia, M.A.; Collins, G.S.; Nielsen, R.O.; Nazemipour, M.; Jewell, N.P.; Altman, D.G.; Campbell, M.J. Checklist for statistical Assessment of Medical Papers: The CHAMP statement. *Br. J. Sports Med.* **2021**, *55*, 1002–1003. [CrossRef]
29. Gilbert, L.K.; Strine, T.W.; Szucs, L.E.; Crawford, T.N.; Parks, S.E.; Barradas, D.T.; Njai, R.; Ko, J.Y. Racial and Ethnic Differences in Parental Attitudes and Concerns About School Reopening During the COVID-19 Pandemic—United States, July 2020. *MMWR Morb. Mortal. Wkly Rep.* **2020**, *69*, 1848–1852. [CrossRef]
30. The Centers for Disease Control and Prevention. Risk for COVID-19 Infection, Hospitalization, and Death by Race/Ethnicity. 1 February 2022. Available online: https://www.cdc.gov/coronavirus/2019-ncov/covid-data/investigations-discovery/hospitalization-death-by-race-ethnicity.html (accessed on 19 February 2022).
31. Domenech Rodríguez, M.M.; Donovick, M.R.; Crowley, S.L. Parenting styles in a cultural context: Observations of "protective parenting" in first-generation Latinos. *Fam. Process* **2009**, *48*, 195–210. [CrossRef] [PubMed]
32. Ayón, C.; Williams, L.R.; Marsiglia, F.F.; Ayers, S.; Kiehne, E. A Latent Profile Analysis of Latino Parenting: The Infusion of Cultural Values on Family Conflict. *Fam. Soc.* **2015**, *96*, 203–210. [CrossRef] [PubMed]

33. Sotomayor-Peterson, M.; Figueredo, A.J.; Christensen, D.H.; Taylor, A.R. Couples' cultural values, shared parenting, and family emotional climate within Mexican American families. *Fam. Process* **2012**, *51*, 218–233. [CrossRef] [PubMed]
34. The New York Times. (n.d.). COVID-19 Updates. Available online: https://www.nytimes.com/interactive/2021/us/covid-cases.html (accessed on 19 February 2022).
35. CNN Politics. The Northeast Leads the Country in Mask Wearing. 2020. Available online: https://www.cnn.com/2020/06/26/politics/maskwearing-coronavirus-analysis/index.html (accessed on 12 February 2022).
36. Coronavirus Resource Center. Understanding Vaccination Process. Available online: https://coronavirus.jhu.edu/vaccines/us-states/ (accessed on 19 February 2022).
37. Henderson, M.B.; Peterson, P.E.; West, M.R. Pandemic parent survey finds perverse pattern: Students are more likely to be attending school in person where Covid is spreading more rapidly. *Educ. Next* **2021**, *21*, 34–48. Available online: https://link.gale.com/apps/doc/A658584616/AONE?u=unlv_main&sid=googleScholar&xid=a895985 (accessed on 18 January 2022).
38. U.S. News. Private School vs. Public School. 2021. Available online: https://www.usnews.com/education/k12/articles/private-school-vs-public-school (accessed on 19 February 2022).
39. Burstein, M.; Ginsburg, G.S.; Tein, J.Y. Parental anxiety and child symptomatology: An examination of additive and interactive effects of parent psychopathology. *J. Abnorm. Child Psychol.* **2010**, *38*, 897–909, Erratum in *J. Abnorm. Child Psychol.* **2010**, *38*, 895. [CrossRef]
40. Ehrler, M.; Werninger, I.; Schnider, B.; Eichelberger, D.A.; Naef, N.; Disselhoff, V.; Kretschmar, O.; Hagmann, C.F.; Latal, B.; Wehrle, F.M. Impact of the COVID-19 pandemic on children with and without risk for neurodevelopmental impairments. *Acta Paediatr.* **2021**, *110*, 1281–1288. [CrossRef]
41. Woodruff-Borden, J.; Morrow, C.; Bourland, S.; Cambron, S. The behavior of anxious parents: Examining mechanisms of transmission of anxiety from parent to child. *J. Clin. Child Adolesc. Psychol.* **2002**, *31*, 364–374. [CrossRef]
42. Borelli, J.L.; Rasmussen, H.F.; John, H.; West, J.L.; Piacentini, J.C. Parental reactivity and the link between parent and child anxiety symptoms. *J. Child Fam. Stud.* **2015**, *24*, 3130–3144. [CrossRef]

healthcare

MDPI

Article

Mental Health of Czech University Psychology Students: Negative Mental Health Attitudes, Mental Health Shame and Self-Compassion

Yasuhiro Kotera [1,*], Denise Andrzejewski [2], Jaroslava Dosedlova [3], Elaina Taylor [4], Ann-Marie Edwards [4] and Chris Blackmore [5]

1 School of Health Sciences, University of Nottingham, Nottingham NG7 2HA, UK
2 Department of Psychology, Middlesex University, Dubai Block 16, Dubai P.O. Box 500697, United Arab Emirates; d.andrzejewski@mdx.ac.ae
3 Department of Psychology, Masaryk University, 602 00 Brno, Czech Republic; dosedlova@mail.muni.cz
4 College of Health, Psychology and Social Care, University of Derby, Derby DE22 1GB, UK; e.taylor3@derby.ac.uk (E.T.); annm.edwards@icloud.com (A.-M.E.)
5 School of Health and Related Research, University of Sheffield, Sheffield S1 4DA, UK; c.m.blackmore@sheffield.ac.uk
* Correspondence: yasuhiro.kotera@nottingham.ac.uk

Abstract: High rates of mental health problems are a growing concern in Czech higher education, negatively impacting students' performance and wellbeing. Despite the serious nature of poor mental health, students often do not seek help because of negative attitudes and shame over mental health problems. Recent mental health research reports self-compassion is strongly associated with better mental health and reduced shame. However, relationships between those constructs remain to be evaluated among Czech students. This study aims to appraise the relationships between mental health problems, negative mental health attitudes, mental health shame, and self-compassion in this population. An opportunity sample of 119 psychology students from a Czech university completed questionnaires regarding these constructs. Correlation, regression, and path analyses were conducted. Mental health problems were positively associated with negative mental health attitudes and shame, and negatively associated with self-compassion. Self-compassion negatively predicted mental health problems, while negative attitudes and shame did not. Last, self-compassion fully mediated the negative attitudes-mental health problems relationship, and partially mediated the shame-mental health problems relationship. Findings suggest self-compassion is essential for mental health in Czech students and associated with negative mental health attitudes and mental health shame. Czech universities can benefit from incorporating self-compassion training into their curricula to protect students' mental health.

Keywords: self-compassion; Czech university students; mental health; mental health attitudes; mental health shame

Citation: Kotera, Y.; Andrzejewski, D.; Dosedlova, J.; Taylor, E.; Edwards, A.-M.; Blackmore, C. Mental Health of Czech University Psychology Students: Negative Mental Health Attitudes, Mental Health Shame and Self-Compassion. *Healthcare* **2022**, *10*, 676. https://doi.org/10.3390/healthcare10040676

Academic Editor: Janet McDonagh

Received: 14 March 2022
Accepted: 30 March 2022
Published: 2 April 2022

1. Introduction

Mental health is high on the national agenda in the Czech Republic. The social and economic transformations that followed the collapse of the socialistic regime took a toll on the mental health of the people of Central and Eastern Europe that persists to this day [1–3]. In the Czech Republic, one in five adults are diagnosed with a mental health illness [4]. Alcohol dependence is almost twice as high as in the rest of Europe [5]. The highest prevalence rates of alcohol dependence (16.64%), as well as mood (7.96%) and anxiety disorders (5.42%), have been found in young people aged 18 to 29 years old, predominantly undergraduate university students [4]. The consequences of poor mental health in university students are higher dropout rates and lower academic achievement [6], which is partially mirrored in lower tertiary qualification attainment (i.e., college, university, and vocational courses) in the Czech Republic [7]. Furthermore, Bobak et al. [8] found a high prevalence of depression among a Czech Republic adult sample and were able to establish an inverse relationship between psychological wellbeing and acquired education: well-educated adults in the Czech Republic tended to have poor psychological wellbeing. This trend is present consistently among Eastern European university students [9]. While the government has recently started reforming mental health care, underfinancing and insufficient legislation [10] are still contributing to the substantial treatment gap in the Czech Republic where 83% of people with a mental disorder need care but have not received it [11,12].

Furthermore, mental health illness and its diagnosis in the historical Soviet discourse has mainly served as an instrument of oppression, and led to inhumane and amoral treatment. These conditions resulted in a stigmatization of psychiatry [1] that continues to foster negative attitudes towards the discipline of mental health at large [13]. As attitudes and beliefs about mental health are formed and maintained through cultural knowledge and perceptions, which are often based largely on historical narratives [14], the higher prevalence of mental health stigma in former socialist societies not only poses an additional barrier to close the treatment gap in the Czech Republic [13,15], but can potentially worsen overall mental health in the region [16].

The detrimental effects of negative mental health attitudes (i.e., believing that having a mental health problem indicates that the person is weak and inadequate) are well known. Research suggests that such attitudes impact negatively on self-esteem [17,18], self-efficacy [19], and physical health [20], and that they are a significant obstacle to treating mental health. These socio-cognitive barriers are among the leading obstacles for help-seeking behaviours, followed by mental health knowledge and awareness [21–26]. While studies have shown that mental health literacy is associated with more positive attitudes towards mental health [27], only 1% of Czech medical students are genuinely interested in pursuing careers in psychiatry [28]. The extent to which negative mental health attitudes and shame in post-Soviet societies, such as the Czech Republic, affect treatment seeking, has not yet been explored in detail.

1.1. Negative Mental Health Attitudes and Shame

It is well established that negative attitudes about mental health can lead to internalisation potentially manifesting in a sense of shame [29–33]. The emotional state of shame is complex and arises when individuals feel that they fall short of internalised socially constructed standards [32,34–36]. Shame involves negative self-evaluations and concerns about the judgements of others, and feelings of regret about one's identity [37,38]. As a marker of psychopathology, shame has been linked to depression [39], anxiety [40], and eating disorders [41]. "Mental health shame" ([42], p. 136)—feeling ashamed for having a mental health problem—is linked in university students to poorer mental health [43] and is especially prominent in students who prepare to enter demanding careers such as business management [44] and health care [45]. Doblytė [13] qualitatively explored feelings of shame regarding mental health problems in an adult Czech sample and observed that shame was a dominant theme for delayed treatment seeking and the adoption of destructive coping strategies to prevent stigmatisation. Though psychopathology is affecting predominately young adults in the Czech Republic [11] and the stigma surrounding mental health in the country remains pervasive [15], the relationship between negative attitudes, mental health shame, and mental health problems has not yet been quantitatively examined in Czech undergraduate students.

1.2. Self-Compassion

Research that focused on mental health improvement and shame reduction has consistently identified self-compassion as a protective factor [31,46–50]. Rooted in the tradition of Buddhism [51], self-compassion is related to practicing kindness towards oneself when facing adversity, acknowledging that struggling and suffering is a shared human condition, and becoming mindful and aware of one's painful thoughts [52]. Self-compassion has been linked to lower rates of depression, anxiety, and stress [52–55], as well as reduced social comparison [56] and self-criticism [31,47,57,58]. It has also been beneficially linked to life satisfaction, happiness, optimism, and overall wellbeing and better mental health [53,54,57]. Some have investigated the moderating role self-compassion plays in psychopathological symptoms such as rumination and stress [59], and self-criticism and depression [57]. However, the mediating role of self-compassion in the relationship between shame and mental health problems have largely only been explored in the context of eating disorders (e.g., refs. [58,60]). Self-compassion has begun to attract attention in the Czech Republic [61]. Montero-Marin et al. [62] suggested that this may be because practicing self-compassion is influenced by cultural values. Most notably, self-compassion is suggested to be inversely related to indulgence and restraint, as outlined by Hofstede's [63] Cultural Dimensions Theory. Like many other Eastern European countries, the Czech culture scores comparably lower in the indulgence domain than other Western societies such as Germany, the United Kingdom, and the United States [64]. Restraint and control of impulses and desires is a governing cultural value in the Czech Republic [63]. Therefore, self-compassion might not be a common trait or known skill in Czech culture and thus is a meaningful area to explore to support the national mental health agenda goals of improving the quality of life of people with mental illnesses and widening access to treatment [10].

1.3. Mental Health and Emotion Regulation

Disorders of distress such as anxiety and depression have been widely linked to emotional dysregulation [65] and often lead to maladaptive coping behaviours such as substance [66,67] and alcohol abuse [68]. Neurophysiological research suggests that there are three main emotion regulation systems [69], namely the threat, drive, and soothing systems [70]. The threat system functions as an alarm apparatus that elicits feelings of anxiety and anger, resulting in protection-seeking behaviours [70]. The drive system is goal oriented and triggers behaviours that bring pleasure [69]. The soothing system focuses on safety, and reduces distress through nurturing and affection [70]. According to Gilbert [71], taken together these emotion regulation systems, if balanced, form the foundation of mental wellbeing, but cause distress and psychopathology if unbalanced. The societal values of restraint and impulse control in Czech culture [64] could lead to imbalances in some individuals, with greater activation of the threat system, and diminished activation of the soothing system [62]. This in turn could explain lower engagement in self-compassionate behaviours. High activation of the threat system could further explain the high prevalence in mental health illness in the Czech Republic. Accordingly, we theorise that Czech individuals might predominantly operate on the threat system, which could be a plausible explanation for the high prevalence of mental health problems in the country. Therefore, if shame and negative attitudes towards mental health are anchored in the threat system then self-compassion as a soothing mechanism should be able to mediate the negative effects of the overstimulated threat system on mental health.

1.4. Aims and Hypotheses

Poor mental health affecting young adults in the Czech Republic can impair educational attainment and quality of life [13]. Recent mental health reform is currently addressing legislative avenues to facilitate accessibility to mental health services and better treatment in the Czech Republic. It is important to consider the cognitive barriers for individuals in accessing support and treatment, which has thus far not been explored in depth.

Given the scarcity of empirical studies investigating this association, the current study aimed to appraise the relationships between mental health problems, negative mental health attitudes, mental health shame, and self-compassion in Czech undergraduate psychology students. We particularly focused on depression, anxiety, and stress as "mental health problems" as these symptoms are most common among the general public and university student populations [72,73]. Additionally, based on Gilbert's [71] emotion regulatory systems, we investigated if self-compassion would be able to mediate the relationship between (H3) negative mental health attitudes and mental health problems, and (H4) mental health shame and mental health problems among Czech university students.

Our guiding hypotheses for the current study were as follows:

H1: *Greater mental health problems are significantly associated with (a) negative mental health attitudes, (b) mental health shame, and (c) self-compassion.*

H2: *Mental health problems are significantly predicted by (a) negative mental health attitudes, (b) mental health shame, and (c) self-compassion.*

H3: *Self-compassion mediates the relationship between negative mental health attitudes and mental health problems.*

H4: *Self-compassion mediates the relationship between mental health shame and mental health problems.*

2. Materials and Methods

2.1. Design

A cross-sectional study design was employed. Ethical approval was granted by the Ethics Panel of the Institute of Psychology, Faculty of Arts, Masaryk University (ID: 035/20).

2.2. Participants

Participants were aged 18 years or older and were undergraduate psychology students studying at a Czech university. Students who were on a break from studies were excluded from the study. Students were invited to take part in the study through a class announcement by programme tutors, who were not the authors of this paper, in February 2020. No incentive for participation was given.

An opportunity sample of 130 students agreed to participate. The final sample consisted of 119 students with an overrepresentation of females (n = 93; 78%) compared to males (n = 20), which is similar to the ratio found in the general psychology population in the Czech Republic [74]. Six students did not disclose their gender. The sample was aged between 19 and 44 (M = 21.87, SD = 3.32 years). The sample satisfied the required sample size of 115 based on statistical a priori power calculations [75]. The majority of students were Czech (n = 98; 82%), and the rest were Slovakian students (n = 21; 18%). The withdrawn 11 students did not report any reason or complaint.

2.3. Measures

Three self-report measures were used in this study (Table 1).

Table 1. Descriptive statistics and correlations between mental health problems, negative mental health attitudes, mental health shame, and self-compassion in Czech psychology students (n = 119).

		M	SD	α	1	2	3	4	5	6
1	Gender (F = 93, M = 20, No Answer = 6)				-					
2	Age (19–44 in our sample)	21.87	3.32		0.25 **	-				
3	Mental Health Problems (0–126)	38.00	20.98	0.90	−0.14	0.02	-			
4	Negative Mental Health Attitudes (0–24)	4.50	4.63	0.86	0.20 *	0.08	0.26 **	-		
5	Mental Health Shame (0–81)	18.34	11.55	0.90	0.13	0.02	0.35 **	0.56 **	-	
6	Self-Compassion (1–5)	3.08	0.60	0.81	0.12	0.002	−0.54 **	−0.25 **	−0.36 **	-

Gender F = 0, M = 1; * $p < 0.05$, ** $p < 0.01$.

Mental health problems were assessed using the Depression Anxiety and Stress Scale 21 (DASS21), a shorter version of the original DASS42 [76]. DASS21 evaluates the levels of depression ("I found it difficult to work up the initiative to do things."), anxiety ("I felt I was close to panic."), and stress ("I felt that I was using a lot of nervous energy."). Participants reflect on the past week and select how much each statement applies to them on a scale of 0 to 3 (0 = "Did not apply to me at all."; 3 = "Applied to me very much, or most of the time."). DASS21 has good reliability: $\alpha \geq 0.87$ [77].

Negative mental health attitudes and mental health shame were evaluated using the Attitudes Towards Mental Health Problems (ATMHP), comprising 35 items on a four-point Likert scale (0 = 'do not agree at all' to 3 = 'completely agree'; a higher score indicates more negative attitudes and stronger shame) [30]. This scale consists of four sections: (i) Community Attitudes and Family Attitudes refer to their perception of how their community and family perceive mental health problems, (ii) External Shame appraises their perception of how their community and family would view them if they had a mental health problem, (iii) Internal Shame considers how they would perceive themselves if they had a mental health problem, and (iv) Reflected Shame examines how their family would be perceived if they had a mental health problem (Family-Reflected Shame), and fears of reflected shame on themselves if a close relative had a mental health problem (Self-Reflected Shame). All of the subscales had good internal consistency (α = 0.85–0.97; [30]). Negative mental health attitudes were calculated by totalling the scores for Community and Family Attitudes, and mental health shame was calculated by totalling the scores for External Shame, Internal Shame, and Reflected Shame.

Self-compassion was assessed using the Self-Compassion Scale-Short Form (SCS-SF). This 12-item self-report measure is a shorter form of the 26-item Self-Compassion Scale [52], responded on a five-point Likert scale (e.g., "When something painful happens I try to take a balanced view of the situation."; 1 = 'Almost never' to 5 = 'Almost always'). The internal consistency was high (α = 0.86) [78].

2.4. Data Analysis

Data was screened for outliers and the assumptions of parametric tests. Pearson correlations between mental health problems, negative mental health attitudes, mental health shame, and self-compassion were evaluated using an SPSS version 26 (IBM, Chicago, IL, USA). Multiple regression analyses were performed to identify significant predictors of mental health problems. Path analysis was undertaken to explore effects of negative mental health attitudes in the relationship between mental health shame and mental health problems. Last, mediation analyses were conducted to examine the impact of negative mental health attitudes on the relationship between mental health shame and mental health problems. For the path analyses, the Process Macro 3 for SPSS (IBM, Chicago, IL, USA) [79] was used, with a setting of 5000 bootstrapping re-samples and bias-corrected 95% confidence intervals (CIs).

3. Results

Three scores in mental health problems and four scores in self-compassion were identified as outliers using the outlier labelling rule [80], so were winsorised [81]. All variables had good reliability (α = 0.81–0.90; Table 1). Because scores on all variables were not normally distributed (Shapiro-Wilk's test, $p < 0.05$), they were square-root-transformed to satisfy the assumption of normality.

3.1. Correlations (H1)

Pearson's correlation was conducted to examine the relationship between mental health problems, negative mental health attitudes, mental health shame, and self-compassion. Mental health problems were significantly positively associated with negative mental health attitudes (r = 0.26) and mental health shame (r = 0.35), and negatively associated with self-compassion (r = −0.54). H1 was supported.

3.2. Regression (H2)

Multiple regression analyses were conducted to appraise the relative contribution of negative mental health attitudes, mental health shame, and self-compassion to mental health problems (Table 2). At step 1, gender and age were entered to statistically adjust for their effects, and at step 2, negative mental health attitudes, mental health shame, and self-compassion were entered. Adjusted coefficient of determination (Adjusted R^2)

was reported. Multicollinearity was not a concern (the variance inflation factor < 10). After adjusting for demographic information, negative mental health attitudes, mental health shame, and self-compassion accounted for an additional 29% of the variance for mental health problems ($p < 0.001$). Negative mental health attitudes ($p = 0.23$) and mental health shame ($p = 0.24$) were not significant predictors for mental health problems. Self-compassion score was a significant negative predictor of the mental health symptom score ($p < 0.001$). A one-unit decrease in the self-compassion score was significantly associated with a 4.25-unit increase in the mental health problems score.

Table 2. Multiple regression examining negative attitudes to mental health, shame, and self-compassion as predictors of mental health symptoms in Czech psychology students ($n = 119$).

	Dependent Variable: Mental Health Problems			
	B	**SE_B**	**β**	**95% CI**
Step 1				
Gender	−0.62	0.43	−14	−1.46, 0.23
Age	0.02	0.06	0.03	−0.10, 0.14
Adj. R^2		0.001		
Step 2				
Gender	−0.61	0.37	−14	−1.35, 0.13
Age	0.05	0.05	0.08	−0.05, 0.16
Negative Mental Health Attitudes	0.16	0.13	0.12	−0.10, 0.42
Mental Health Shame	0.14	0.12	0.12	−0.10, 0.38
Self-Compassion	−4.25	0.80	−0.46 ***	−5.84, −2.66
Adj. R^2 Δ		0.29 ***		

Gender F = 0, M = 1; *** $p < 0.001$.

3.3. Mediation of Self-Compassion on Negative Mental Health Attitudes—Mental Health Problems (H3)

Path analysis was conducted, using Model 4 in the Process Macro parallel mediation model [79], to examine whether self-compassion (mediator variable) mediated the relationship between negative mental health attitudes (predictor variable) and mental health problems (outcome variable).

Self-compassion mediated the relationship between negative mental health attitudes and mental health problems (Figure 1). The total effect of negative mental health attitudes on mental health problems, including self-compassion, was significant, $b = 0.36$, $t(117) = 2.92$, $p = 0.004$, CI [0.12, 0.60]. The direct effect of negative mental health attitudes on mental health problems was not significant, $b = 0.18$, $t(116) = 1.66$, $p = 0.10$, CI [−0.04, 0.40]. The indirect effect of negative mental health attitudes on mental health problems, controlling for self-compassion, was significant, $b = 0.18$, CI [0.06, 0.32]. H3 was supported by a full mediation.

3.4. Mediation of Self-Compassion on Mental Health Shame—Mental Health Problems (H4)

Last, another path analysis was conducted, using Model 4 in the Process Macro (parallel mediation model [82]). Mental health shame was entered as a mediator variable, instead of negative mental health attitudes. Self-compassion (mediator variable) and mental health problems (outcome variable) remained the same.

Self-compassion partially mediated the relationship between mental health shame and mental health problems (Figure 2). The total effect of mental health shame on mental health problems, including self-compassion, was significant, $b = 0.42$, $t(117) = 3.99$, $p = 0.0001$, CI [0.21, 0.62]. The direct effect of mental health shame on mental health problems was also significant, $b = 0.21$, $t(116) = 2.14$, $p = 0.03$, CI [0.02, 0.40]. The indirect effect of mental health shame on mental health problems, controlling for self-compassion, was significant, $b = 0.21$, CI [0.09, 0.34]. H4 was supported by a partial mediation.

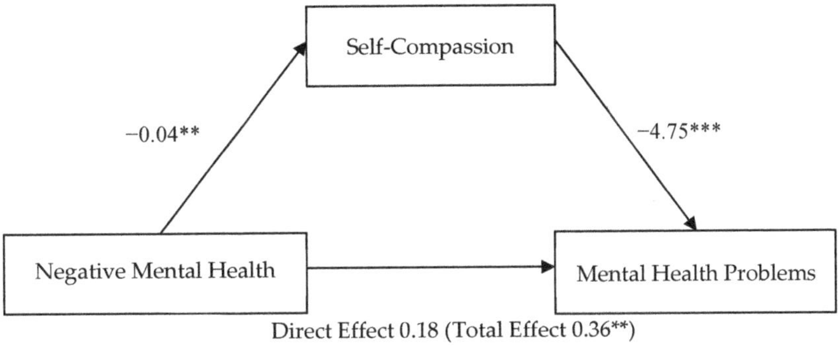

Figure 1. Parallel Mediation Model: Negative Mental Health Attitudes as a predictor of Mental Health Problems, mediated by Self-Compassion. Note: The confidence interval for the indirect effect is a BCa bootstrapped CI based on 5000 samples. Direct effect (total effect). Values attached to arrows are coefficients indicating impacts. ** $p < 0.01$, *** $p < 0.001$.

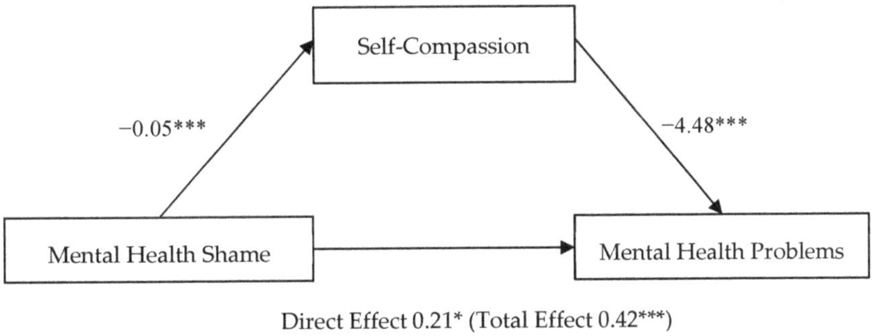

Figure 2. Parallel Mediation Model: Mental Health Shame as a predictor of Mental Health Problems, mediated by Self-Compassion. Note: The confidence interval for the indirect effect is a BCa bootstrapped CI based on 5000 samples. Direct effect (total effect). Values attached to arrows are coefficients indicating impacts. * $p < 0.05$, *** $p < 0.001$.

4. Discussion

Our aim was to explore how negative mental health attitudes, mental health shame, and self-compassion are associated with psychopathological symptoms. We hypothesised that all variables of interest would be significantly associated and predictive of mental health problems. While negative attitudes and mental health shame were significantly associated with mental health problems, they showed no significant predictive power for the mental health problems (H1ab, H2ab). On the other hand, self-compassion was both a significant correlate with, and predictor of mental health problems (H1c, H2c). Mediation analyses also indicated self-compassion fully mediated the relationship between negative mental health attitudes and mental health problems (H3), and partially mediated the relationship between mental health shame and mental health problems (H3, H4). Overall, the importance of self-compassion to mental health was demonstrated in all analyses.

Our findings support and add to the empirical literature of the beneficial relationship between practicing self-compassion and better mental health. Gilbert and Procter [31] argue that self-compassion can reduce feelings of embarrassment and humiliation, which provides further insight into the study findings. This is in line with earlier findings that self-compassion mediated the relationship between shame and negative body image [83]. Self-compassion has also been found to mediate the relationship between shame and symptoms of depression [84] and students' psychological health [85]. This research supports the beneficial role of self-compassion on shame reduction [31,42,45,48]. Though mental health shame is often a by-product of socially constructed attitudes and beliefs about mental health [14], our findings indicate that despite their strong relationship in our sample, these negative attitudes do not directly affect mental health but somewhat compromise the positive impact of self-compassion on mental health.

Furthermore, in our sample, while negative mental health attitudes and mental health shame were significantly correlated with mental health problems, these did not significantly predict mental health problems. This stands in contrast to some earlier findings [43], but echoes others [33]. A plausible explanation for this could be that perceived attitudes and mental health shame are not precursors of mental health problems but rather co-occurrences and consequences. Therefore, mental health shame possibly only arises once mental health problems are experienced. Mental health shame, as measured in the current study, takes into account external, reflective, and internal shame of having mental health issues [36,86,87], hence it is possible that one's own and the perceived attitudes of others towards mental health only become relevant when one becomes a potential target of these beliefs. There are some findings suggesting that anticipated stigma is significantly linked to higher rates of mental distress [88]. Indeed, this cross-sectional study did not appraise the causality, but our findings indicate a possible interpretation for non-predictive, yet correlative relationships of negative mental health attitudes and mental health shame with mental health problems. Further research is needed to investigate the directions of these relationships through longitudinal observation.

Applied to the context of mental health in the Czech Republic, our current findings indicate that the pervasive negative attitudes associated with mental health are strongly linked to feelings of shame, which have a direct effect on mental health. These attitudes and feelings of shame may create cognitive barriers for help-seeking behaviours that prevent individuals in distress from accessing mental health services and receiving treatment. It seems that the historical demonised image of psychiatry in former socialist societies [1] continues to cloud the cultural perception of mental health and therefore continues to fuel scepticism and shame [13]. This may be potentially aggravated and maintained by feelings of inferiority and elevated status anxiety in the country [89]. With this persistent fear of losing one's standing, the threat system seems constantly activated, and with a culturally inclined tendency to restrain from pleasures and soothing behaviours [62,64], the emotion-regulation systems may be unbalanced. This imbalance could be a plausible explanation for the elevated prevalence of mental distress [71] in the Czech Republic [4], and addressing the same might be a feasible avenue to explore in the reform of mental health. Given that poor mental health is a nationally recognised problem, the importance of self-compassion needs to be emphasised in Czech higher education. Training that targets self-compassion would be helpful for the mental health of Czech students. Considering the heightened anxiety associated with the transition to university [90], the effects of such training may be maximised during the orientation/induction session.

4.1. Limitations

While the current study has added to the understanding of the relationship between negative mental health attitudes, mental health shame, self-compassion and mental health problems, it is not without limitations. One of the key limitations would be that our sample was recruited at one university in the Czech Republic. Moreover, we recruited psychology students, many of whom were female students. Research has shown that mental health literacy [91,92], as well as being female, is associated with less stigma and negative attitudes toward mental health [32]. These might have caused sample biases. Therefore, a more diverse sample of students needs to be involved. Second, self-report measures might compromise the accuracy of reporting because of response biases [93]. Moreover, process-oriented constructs such as self-compassion may require closer evaluation [94,95]. Third, the causal direction of these relationships has not been examined. Longitudinal studies would be helpful to elucidate the temporal patterning of the relationships identified in this study and may help develop more effective interventions.

4.2. Conclusions

In conclusion, this study highlights the relationship between greater self-compassion and better mental health in university students. Practicing self-compassion, particularly in student samples who experience high levels of stress and who may be at greater risk of developing mental health illness, may be beneficial. Future research should examine ways in which positive psychology techniques can be introduced into the university curriculum. Longitudinal research should also be conducted to examine the causal relationships between negative attitudes and shame towards mental health, self-compassion and poor mental health.

Author Contributions: Conceptualization, Y.K. and J.D.; methodology, Y.K.; software, Y.K.; validation, D.A., E.T. and A.-M.E.; formal analysis, Y.K.; investigation, Y.K. and J.D.; resources, J.D.; data curation, J.D.; writing—original draft preparation, Y.K., D.A. and E.T.; writing—review and editing, A.-M.E. and C.B.; project administration, Y.K. All authors have read and agreed to the published version of the manuscript.

Funding: This research received no external funding.

Institutional Review Board Statement: The study was conducted according to the guidelines of the Declaration of Helsinki, and approved by Ethical approval was granted by the Ethics Panel of the Institute of Psychology, Faculty of Arts, Masaryk University (ID: 035/20; 28 January 2020).

Informed Consent Statement: Informed consent was obtained from all subjects involved in the study.

Data Availability Statement: The data presented in this study are available on request from the corresponding author. The data are not publicly available due to ethical restrictions.

Conflicts of Interest: The authors declare no conflict of interest.

References

1. Raikhel, E.; Bemme, D. Postsocialism, the Psy-Ences and Mental Health. *Transcult. Psychiatry* **2016**, *53*, 151–175. [CrossRef] [PubMed]
2. Skultans, V. From Damaged Nerves to Masked Depression: Inevitability and Hope in Latvian Psychiatric Narratives. *Soc. Sci. Med.* **2003**, *56*, 2421–2431. [CrossRef]
3. Skultans, V. The Appropriation of Suffering: Psychiatric Practice in the Post-Soviet Clinic. *Theory Cult. Soc.* **2007**, *24*, 27–48. [CrossRef]
4. Formánek, T.; Kagström, A.; Cermakova, P.; Csémy, L.; Mladá, K.; Winkler, P. Prevalence of Mental Disorders and Associated Disability: Results from the Cross-Sectional CZEch Mental Health Study (CZEMS). *Eur. Psychiatry* **2019**, *60*, 1–6. [CrossRef] [PubMed]
5. Wittchen, H.U.; Jacobi, F.; Rehm, J.; Gustavsson, A.; Svensson, M.; Jönsson, B.; Olesen, J.; Allgulander, C.; Alonso, J.; Faravelli, C.; et al. The Size and Burden of Mental Disorders and Other Disorders of the Brain in Europe 2010. *Eur. Neuropsychopharmacol.* **2011**, *21*, 655–679. [CrossRef] [PubMed]
6. Poh Keong, P.; Chee Sern, L.; Foong, M.; Ibrahim, C. The Relationship between Mental Health and Academic Achievement among University Students—A Literature Review. *Second Int. Conf. Glob. Trends Acad. Res.* **2015**, *2*, 755–764.
7. Organisation for Economic Co-Operation and Development Czech Republic Country Note. Available online: http://www.oecd.org/education/Czech%20Republic-EAG2014-Country-Note.pdf (accessed on 14 March 2022).
8. Bobak, M.; Pikhart, H.; Pajak, A.; Kubinova, R.; Malyutina, S.; Sebakova, H.; Topor-Madry, R.; Nikitin, Y.; Marmot, M. Depressive Symptoms in Urban Population Samples in Russia, Poland and the Czech Republic. *Br. J. Psychiatry J. Ment. Sci.* **2006**, *188*, 359–365. [CrossRef] [PubMed]
9. Mikolajczyk, R.T.; Maxwell, A.E.; El Ansari, W.; Naydenova, V.; Stock, C.; Ilieva, S.; Dudziak, U.; Nagyova, I. Prevalence of Depressive Symptoms in University Students from Germany, Denmark, Poland and Bulgaria. *Soc. Psychiatry Psychiatr. Epidemiol.* **2008**, *43*, 105–112. [CrossRef]
10. Pec, O. Mental Health Reforms in the Czech Republic. *BJPsych Int.* **2019**, *16*, 4–6. [CrossRef]
11. Kagstrom, A.; Alexova, A.; Tuskova, E.; Csajbók, Z.; Schomerus, G.; Formanek, T.; Mladá, K.; Winkler, P.; Cermakova, P. The Treatment Gap for Mental Disorders and Associated Factors in the Czech Republic. *Eur. Psychiatry* **2019**, *59*, 37–43. [CrossRef] [PubMed]

12. Kale, R. The Treatment Gap. *Epilepsia* **2002**, *43*, 31–33. [CrossRef] [PubMed]
13. Doblytė, S. Shame in a Post-Socialist Society: A Qualitative Study of Healthcare Seeking and Utilisation in Common Mental Disorders. *Sociol. Health Illn.* **2020**, *42*, 1858–1872. [CrossRef] [PubMed]
14. Choudhry, F.R.; Mani, V.; Ming, L.; Khan, T.M. Beliefs and Perception about Mental Health Issues: A Meta-Synthesis. *Neuropsychiatr. Dis. Treat.* **2016**, *12*, 2807–2818. [CrossRef] [PubMed]
15. Winkler, P.; Mladá, K.; Janoušková, M.; Weissová, A.; Tušková, E.; Csémy, L.; Evans-Lacko, S. Attitudes towards the People with Mental Illness: Comparison between Czech Medical Doctors and General Population. *Soc. Psychiatry Psychiatr. Epidemiol.* **2016**, *51*, 1265–1273. [CrossRef]
16. Weiss, M.G.; Ramakrishna, J.; Somma, D. Health-Related Stigma: Rethinking Concepts and Interventions. *Psychol. Health Med.* **2006**, *11*, 277–287. [CrossRef] [PubMed]
17. Link, B.G.; Struening, E.L.; Neese-Todd, S.; Asmussen, S.; Phelan, J.C. Stigma as a Barrier to Recovery: The Consequences of Stigma for the Self-Esteem of People with Mental Illnesses. *Psychiatr. Serv.* **2001**, *52*, 1621–1626. [CrossRef] [PubMed]
18. Rodrigues, S.; Serper, M.; Novak, S.; Corrigan, P.; Hobart, M.; Ziedonis, M.; Smelson, D. Self-Stigma, Self-Esteem, and Co-Occurring Disorders. *J. Dual Diagn.* **2013**, *9*, 129–133. [CrossRef]
19. Watson, A.C.; Corrigan, P.; Larson, J.E.; Sells, M. Self-Stigma in People With Mental Illness. *Schizophr. Bull.* **2007**, *33*, 1312–1318. [CrossRef]
20. Sickel, A.E.; Seacat, J.D.; Nabors, N.A. Mental Health Stigma: Impact on Mental Health Treatment Attitudes and Physical Health. *J. Health Psychol.* **2019**, *24*, 586–599. [CrossRef]
21. Corrigan, P.W.; Druss, B.G.; Perlick, D.A. The Impact of Mental Illness Stigma on Seeking and Participating in Mental Health Care. *Psychol. Sci. Public Interest J. Am. Psychol. Soc.* **2014**, *15*, 37–70. [CrossRef]
22. Eisenberg, D.; Golberstein, E.; Gollust, S.E. Help-Seeking and Access to Mental Health Care in a University Student Population. *Med. Care* **2007**, *45*, 594–601. [CrossRef] [PubMed]
23. Kopinak, J.; Berisha, B.; Mursali, B. An Investigation into the Health of a Representative Sample of Adults in Kosovo. *J. Humanit. Assist.* **2001**, *17*, 1–7.
24. Lannin, D.G.; Vogel, D.L.; Brenner, R.E.; Abraham, W.T.; Heath, P.J. Does Self-Stigma Reduce the Probability of Seeking Mental Health Information? *J. Couns. Psychol.* **2016**, *63*, 351–358. [CrossRef]
25. Shannon, P.J.; Wieling, E.; Simmelink-McCleary, J.; Becher, E. Beyond Stigma: Barriers to Discussing Mental Health in Refugee Populations. *J. Loss Trauma* **2015**, *20*, 281–296. [CrossRef]
26. Ting, L. Depressive Symptoms in a Sample of Social Work Students and Reasons Preventing Students from Using Mental Health Services: An Exploratory Study. *J. Soc. Work Educ.* **2011**, *47*, 253–268. [CrossRef]
27. Milin, R.; Kutcher, S.; Lewis, S.P.; Walker, S.; Wei, Y.; Ferrill, N.; Armstrong, M.A. Impact of a Mental Health Curriculum on Knowledge and Stigma Among High School Students: A Randomized Controlled Trial. *J. Am. Acad. Child Adolesc. Psychiatry* **2016**, *55*, 383–391.e1. [CrossRef]
28. Janoušková, M.; Formánek, T.; Bražinová, A.; Mílek, P.; Alexová, A.; Winkler, P.; Motlová, L.B. Attitudes towards People with Mental Illness and Low Interest in Psychiatry among Medical Students in Central and Eastern Europe. *Psychiatr. Q.* **2021**, *92*, 407–418. [CrossRef]
29. Corrigan, P.W.; Watson, A.C.; Miller, F.E. Blame, Shame, and Contamination: The Impact of Mental Illness and Drug Dependence Stigma on Family Members. *J. Fam. Psychol.* **2006**, *20*, 239–246. [CrossRef]
30. Gilbert, P.; Bhundia, R.; Mitra, R.; McEwan, K.; Irons, C.; Sanghera, J. Cultural Differences in Shame-Focused Attitudes towards Mental Health Problems in Asian and Non-Asian Student Women. *Ment. Health Relig. Cult.* **2007**, *10*, 127–141. [CrossRef]
31. Gilbert, P.; Procter, S. Compassionate Mind Training for People with High Shame and Self-Criticism: Overview and Pilot Study of a Group Therapy Approach. *Clin. Psychol. Psychother.* **2006**, *13*, 353–379. [CrossRef]
32. Hampton, N.Z.; Sharp, S.E. Shame-Focused Attitudes Toward Mental Health Problems: The Role of Gender and Culture. *Rehabil. Couns. Bull.* **2014**, *57*, 170–181. [CrossRef]
33. Kotera, Y.; Green, P.; Sheffield, D. Mental Health Attitudes, Self-Criticism, Compassion and Role Identity among UK Social Work Students. *Br. J. Soc. Work* **2019**, *49*, 351–370. [CrossRef]
34. Gilbert, P. Evolution, Social Roles, and the Differences in Shame and Guilt. *Soc. Res. Int. Q.* **2003**, *70*, 1205–1230.
35. Kim, S.; Thibodeau, R.; Jorgensen, R.S. Shame, Guilt, and Depressive Symptoms: A Meta-Analytic Review. *Psychol. Bull.* **2011**, *137*, 68–96. [CrossRef] [PubMed]
36. Tangney, J.P. Assessing Individual Differences in Proneness to Shame and Guilt: Development of the Self-Conscious Affect and Attribution Inventory. *J. Pers. Soc. Psychol.* **1990**, *59*, 102–111. [CrossRef] [PubMed]
37. Benetti-McQuoid, J.; Bursik, K. Individual Differences in Experiences of and Responses to Guilt and Shame: Examining the Lenses of Gender and Gender Role. *Sex Roles* **2005**, *53*, 133–142. [CrossRef]
38. Tangney, J.P.; Stuewig, J.; Mashek, D.J. Moral Emotions and Moral Behavior. *Annu. Rev. Psychol.* **2007**, *58*, 345–372. [CrossRef]
39. Matos, M.; Pinto-Gouveia, J. Shame as a Traumatic Memory. *Clin. Psychol. Psychother.* **2010**, *17*, 299–312. [CrossRef]
40. Gilbert, P. The Relationship of Shame, Social Anxiety and Depression: The Role of the Evaluation of Social Rank. *Clin. Psychol. Psychother.* **2000**, *7*, 174–189. [CrossRef]

41. Caglar-Nazali, H.P.; Corfield, F.; Cardi, V.; Ambwani, S.; Leppanen, J.; Olabintan, O.; Deriziotis, S.; Hadjimichalis, A.; Scognamiglio, P.; Eshkevari, E.; et al. A Systematic Review and Meta-Analysis of 'Systems for Social Processes' in Eating Disorders. *Neurosci. Biobehav. Rev.* **2014**, *42*, 55–92. [CrossRef]
42. Kotera, Y.; Green, P.; Sheffield, D. Mental Health Shame of UK Construction Workers: Relationship with Masculinity, Work Motivation, and Self-Compassion. *Rev. Psicol. Trab. Las Organ.* **2019**, *35*, 135–143. [CrossRef]
43. Arimitsu, K. The Relationship of Guilt and Shame to Mental Health. [The Relationship of Guilt and Shame to Mental Health.]. *Jpn. J. Health Psychol.* **2001**, *14*, 24–31. [CrossRef]
44. Kotera, Y.; Conway, E.; Van Gordon, W. Mental Health of UK University Business Students: Relationship with Shame, Motivation and Self-Compassion. *J. Educ. Bus.* **2019**, *94*, 11–20. [CrossRef]
45. Kotera, Y.; Cockerill, V.; Chircop, J.G.E.; Forman, D. Mental Health Shame, Self-compassion and Sleep in UK Nursing Students: Complete Mediation of Self-compassion in Sleep and Mental Health. *Nurs. Open* **2021**, *8*, 1325–1335. [CrossRef]
46. Kotera, Y.; Green, P.; Sheffield, D. Mental Health of Therapeutic Students: Relationships with Attitudes, Self-Criticism, Self-Compassion, and Caregiver Identity. *Br. J. Guid. Couns.* **2019**. [CrossRef]
47. Neely, M.E.; Schallert, D.L.; Mohammed, S.S.; Roberts, R.M.; Chen, Y.-J. Self-Kindness When Facing Stress: The Role of Self-Compassion, Goal Regulation, and Support in College Students' Well-Being. *Motiv. Emot.* **2009**, *33*, 88–97. [CrossRef]
48. Neff, K.D.; Kirkpatrick, K.L.; Rude, S.S. Self-Compassion and Adaptive Psychological Functioning. *J. Res. Personal.* **2007**, *41*, 139–154. [CrossRef]
49. Sirois, F.M.; Kitner, R.; Hirsch, J.K. Self-Compassion, Affect, and Health-Promoting Behaviors. *Health Psychol.* **2015**, *34*, 661–669. [CrossRef]
50. King, R.; Gaerlan, M.J.M. High Self-Control Predicts More Positive Emotions, Better Engagement, and Higher Achievement in School. *Eur. J. Psychol. Educ.* **2014**, *29*, 81–100. [CrossRef]
51. Neff, K.D. Self-Compassion, Self-Esteem, and Well-Being: Self-Compassion, Self-Esteem, and Well-Being. *Soc. Personal. Psychol. Compass* **2011**, *5*, 1–12. [CrossRef]
52. Neff, K.D. The Development and Validation of a Scale to Measure Self-Compassion. *Self Identity* **2003**, *2*, 223–250. [CrossRef]
53. MacBeth, A.; Gumley, A. Exploring Compassion: A Meta-Analysis of the Association between Self-Compassion and Psychopathology. *Clin. Psychol. Rev.* **2012**, *32*, 545–552. [CrossRef] [PubMed]
54. Ying, Y.W. Contribution of Self-Compassion to Competence and Mental Health in Social Work Students. *J. Soc. Work Educ.* **2009**, *45*, 309–323. [CrossRef]
55. Kotera, Y.; Dosedlova, J.; Andrzejewski, D.; Kaluzeviciute, G.; Sakai, M. From Stress to Psychopathology: Relationship with Self-Reassurance and Self-Criticism in Czech University Students. *Int. J. Ment. Health Addict.* **2021**. [CrossRef] [PubMed]
56. Neff, K.D.; Vonk, R. Self-Compassion Versus Global Self-Esteem: Two Different Ways of Relating to Oneself. *J. Pers.* **2009**, *77*, 23–50. [CrossRef] [PubMed]
57. Braehler, C.; Gumley, A.; Harper, J.; Wallace, S.; Norrie, J.; Gilbert, P. Exploring Change Processes in Compassion Focused Therapy in Psychosis: Results of a Feasibility Randomized Controlled Trial. *Br. J. Clin. Psychol.* **2012**, *52*, 199–214. [CrossRef] [PubMed]
58. Liss, M.; Erchull, M.J. Not Hating What You See: Self-Compassion May Protect against Negative Mental Health Variables Connected to Self-Objectification in College Women. *Body Image* **2015**, *14*, 5–12. [CrossRef] [PubMed]
59. Samaie, G.; Farahani, H.A. Self-Compassion as a Moderator of the Relationship between Rumination, Self-Reflection and Stress. *Procedia-Soc. Behav. Sci.* **2011**, *30*, 978–982. [CrossRef]
60. Wollast, R.; Riemer, A.; Sarda, E.; Wiernik, B.; Klein, O. How Self-Compassion Moderates the Relation Between Body Surveillance and Body Shame Among Men and Women. *Mindfulness* **2020**, *11*, 2298–2313. [CrossRef]
61. Benda, J.; Reichová, A. Psychometrické Charakteristiky České Verze Self-Compassion Scale (SCS-CZ). [Psychometric Characteristics of the Czech Version of the Self-Compassion Scale (SCS-CZ).]. *Československ. Psychol. Časopis Psychol. Teor. Praxi* **2016**, *60*, 120–136.
62. Montero-Marin, J.; Kuyken, W.; Crane, C.; Gu, J.; Baer, R.; Al-Awamleh, A.A.; Akutsu, S.; Araya-Véliz, C.; Ghorbani, N.; Chen, Z.J.; et al. Self-Compassion and Cultural Values: A Cross-Cultural Study of Self-Compassion Using a Multitrait-Multimethod (MTMM) Analytical Procedure. *Front. Psychol.* **2018**, *9*, 2638. [CrossRef] [PubMed]
63. Hofstede, G.H. *Culture's Consequences: Comparing Values, Behaviors, Institutions, and Organizations across Nations*, 2nd ed.; Sage Publications: Thousand Oaks, CA, USA, 2001; ISBN 978-0-8039-7323-7.
64. Hofstede Insights. Available online: https://www.hofstede-insights.com/country-comparison/czech-republic,germany,the-uk,the-usa/ (accessed on 23 December 2020).
65. Suveg, C.; Morelen, D.; Brewer, G.A.; Thomassin, K. The Emotion Dysregulation Model of Anxiety: A Preliminary Path Analytic Examination. *J. Anxiety Disord.* **2010**, *24*, 924–930. [CrossRef] [PubMed]
66. Carver, C.S.; Scheier, M.F.; Weintraub, J.K. Assessing Coping Strategies: A Theoretically Based Approach. *J. Pers. Soc. Psychol.* **1989**, *56*, 267–283. [CrossRef]
67. Lazarus, R.S.; Folkman, S. *Stress, Appraisal, and Coping*; 11. [print.]; Springer: New York, NY, USA, 1984; ISBN 978-0-8261-4191-0.
68. Sher, K.J.; Grekin, E.R. Alcohol and Affect Regulation. In *Handbook of Emotion Regulation*; Gross, J.J., Ed.; Guilford Press: New York, NY, USA, 2007; pp. 560–580.
69. Depue, R.A.; Morrone-Strupinsky, J.V. A Neurobehavioral Model of Affiliative Bonding: Implications for Conceptualizing a Human Trait of Affiliation. *Behav. Brain Sci.* **2005**, *28*. [CrossRef] [PubMed]

70. Gilbert, P. Evolutionary Approaches to Psychopathology: The Role of Natural Defences. *Aust. N. Z. J. Psychiatry* **2001**, *35*, 17–27. [CrossRef] [PubMed]
71. Gilbert, P. *Compassion: Conceptualisations, Research and Use in Psychotherapy*, 1st ed.; Routledge: New York, NY, USA, 2005; ISBN 978-1-58391-982-8.
72. Hsiung, D.-Y.; Tsai, C.-L.; Chiang, L.-C.; Ma, W.-F. Screening Nursing Students to Identify Those at High Risk of Poor Mental Health: A Cross-Sectional Survey. *BMJ Open* **2019**, *9*, e025912. [CrossRef]
73. Mirzaei, M.; Yasini Ardekani, S.M.; Mirzaei, M.; Dehghani, A. Prevalence of Depression, Anxiety and Stress among Adult Population: Results of Yazd Health Study. *Iran. J. Psychiatry* **2019**, *14*, 137. [CrossRef]
74. Czech Ministry of Education. *University Students by Groups of Fields and Sex in 2018 [Studenti Vysokých Škol Podle Skupin Oborů a Pohlaví v Roce 2018]*; Czech Ministry of Education: Prague, Czech Republic, 2019.
75. Faul, F.; Erdfelder, E.; Buchner, A.; Lang, A.-G. Statistical Power Analyses Using G*Power 3.1: Tests for Correlation and Regression Analyses. *Behav. Res. Methods* **2009**, *41*, 1149–1160. [CrossRef] [PubMed]
76. Lovibond, P.F.; Lovibond, S.H. The Structure of Negative Emotional States: Comparison of the Depression Anxiety Stress Scales (DASS) with the Beck Depression and Anxiety Inventories. *Behav. Res. Ther.* **1995**, *33*, 335–343. [CrossRef]
77. Antony, M.M.; Bieling, P.J.; Cox, B.J.; Enns, M.W.; Swinson, R.P. Psychometric Properties of the 42-Item and 21-Item Versions of the Depression Anxiety Stress Scales in Clinical Groups and a Community Sample. *Psychol. Assess.* **1998**, *10*, 176–181. [CrossRef]
78. Raes, F.; Pommier, E.; Neff, K.D.; Van Gucht, D. Construction and Factorial Validation of a Short Form of the Self-Compassion Scale. *Clin. Psychol. Psychother.* **2011**, *18*, 250–255. [CrossRef] [PubMed]
79. Hayes, A.F. *Introduction to Mediation, Moderation, and Conditional Process Analysis: A Regression-Based Approach*; The Guilford Press: New York, NY, USA, 2013; ISBN 978-1-60918-230-4.
80. Hoaglin, D.C.; Iglewicz, B. Fine-Tuning Some Resistant Rules for Outlier Labeling. *J. Am. Stat. Assoc.* **1987**, *82*, 1147–1149. [CrossRef]
81. Tukey, J.W. The Future of Data Analysis. *Ann. Math. Stat.* **1962**, *33*, 1–67. [CrossRef]
82. Hayes, A.F. PROCESS: A Versatile Computational Tool for Observed Variable Mediation, Moderation, and Conditional Process Modeling [White Paper]. Available online: http://www.afhayes.com/public/process2012.pdf (accessed on 12 March 2022).
83. Ferreira, C.; Pinto-Gouveia, J.; Duarte, C. Self-Compassion in the Face of Shame and Body Image Dissatisfaction: Implications for Eating Disorders. *Eat. Behav.* **2013**, *14*, 207–210. [CrossRef] [PubMed]
84. Zhang, H.; Carr, E.R.; Garcia-Williams, A.G.; Siegelman, A.E.; Berke, D.; Niles-Carnes, L.V.; Patterson, B.; Watson-Singleton, N.N.; Kaslow, N.J. Shame and Depressive Symptoms: Self-Compassion and Contingent Self-Worth as Mediators? *J. Clin. Psychol. Med. Settings* **2018**, *25*, 408–419. [CrossRef]
85. Fong, M.; Loi, N.M. The Mediating Role of Self-Compassion in Student Psychological Health: Self-Compassion and Psychological Health. *Aust. Psychol.* **2016**, *51*, 431–441. [CrossRef]
86. Andrews, B.; Qian, M.; Valentine, J.D. Predicting Depressive Symptoms with a New Measure of Shame: The Experience of Shame Scale. *Br. J. Clin. Psychol.* **2002**, *41*, 29–42. [CrossRef]
87. Andrews, B.; Hunter, E. Shame, Early Abuse, and Course of Depression in a Clinical Sample: A Preliminary Study. *Cogn. Emot.* **1997**, *11*, 373–381. [CrossRef]
88. Quinn, D.M.; Williams, M.K.; Quintana, F.; Gaskins, J.L.; Overstreet, N.M.; Pishori, A.; Earnshaw, V.A.; Perez, G.; Chaudoir, S.R. Examining Effects of Anticipated Stigma, Centrality, Salience, Internalization, and Outness on Psychological Distress for People with Concealable Stigmatized Identities. *PLoS ONE* **2014**, *9*, e96977. [CrossRef]
89. Layte, R.; Whelan, C.T. Who Feels Inferior? A Test of the Status Anxiety Hypothesis of Social Inequalities in Health. *Eur. Sociol. Rev.* **2014**, *30*, 525–535. [CrossRef]
90. Carr, S.; Colthurst, K.; Coyle, M.; Elliott, D. Attachment Dimensions as Predictors of Mental Health and Psychosocial Well-Being in the Transition to University. *Eur. J. Psychol. Educ.* **2013**, *28*, 157–172. [CrossRef]
91. Cheng, H.-L.; Wang, C.; McDermott, R.C.; Kridel, M.; Rislin, J.L. Self-Stigma, Mental Health Literacy, and Attitudes Toward Seeking Psychological Help. *J. Couns. Dev.* **2018**, *96*, 64–74. [CrossRef]
92. Rüsch, N.; Müller, M.; Ajdacic-Gross, V.; Rodgers, S.; Corrigan, P.W.; Rössler, W. Shame, Perceived Knowledge and Satisfaction Associated with Mental Health as Predictors of Attitude Patterns towards Help-Seeking. *Epidemiol. Psychiatr. Sci.* **2014**, *23*, 177–187. [CrossRef] [PubMed]
93. Kotera, Y.; Laethem, M.; Ohshima, R. Cross-Cultural Comparison of Mental Health between Japanese and Dutch Workers: Relationships with Mental Health Shame, Self-Compassion, Work Engagement and Motivation. *Cross Cult. Strateg. Manag.* **2020**. ahead-of-print. [CrossRef]
94. Kotera, Y.; Sheffield, D. Revisiting the Self-Compassion Scale-Short Form: Stronger Associations with Self-Inadequacy and Resilience. *SN Compr. Clin. Med.* **2020**, *2*, 761–769. [CrossRef]
95. Redden, E.K.; Bailey, H.N.; Katan, A.; Kondo, D.; Czosniak, R.; Upfold, C.; Newby-Clark, I.R. Evidence for Self-Compassionate Talk: What Do People Actually Say? *Curr. Psychol.* **2021**, 1–17. [CrossRef]

healthcare

Article

Attitudes and Beliefs of Portuguese and American Nursing Students about Patients' Sexuality

Margarida Sim-Sim [1,†], Vicki Aaberg [2,†], Hélia Dias [3,†], Ermelinda Caldeira [1,†], Cinzia Gradellini [4,*,†], Daniela Mecugni [4,†] and Sagrario Gomez-Cantarino [5,6,†]

1 Nursing Department, University of Evora, Comprehensive Health Research, Integrated Researcher (CHRC), Escola Superior de Enfermagem S. João de Deus, 7000-811 Evora, Portugal; msimsim@uevora.pt (M.S.-S.); ecaldeira@uevora.pt (E.C.)
2 School of Health Sciences, Seattle Pacific University, Seattle, WA 98119, USA; aaberv@spu.edu
3 Instituto Politecnico de Santarém, Escola Superior de Saúde, 2005-075 Santarém, Portugal; helia.dias@essaude.ipsantarem.pt
4 EdSex Project, University of Modena and Reggio Emilia, Azienda Unità Sanitaria Locale-IRCCS of Reggio Emilia, 42122 Reggio Emilia, Italy; daniela.mecugni@unimore.it
5 Faculty of Physiotherapy and Nursing, Toledo Campus, University of Castilla-La Mancha, Avda Carlos III, s/n, 45071 Toledo, Spain; sagrario.gomez@uclm.es
6 Health Sciences Research Unit: Nursing (UICISA: E), The Nursing School of Coimbra (ESEnfC), 3004-011 Coimbra, Portugal
* Correspondence: cinzia.gradellini@ausl.re.it; Tel.: +34-39-0522-522514
† EdSex Projec.

Abstract: Nursing school graduates must be prepared to interact comfortably and effectively with patients about their sexual health. This study analyses the attitudes and beliefs about patient sexuality held by Portuguese and American nursing students. Objective: In Portuguese and American nursing students, (1) we analyzed students' attitudes and beliefs towards sexuality using the Sexuality Attitudes and Beliefs Survey (SABS); (2) we identified nationality, socio-demographic information, and affective-sexual beliefs and attitudes. Method: Quantitative, cross-sectional study; convenience sample of 296 students (63.2% Portuguese; 36.8% American); mean age: 21.9 years (SD = 3.12); two-way ANOVA and multiple correspondence analyses were performed. Results: Attitudes and beliefs toward sexuality: Portuguese women are more liberal than men, contrary to American students. Among both nationalities, participants with multiple sexual partners held more conservative attitudes. Sexual orientation: bisexual American students and homosexual Portuguese students are conservative. Multiple correspondence analysis revealed two profiles: (1) Portuguese students: liberal-tolerant in attitudes towards patient sexuality, live with family/roommate, 18 to 21 years old, no or one sexual partner; (2) US students: traditionalist attitudes towards patient sexuality, share house, 22 and 23 years old, multiple partners. Conclusion: Human sexuality must be addressed in nursing education curricula.

Keywords: nursing; health personnel; clinical competence; sexuality; intervention

Citation: Sim-Sim, M.; Aaberg, V.; Dias, H.; Caldeira, E.; Gradellini, C.; Mecugni, D.; Gomez-Cantarino, S. Attitudes and Beliefs of Portuguese and American Nursing Students about Patients' Sexuality. *Healthcare* **2022**, *10*, 615. https://doi.org/10.3390/healthcare10040615

Academic Editors: Yasuhiro Kotera and Elaina Taylor

Received: 14 February 2022
Accepted: 22 March 2022
Published: 25 March 2022

1. Introduction

Nursing students, while studying to work in their profession, perform care interactions in a real context. They deal with the patient's body and intimate issues and must engage with expressions of people's sexuality.

At the beginning of their undergraduate education, students' knowledge about sexuality is rooted in primary sources such as family, friends and the media [1,2]. At this stage, attitudes and beliefs tend to express traditionalism [3,4]. The social and family environment from which they come plays a determining role by offering values and life perspectives. Students from conservative families or those belonging to religious denominations are more traditionalist [2,5].

Sexual health is an important area for holistic care [3] and requires intentional training in nursing curricula [3,6]. The undergraduate curricula of European and American nursing education programs cover similar subjects, including sexual orientation, gender and identity [6]. However, as the inclusion of sexuality issues is controversial in some settings, the need for sexuality content in nursing education is clear and demands consistent use of pedagogical curricular strategies for effective inclusion in all settings [6–10].

Nursing students consider addressing sexuality a requirement of the profession [1,11,12]. In addition, teachers emphasize the need to educate students on positive attitudes in order to be more effective in dealing with patients [10,13]. Some students criticize the curricula, stating that there is little theoretical preparation and little training for the clinical experiences [3,9], especially with regard to lesbian, gay, bisexual, and transgender (LGBTQ) issues [1,14]. In clinical practice, they report a lack of time to address patients' sexual concerns and find that patients do not expect nurses or students to address sexual issues [7,15]. Studies reveal that students feel anxiety and worry about how they may be viewed by their colleagues when deciding to address patients regarding sexuality [12]. The students' resistance or difficulty worsens when the patient has a non-heterosexual orientation [16]. Although they may hold favourable attitudes, they feel discomfort and are reluctant to initiate the subject [1,5,11,16].

In Portugal, undergraduate nursing students' attitudes are less permissive and more conservative compared to students studying other health professions [17]. On the other hand, American studies identify more favourable attitudes and better knowledge and self-efficacy after specific training of undergraduate students in sexual health [18]. When the same instrument was applied to younger (accelerated program) American students, they held more conservative attitudes and beliefs than older (traditional program) undergraduate students in addressing the sexual concerns of patients [15]. In nursing students, research has used several instruments [17,19] for analysing attitudes, opinions and beliefs towards patient sexuality. However, it is rare that the same study takes place in samples of different countries and in diverse languages. In fact, the application of a single instrument, validated for different cultures, guarantees a reliable analysis of the construct. The objectives are (1) to analyse the attitudes and beliefs towards sexuality of the patient through the application of the Sexuality Attitudes and Beliefs Survey (SABS) and (2) to identify the profile of the students, considering characteristics such as nationality and socio-demographic and affective-sexual data.

2. Materials and Methods

This is a quantitative study of inferential character and uses a transversal approach with a convenience sample, with the participation of 296 undergraduate nursing students. Portuguese students attended state schools and American students attended private schools. Data from the Portuguese students were collected online, through the Lime-survey platform, by sending 340 emails to students from Évora and 375 emails to students from Santarém. The data from the American students, located in Seattle, were collected on paper from 135 students.

2.1. Data Collection Instrument

The instrument applied consisted of two sections: (a) sociodemographic data and (b) SABS scale in the original version [19] or in the version validated for Portuguese [20]. The scale is unidimensional, and the original form consists of a set of 12 items with a Cronbach's alpha coefficient of 0.75 and retest of 0.82 [19]. In the Portuguese version, the observation of the psychometric properties implied the removal of item 3, thus resulting in 11 items. In this version, the Cronbach's alpha coefficient was 0.72 and the test-retest was 0.80 [20]. In the current study, the subsample of Portuguese students (11 items in the SABS scale) shows a Cronbach's alpha value of 0.67, and the subsample of American students (12 items in the SABS scale) demonstrates a Cronbach's alpha value of 0.66.

The items are presented on a six-position scale, ranging between 1 (i.e., strongly disagree) and 6 (i.e., strongly agree). Items 1, 2, 4, 6, 8, 10 and 12 are reverse-formulated and are scored accordingly. Higher scores indicate stronger and more limiting attitudes and beliefs about addressing sexuality with the patient. In the current study, the response score on the SABS was analysed using the mean.

2.2. Ethical Procedures

The invitation to participate was presented face to face in the classroom, at a date prior to data collection. In the data collection instrument, the first page of both the online and paper versions presented the informed consent. In view of the student's inability to manually sign the online version, the first question was mandatory to answer. He/she could not progress to the questionnaire without explicitly ticking (i.e., yes versus no) to confirm consent. In the paper version, consent was signed in person. The questionnaires were anonymised, with no elements that could link the respondent to the response.

The study was submitted to and approved by the Ethics Committee for Research in the Areas of Human Health and Welfare of the University of Évora (Registration No. 18175).

2.3. Data Analysis

Statistical analysis was performed using the Statistical Package for the Social Sciences (IBM SPSS Statistics for Windows, Version 24®). The distribution of the sample, according to nationality showed normality (KS = 0.055; df = 187; p = 0.200; KS = 0.054; df = 109; p = 0.200) in the Portuguese and Americans, respectively. We opted for a parametric test approach, performing two-factor ANOVA analysis of variance and multiple correspondence analysis. Values are presented in mean and standard deviation, and the significance level is $p < 0.05$.

3. Results

A total of 296 students participated, of whom 187 (63.2%) were Portuguese and 109 (36.8%) American. The response rate in Évora was 28.5% (i.e., 340 sent vs. 97 completed responses), in Santarém was 24% (i.e., 375 sent vs. 90 completed responses) and in Seattle was 80.7% (i.e., 135 invited vs. 109 completed).

The age of participants ranged from 18 to 40 years, with a mean of 21.9 years (SD = 3.12) and a mode of 22 years. The representation of males compared to females was lower in both schools (n = 25: 13.4% vs. n = 162: 86.6% in the Portuguese; n = 15: 13.8% vs. n = 94: 86.2% in the Americans). The majority were in the 22–23 age group (n = 115: 38.9%). Regarding current residence, most Portuguese lived with their parents (n = 70: 37.4%), while the Americans lived with their peers in a rented house (n = 38: 34.9%).

Regarding sexual orientation, most defined themselves as heterosexual (n = 274: 92.6%) in the total sample. The majority of the students stated that they currently had a sexual partner (n = 152: 51.4%), while 76 (25.7%) stated they were currently not in a relationship and 64 (21.6%) had never had an affective-sexual relationship. The sociodemographic, sexual orientation and affective-sexual relationship data are shown in Table 1.

Through a two-factor ANOVA (gender and nationality), having verified the equality of variances by Levene's test (p = 0.261), it was observed, without significant differences ($F_{(3:292)}$ = 2.119; p = 0.098), that American students had higher mean scores on the SABS (M = 2.91; SD = 0.081) than Portuguese students (M = 2.78; SD = 0.063). Among Portuguese students, females show less conservative attitudes and beliefs towards the sexuality of the patient (M = 2.77; SD = 0.057) than males (M = 2.80; SD = 0.667). However, in American students, males are more liberal in their approach to patient sexuality than females (M = 2.86; SD = 0.410 vs. M = 2.95; SD = 0.608), according to Figure 1.

A two-way ANOVA test was performed, using the SABS scale and the students' nationality and the affective relationships they reported. The homogeneity of variances was verified in Levene's test (p = 0.945). It was observed without significant differences ($F_{(7:288)}$ = 1.239; p = 0.281), that American students held numerically higher averages in the SABS, that is, more limiting attitudes and beliefs in the approach to sexuality of the patient compared to the Portuguese students (M = 2.98; SD = 0.113 vs. M = 2.84; SD = 0.111). On the other hand, in both nationalities, the most conservative participants in their approach to patient sexuality were those with multiple sexual partners (M = 3.08; SD = 0.415 American; M = 3.00, SD = 0.415 Portuguese). The more liberal participants revealed having only one sexual partner (M = 2.86, SD = 0.087 and M = 2.74, SD = 0.057) (Figure 2).

Table 1. Characteristics of the participants.

Category	Characteristic	Portuguese	American	Total
	n (%)	*n* (%)	*n* (%)	
Sex	Male	40 (13.5)	25 (13.4)	15 (13.8)
	Female	256 (86.5)	162 (86.6)	94 (86.2)
Age	18–19 years old	51 (17.2)	45 (24.1)	6 (5.5)
	20–21 years old	92 (31.1)	57 (30.5)	35 (32.1)
	22–23 years old	115 (38.9)	51 (27.3)	64 (58.7)
	23–40 years old	38 (12.8)	34 (18.2)	4 (3.7)
Living at	Family home	85 (28.7)	70 (37.4)	15 (13.8)
	Academic residency	53 (17.9)	22 (11.8)	31 (28.4)
	Rented room	65 (22.0)	47 (25.1)	18 (16.5)
	Rented house with colleagues	67 (2.6)	29 (15.5)	38 (34.9)
	Other	26 (8.8)	19 (10.2)	7 (6.4)
Sexual Orientation	Heterosexual	274 (92.6)	171 (91.4)	103 (94.5)
	Homosexual	9 (3.0)	6 (3.2)	3 (2.8)
	Bisexual	10 (3.4)	7 (3.7)	3 (2.8)
	Not defined	3 (1.0)	3 (1.6)	-
Sexual partner	Never had sex partner	64 (21.6)	25 (13.4)	39 (35.8)
	No sex partner now	76 (25.7)	53 (28.3)	23 (21.1)
	One sex partner now	152 (51.4)	107 (57.2)	45 (41.3)
	Multiple sex partners	4 (1.4)	2 (1.1)	2 (1.8)
Total		296	187	109

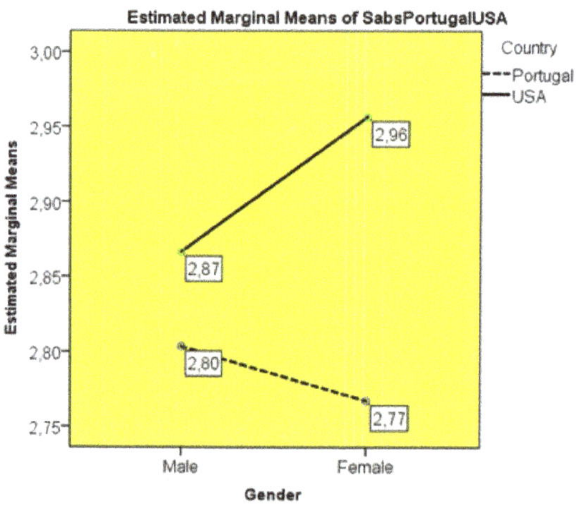

Figure 1. SABS considering students' nationality and gender.

Considering the students' sexual orientation (heterosexual, homosexual and bisexual), excluding the three participants who did not define themselves, a new two-factor ANOVA test (i.e., nationality and sexual orientation) was performed. Levene's test showed equality of variances ($p = 0.086$). Without significant differences ($F_{(5:287)} = 1.906$; $p = 0.093$), American students showed numerically higher means in the attitudes and beliefs towards sexuality of the patient when compared to the Portuguese (M = 2.863, SD = 0.160 vs. M = 2.772, SD = 0.109). It was observed that the most conservative about patient sexuality were the American students with a bisexual orientation (M = 3.19, SD = 0.337) and the least traditional were those who declared themselves homosexual (M = 2.44, SD = 0.337). In contrast, in the Portuguese, the most conservative were the students with a homosexual orientation (M = 2.86, SD = 0.238) (Figure 3).

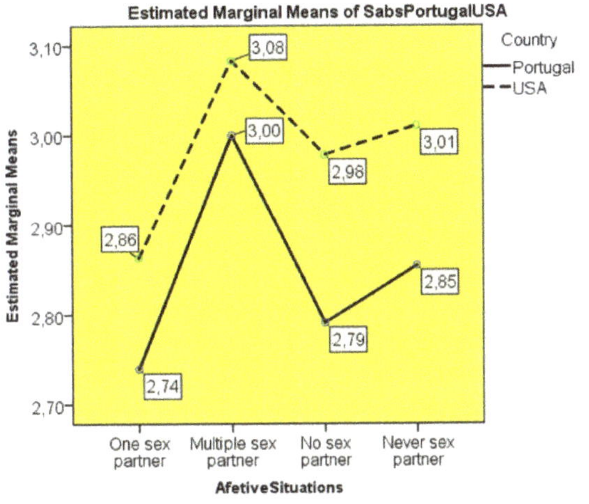

Figure 2. ABS considering students' nationality and affective relationship.

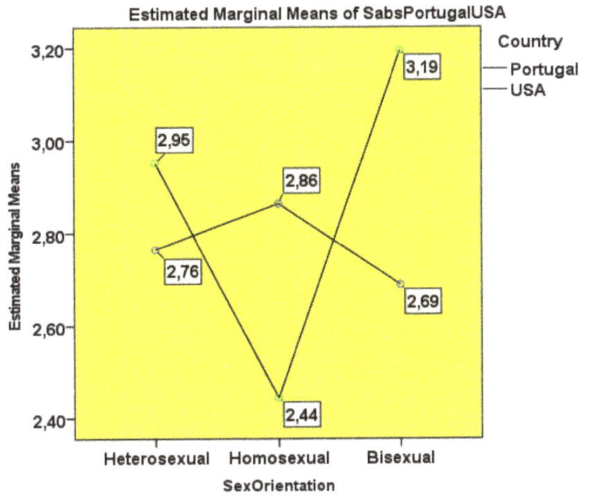

Figure 3. SABS considering students' nationality and sexual orientation.

One focus of this study was to identify the profile of nursing students related to attitudes toward patient sexuality, considering underlying characteristics like nationality, sociodemographic data (age, gender, type of residence where they live) and affective relationships. A multiple correspondence analysis (MCA) was developed to process the responses.

Before starting the analysis, the maximum and minimum scores on the SABS scale (Min = 1.36 and Max = 4.64) were considered, and three groups with the same range were formed, with cut-off points at 2.46 and 3.55. The SABS was thus defined as the liberal group (scale score between 1.36 and 2.45), the tolerant group (scale score between 2.46 and 3.54) and the traditionalist group (scale score between 3.55 and 4.64).

In the first exploration, the maximum number of dimensions (i.e., nine dimensions) was used to observe the behaviour of the inertia values (i.e., eigenvalues), according to Table 2. The first two dimensions have supremacy, with higher eigenvalues (i.e., 1.907, 1.371). A two-dimensional solution was then used. The total inertia became 0.546. The variance explained for Dimension 1 is 31.78% and for Dimension 2 is 22.85%.

Table 2. Two-dimensional solution.

Model Summary				
		Variance Accounted for		
Dimension	**Cronbach's Alpha**	**Total (Eigenvalue)**	**Inertia**	**% of Variance**
1	0.571	1907	0.318	31.778
2	0.325	1371	0.229	22.854
Total		3278	0.546	
Mean	0.468 [a]	1639	0.273	27.316

[a] Mean Cronbach's Alpha is based on the mean Eigenvalue.

Discrimination Measures

The discrimination measures of the variables residence, affective relationship and nationality have higher values in Dimension 1 (i.e., respectively, 0.571, 0.337 and 0.540).

In Dimension 2, the discrimination measures of the variables SAB groups and age groups have higher values (i.e., SABS groups = 0.159; Age Groups = 0.563). The gender indicator is not relevant in any of the dimensions, as revealed in Table 3, and in the observation of the discrimination measures chart, since the sex variable is very close to the origin (Figure 4).

Table 3. Discrimination measures of variables entered in the 2-dimensional model.

Discrimination Measures			
	Dimension		
	1	2	Mean
ResidNow	0.571	0.376	0.474
AfectiveRelationship	0.337	0.186	0.262
Country	0.540	0.087	0.314
AgeGroups	0.375	0.563	0.469
SABSgroups	0.082	0.159	0.121
Gender	0.000	0.001	0.000
Active Total	1.907	1.371	1.639
% of Variance	31.778	22.854	27.316

If we remove the gender variable from the analysis, the distribution and grouping of the categories suggests that the American students have a traditionalist profile manifested by students aged 22–23 years, living in rented accommodation with colleagues and with multiple affective relationships. In the Portuguese students, the profile oscillates between liberal and tolerant attitudes toward engaging with the patients' sexuality, falling into lower age groups, living in a family house or rented room, with no current affective-sexual relationship or with an exclusive partner (Figure 5).

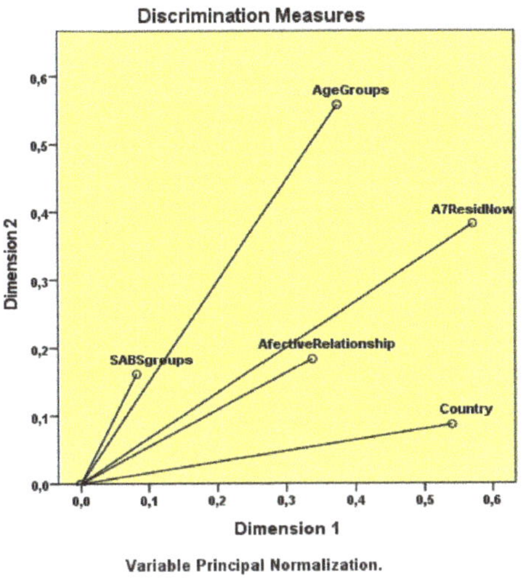

Figure 4. Discriminant measures.

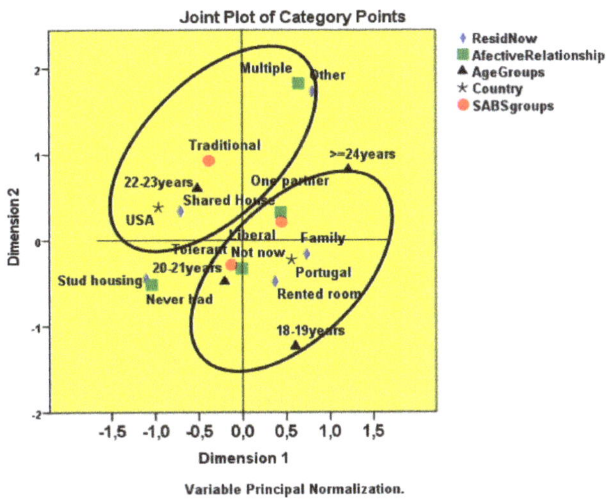

Figure 5. Joint plot of category points resulting from the multiple analysis.

4. Discussion

The objective of this study was to examine Portuguese and American nursing students' attitudes toward patients' sexuality and define their profile. The different proportion of answers obtained in the Portuguese versus American institutions reveals the influence that the collection method may have on the final data set. The response rates in the online application coincide with previous studies, which refer to a representation between 20% and 47%, while in the collection via paper, the sample is higher than those found in the literature, which vary between 32.6% and 75% [21]. Although the use of the internet increases the dissemination to and access of potential respondents, as well as the suitability for personal schedules and privacy, compared to the classroom during school hours, the response rates do not seem to reflect these advantages. The overload of requests for collaboration via the internet may be a demotivating factor. Perhaps the direct and personal stimulus of the researcher, providing a paper questionnaire and requesting collaboration from the students, is the cause of a more successful response rate. On the other hand, the online data collection route should be cultivated, as it favours the preservation of the environment and the dematerialisation of communication, and it saves financial resources. In fact, in addition to the non-countable expenses to the environment, the paper-based application would be a considerable burden on Portuguese students.

Gender representation is similar in both sets of participants, favouring women. This was expected, as nursing is a career widely chosen by women. In this profession, the nature of caring imprints social attributes associated with femininity. In addition, there may be higher social expectation for women to be nurses, thereby feminising the profession [22]. The lower number of male students may be due to ambivalence between the call to the profession and the construct of professional identity. Some studies show that there are barriers to student development, including the lack of a clear and acceptable public image, which suggests the idea that the option for nursing is not appropriate for men [23]. Gender stereotypes in professions may induce biases, since performance expectations are misrepresented. Not only in the population but also among nursing students, there are gender stereotypes [22] that can negatively affect students' professional paths. Professions that, like nursing, are associated overwhelmingly with one gender present a risk that some members of the profession may be devalued when the traditional gender boundaries are broken [24].

Although heterosexuality is the orientation most frequently mentioned by the participants, the fact that some students reported other orientations reveals that they are in the process of recognizing and making known the way they feel and live their sexuality. The entry into university approximately coincides with the leaving of the parents' home, a phase of searching for adult identity in which young people tend to free themselves from family norms. In this context, a positive view of orientations other than heterosexuality grows and becomes more acceptable [25] and feelings of self-marginalisation can be reduced.

The affective-sexual relationships recognised by the participants, in the sense of romantic relationships, follow the juvenile framework, when the adolescent crisis is overcome [26]. The type of involvements and the expressions of gender identity in the current study correspond to those identified in the literature [26,27]. In fact, during university education there are typical behaviours (i.e., exclusivity, short relationships, sequential monogamies, multiple partners), which may reflect development, opportunities or experimentation.

4.1. Gender vs. Nationality in the SABS

The SABS results in the two-way ANOVA between the factors of sex and nationality point to a difference in female students, with American students demonstrating the most conservative results, and Portuguese female students demonstrating the most liberal. The interpretation requires care to avoid xenophobic allocations or approximations to nationalism, machismo or marianism. The similar SABS scores for Portuguese and American men can be interpreted from a gender perspective, as men may be less subject to social influences than women. In fact, the role norms with which men and women from different cultural, ethnic and religious traditions invest in care are different [28]. The results seem to be rooted in the pathway of nursing education. The admission of more male students, older people, individuals with different marital status and families, and emigrants or their descendants has broken the conventional female profile [28]. Moreover, although the profession retains a female majority, the sexual division of care is blurring [29]. The higher score of American women in the SABS score suggests that private university institutions may favour a more conservative education. In the Portuguese polytechnic environment, there may be greater openness in this area of patient care. Controversy, however, may still reside in the eventual perception of the students regarding the client's responses, since in some cultures, men prefer male caregivers, while women prefer female [28,29]. A bias related to social desirability may still be present, as American and Portuguese students responded to what is expected of the good caregiver [29].

4.2. Affective-Sexual Relationship vs. Nationality

The SABS results in the two-way ANOVA between the factors of type of affective-sexual relationship and nationality suggest the dominance of a hidden curriculum, prior and simultaneous to the nursing training process, to which both samples are exposed. There may be social stereotypes in which both the experience of one's own sexuality and opinions about the sexuality of others are rooted. On the other hand, this suggests that in spite of the sociocultural surroundings, there is an individual and singular path in the development of the student. Not having experienced affective-sexual relationships and not having a partner or being in an exclusive relationship seem to transmit a greater openness to the sexuality of the others and to consider the patient as a total being. The participants with multiple partners, on the other hand, by showing more barriers, suggest less maturity, and less affective-emotional development. Perhaps these individuals are still trapped in a late adolescence, which prolongs experimentation [26].

4.3. Sexual Orientation vs. Nationality

Students coming from conservative, religious backgrounds are more traditionalist [2,5] in their attitudes toward sexuality. The relationship between conservative religious beliefs and attitudes and beliefs toward patient sexuality need to be explored in further detail. These authors describe the results obtained in the ANOVA with the two factors of sexual orientation and nationality.

4.4. The SABS Profile in Portuguese and American Students

The MCA revealed relations of interdependence, and thus profiling, between the student attitudes to addressing sexuality in patient care and a series of sociodemographic variables, including affective relationship, nationality, place of residence and age. The MCA, although with exploratory purposes, was useful in this analysis, as it involved multifaceted structures rooted in the cultural environment [30]. It allowed us to observe the association between the categories of different variables, thus defining profiles. This is useful for understanding the vision of the participants and for emphasizing to teachers and clinical supervisors the students' needs for guidance in the care of patients as unique and total beings who live their sexuality independently of their health or illness status. It is important to emphasize the patient's sexuality as a human dimension in the teaching–learning process of nursing students, both in theoretical and clinical approaches [15]. The identified profiles highlight aspects that call for teaching strategies so that students can develop. These profiles are consistent with studies on university students from the American continent, where uninhibitedness (i.e., multiple partners) is evident when it comes to personal sexuality [31] but, contradictorily, the same individuals report more conservative sexual attitudes with regard to patient care, suggesting that their attitude is rooted in their sociocultural background.

With regard to the Portuguese students, the profile suggests a liberal attitude. However, they are younger students and thus more susceptible to prescriptive teaching models with a biomedical root. In Portugal, although under the European guidelines, it is not clear in the undergraduate nursing study plans which subjects include the patient's sexuality as an explicit and concrete teaching subject. The approach to the sexuality of the patient mainly involves strategies based on the transmission of anatomical and physiological knowledge. In fact, the profile of the Portuguese students participating in this study is in agreement with semantic representations of sexuality, which are oriented toward the normative in the psychophysiological dimensions [32–34].

We may suspect that in both profiles the predisposition to approach the sexuality of the patient reverts into postponement or is even contradictory. Although the students identify their roles, they hesitate to follow through [35,36]. Considering sexuality as an area of care, but having reservations and subsequently not carrying out the care, underlines beliefs or barriers.

5. Conclusions

Despite the current eroticized society, nursing students' attitudes towards patient sexuality seem to be anchored in traditionalism. On the other hand, a certain cleavage seems to emerge between the recognition of the patient's sexuality as an area to be considered in care provision and the predisposition to address it in clinical practice. In fact, the profile of both groups of students reveals openness to providing this care yet hesitation to do so.

Although the students identify their role, they hesitate to play it. It will be important to think of undergraduate training programmes in nursing as the beginning of the construction of competencies, so that the sexuality of the patient is assumed as a dimension of human life. Such programmes may need evaluation indicators oriented toward the theme of sexuality and strategies for the incorporation of this content. These strategies may include small group work with discussion of cases or dilemmas, interview training for data collection, simulation, autoscopies, and role playing under the guidance of a supervisor.

Considering sexuality as an area of care, but having reservations and not following through on the care, reveals underlying beliefs or barriers that, in the profession, are contradictory, since nursing mandates holistic care. The inclusion of skills to learn to care for the patient in this area seems urgent. The current results show the need for formative programs where the student's individual, social and cultural dimensions should be examined for the influence they may have on students' ability to engage in the topic of sexuality with patients.

6. Limitations

The non-randomised sample prevents generalisation of the results. Data collection through two formats (online and face to face) may imprint bias. Another weakness of the current study is the heterogeneity of the sample, with an age range of 18–40 years and in various years of training.

Author Contributions: Conceptualization, M.S.-S., S.G.-C. and C.G.; data curation, V.A. and S.G.-C.; formal analysis, M.S.-S.; funding acquisition, H.D. and D.M.; project administration, M.S.-S. and E.C.; software, M.S.-S.; supervision, V.A.; validation, M.S.-S. and S.G.-C.; writing—original draft, V.A., S.G.-C., H.D., E.C., D.M. and C.G.; writing—review and editing, H.D., E.C., D.M. and C.G. All authors have read and agreed to the published version of the manuscript.

Funding: This research received no external funding.

Institutional Review Board Statement: The study was submitted to the Ethics Committee for Research in the Areas of Human Health and Welfare of the University of Évora, resulting in approval (Registration No. 18175).

Informed Consent Statement: Informed consent was obtained from all subjects involved in the study.

Data Availability Statement: Not applicable.

Conflicts of Interest: The authors declare no conflict of interest.

References

1. Evcili, F.; Demirel, G. Sexual myths of midwifery and nursing students and their attitude regarding the assessment of sexual health. *J. Hum. Sci.* **2018**, *15*, 2280–2288. [CrossRef]
2. Papaharitou, S.; Nakopoulou, E.; Moraitou, M.; Tsimtsiou, Z.; Konstantinidou, E.; Hatzichristou, D. Exploring sexual attitudes of students in health professions. *J. Sex. Med.* **2008**, *5*, 1308–1316. [CrossRef] [PubMed]
3. Pengpid, S.; Peltzer, K. Associations of physical partner violence and sexual violence victimization on health risk behaviours and mental health among university students from 25 countries. *BMC Public Health* **2020**, *20*, 937. [CrossRef] [PubMed]
4. Atlý Özbaþ, A.; Gürhan, N.; Duyan, V. An Evaluation of Sexual Attitudes of Nursing Students in Turkey. *J. Psychiatr. Nurs.* **2016**, *7*, 114–120. [CrossRef]
5. Bal, M.D.; Sahiner, N.C. Turkish Nursing Students' Attitudes and Beliefs Regarding Sexual Health. *Sex. Disabil.* **2015**, *33*, 223–231. [CrossRef]
6. Cappiello, J.; Coplon, L.; Carpenter, H. Systematic Review of Sexual and Reproductive Health Care Content in Nursing Curricula. *J. Obstet. Gynecol. Neonatal Nurs.* **2017**, *46*, e157–e167. [CrossRef]

7. Cesnik, V.M.; Zerbini, T. Sexuality education for health professionals: A literature review. *Estud. Psicol. Camp.* **2017**, *34*, 161–172. [CrossRef]
8. Carteiro, D.; Néné, M. A importância da formação na área da sexualidade em enfermagem. *Rev. Assoc. Port. Enferm. Obs.* **2015**, *16*, 4–8.
9. Figueiroa, M.d.N.; Menezes, M.L.N.d.; Monteiro, E.M.L.M.; Andrade, Â.R.L.D.; Fraga, D.P.F.; Oliveira, M.V.D. A formação relacionada com a sexualidade humana na perceção de estudantes de enfermagem. *Rev. Enferm. Ref.* **2017**, *4*, 21–30. [CrossRef]
10. Aaberg, V. The state of sexuality education in baccalaureate nursing programs. *Nurse Educ. Today* **2016**, *44*, 14–19. [CrossRef]
11. Blakey, E.P.; Aveyard, H. Student nurses' competence in sexual health care: A literature review. *J. Clin. Nurs.* **2017**, *26*, 3906–3916. [CrossRef] [PubMed]
12. Kong, S.K.; Wu, L.H.; Loke, A.Y. Nursing students' knowledge, attitude and readiness to work for clients with sexual health concerns. *J. Clin. Nurs.* **2009**, *18*, 2372–2382. [CrossRef] [PubMed]
13. Sung, S.C.; Huang, H.C.; Lin, M.H. Relationship between the knowledge, attitude, and self-efficacy on sexual health care for nursing students. *J. Prof. Nurs.* **2015**, *31*, 254–261. [CrossRef] [PubMed]
14. Greene, M.Z.; France, K.; Kreider, E.F.; Wolfe-Roubatis, E.; Chen, K.D.; Wu, A.; Yehia, B.R. Comparing medical, dental, and nursing students' preparedness to address lesbian, gay, bisexual, transgender, and queer health. *PLoS ONE* **2018**, *13*, e0204104. [CrossRef] [PubMed]
15. Magnan, M.A.; Norris, D.M. Nursing students' perceptions of barriers to addressing patient sexuality concerns. *J. Nurs. Educ.* **2008**, *47*, 260–268. [CrossRef]
16. Richardson, B.P.; Ondracek, A.E.; Anderson, D. Do student nurses feel a lack of comfort in providing support for Lesbian, Gay, Bisexual or Questioning adolescents: What factors influence their comfort level? *J. Adv. Nurs.* **2017**, *73*, 1196–1207. [CrossRef]
17. Azevedo, J.; Palha, A.; Esteves, M.; Coelho, M.V. P02–197 Attitudes of medicine and nursing students towards the sexuality of older people. *Eur. Psychiatry* **2009**, *24*, S887. [CrossRef]
18. Aaberg, V. Increasing Student Knowledge, Attitudes, and Self-Efficacy in Sexuality Education. *J. Obstet. Gynecol. Neonatal Nurs.* **2019**, *48*, S76. [CrossRef]
19. Reynolds, K.E.; Magnan, M.A. Nursing attitudes and beliefs toward human sexuality: Collaborative research promoting evidence-based practice. *Clin. Nurse Spec.* **2005**, *19*, 255–259. [CrossRef]
20. Dias, H.; Sim-Sim, M. Validação para a população portuguesa do Sexuality Attitudes and Beliefs Survey (SABS). *Acta Paul. Enferm.* **2015**, *28*, 196–201. [CrossRef]
21. Nulty, D.D. The adequacy of response rates to online and paper surveys: What can be done? *Assess. Eval. High. Educ.* **2008**, *33*, 301–314. [CrossRef]
22. Thurston, M.M.; Chesson, M.M.; Harris, E.C.; Ryan, G.J. Professional Stereotypes of Interprofessional Education Naive Pharmacy and Nursing Students. *Am. J. Pharm. Educ.* **2017**, *81*, 84. [CrossRef] [PubMed]
23. Zamanzadeh, V.; Valizadeh, L.; Negarandeh, R.; Monadi, M.; Azadi, A. Factors influencing men entering the nursing profession, and understanding the challenges faced by them: Iranian and developed countries' perspectives. *Nurs. Midwifery Stud.* **2013**, *2*, 49–56. [CrossRef] [PubMed]
24. Heilman, M.E. Gender stereotypes and workplace bias. *Res. Organ. Behav.* **2012**, *32*, 113–135. [CrossRef]
25. Feng, Y.; Lou, C.; Gao, E.; Tu, X.; Cheng, Y.; Emerson, M.R.; Zabin, L.S. Adolescents' and young adults' perception of homosexuality and related factors in three Asian cities. *J. Adolesc. Health* **2012**, *50* (Suppl. S3), S52–S60. [CrossRef]
26. Gómez-López, M.; Viejo, C.; Ortega-Ruiz, R. Well-Being and Romantic Relationships: A Systematic Review in Adolescence and Emerging Adulthood. *Int. J. Environ. Res. Public Health* **2019**, *16*, 2415. [CrossRef] [PubMed]
27. Baptiste-Roberts, K.; Oranuba, E.; Werts, N.; Edwards, L.V. Addressing Health Care Disparities among Sexual Minorities. *Obstet. Gynecol. Clin. N. Am.* **2017**, *44*, 71–80. [CrossRef]
28. Bednarz, H.; Schim, S.; Doorenbos, A. Cultural diversity in nursing education: Perils, pitfalls, and pearls. *J. Nurs. Educ.* **2010**, *49*, 253–260. [CrossRef]
29. Dias, H.; Lima, L. Do ser mulher ao ser enfermeira: Cuidar na perspetiva de género. *Psicol. Teor. Investig. Prática* **2003**, *2*, 205–231.
30. Fennell, R.; Grant, B. Discussing sexuality in health care: A systematic review. *J. Clin. Nurs.* **2019**, *28*, 3065–3076. [CrossRef]
31. Döring, N.; Daneback, K.; Shaughnessy, K.; Grov, C.; Byers, E.S. Online Sexual Activity Experiences among College Students: A Four-Country Comparison. *Arch. Sex. Behav.* **2017**, *46*, 1641–1652. [CrossRef] [PubMed]
32. Dias, H.; Sim-Sim, M. Representações sobre a sexualidade dos estudantes do 1° ano do Curso de Enfermagem: Um estudo exploratório. *Rev. UIIPS* **2015**, *3*, 385–395.
33. Yáñez-Araque, B.; Gómez-Cantarino, S.; Gutiérrez-Broncano, S.; López-Ruiz, V.R. Examining the Determinants of Healthcare Workers' Performance: A Configurational Analysis during COVID-19 Times. *Int. J. Environ. Res. Public Health* **2021**, *18*, 5671. [CrossRef]
34. Sim-Sim, M.; Aaberg, V.; Gómez-Cantarino, S.; Dias, H.; Caldeira, E.; Soto-Fernandez, I.; Gradellini, C. Sexual Quality of Life-Female (SQoL-F): Cultural Adaptation and Validation of European Portuguese Version. *Healthcare* **2021**, *10*, 255. [CrossRef] [PubMed]

35. Cantarino, S.G.; Pinto, J.M.; Fabião, J.A.; García, A.M.; Abellán, M.V.; Rodrigues, M.A. The Importance of Religiosity/Spirituality in the Sexuality of Pregnant and Postpartum Women. *PLoS ONE* **2016**, *16*, e0156809. [CrossRef] [PubMed]
36. Ebert, J.F.; Huibers, L.; Christensen, B.; Christensen, M.B. Paper- or Web-Based Questionnaire Invitations as a Method for Data Collection: Cross-Sectional Comparative Study of Differences in Response Rate, Completeness of Data, and Financial Cost. *J. Med. Internet. Res.* **2018**, *20*, e24. [CrossRef] [PubMed]

healthcare

MDPI

Article

The Nine-Item Internet Gaming Disorder Scale (IGDS9-SF): Its Psychometric Properties among Sri Lankan Students and Measurement Invariance across Sri Lanka, Turkey, Australia, and the USA

Amira Mohammed Ali [1], Rasmieh Al-Amer [2,3], Maha Atout [4], Tazeen Saeed Ali [5], Ayman M. Hamdan Mansour [6], Haitham Khatatbeh [7], Abdulmajeed A. Alkhamees [8,*] and Amin Omar Hendawy [9,10]

1 Department of Psychiatric Nursing and Mental Health, Faculty of Nursing, Alexandria University, Smouha, Alexandria 21527, Egypt; mercy.ofheaven2000@gmail.com
2 Faculty of Nursing, Isra University, Amman 11953, Jordan; r.al-amer@outlook.com
3 School of Nursing and Midwifery, Western Sydney University, Penrith, NSW 2751, Australia
4 School of Nursing, Philadelphia University, Amman 19392, Jordan; m.atout@philadelphia.edu.jo
5 School of Nursing and Midwifery, Aga Khan University, Karachi 3500, Pakistan; tazeen.ali@aku.edu
6 Department of Psychiatric and Mental Health Nursing, School of Nursing, The University of Jordan, Amman 11942, Jordan; a.mansour@ju.edu.jo
7 Faculty of Health Sciences, Doctoral School of Health Sciences, University of Pécs, 7621 Pécs, Hungary; khatatbeh.haitham@etk.pte.hu
8 Department of Medicine, Unayzah College of Medicine and Medical Sciences, Qassim University, Unayzah, Al Qassim 52571, Saudi Arabia
9 Department of Biological Production, Tokyo University of Agriculture and Technology, Tokyo 183-8509, Japan; amin.hendawy@gmail.com
10 Department of Animal and Poultry Production, Faculty of Agriculture, Damanhour University, Damanhour 22516, Egypt
* Correspondence: a.alkhamees@qu.edu.sa; Tel.: +966-06-3800916

Citation: Ali, A.M.; Al-Amer, R.; Atout, M.; Ali, T.S.; Mansour, A.M.H.; Khatatbeh, H.; Alkhamees, A.A.; Hendawy, A.O. The Nine-Item Internet Gaming Disorder Scale (IGDS9-SF): Its Psychometric Properties among Sri Lankan Students and Measurement Invariance across Sri Lanka, Turkey, Australia, and the USA. *Healthcare* **2022**, *10*, 490. https://doi.org/10.3390/healthcare10030490

Academic Editors: Yasuhiro Kotera and Elaina Taylor

Received: 9 February 2022
Accepted: 5 March 2022
Published: 7 March 2022

Abstract: The prevalence of internet gaming disorders (IGD) is considerably high among youth, especially with the social isolation imposed by the ongoing COVID-19 pandemic. IGD adversely affects mental health, quality of life, and academic performance. The Internet Gaming Disorder Scale (IGDS9-SF) is designed to detect IGD according to DSM-IV diagnostic criteria. However, inconsistent results are reported on its capacity to diagnose IGD evenly across different cultures. To ensure the suitability of the IGDS9-SF as a global measure of IGD, this study examined the psychometric properties of the IGDS9-SF in a sample of Sri Lankan university students (N = 322, mean age = 17.2 ± 0.6, range = 16–18 years, 56.5% males) and evaluated its measurement invariance across samples from Sri Lanka, Turkey, Australia, and the USA. Among Sri Lankan students, a unidimensional structure expressed good fit, invariance across different groups (e.g., gender, ethnicity, and income), adequate criterion validity (strong correlation with motives of internet gaming, daily gaming duration, and sleep quality), and good reliability (alpha = 0.81). Males and online multiplayers expressed higher IGD levels, greater time spent gaming, and more endorsement of gaming motives (e.g., Social and Coping) than females and offline players. Across countries, the IGDS9-SF was invariant at the configural, metric, and scalar levels, although strict invariance was not maintained. The lowest and highest IGD levels were reported among Turkish and American respondents, respectively. In conclusion, the IGDS9-SF can be reliably used to measure IGD among Sri Lankan youth. Because the scale holds scalar invariance across countries, its scores can be used to compare IGD levels in the studied countries.

Keywords: coronavirus disease 2019/COVID-19; Internet Gaming Disorder Scale 9—Short Form (IGDS9-SF); university students; factorial structure/psychometric properties/structural validity/validation; cultur*/collectivisti*/individualis*; invariance; gender; game type

1. Introduction

Internet gaming is commonly used as a recreational activity among children, adolescents, and young adults [1]. With the expansion of internet technology, internet gaming has been employed to serve educational purposes, promote physical activity, develop cognitive skills and for therapeutic actions (e.g., reasoning, spatial awareness, and problem-solving) [2]. While most industries were negatively affected during the coronavirus disease 2019 (COVID-19) pandemic, the gaming industry has considerably flourished worldwide, with increased gaming time spent by previous users, numbers of new users, female users, and traffic in online mobile gaming. Indeed, the World Health Organization's (WHO) collaborative campaign (#PlayApartTogether) has been employed to promote online gaming as a method of fostering socialization while maintaining spatial distancing to prevent infection spread [2–4]. Increased risk for excessive gaming and increased screen use time among youth during COVID-19 may be a method to compensate for negative emotions associated with social isolation/being homebound, lack of meaningful activities (e.g., due to closure of schools and workplaces), inability to participate in entertainment activities that were available before the pandemic (e.g., clubs and cinemas), and COVID-19 burnout [3–6].

Children and adolescents are particularly vulnerable to developing maladaptive patterns of excessive or problematic gaming [3]. Internet Gaming Disorder (IGD) is defined as a behavioral pattern of persistent and recurrent involvement in online and offline games, resulting in remarkable distress and impairment in essential life activities (work and study) for a period of 12 months or more [7,8]. The global prevalence of IGD ranges between 0.7% and 15.6% [9,10]. In a Chinese study comprising 2863 school children, 83.0% played video games during the COVID-19 pandemic. Excessive and pathological gaming were evident in 20.9% and 5.3% of the participants [11].

Factors associated with increased IGD during COVID-19 include male gender, young age, loneliness, lack of parental support and/or supervision, mental health problems, and low socio-economic status [1,3,11,12]. Internet gaming among youth is associated with a plethora of negative consequences: poor health-related quality of life, sleep disturbances, impaired life skills, low self-esteem, concentration problems, poor communication skills, higher social distress, poor real-life relationships, loneliness, aggression, poor academic/work performance, poor impulse control, and tendency toward psychopathology [2,13–16]. Persistent or excessive gaming may increase the risk of more serious mental health problems [4]. Suicide has been reported among adolescents and emerging adults with psychological predispositions who played videogames that take up many hours a day such as PlayerUnknown's Battlegrounds (PUBG) [12].

The Diagnostic and Statistical Manual of Mental Disorders (DSM–5) classifies IGD in Section III—disorders requiring further investigation [8,14]. According to the DSM-5, meeting five out of nine normative symptomatic criteria is sufficient for establishing IGD as a diagnosis: (1) preoccupation with playing on the internet/digital games; (2) withdrawal symptoms when internet games are not available; (3) tolerance noted by increased time spent in gaming; (4) relapse noted by failed attempts to quit gaming; (5) loss of interest in other previous hobbies/entertainment behaviors because of, and with the exception of, online games; (6) continued and excessive use of online games despite knowledge of the psychosocial problems it causes; (7) deception of relatives, therapists, or other people about the amount of time spent in gaming; (8) mood modification is noted by use of online games to escape or mitigate negative emotions; and, (9) losing significant interpersonal relationships, work and educational or professional opportunities as a result of participating in internet gaming [7,9]. Because research on IGD is relatively recent, these nine criteria have been described based on existing research on pathological gambling and substance use disorder. Therefore, IGD measurement may be associated with methodological issues i.e., in relation to the definition and presentation of IGD [8]. Among 18 measures designed to assess IGD, the Internet Gaming Disorder Scale 9–Short-Form (IGDS9-SF), a brief form of the IGDS, has been designed to detect all nine diagnostic criteria of IGD [8,16].

The IGDS9-SF has been translated into many languages, including Italian and Albanian [8], Korean [15], Spanish [14], Polish [17], Turkish [10], Arabic [18], and Portuguese [19]. Although numerous studies report good reliability and validity of the IGDS9-SF in samples from Western and developed countries, its psychometric evaluation in various cultural contexts is limited, which may restrict its use as a global measure of IGD [8,17].

Cultural orientations may considerably influence the way through which individuals respond to a symptom scale. For instance, collectivistic cultures put a great appreciation for group values and norms while conditions entailing deviation from group norms (e.g., mental disorders such as IGD) can be stigmatized [20]. Therefore, collectivistic individuals tend to express IGD scores close to the mean, resulting in a minimal range of IGDS9-SF item responses. On the other hand, individualistic cultures entail appreciation of individual goals, values, competition and achievement as a base of social hierarchies. Therefore, individualistic persons tend to compete in gaming in order to achieve higher rankings [8,21].

Measurement invariance is frequently tested to ensure the usability of a measure for comparing the levels of a latent construct across different groups (e.g., cultures, ethnicities, genders, and age) [22,23]. Measurement invariance is assessed at four levels: configural, metric, scalar, and strict. Ideally, an invariant scale successfully reflects similar conceptualization of the underlying latent structure among groups (usually by reporting the same observed scores), similar degree of endorsement of items, and a capacity to objectively compare scale mean scores among groups. Comparisons based on non-invariant measures, especially at the scalar level, are likely to be invalid because group scores are confounded by differences in scaling properties across groups [8,22]. Although all previous findings support the unidimensional structure of the IGDS9-SF, some degree of measurement non-invariance has been expressed, especially in studies comprising English-speaking countries (e.g., Australia, United Kingdome (UK), and the United States of America (USA)) and non-English-speaking samples e.g., Polish and Indian [17,21]. Even among English-speakers from the USA and Australia, investigations of time invariance (three months) uncovered partial metric and scalar non-invariance among Australian gamers [24].

Given the widespread incidence of IGD and related mental health and academic adverse effects among youth, careful identification and proper management of IGD in this group may have implications for preventing/mitigating psychiatric comorbidities [23,25]. Lack of validation of IGD measures in developing countries such as Sri Lanka represents a challenge for IGD detection and treatment. It is not clear if IGD levels among youth from Sri Lanka can be compared with IGD levels among countries with evolving economies such as Turkey and developed countries such as Australia and the USA. These four countries do not only vary according to their economy but also according to the dominant cultural orientations. To fill the gap, the current study aimed to examine the structure, invariance, and criterion validity of the IGD9-SF among university students from Sri Lanka. We hypothesized that the IGD9-SF would express a unidimensional structure that would be invariant across different groups (e.g., gender, ethnicity, and income). We also expected that the IGD9-SF would strongly correlate with average daily gaming time, different motives of gaming, as well as sleep quality and quantity. In addition, we expected that the IGD9-SF would express some none-invariance across countries (Sri Lanka, Turkey, Australia, and the USA).

2. Materials and Methods

2.1. Study Design, Participants, and Procedure

This cross-sectional study is a secondary analysis based on three publicly accessible datasets. The first dataset is affiliated with University of Colombo Faculty of Medicine, Sri Lanka [26] and it is associated with a published study [27]. This dataset comprises a sample of advanced level university students obtained through random cluster sampling from four schools of the Colombo Educational Zone in Sri Lanka. This sample was used to examine the psychometric properties of the IGDS9-SF in Sri Lanka [27]. The sociodemographic, academic, and gaming characteristics of this sample are shown in Table 1.

Table 1. Sociodemographic, academic, and gaming characteristics of university students from Sri Lanka.

Participant Characteristics	(N = 322)
Age in years mean (SD)	17.2 ± 0.6
Gender	
Males	182 (56.5%)
Females	140 (43.5%)
Ethnicity	
Sinhala	249 (77.3%)
Others	73 (22.7%)
Language used	
Sinhala	152 (47.2%)
English	170 (52.8%)
Major	
Physical education	88 (27.3%)
Commerce	106 (32.9%)
Biology	67 (20.8%)
Arts	51 (15.9%)
Others	10 (3.1%)
Monthly income per household (SLR) ▲	
<100,000	124 (38.5%)
<200,000	106 (32.9%)
>200,000	92 (28.6%)
Gaming Type	
Offline single player	109 (33.9)
Online multiplayer	164 (50.9)
Online single player	49 (15.2)
Gaming hours/day	
One hour or less	172 (53.4)
Two to three hours	91 (28.3)
More than three hours	59 (18.3)
Device	
Mobile	206 (64.0%)
Others	116 (36.0%)
Sleep hours/day	
Five hours or less	102 (31.7%)
Six hours or more	220 (68.3.2%)

SLR: Sri Lankan rupee; ▲: one SLR is equal to $0.0049 or €0.0044.

Invariance of the IGDS9-SF across countries is based on data of the IGDS9-SF only from an international sample, which was integrated from three publicly accessible datasets, including the current sample from Sri Lanka [26]. It also included a sample from Turkey comprising 244 university students who reported playing digital games. Males and females were almost equally represented (N =113, 46.3%) and (N = 131, 53.7%), respectively [28]. No further details are available about the characteristics of those students. The third dataset comprised online gamers from Australia (N = 738, mean age = 25.8 ± 7.6, range = 18–72 years, 49.3% females, 71.7% employed, 32.7% students) and the USA (N = 222, mean age = 27.0 ± 8.0, range = 18–63 years, 54.1% females, 70.3% employed, 39.2% students) [29]. Data collection was obtained through an anonymous online survey conducted through SurveyGizmo, and an ethical approval for data collection was issued by the ethics committee of Cairnmillar Institute. Further details on the characteristics of the participants from those countries are reported in detail elsewhere [30]. Because all the datasets are shared under the terms of creative common license (CC BY 4.0) [26,28] or are in the public domain [29], we did not obtain an ethical approval for the current study.

2.2. Data Collection Measures

The questionnaire addressed to students from Sri Lanka included a personal information form inquiring about students' age, gender, academic major, ethnicity, and gaming experience (hours, type of gaming, age of start). It also comprised other measures, including the Internet Gaming Disorder Scale 9—Short Form (IGDS9-SF), a brief measure of the severity of IGD symptoms [31]. The scale consists of nine items. Each item is rated on a 5-point Likert scale ranging from (1 = never) to (5 = very often). The maximum and minimum scores of the IGDS9-SF are 9 and 45. Higher scores reflect higher levels of problematic internet gaming [31]. The IGDS9-SF was administered both in Sinhala and in English. Its reliability in this Sri Lankan sample is good (coefficient alpha = 0.81).

Motives for Online Gaming Questionnaire (MOGQ), a scale that comprises 27 items, which measure seven major gaming motives: Social (building and maintaining social relationships), Escape (escaping from reality), Coping (coping with stress and distress), Competition (challenging and competing with others), Skill Development (attention and coordination), Fantasy (in-game identities and experience), and Recreation (entertainment and enjoyment). Items are rated on a five-point Likert scale (1 = almost never/never) to (5 = almost always/always) [32]. Its reliability in the current sample is excellent (coefficient alpha = 0.92).

The Single Item Sleep Quality Scale (SISQ) was used to evaluate sleep quality. The response is rated on a scale from 1 to 10, with higher scores indicating better sleep quality [33].

Self-esteem has been assessed by a single question prompting the participants to rate their self-esteem on a 5-point Likert (1 = very poor) to (5 = excellent).

2.3. Statistical Analysis

Checking the original Sri Lankan dataset, which comprised 395 responses, for missing data revealed that seventy-three responses had missing data on the IGDS9-SF. Therefore, they were excluded from the analysis, ending with a final sample of 322 respondents—response rate = 82%. The distribution of the IGDS9-SF and MOGQ was examined using Shapiro-Wilk test. Quantitative variables with normal distribution were described by mean and standard deviation while those with non-normal distribution were described using median and interquartile range (IQR: 25–75%). Categorical variables were described using number and percentage.

Based on the literature, the unidimensional structure of the IGDS9-SF was examined in the Sri Lankan sample by confirmatory factor analysis (CFA) using the maximum likelihood method of estimation with bootstrap that generates 2000 random replications. Multigroup CFA was conducted to examine invariance of the IGDS9-SF across countries (Sri Lanka, Turkey, Australia, and the USA) in the international sample as well as across groups of gender, ethnicity, and language used to complete the questionnaire, income, device, game type, and academic major in the Sri Lankan sample (groups are shown in Table 1).

The chi square (χ^2) index is largely dependent on sample size, and well-fitting models with minor misspecifications may be disqualified based on a significant χ^2 [25]. Meanwhile, absolute fit indices represent more reliable indicators of model fit because they are sample-size independent. Therefore, we considered model fit in CFA/multigroup CFA to be good or acceptable based on a Comparative Fit Index (CFI) and Tucker–Lewis Index (TLI) equal to or above 0.95 and 0.90, respectively, along with a root mean square error of approximation (RMSEA) and standardized root-mean-square residual (SRMR) less than 0.06 and 0.08, respectively [34].

The internal consistency of the IGDS9-SF was examined by coefficient alpha, alpha-if-item deleted, and item-total correlations. Its criterion validity was evaluated by Spearman's r correlations with the MOGQ, number of sleeping hours, SISQ, and single-item measure of self-esteem in the Sri Lankan sample. Additionally, a Mann Whitney U test and Kruskal Wallis test were used to examine differences in the number of gaming hours as well as key constructs measured by MOGQ among groups of gender and game type since differences in IGD were depicted in these groups, and they expressed a tendency toward non-invariance in multigroup CFA. Statistical analyses were conducted in SPSS and Amos, and significance was considered at a probability level less than 0.05 in two-tailed tests.

3. Results

3.1. *Confirmatory Factor Analysis and Invariance Analysis of the Internet Gaming Disorder Scale 9—Short Form*

As shown in Table 2, the IGDS9-SF expressed excellent fit among students from Sri Lanka. According to Supplementary Table S1, the scale expressed invariance at the configural, metric, and scalar levels among Sri Lankan students across groups of gender, ethnicity, language used to complete the questionnaire, income, game type, academic major, and device used for gaming. However, there was a tendency toward configural non-invariance across groups of monthly income (ΔCFI = 0.022, ΔTLI = 0.020) and strict non-invariance across groups of gender and game type (ΔCFI = 0.042, 0.028; ΔTLI = 0.037, 0.020). However, ΔRMSEA was within the acceptable range in all tests, which supports invariance of the IGDS9-SF.

Table 2. Goodness-of-fit indices for the one-factor structure of the Internet Gaming Disorder Scale 9—Short form (IGDS9-SF) among university students from Sri Lanka, its invariance across countries, normality tests, and internal consistency.

Groups	Invariance Levels	χ^2	df	p	$\Delta\chi^2$	Δdf	p ($\Delta\chi^2$)	CFI	ΔCFI	TLI	ΔTLI	RMSEA	ΔRMSEA	SRMR	W ▲	Coefficient Alpha
Countries	Sri Lanka	41.85	27	0.034				0.978		0.970		0.041		0.0357	0.961	0.811
	Turkey	73.68	26	0.001				0.967		0.954		0.076		0.0353	0.838	0.902
	Australia	109.95	25	0.001				0.965		0.950		0.068		0.0359	0.950	0.862
	USA	113.02	25	0.001				0.967		0.953		0.069		0.0350	0.965	0.876
	Configural	351.23	108	0.001				0.956		0.941		0.038		0.0357		
	Metric	459.81	132	0.001	108.578	24	0.001	0.940	0.016	0.935	0.006	0.040	−0.002	0.0577		
	Scalar	473.97	135	0.001	14.163	3	0.003	0.938	0.002	0.934	0.001	0.041	−0.001	0.0735		
	Strict	1307.43	162	0.001	833.458	27	0.001	0.791	0.147	0.814	0.120	0.068	0.027	0.1086		

χ^2: chi-square; df: degrees of freedom; CFI: comparative fit index; TLI: Tucker–Lewis index; RMSEA: root mean square error of approximation; SRMR: standardized root mean residual; ▲: Shapiro–Wilk W test with all p values < 0.01; values in boldface indicate variance.

The IGDS9-SF expressed good fit in the Turkish, Australian, and American subsamples as well (Table 2). It was invariant at the configural, metric and scalar levels across participants from the four countries. However, strict invariance was not maintained (ΔCFI = 0.147, ΔTLI = 0.120, ΔRMSEA = 0.027) due to minor misspecifications in the covariances of a few of the items among participants from Turkey, Australia, and the USA (Figure 1). A Kruskal Wallis test revealed significant differences in the levels of IGD across countries (H (3) = 185.99, p = 0.001), with the lowest levels reported in participants from Turkey (median (IQR) = 13.0 (10.0–18.0)) and Sri Lanka (median (IQR) = 18.0 (13.0–22.0)) while the highest levels were reported in participants from the USA (median (IQR) = 21.0 (17.0–27.0)) and Australia (median (IQR) = 20.0 (16.0–25.0)). The Shapiro Wilk W test showed that the IGDS9-SF demonstrates a similar distribution in all the samples. Reliability tests showed that the internal consistency of the scale in the international subsamples ranged between very good and excellent (Table 2).

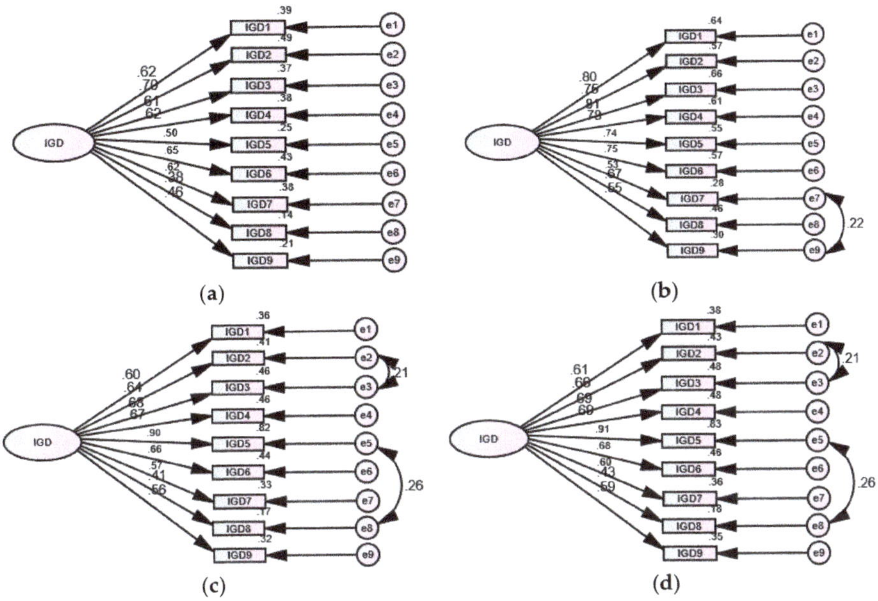

Figure 1. Factor structure of the Internet Gaming Disorder 9—Short Form (IGDS9-SF) among participants from Sri Lanka (**a**), Turkey (**b**), Australia (**c**), and the USA (**d**).

3.2. Reliability and Criterion Validity of the Internet Gaming Disorder Short-Form 9

Among Sri Lankan students, the IGDS9-SF expressed adequate internal consistency (coefficient alpha = 0.81), item-total correlations ranging between 0.341 and 0.611, and alpha if item deleted ranging between 0.779 and 0.816. It also demonstrated adequate criterion validity by exhibiting strong positive correlation with all motives of gaming measured by the MOGQ; the highest correlations were demonstrated with the dimensions of Escape and Coping (Table 3). It also correlated with the number of gaming hours, and the Kruskal Wallis test revealed significantly higher scores of the IGDS9-SF among those with daily gaming for more than three hours (H (2) = 88.8, p = 0.001). The IGDS9-SF did not correlate with the number of sleep hours (r = −0.051, p = 0.360), but it was negatively correlated with sleep quality (r = −0.120, p = 0.05). It negatively correlated with the single measure of self-esteem, albeit non-significantly.

The IGDS9-SF strongly correlated with Gender. The Mann Whitney U test revealed significantly higher IGD levels among males than females (U = 9261.0, z = −4.21, p = 0.001). Males recorded more daily time (>3 h) spent in gaming than females (χ^2 (1) = 32.19, p = 0.001). The Mann Whitney U test also revealed significantly higher scores of Social, Escape, Coping, Fantasy, and Recreation motives among males than females (U = 9724.0, 10,934.5, 10,987.0, 10,619.0, 8382.0; z = −3.66, −2.19, −2.12, −2.59, −5.29; p = 0.001, 0.028, 0.034, 0.010, 0.001). IGD correlated with ethnicity, and the Mann Whitney U test revealed significantly higher IGD among Sinhala students compared with students from other ethnicities (U =7 446.0, z = −2.35, p = 0.019).

Table 3. Descriptive statistics of and correlation of the Internet Gaming Disorder Scale 9—Short form (IGDS9-SF) with criterion variables among university students from Sri Lanka.

Variables	1	2	3	4	5	6	7	8	9	10	11	12
1. IGDS9-SF	-											
2. Gender	0.235 **	-										
3. Ethnicity	0.131 *	0.288 **	-									
4. Sleep quality	−0.120 *	0.082	0.000	-								
5. Gaming hours/day	0.532 **	0.312 **	0.170 **	−0.066	-							
6. Self-esteem	−0.104	0.084	0.030	0.193 **	-.008	-						
7. Social	0.514 **	0.204 **	−0.014	−0.038	0.463 **	0.033	-					
8. Escape	0.616 **	0.123 *	0.043	−0.161 **	0.383 **	−0.122 *	0.537 **	-				
9. Competition	0.367 **	0.085	0.028	−0.059	0.340 **	0.017	0.495 **	0.452 **	-			
10. Coping	0.513 **	0.119 *	−0.042	−0.134 *	0.295 **	−0.149 **	0.503 **	0.708 **	0.426 **	-		
11. Skill development	0.398 **	0.060	−0.098	−0.030	0.334 **	−0.057	0.552 **	0.538 **	0.534 **	0.590 **	-	
12. Fantasy	0.451 **	0.145 **	0.073	−0.074	0.321 **	−0.017	0.469 **	0.604 **	0.459 **	0.505 **	0.497 **	-
13. Recreation	0.389 **	0.295 **	0.098	0.020	0.448 **	−0.024	0.464 **	0.254 **	0.356 **	0.439 **	0.418 **	0.322 **
Median	18.0	-	-	7.0	-	4.0	8.0	7.0	9.0	9.0	8.0	7.0
IQR (Q1–Q3)	13.0–22.0	-	-	5.0–8.0	-	3.0–4.0	5.0–8.0	5.0–10.0	6.0–12.0	7.0–12.0	5.8–12.0	4.0–10.0

*: $p < 0.05$; **: $p < 0.01$.

Despite a tendency toward non-invariance across income groups, no significant differences between groups in the scores of the IGDS9-SF were noted among income groups (H (2) = 0.72, $p = 0.696$). Significant differences in IGD were recorded among game types, with the lowest scores reported among single offline players and the highest scores expressed among online multiplayers (H (2) = 9.70, $p = 0.008$). Multiple online players (28.0%) spent more than three hours gaming a day compared with 11.0% of offline players and 2.0% of single online players (χ^2 (4) = 41.54, $p = 0.001$). They also exhibited significantly higher levels of Social, Competition, Skill development, and Recreation motives (H (2) = 47.49, 20.06, 11.83, 18.99; all p values = 0.001).

4. Discussion

Concerns about psychometric equivalence of IGD measures in different parts of the world represent a challenge for adequate identification of IGD in different cultural contexts [24]. The psychometric properties of the IGDS9-SF have been largely tested in English-speaking, European, and a small number of less developed countries. Accordingly, the current study complements existing knowledge by examining the psychometric properties of this scale among respondents from Sri Lanka and evaluating its measurement invariance across four international groups from distinct cultural backgrounds.

Consistent with previous studies, data obtained from the Sri Lankan and international samples expressed good fit of the single factor structure of the IGDS9-SF, with all items adequately loading on this factor (Table 2, Figure 1). Multigroup CFA revealed non-invariance of this measure at the configural, metric, and scalar levels across the international participants as well as across various groups in the Sri Lanka sample. However, the IGDS9-SF did not hold strict invariance in the international sample, and there was a tendency toward strict non-invariance across groups of gender and game type in the Sri Lankan sample. Obviously, slight improvements in the fit of the IGDS9-SF were attained by correlating the error terms of item 7 (deceiving others) and item 9 (jeopardizing relationships) in the Turkish sample. Among Turkish adolescents, excessive social media use is associated with an interplay between family life satisfaction and social connectedness [35]. Therefore, the interaction between items of deception and jeopardizing relationships may reflect on a subtle relation factor underlying IGD in the Turkish context. On the other hand, correlating the error of item 5 (loss of interest) with item 8 (escape) and the errors of item 2 (withdrawal—Irritability when reducing or stopping use) and item 3 (tolerance—need to spend more time gaming) improved the fit in the Australian and American samples. It seems that gaming is adopted as a measure to escape negative emotions (loss of interest) in the Australian and American contexts. Meanwhile, the mood-modifying effect that results from achieving status or progress in online gaming diminishes in persons with prolonged engagement [36]. Therefore, loss of interest in previous gaming activities may derive tolerance—gaming for a longer time to achieve the previous satisfactory effect. Accordingly, lack of satisfaction of gamers' emotional needs may be related to irritability upon reducing/stopping (withdrawal). Cumulative knowledge shows variations in the internal and external events which shape the psychosocial background of IGD among individuals meeting the diagnostic criteria of IGD [37]. Thus, further investigations are needed to explore different dynamics underlying IGD in the cultures addressed in our study and whether they affect people's responses to the items of the scale.

Although the loadings of items 2 and 3 were strong in our American and Australian samples, they contributed to metric non-invariance in Australian respondents in a previous assessment of time invariance across Australian and American gamers. In the same Australian sample, items 4, 6, 8, and 9 were involved in partial scalar non-invariance of the scale [24]. Likewise, the single IGD factor was replicated in Polish gamers; however, poor fit was expressed by item 6 (continuation), item 7, and item 8 [17]. In another study comprising participants from the UK, USA, and India, cross-country variations were noted for items pertaining to preoccupation, tolerance, deception, escape, impairment in daily activities, and lack of control [21]. These findings generally suggest that the IGDS9-SF possibly involves a general component of high time and energy investment into IGD, in addition to the specific core components of IGD [38,39].

Configural and metric invariance of the IGDS9-SF across countries in our study indicates that IGD is similarly conceptualized by participants from poor countries (e.g., Sri Lanka), countries with evolving economies (e.g., Turkey), and developed countries (Australia and the USA). Strict non-invariance is rarely achieved while one third of the commonly used psychometric measures exhibit partial non-invariance [22,40]. Because the scale demonstrated scalar invariance across countries, the findings of this study show that the IGDS9-SF can be reliably used to compare IGD levels in those countries [40]. Significant differences in the level of IGD were noted among the international groups, with the highest occurrence reported among participants from the USA and Australia. On the contrary, Turkish students reported the lowest level followed by Sri Lankan students. In accordance, aggregate dates show that the levels of IGD and social media abuse are evidently higher in Europe and America than Asia [41]. These findings lend further support to previous studies reporting more pathologized IGD scores less close to the mean among respondents from individualistic cultural orientations such as the USA and close to normal scores among collectivistic or less individualistic countries (e.g., Turkey and the UK) [10,21]. It is worthy to note that the IGDS9-SF was examined in India before. However, it was administered in English [21], with a possibility that the scale may demonstrate different properties if it was administered among Indians who speak only local languages. In the present study, the IGD-SF9 was presented both in English and Sinhala, and it maintained perfect invariance across languages used, with no significant IGD differences between Sinhala and respondents from other ethnicities. Given the proximity as well as the cultural and geographical similarities between India and Sri Lanka, the IGDS9-SF seems to be a suitable measure in this region of south Asia.

Non-invariance of the IGDS9-SF across different groups (e.g., gender, ethnicities, and game types) denotes that the IGDS9-SF operates evenly as an IGD measure among youth from different social backgrounds in Sri Lanka. However, IGD significantly correlated with gender, with males expressing significantly higher IGD levels, IGD motives, and gaming time than females. This finding is consistent with reports of a recent meta-analysis comprising studies across 22 countries, which confirms that men exhibit significantly higher levels of IGD than women. On the other hand, women express higher levels of excessive social media abuse than men [41]. Increased risk for IGD among males is likely to be attributed to gender differences in the activation and connectivity of brain regions associated with the mesocorticolimbic reward system [37]. While females retreat to social networking to meet their need to relate [39], it seems that males retreat to gaming to satisfy their need to relate, as noted by significantly higher levels of Social motives among males in the present study. Moreover, online multiplayers were a majority in this study (Table 1); they also expressed the greatest time spent gaming, as well as the highest levels of IGD and gaming motives (Social, Competition, Skill development, and Recreation), particularly compared to offline single players. A qualitative investigation shows that distinctions between online and offline gaming can be largely shaped by the development of relationships, norms, and expectations—a person frequently plays with 'people like me' [42]. Overall, males engage in gaming more than females, engage more in online multi-playing, and endorse more Social and Competition motives, denoting that gaming is probably employed to meet social needs.

5. Strength and Limitations

This study has the strength of integrating public data to examine the psychometric properties of the IGDS9-SF in Sri Lanka and its measurement invariance across four countries with different cultural backgrounds. However, the study entails numerous limitations that must be acknowledged. The use of public data in the analysis makes us unable to answer pivotal questions. For example, it is not clear how the IGDS9-SF was translated into Sinhala and whether it was back translated into English before data collection. Priori calculations of sample sizes were not performed. Details on the number of people initially contacted as well as the specific sampling method are not available. The cross-sectional design precluded the evaluation of test-retest reliability of the scale. Lack of sociodemographic characteristics in the international sample makes it impossible to examine invariance of the scale across specific groups (e.g., gender, age, and education). The reported differences in IGD across countries may not be sufficiently accurate. This is because participants in the American and Australian samples were recruited from online gaming platforms. On the other hand, gaming was self-reported by the Sri Lankan and Turkish participants—a possibility of reporting bias. Future studies may remedy the flaws implicated in the present study.

6. Conclusions

The IGDS9-SF expressed good fit and invariance across different groups, along with satisfactory levels of reliability and criterion validity in Sri Lankan students. Males and online multiplayers spent more time playing and expressed higher levels of IGD and playing motives. Significantly higher levels of the social dimension of the MOGQ in both groups suggest that gaming is used to meet the need to relate. Future investigations of relevant factors (e.g., family relations, social connectedness, social skills/competence) may provide further explanations of the use of gaming by males to satisfy their social needs.

In addition to demonstrating good fit of the unidimensional structure among the Turkish, Australian, and American subsamples, the IGDS9-SF also expressed configural, metric, and scalar invariance across countries. Therefore, this scale may be reliably used to compare IGD levels in those countries. The covariances of error terms of items related to deceiving others and jeopardizing relationships in the Turkish sample suggest the involvement of social factors in IGD in this culture. On the other hand, covariances between the error terms of items 5 and 8 as well as items 2 and 3 in the Australian and American subsamples indicate that gaming is employed by respondents from those countries to escape negative emotions. However, failure of gaming to resolve negative emotions may contribute to continued gaming (tolerance) and withdrawal symptoms upon sudden cessation. Future studies examining cultural invariance of the IGDS9-SF may need to identify the dynamics underling IGD in different countries and whether they may affect individual responses to the items of the IGDS9-SF.

Supplementary Materials: The following supporting information can be downloaded at: https://www.mdpi.com/article/10.3390/healthcare10030490/s1, Table S1: Invariance of IGDS9-SF across different student groups in the Sri Lankan sample.

Author Contributions: Conceptualization, A.M.A., A.O.H. and R.A.-A.; methodology, A.M.A., A.M.H.M. and H.K.; software A.M.A. and T.S.A.; validation, A.M.A., A.O.H. and R.A.-A.; formal analysis, A.M.A., M.A. and A.O.H.; resources, M.A. and A.A.A.; data curation, T.S.A.; writing—original draft preparation, A.M.H.M., H.K. and A.A.A.; writing—review and editing, A.M.A. and R.A.-A.; project administration, A.M.A. All authors have read and agreed to the published version of the manuscript.

Funding: This research received no external funding.

Institutional Review Board Statement: This study is based on publicly accessible datasets. Available reports based on these data state compliance with ethical research standards (e.g., ethical approval and informed consent). Apart from those public data, no new data were collected in the present study. Therefore, ethical approval has not been obtained.

Informed Consent Statement: Not applicable.

Data Availability Statement: The datasets used to produce the current article are publicly available in Mendeley at: https://data.mendeley.com/datasets/8r2jgm6ygh/1 [26], https://data.mendeley.com/datasets/k698sznwf6/3 [28], and in DANS at: https://easy.dans.knaw.nl/ui/datasets/id/easy-dataset:162945 [29], (accessed on 2 August 2021).

Acknowledgments: The researchers would like to thank the Deanship of Scientific Research, Qassim University for funding the publication of this project.

Conflicts of Interest: The authors declare that they have no conflicts of interest.

References

1. Dong, H.; Yang, F.; Lu, X.; Hao, W. Internet Addiction and Related Psychological Factors Among Children and Adolescents in China during the Coronavirus Disease 2019 (COVID-19) Epidemic. *Front. Psychiatry* **2020**, *11*, 751. [CrossRef] [PubMed]
2. Amin, K.P.; Griffiths, M.D.; Dsouza, D.D. Online Gaming During the COVID-19 Pandemic in India: Strategies for Work-Life Balance. *Int. J. Ment. Health Addict.* **2020**, *20*, 296–302. [CrossRef] [PubMed]
3. Paschke, K.; Austermann, M.I.; Simon-Kutscher, K.; Thomasius, R. Adolescent gaming and social media usage before and during the COVID-19 pandemic. *Sucht* **2021**, *67*, 13–22. [CrossRef]
4. Xu, S.; Park, M.; Kang, U.G.; Choi, J.S.; Koo, J.W. Problematic Use of Alcohol and Online Gaming as Coping Strategies During the COVID-19 Pandemic: A Mini Review. *Front. Psychiatry* **2021**, *12*, 685964. [CrossRef] [PubMed]
5. Lemenager, T.; Neissner, M.; Koopmann, A.; Reinhard, I.; Georgiadou, E.; Müller, A.; Kiefer, F.; Hillemacher, T. COVID-19 Lockdown Restrictions and Online Media Consumption in Germany. *Int. J. Environ. Res. Public Health* **2020**, *18*, 14. [CrossRef] [PubMed]
6. Ali, A.M.; Alkhamees, A.A.; Elhay, E.S.A.; Taha, S.M.; Hendawy, A.O. COVID-19-related psychological trauma and psychological distress among community-dwelling psychiatric patients: People struck by depression and sleep disorders endure the greatest burden. *Front. Public Health* **2022**. [CrossRef]
7. Gomez, R.; Stavropoulos, V.; Beard, C.; Pontes, H.M. Item Response Theory Analysis of the Recoded Internet Gaming Disorder Scale-Short-Form (IGDS9-SF). *Int. J. Ment. Health Addict.* **2019**, *17*, 859–879. [CrossRef]
8. Stavropoulos, V.; Beard, C.; Griffiths, M.D.; Buleigh, T.; Gomez, R.; Pontes, H.M. Measurement Invariance of the Internet Gaming Disorder Scale–Short-Form (IGDS9-SF) Between Australia, the USA, and the UK. *Int. J. Ment. Health Addict.* **2018**, *16*, 377–392. [CrossRef]
9. Bernaldo-de-Quirós, M.; Labrador-Méndez, M.; Sánchez-Iglesias, I.; Labrador, F.J. Measurement instruments of online gaming disorder in adolescents and young people according to DSM-5 criteria: A systematic review. *Adicciones* **2020**, *32*, 291–302. [CrossRef]
10. Evren, C.; Dalbudak, E.; Topcu, M.; Kutlu, N.; Evren, B.; Pontes, H.M. Psychometric validation of the Turkish nine-item Internet Gaming Disorder Scale-Short Form (IGDS9-SF). *Psychiatry Res.* **2018**, *265*, 349–354. [CrossRef]
11. Zhu, S.; Zhuang, Y.; Lee, P.; Li, J.C.; Wong, P.W.C. Leisure and Problem Gaming Behaviors Among Children and Adolescents During School Closures Caused by COVID-19 in Hong Kong: Quantitative Cross-sectional Survey Study. *JMIR Serious Games* **2021**, *9*, e26808. [CrossRef] [PubMed]
12. Mamun, M.A.; Ullah, I.; Usman, N.; Griffiths, M.D. PUBG-related suicides during the COVID-19 pandemic: Three cases from Pakistan. *Perspect. Psychiatr. Care* **2020**. [CrossRef] [PubMed]
13. Wu, A.M.; Lai, M.H.; Yu, S.; Lau, J.T.; Lei, M.W. Motives for online gaming questionnaire: Its psychometric properties and correlation with Internet gaming disorder symptoms among Chinese people. *J. Behav. Addict.* **2017**, *6*, 11–20. [CrossRef] [PubMed]

14. Beranuy, M.; Machimbarrena, J.M.; Vega-Osés, M.A.; Carbonell, X.; Griffiths, M.D.; Pontes, H.M.; González-Cabrera, J. Spanish Validation of the Internet Gaming Disorder Scale–Short Form (IGDS9-SF): Prevalence and Relationship with Online Gambling and Quality of Life. *Int. J. Environ. Res. Public Health* **2020**, *17*, 1562. [CrossRef] [PubMed]
15. Kim, B.N.; Ko, H. Psychometric Properties of the Nine-Item Korean Internet Gaming Disorder Scale: Short Form. *Cyberpsychol. Behav. Soc. Netw.* **2020**, *23*, 854–859. [CrossRef]
16. Lei, W.; Liu, K.; Zeng, Z.; Liang, X.; Huang, C.; Gong, K.; He, W.; Xiang, B.; Zhang, J.; Zheng, X.; et al. The psychometric properties of the Chinese version internet gaming disorder scale. *Addict. Behav.* **2020**, *106*, 106392. [CrossRef]
17. Schivinski, B.; Brzozowska-Woś, M.; Buchanan, E.M.; Griffiths, M.D.; Pontes, H.M. Psychometric assessment of the Internet Gaming Disorder diagnostic criteria: An Item Response Theory study. *Addict. Behav. Rep.* **2018**, *8*, 176–184. [CrossRef]
18. Hawi, N.S.; Samaha, M. Validation of the Arabic Version of the Internet Gaming Disorder-20 Test. *Cyberpsychol. Behav. Soc. Netw.* **2017**, *20*, 268–272. [CrossRef]
19. Pontes, H.M.; Griffiths, M.D. Portuguese Validation of the Internet Gaming Disorder Scale-Short-Form. *Cyberpsychol. Behav. Soc. Netw.* **2016**, *19*, 288–293. [CrossRef]
20. Ali, A.M.; Hori, H.; Kim, Y.; Kunugi, H. The Depression Anxiety Stress Scale 8-items expresses robust psychometric properties as an ideal shorter version of the Depression Anxiety Stress Scale 21 among healthy respondents from three continents. *Front. Psychol.* **2022**. [CrossRef]
21. Pontes, H.M.; Stavropoulos, V.; Griffiths, M.D. Measurement Invariance of the Internet Gaming Disorder Scale-Short-Form (IGDS9-SF) between the United States of America, India and the United Kingdom. *Psychiatry Res.* **2017**, *257*, 472–478. [CrossRef] [PubMed]
22. Ali, A.M.; Hendawy, A.O.; Ahmad, O.; Sabbah, H.A.; Smail, L.; Kunugi, H. The Arabic version of the Cohen perceived stress scale: Factorial validity and measurement invariance. *Brain Sci.* **2021**, *11*, 419. [CrossRef] [PubMed]
23. Ali, A.M.; Alkhamees, A.A.; Hori, H.; Kim, Y.; Kunugi, H. The Depression Anxiety Stress Scale 21: Development and Validation of the Depression Anxiety Stress Scale 8-item in Psychiatric Patients and the General Public for Easier Mental Health Measurement in a Post-COVID-19 World. *Int. J. Environ. Res. Public Health* **2021**, *18*, 10142. [CrossRef]
24. Stavropoulos, V.; Bamford, L.; Beard, C.; Gomez, R.; Griffiths, M.D. Test-Retest Measurement Invariance of the Nine-Item Internet Gaming Disorder Scale in Two Countries: A Preliminary Longitudinal Study. *Int. J. Ment. Health Addict.* **2019**. [CrossRef]
25. Ali, A.M.; Hendawy, A.O.; Almarwani, A.M.; Alzahrani, N.; Ibrahim, N.; Alkhamees, A.A.; Kunugi, H. The Six-item Version of the Internet Addiction Test: Its development, psychometric properties, and measurement invariance among women with eating disorders and healthy school and university students. *Int. J. Environ. Res. Public Health* **2021**, *18*, 12341. [CrossRef] [PubMed]
26. Manchanayake, M. Data For: Internet Gaming Disorder: An Emerging Public Health Concern Among an Advanced Level Student Population from Colombo, Sri Lanka. *Mendeley Data* **2021**, *Version 1*. [CrossRef]
27. Manchanayake, M.; Malsirini, T.G.M.; Vithanage, M.; Jayawardene, D.M.S. Internet Gaming Disorder: Prevalence and Correlates Among an Advanced Level Student Population from the Colombo Educational Zone. In Proceedings of the 26th Annual Academic Sessions of the College of Community Physicians, Colombo, Sri Lanka, 14 July 2021.
28. Demirbaş-Çelik, N.; Yeşilyurt, F.; Çelik, B. The Mediating Role of Life Satisfaction, Internet Gaming Disorder, and Negative Affect within the Relationship of Basic Psychological Needs Satisfaction and Depression. *Mendeley Data* **2021**, *V3*. [CrossRef]
29. Burleigh, T.L. IGD_Culture. *Data Arch. Netw. Serv. (DANS)* **2019**, *V1*. [CrossRef]
30. Andreetta, J.; Teh, M.J.; Burleigh, T.L.; Gomez, R.; Stavropoulos, V. Associations between comorbid stress and Internet Gaming Disorder symptoms: Are there cultural and gender variations? *Asia-Pac. Psychiatry Off. J. Pac. Rim Coll. Psychiatr.* **2020**, *12*, e12387. [CrossRef]
31. Pontes, H.M.; Griffiths, M.D. Measuring DSM-5 internet gaming disorder: Development and validation of a short psychometric scale. *Comput. Hum. Behav.* **2015**, *45*, 137–143. [CrossRef]
32. Demetrovics, Z.; Urbán, R.; Nagygyörgy, K.; Farkas, J.; Zilahy, D.; Mervó, B.; Reindl, A.; Ágoston, C.; Kertész, A.; Harmath, E. Why do you play? The development of the motives for online gaming questionnaire (MOGQ). *Behav. Res. Methods* **2011**, *43*, 814–825. [CrossRef]
33. Snyder, E.; Cai, B.; DeMuro, C.; Morrison, M.F.; Ball, W. A New Single-Item Sleep Quality Scale: Results of Psychometric Evaluation in Patients With Chronic Primary Insomnia and Depression. *J. Clin. Sleep Med.* **2018**, *14*, 1849–1857. [CrossRef]
34. Ali, A.M.; Hori, H.; Kim, Y.; Kunugi, H. Predictors of nutritional status, depression, internet addiction, Facebook addiction, and tobacco smoking among women with eating disorders in Spain. *Front. Psychiatry* **2021**, *12*, 2001. [CrossRef]
35. Savci, M.; Akat, M.; Ercengiz, M.; Griffiths, M.D.; Aysan, F. Problematic Social Media Use and Social Connectedness in Adolescence: The Mediating and Moderating Role of Family Life Satisfaction. *Int. J. Ment. Health Addict.* **2020**. [CrossRef]
36. King, D.L.; Herd, M.C.E.; Delfabbro, P.H. Tolerance in Internet gaming disorder: A need for increasing gaming time or something else? *J. Behav. Addict.* **2017**, *6*, 525–533. [CrossRef]
37. Paulus, F.W.; Ohmann, S.; von Gontard, A.; Popow, C. Internet gaming disorder in children and adolescents: A systematic review. *Dev. Med. Child Neurol.* **2018**, *60*, 645–659. [CrossRef]
38. Atroszko, P.A.; Balcerowska, J.M.; Bereznowski, P.; Biernatowska, A.; Pallesen, S.; Schou Andreassen, C. Facebook addiction among Polish undergraduate students: Validity of measurement and relationship with personality and well-being. *Comput. Hum. Behav.* **2018**, *85*, 329–338. [CrossRef]

39. Ali, A.M.; Hendawy, A.O.; Elhay, E.S.A.; Ali, E.M.; Alkhamees, A.A.; Kunugi, H.; Hassan, N.I. The Bergen Facebook Addiction Scale: Its psychometric properties and invariance among women with eating disorders. *BMC Women's Health* **2022**. *under review*.
40. Putnick, D.L.; Bornstein, M.H. Measurement Invariance Conventions and Reporting: The State of the Art and Future Directions for Psychological Research. *Dev. Rev. DR* **2016**, *41*, 71–90. [CrossRef]
41. Su, W.; Han, X.; Yu, H.; Wu, Y.; Potenza, M.N. Do men become addicted to internet gaming and women to social media? A meta-analysis examining gender-related differences in specific internet addiction. *Comput. Hum. Behav.* **2020**, *113*, 106480. [CrossRef]
42. Eklund, L. Bridging the online/offline divide: The example of digital gaming. *Comput. Hum. Behav.* **2015**, *53*, 527–535. [CrossRef]

healthcare MDPI

Article

Psychological Impacts of the COVID-19 Pandemic on Rural Physicians in Ontario: A Qualitative Study

Anchaleena Mandal [1,*] and Eva Purkey [1,2]

1 School of Medicine, Queen's University, Kingston, ON K7L 3N6, Canada; eva.purkey@queensu.ca
2 Department of Family Medicine, Queen's University, Kingston, ON K7L 3G2, Canada
* Correspondence: anchaleena.mandal@queensu.ca

Abstract: Frontline rural physicians in Canada are vulnerable to the psychological impacts of the COVID-19 pandemic considering their high pre-pandemic burnout rates as compared to their urban counterparts. This study aims to understand the psychological impacts of the COVID-19 pandemic on rural family physicians engaged in full-time primary care practice in Ontario and the stressors behind any identified challenges. Recruitment combined purposive, convenience, and snowball sampling. Twenty-five rural physicians participated in this study. Participants completed a questionnaire containing Patient Health Questionnaire-2 (depression), General Anxiety Disorder-2 (anxiety), and Perceived Stress Scale-4 (stress) screening as well as questions exploring self-reported perceptions of change in their mental health, followed by a semi-structured virtual interview. Quantitative data showed an overall increase in self-reported depression, anxiety, and stress levels. Thematic analysis revealed seven qualitative themes including the positive and negative psychological impacts on rural physicians, as well as the effects of increased workload, infection risk, limited resources, and strained personal relationships on the mental health of rural physicians. Coping techniques and experiences with physician wellness resources were also discussed. Recommendations include establishing a rapid locum supply system, ensuring rural representation at decision-making tables, and taking an organizational approach to support the mental health of rural physicians.

Keywords: COVID-19; mental health; psychological impact; pandemic; epidemic; primary care physicians; coronavirus disease; rural

Citation: Mandal, A.; Purkey, E. Psychological Impacts of the COVID-19 Pandemic on Rural Physicians in Ontario: A Qualitative Study. *Healthcare* **2022**, *10*, 455. https://doi.org/10.3390/healthcare10030455

Academic Editor: Roopma Wadhwa

Received: 22 November 2021
Accepted: 23 February 2022
Published: 28 February 2022

1. Introduction

The Coronavirus Disease 2019 (COVID-19) pandemic has introduced additional burdens on the publicly funded Canadian healthcare system, alongside great uncertainty and personal health risk for people working within the healthcare system. This has resulted in significant psychological impacts on frontline healthcare workers (HCW), who experience high levels of work-related stress even under normal circumstances [1]. This is particularly true for rural and remote communities, comprising approximately 21% of the Canadian population, which are primarily serviced by general practitioners [2–4]. While one in four Canadian physicians and residents reported high levels of burnout before the pandemic, the available pre-pandemic literature also shows profoundly reduced wellness, increased burnout, decreased job satisfaction, and lower retention rates of rural physicians as compared to their urban counterparts [3–5]. This can be attributed to limited resources, longer work hours, reduced access to specialty consultation, lack of privacy from patients outside of the clinical context, difficulty in taking time off, and increased social isolation among other factors [3,4,6].

Global research during the COVID-19 pandemic and the previous SARS pandemic shows higher rates of depression, anxiety, fear of infection, stress, burnout, isolation, poor sleep quality, exhaustion, psychiatric disorders, and even suicide of HCWs, including physicians [7–20]. The current international literature on rural HCWs shows similar worsening mental health during the COVID-19 pandemic [11,21,22]. The consequences of excessive and prolonged stressors can be detrimental to the quality of care provided, resulting in reduced patient satisfaction, compassion fatigue, lower productivity, absences, and higher turnover rates [23,24]. Increased rates of medical errors, breach of protocols and guidelines, and patient safety concerns are also a risk [23,24]. Supporting the mental health of physicians is, therefore, also vital for sustaining an effective response from the medical workforce during the pandemic, particularly in underserviced rural areas.

In October 2020, the Canadian Medical Association (CMA) responded to the mental health crisis of physicians by introducing a Physician Wellness Initiative [25]. Other organizations, such as the Society of Rural Physicians of Canada (SRPC) and Canadian Medical Protective Association (CMPA), posted lists of online COVID-19 and mental health resources [26,27]. Some hospitals have developed "grief circles" and response teams to provide urgent support to distressed physicians [28,29]. Nevertheless, in February 2021, a survey by the CMA showed that 64% of respondents were still experiencing anxiety and reported that their fatigue had increased by 69% within the first year of the pandemic [30]. This may be because most of these services do not address the underlying challenges, and physicians may be reluctant to disclose mental health distress due to the fear of stigma [1]. Furthermore, there are little to no physician wellness services that specifically address the unique psychological challenges of rural doctors. There is also a paucity of pre-pandemic research on rural physician wellness in Canada. Therefore, this paper will examine the psychological impacts of the COVID-19 pandemic on rural physicians from the province of Ontario within Canada and the stressors behind any identified challenges.

2. Methods

2.1. Participants

This study recruited participants from the province of Ontario, which had the highest total number of COVID-19 cases in Canada at the time, the fewest hospital beds per capita, a severe shortage of staff, and a greater number of rural residents than any of the other provinces [7,31,32]. This study uses Statistics Canada's definition of rural, which is defined as communities with less than 10,000 people and are located outside of the commuting zones of larger urban centers [33].

Recruitment combined purposive, convenience, and snowball sampling between May and early August 2021. A recruitment poster and letter were advertised through the SRPC, on social media, and by contacting rural hospitals and affiliated rural sites of medical schools in Ontario. The inclusion criteria included physicians with College of Family Physicians of Canada certification, working in rural Ontario since at least March 2020, and having a full-time primary care practice with regular duties in at least two clinical settings (i.e., emergency department, clinics, hospitals, long-term care, etc.). Participants were recruited to the point of thematic saturation, when data collection no longer resulted in the identification of new themes [34]. This was confirmed when researchers started to hear the same comments again and again, and there were no new codes. Ethics approval was obtained through the Queen's University Health Sciences and Affiliated Teaching Hospitals Research Ethics Board. Participants were provided a $50 SRPC gift card honorarium.

2.2. Data Collection and Analysis

Once participants were verified as fitting the inclusion criteria, they were emailed a brief online questionnaire, which collected demographic information, a Generalized Anxiety Disorder (GAD)-2 score, a Patient Health Questionnaire (PHQ)-2 score, a Perceived Stress Scale (PSS)-4 score, and self-reported perceptions of change in their mental health compared to before the pandemic [35–37]. The purpose of collecting these quantitative descriptive data was to understand the current mental health status of the participants. GAD-2 is a two-item tool with a cut-off at $\geq$3 points that rapidly screens for Generalized Anxiety Disorder (sensitivity 76%, specificity 81%) [35]. Similarly, the PHQ-2 is a 2-item screening tool for the presence of depression (sensitivity 97%, specificity 67%) with a cut-off at ≥3 points [36]. Finally, the 4-item PSS-4 is the most widely used psychological instrument for measuring the perception of stress and has an internal reliability of r = 0.60 [37].

The questionnaire was followed by an approximately 60-min one-on-one semi-structured interview over Zoom video conferencing or phone call. The interview guide included seven questions and associated prompts that were developed through an extensive review of the literature on the topic of interest. Please see Appendix A Interview Guide for a copy of the interview guide.

The qualitative data of this study was analyzed using thematic analysis, which allows for the identification of recurring ideas (referred to as themes) in a data set [38,39]. Data analysis was conducted by the primary researcher (AM). Interviews were audio-recorded and transcribed verbatim. Interview transcripts were imported into NVivo12 (QSR International, Burlington, MA, USA). All data were read in its entirety, encoded into twenty-four codes, and key conceptual themes were searched, reviewed, and defined by AM. Themes and codes were discussed iteratively with the research supervisor (EP) to ensure validity. An external family physician in rural Ontario was consulted regarding the research findings and recommendations to check for relevance and sensitivity to the rural physician community.

3. Results

3.1. Quantitative Findings: Demographic and Questionnaire Results

Twenty-five rural physicians participated in this study. Please reference Tables 1–3. Most participants were between the ages of 35 and 49 years. There was a relatively equal distribution between participants from Southern and Northern Ontario as well as between male and female gender, with one participant who identified as trans, gender fluid, or non-binary. The majority of participants (72%) scored between 6 and 10 on the PSS-4 screening, indicating moderate stress levels. Although 16% screened positive for depression in PHQ-2 and 44% screened positive for anxiety in GAD-2 at the time of the study, the majority of the participants self-reported increased levels of perceived depression (48%), anxiety (72%), and stress (92%), as well as a decreased sense of overall mental wellbeing (80%) compared to before the pandemic. Full results of PHQ-2, GAD-2, and PSS-4 are reported in Table A1.

Table 1. Demographics of participants.

Variable	Response % (*n* = 25)	Variable	Response % (*n* = 25)
Age, years		*No. of physicians in practice*	
18 to 34	20.0 (5)	<5	00.0 (0)
35 to 49	48.0 (12)	5 to 10	40.0 (10)
50 to 64	28.0 (7)	10 to 20	36.0 (9)
65 and older	4.0 (1)	>20	24.0 (6)
Gender		*Medical setting of practice*	
Female	52.0 (13)	Clinics	92.0 (23)
Male	44.0 (11)	Emergency	88.0 (22)
Trans/Gender Fluid/Non-binary	4.0 (1)	Hospitalist	96.0 (24)
Other (No comments)		Obstetrics	20.0 (5)
Location		Public Health	8.0 (2)
Northern Ontario	44.0 (11)	Anesthesia	00.0 (0)
Southern Ontario	56.0 (14)		
Years of practice in Ontario		Other (Comments include indigenous communities and reserves, home visits, long-term care, palliative care, MAID, surgical assist, and teaching)	
1 to 5	24.0 (6)		
5 to 10	20.0 (5)		
10 to 20	40.0 (10)		
>20	16.0 (4)		

Table 2. PHQ-2 (depression), GAD-2 (anxiety), and PSS-4 (stress) screening results.

Variable	Response % (n = 25)
PHQ-2	
Screened positive for depression	16.0 (4)
Screened negative for depression	84.0 (21)
GAD-2	
Screened positive for anxiety	44.0 (11)
Screened negative for anxiety	56.0 (14)
PSS-4 *	
≤5 (low stress)	24.0 (6)
6–10 (moderate stress)	72.0 (18)
≥11 (high stress)	4.00 (1)

* Please note that PSS-4 results were divided into low, moderate, and high stress groups based on data distribution.

Table 3. Self-reported perceived change in mental health of participants from before the pandemic compared to the present.

Variable	Response % (n = 25)	Variable	Response % (n = 25)
Level of anxiety		*Level of stress*	
Increased	72.0 (18)	Increased	92.0 (23)
Stayed the same	28.0 (7)	Stayed the same	8.0 (2)
Decreased	00.0 (0)	Decreased	00.0 (0)
Not sure	00.0 (0)	Not sure	00.0 (0)
Level of depression		*Level of overall mental wellbeing*	
Increased	48.0 (12)	Improved	4.0 (1)
Stayed the same	48.0 (12)	Stayed the same	16.0 (4)
Decreased	00.0 (0)	Worsened	80.0 (20)
Not sure	4.0 (1)	Not sure	00.0 (0)

3.2. Qualitative Findings

From the qualitative data, seven main themes emerged (Figure 1) that are organized into three broad categories. Please refer to Tables A2–A4 for quotes relevant to each theme.

3.2.1. Category One: Impacts on Psychological Wellbeing

Theme 1: Psychological Impacts of the COVID-19 Pandemic on Rural Physicians

Although rural Ontario generally had delayed onset and a low number of COVID-19 cases due to geographical isolation and eventual high vaccination rates, participants described how the trickle-down effects on the rural healthcare system continued to impact the psychological wellbeing of most rural physicians to varying degrees. Almost all participants reported that their mental health had worsened at the beginning of the pandemic. However, over time, as they learned to adapt, their mental health either stayed the same or seemed to improve, although still at an overall lower level than their baseline from before the pandemic.

Beyond work, many physicians struggled to balance family obligations, social isolation, and personal struggles such as loss of loved ones, going through a divorce, managing personal/family member health issues, and worrying about loved ones in long-term care homes. Common words used by participants to describe their mental health included frustration, anxiety, stress, tired, anger, irritable, hopeless, drowning, defensive, burnout, COVID-fatigue, depression, disappointed, emotional, low mood, and numb.

Earlier in the pandemic, anxiety largely stemmed from the uncertainty surrounding the virus. Many local public health projections predicted massive outbreaks, which generated fear of the worst-case scenarios and "feeling on edge" in understaffed and low resourced rural centers. This anxiety manifested as "doom and gloom" conversations, feeling behind on scientific readings, feeling underprepared for work, as well as worrying about long-term impacts on personal life, hospital backlog, the economy, etc.

Outbreaks in rural communities triggered anxiety, a sense of demoralization, and feelings of helplessness, and several participants found it stressful to deal with the medical complications of decompensating COVID-19 patients in a low-resource setting. Some participants also resented that public health protocols for COVID-19 patient care undermined the dignity of the patient and restricted visits from family during end-of-life care. The death of a COVID-19 patient was personally difficult for many rural physicians, as individuals in rural communities are usually intimately familiar with each other, and physicians know them as more than just patients.

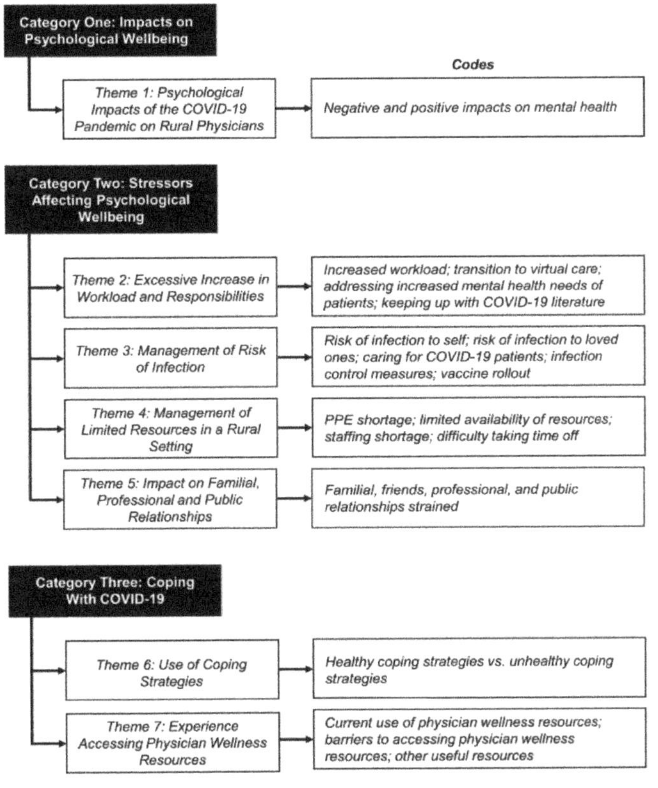

Figure 1. Qualitative thematic analysis with categories, themes, and associated codes.

As the pandemic progressed, feelings of burnout evoked self-questioning about the purpose of working when they were unable to enjoy their personal lives, especially in younger physicians who may place higher value on work–life balance. They reported feeling like there was no end in sight and not having a life outside of work. Interestingly, previous experience of the SARS pandemic appeared to have helped older physicians to mentally prepare themselves for the COVID-19 pandemic.

A few participants identified an overwhelming feeling of constant generalized anger at the beginning of the pandemic relating to their stressful workload and dissatisfaction with the government, public reaction, and vaccine rollout. This often manifested as easily losing patience with uncooperative patients, emotional outbursts, and strained relationships with colleagues and family.

Interestingly, despite adverse psychological impacts, some participants also felt that it was exciting to deal with a new challenge through teamwork with their colleagues and rewarding to be able to use their skills to help their community. This sparked feelings of "rising to the challenge" and "humble gratitude."

3.2.2. Category Two: Stressors Affecting Psychological Wellbeing

Theme 2: Excessive Increase in Workload and Responsibilities

Planning and preparations for the COVID-19 pandemic, such as setting up assessment centers, restructuring teams, practicing protected code blue simulations, modifying ventilator capacity, etc., were a substantial source of stress for rural family physicians. This is because, unlike urban centers, rural generalists work across multiple practice settings and, thus, were responsible for enforcing COVID-19 protocols across several departments. Participants described it as "scrambling" to adjust to the new practice and protocols, and many physicians returned early from their vacation to assist.

Initially, clinic and emergency department workload decreased as patients delayed accessing healthcare due to fear of contracting COVID-19, but the workload quickly spiked as patients started coming with acute presentations requiring a higher level of care. The creation of new roles in the assessment center, vaccination clinics, and additional services added strain on the already understaffed system. Physicians now also had to cover for their colleagues on short notice while the colleague was awaiting swab results or recovering from illness. Workload and stress further increased during peaks of COVID-19 outbreaks, and the transfer of complex or ventilated patients to overwhelmed tertiary centers was difficult to organize. Many participants reported greater anxiety because of the unpredictability of their workload with the changing nature of the pandemic.

With the transition to virtual care, physicians were now taking on more patients per day and, thus, having to do more paperwork. Their workload also substantially increased as they were seeing some patients twice, virtually followed by in-person. Concerns of the physicians included doing more investigations to compensate for the lack of a physical exam, fearing that they were providing inferior quality of care virtually, and frustration dealing with patients who were resistant to this change. However, one participant reported enjoying virtual care as they were able to work from the comfort of their home.

Most participants expressed empathy for the increased mental health challenges and struggles of patients during the pandemic for reasons such as unemployment, isolation, difficulty finding childcare, etc. In response, some clinics extended their appointment duration to accommodate mental health discussions. However, some participants disclosed that having to deal with so many mental health appointments contributed to second-hand anxiety, compassion fatigue, and/or exhaustion.

The administrative workload also increased, with more communications, emails, paperwork, as well as investigation results to review. Moreover, there was a drastic increase in the number of meetings for COVID-19-related preparations and decision-making or policy making, in turn adding multiple hours to their workweek.

All participants reported feeling a sense of duty to keep up with the rapidly evolving COVID-19 literature, protocols, and guidelines to provide evidence-based patient care or to develop policies for their local hospital, often spending hours and working late at night to catch up. The majority expressed feeling information overload, cognitive fatigue, and exhaustion over time. Differing and conflicting messaging from various sources and lack of relevance to the rural context created confusion, prompting some participants to selectively review a few reliable information sources or to only consume the literature that was practical for their daily use. Frustration was also augmented by political influence on public health decisions. It was especially stressful for physicians who were responsible for translating the evolving literature to the public on behalf of their hospital. As information rapidly changed, most notably during the AstraZeneca vaccination controversy, many physicians felt that the public lost confidence in the guidance of physicians, and many physicians in turn lost trust in government messaging, which one participant described as "shady" and "lacking transparency" [40].

Out of professional responsibility, commitment to their community drove rural physicians to take on more responsibilities and workload. With time, this led to burnout, and some physicians gave up leadership positions and refused more work. New grads reported burning out faster and feeling under-skilled. On the other hand, provincial restrictions eventually made travel more difficult for locum physicians, so some of them chose to either spend longer periods of time working in a community or to settle down in a permanent practice. Because they were no longer able to locum for short periods and take immediate breaks, taking on full or longer practice increased their workload and fatigue. Locum physicians who continued to travel expressed frustration with multiple quarantines, swab testing, and differences in regional protocols.

Theme 3: Management of Risk of Infection

In general, younger physicians were less anxious about personal infection risk compared to older physicians, who felt more vulnerable and some of whom feared death due to their older age and co-morbidities. Personal fear stemmed from insufficient knowledge about the virus, not having adequate personal protective equipment (PPE), and lower intubation proficiency among other factors. Physicians with young children were particularly anxious about what would happen to their children if they died from COVID-19. Some physicians reported that they had reviewed their will, revised advanced directives, and shared account passwords with their partner in apprehension of their death. Nevertheless, most participants reported feeling desensitized to the risk of infection over time, and vaccines provided them with a greater sense of safety.

Many participants feared bringing home COVID-19 to their family or spreading it to vulnerable patients. A few participants even experienced stigma from family members. As a result, some physicians reduced their contacts, temporarily lived separately from their family, avoided entering nursing homes unless necessary, strictly adhered to infection control and PPE protocols, wore scrubs at work, and/or developed a sanitization ritual after coming home from work (i.e., washing clothes and showering before interacting with family). Several participants reported choosing to stay separate from their family during self-isolation while awaiting swab results, which was emotionally difficult but relatively easy to do in a larger rural home.

Participants who had a greater sense of personal risk of infection also expressed frustration with individuals who did not strictly follow COVID-19 protocols, demonstrated negligence, or had a lower perception of infection risk. Some physicians reported feeling a responsibility to gently remind colleagues and patients to wear masks (like a "PPE police") and worrying that other staff may become lax with protocols over time.

Most COVID-19 patients were managed at home or transferred to tertiary care center intensive care units (ICU) due to lack of infrastructure to care for complex conditions in rural centers. Many participants did not know that they had cared for a COVID-19 patient until later when swab results returned. These participants often reported more post-exposure anxiety due to fear of not having taken enough precautions at the time. Participants also reported that as rural generalists are not as experienced with intubating as anesthesiologists in urban centers, they felt greater anxiety with intubations and other aerosol generating procedures that would increase the risk of infection for healthcare providers.

Infection control measures and cleaning protocols fostered an increased sense of safety. However, after a while, the majority of participants found these protocols to be an additional annoyance, such as having to screen patients, skin irritations from cleaning products, putting on full PPE before seeing every patient, waiting for rooms to dry after sanitization, not being able to eat at their desk or in a break room, as well as entering and exiting through different doors. Protected code-blue barriers also made it more difficult to manipulate equipment and provide quality care.

Although healthcare professionals were labelled as the priority during the COVID-19 vaccine rollout, most participants reported that the general public in several urban centers were receiving the vaccine before them. Rural physicians felt that their risk of infection was "ignored" and felt insulted and undervalued by not being prioritized for vaccines despite the vulnerability of short-staffed rural centers. They were also frustrated by the variation in rollout between communities, the redirection of vaccines away from rural communities, or receiving an insufficient vaccine supply for their population. Several participants reported feeling frustrated by the lack of cooperation of health units to plan for vaccine clinics and later being instructed to set-up a vaccine clinic at the last minute.

Theme 4: Management of Limited Resources in a Rural Setting

The global shortage of PPE at the beginning of the pandemic also impacted unprotected frontline rural physicians, causing much stress and anxiety. Some participants felt neglected by local integrated health networks (LIHN), which are regional organizations responsible for planning, integrating, and funding local health services in Ontario, as they did not provide PPE for community physicians [41]. At some rural hospitals, there was also conflict over the banned use of personally supplied masks at the beginning of the pandemic. To mitigate the PPE shortage issue, many rural centers collected donations from the community and sub-optimal handmade PPE by local sewers, retrieved expired PPE stock from past Ebola, H1N1, and SARS precautions, and reused PPE when possible. No participants reported running out of PPE, but most reported having to use their reserve sparingly until supplies improved.

Use of PPE also had its negatives such as mask acne, headaches from visors, visors steaming up, being dehydrated, and sweating in PPE, especially during hot summers. PPE also made it more difficult to communicate with patients who lip read or are hard of hearing and decreased the human touch in patient care. Nevertheless, most participants did not mind PPE as part of their job and felt safe and grateful to have it.

The limited COVID-19 swab testing capacity and long result turnaround times earlier in the pandemic disproportionately affected short-staffed rural communities where they were burning through PPE and losing potentially exposed staff while awaiting test results. Unprotected frontline physicians also felt unsafe not knowing if their patient had COVID-19 due to limited testing capacity. Moreover, many rural hospitals have access to only one or two transport ventilators and limited oxygen supply, so participants were fearful of running out of these resources if they were unable to transfer ventilated patients out to overwhelmed tertiary care centers during a local COVID-19 outbreak. This worry was compounded by the limited training of rural nurses to care for complex ventilated patients. Interestingly, limited resources prompted many rural physicians to focus more on the patient's goals of care, such as providing palliative care at home as per the patient's request instead of seeking aggressive care in a tertiary care center ICU where they might die alone or contract COVID-19 if ill with something else.

On top of the chronic shortage of human resources, the creation of new roles and clinics during the pandemic required rural family physicians to take on extra shifts, cover for quarantining colleagues, work "double duty" (cover two services concurrently), and be deployed to ICUs with minimal training. Many older physicians were unable to retire as they could not find someone to take over the care of their patients. Moreover, taking time off was difficult as there was a profound shortage of locum physicians, and permanent physicians felt guilty for creating additional burden on colleagues to cover them. As a shortage of staffing threatened some rural emergency departments to close, physicians felt an obligation to continue working to ensure continuity of healthcare services in their community. Even if they were able to take time off, their break was fragmented by being called in to work extra shifts, completing paperwork, and other duties. In other words, there is a culture in rural centers to constantly be on-call if within town, even during time off. Participants explained that true time off requires leaving town, which was not possible due to COVID-19 travel restrictions. Similarly, taking time off did not feel worthwhile for some, as they were unable to visit loved ones in other communities.

Over time, overworked rural physicians started feeling the effects of burnout. The provincial government provided funding to incentivize rural physicians to continue working and keep emergency departments open, but almost all participants revealed that they would prefer more time off to care for their physical and mental health needs rather than earning more money.

Theme 5: Impact on Familial, Professional, and Public Relationships

The social life of rural physicians became limited during the pandemic due to public health restrictions. At work, socializing opportunities such as eating and sharing food with colleagues were halted. Virtual care and decreased face-to-face time with patients and colleagues caused increased professional isolation. Participants who were locums reported that travel restrictions prompted some of them to settle in a community, where they struggled to develop a new social circle due to social restrictions. This enhanced their sense of isolation as their main social supports were residing in distant locations. One participant even contemplated terminating their contract to be with their loved ones. Physicians who were single, unable to date, and living alone also reported feeling a great sense of isolation. However, at the same time, participants reported feeling an increased sense of camaraderie with their rural colleagues as they worked together through the various challenges presented by the pandemic.

The impact on familial relationships varied. Most participants reported that their partner was very supportive, and that having a partner who took care of home demands was helpful. Common regrets included not spending enough time with their partner, being cranky, and feeling like their partner did not understand a physician's stress. Rural physician couples or couples with both individuals working in healthcare struggled with being too busy with work, having differing views about the pandemic, talking about COVID-19 all the time, or staying separate from each other so that the hospital did not lose two doctors to illness at the same time. Virtual schooling or homeschooling allowed physician parents to spend more time with their kids at home, although some participants regretted being too busy to see their child's growth or being irritable with their children. Physicians with young children reported more burnout as they were either working or they were parenting at home, with no time to care for their own needs. Other common concerns included difficulty finding childcare, worrying about children's mental health and safety, and worrying about the quality of virtual education.

Most rural physicians limited in-person interactions with extended family and friends due to pandemic restrictions and to protect elderly/vulnerable family members. While most stayed connected virtually, some met with friends in outdoor settings such as on hikes, picnics, or sitting around a fire. A few participants expressed being too busy to stay connected, or they avoided contact to limit COVID-19-related discussions, or the frustration of their work stresses not being understood by others. Participants also reported feeling judgmental towards those who did not follow public health guidelines, chose not to get vaccinated, or did not share the same sense of risk of infection or view about the pandemic as them.

At the beginning of the pandemic, most rural physicians reported feeling grateful and motivated by the appreciation from the public for being "healthcare heroes." However, a few participants felt it was misplaced since it was their job and other essential workers, such as grocery store employees, deserved more appreciation. One participant also reported feeling less like a hero and more like "cannon fodder" earlier in the pandemic and would have preferred to receive proper PPE and supports instead of the hero label. Later, participants reported feeling like scapegoats for the public's exhaustion with the pandemic, manifested as increased rudeness, frustration, and lack of cooperation from patients. Rural physicians also felt disheartened by anti-vaccination discussions, anti-masking protests, and patients trusting "internet doctors" over their own evidence-based recommendations. Most physicians tried to avoid social media; people were always asking for their opinion, social media attacks in rural communities are often personal where HCWs are unable to defend themselves, and differences in opinion about the pandemic often shattered relationships.

3.2.3. Category Three: Coping with COVID-19

Theme 6: Use of Coping Strategies

The long-term nature of the COVID-19 pandemic posed a challenge for rural physicians to adapt and use their usual healthy coping strategies. Due to restrictions and closures, leisure travel and most recreational activities providing a distraction from work were no longer an option.

While rural physicians initially struggled to cope, the majority of the participants of this study eventually learned to develop healthy coping techniques to retain some consistency in life during their limited free time. One common method was to find ways to distract oneself from work such as limiting COVID-19-related discussions and consumption of news, gardening, practicing music, watching movies, reading books, and doing outdoor activities. Some physicians engaged in activities that increased their mindfulness, positivity, and self-confidence. This included meditation, journaling, self-reflection, therapeutic letter writing, focusing on things one can change rather than things one cannot, celebrating "small wins" from one's efforts, and educating oneself about COVID-19. Many participants also worked on strategies to improve self-care, balance relationships, and follow a healthy lifestyle such as by exercising, having a reasonable sleep schedule, connecting with loved ones, learning to incorporate breaks, not doing work outside of business hours, and relying on their partner to identify limits. When possible, some participants visited their social supports in other communities to recharge their mental health.

Individuals with unhealthy coping strategies often reported de-stressing by crying, being too busy or exhausted to apply the usual coping strategies, forcing themselves to continue functioning for their loved ones, and repressing their emotions. These individuals were also more likely to have a worsening diet, do less exercise, feel too tired to do activities that they enjoyed, or lose interest in pursuing their hobbies. Most participants reported that their sleep schedule and quality was impacted by more night shifts, staying up late to catch up on paperwork, insomnia from increased anxiety (both generalized and COVID-19-related), and COVID-19-related dreams or nightmares. A few participants also reported drinking more alcohol than usual and using dark humor to reduce their stress.

Theme 7: Experience Accessing Physician Wellness Resources

When all participants were asked if they were familiar with or had used any physician wellness resources offered by any professional associations or organizations, only eight participants responded affirmatively. Of them, four individuals reported satisfactory experiences specifically with the Ontario Medical Association's (OMA) wellness programs and would recommend it to their colleagues, including the OMA Wellness Support Line, positive messages texting pilot project, and the physician leadership support groups.

A major barrier to accessing physician wellness services by rural physicians was reported as being too busy with their increased pandemic workload to read promotional emails on wellness resources, to explore resources online, or to even take time off to focus on their mental health. Participants also felt overwhelmed by the vast quantity of resources and emails from various sources and did not have time to comb through them all. Furthermore, negative past experiences with mental health services deterred many rural physicians from seeking support. For example, six participants reported having tried the OMA wellness program or other help lines in the past but did not find it useful or knew a colleague who received suboptimal support. Five participants expressed wanting supports tailored specifically for rural physicians to avoid urban-based "geographical narcissism" from province-wide programs. Fear of repercussions on their professional license and insurance, as well as fear of stigma from colleagues, family and the community were also described.

Beyond physician wellness services, while some participants felt uncomfortable reaching out to local mental health services due to the lack of anonymity in the rural setting, five participants reported having a personal counsellor or therapist from outside their community. They expressed enjoying the validation and the personalized service they received, including for their non-professional problems. A few participants were hesitant to seek support from colleagues due to the fear of creating an added burden for them, but most participants reported enjoying support from peers and community mentors with whom they had a deeper connection and a shared rural context, compared to a stranger from a help line. Participants cited physician support groups, the SRPC RuralMed listserv, and WhatsApp group chats with colleagues as useful resources during the pandemic.

Almost all the participants opined that despite all of these resources, improving workforce capacity through increased locum coverage or permanent physician staffing was the most important need of rural physicians.

4. Discussion

From the literature, it was hypothesized by the researchers that rural physicians would have a high level of emotional distress during the COVID-19 pandemic [7–22]. Interestingly, PHQ-2, GAD-2, and PSS-4 positive screening rates (Table 2) demonstrated lower levels of distress than the self-reported perceived change in depression, anxiety, and stress levels of the participants compared to their pre-pandemic baseline (Table 3). This can be attributed to the fact that this study was conducted more than a year after the start of the COVID-19 pandemic, but the objective screening tools only looked at the past two weeks to one month. Thus, the questions in Table 3 were a much more sensitive tool for gauging the longitudinal decline in their mental wellness. As discussed in theme 1, most participants agreed that their responses earlier in the pandemic would have reflected greater distress because they learned to cope with this "new normal" over time despite their overall decline in mental health. This reflects the resiliency and adaptability of rural physicians in the face of adversity and the positive impact of adopted healthy coping strategies. However, resilience can only take an individual so far before systemic change is required to resolve their ongoing stressors and prevent long-term psychological harm.

The findings of this study are consistent with the global literature, although the context behind these stressors is different for rural Canadian physicians [7–22]. Most psychological stressors identified in this study, particularly in theme 2 and 4, stemmed from previously prevailing systemic challenges within the rural healthcare system that were exacerbated by the pandemic [2–4].

This study identified increased workload and staffing shortage as the major stressors that prevented participants from finding time for self-care or taking time off during the pandemic. Even before the pandemic, it was difficult to take time off and arrange for coverage by locums or colleagues due to the chronic shortage of physicians [42]. This contributes to burnout, which was further complicated by the fact that physicians in general are more likely to reach out for mental wellness supports when they are in an acute dire crisis rather than earlier [43].

Health Force Ontario, a marketing and recruitment agency under the provincial government, currently offers centralized and coordinated assistance for locum physician placement across the province through Ontario Physician Locum Programs (including the Rural Family Medicine Locum Program, Emergency Department Locum Program and Northern Specialist Locum Program) to which short-staffed hospitals, individual physicians seeking support and interested locums seeking employment can register [44]. While there is adequate locum supply through these programs, most of them are not available on an urgent basis. Moreover, many employers usually do not arrange locum coverage for their staff and the onus is put on the respective physician. For a physician in crisis, however, the multi-step application process may be cumbersome and time-consuming.

Based on the request of the participants, aside from ongoing efforts to increase recruitment and retention of the rural medical workforce, the researchers recommend the establishment of a rapid, easy-to-access locum coverage system to allow acutely struggling rural physicians to take a much-needed break without worrying about finding someone to cover their practice. This time off will also enable them to access mental health supports and attend to their basic physical needs, so that they can return to work in a better mental health state and provide safe patient care. This is in alignment with the literature on the Physician Wellness Hierarchy, which prioritizes attending to a physician's physical and mental needs first for them to be able to deliver quality patient care [45]. A logistical model such as the Ontario Physician Locum Programs may be considered but with a simplified application process and standby locums who can emergently ensure continuity of primary medical care in the community.

Most of the stressors identified in this study can be addressed by system-level changes. The CMA positions that health care employers have an ethical responsibility to address physician wellness and remove any occupational and personal barriers to positive physician health [46]. There is also evidence that employer support can be a protective factor for physician mental health during pandemics [1]. An organizational approach, where the employer and hospital leadership are empowered to cultivate a healthy professional culture may, therefore, be considered to address stressors faced by rural physicians. This will also ensure that the approach is relevant to the local context of the physician. Further, employers can play a role in countering mental health stigma in the workplace by fostering open communication; challenging unhealthy attitudes, for example, by ensuring vacation time is protected; hosting an educational campaign about the true impacts of seeking mental health support on licensure and insurance; providing options to seek mental health supports locally or external of their community [1]. When physicians perceive that their efforts are recognized and reciprocated by employers and authorities in these ways, their psychological outcomes, burnout rates, motivation, and morale can be significantly improved [47–49].

Finally, during the COVID-19 pandemic, many new urban-centered decision-making committees were established on short notice to develop COVID-19 management policies, but rural representation was not always considered on these teams. As articulated by the study participants, this led to many unique rural challenges being unaddressed, including inequities in access to healthcare resources such as PPE, oxygen, ventilators, and vaccines, which caused increased anxiety for rural physicians and threatened the stability of fragile low-staffed rural healthcare centers. Participants also reported feeling ignored and unvalued by urban-based policy makers. While the *Equity and Diversity in Medicine* policy by the CMA recognizes geographical minorities, it is not as commonly applied in practice compared to other minority populations [50]. Urban-based policymakers also often try to implement urban interventions to rural areas without understanding that it may not always be successful in the rural context [51]. If rural physicians are represented at decision-making tables, they can effectively guide discussions and inform decisions on issues relevant to rural communities. To improve rural representation at the provincial and national level, greater advocacy for the recognition of rural as a minority group is warranted, so that future policies and planning will be inclusive of the rural context and the voice of rural physicians is heard.

5. Limitations

A major limitation of this study is its small sample size, which may lower the statistical power of the study. As anticipated, it was difficult to recruit participants who had enough time and energy to contribute to this study due to their heavy workload, burnout, and family responsibilities. This may also explain why there was no participants from communities with less than five physicians.

Participant bias may exist in this study as physicians who are distressed or have strong opinions on the topic are more likely to volunteer as a participant. Further, while the demographic distribution of the participants in this study was diverse, there are certainly many physician perspectives from rural Ontario that were missed. As healthcare is under provincial jurisdiction in Canada, the findings of our study may also not be generalizable to the experience of rural physicians in the rest of Canada or to rural physicians in other countries, despite some similarities.

Finally, thematic saturation was not quantitatively assessed in this study due to the challenges of measuring the degree of saturation statistically. Instead, this study determined thematic saturation through the widely accepted empirical approach proposed by Glaser and Strauss (1967) [52]. Despite its popularity, there are a few limitations that are inherent to this approach. It is dependent on the researcher's subjective judgement to determine when thematic saturation has been achieved. Moreover, an observation that does not contribute new themes cannot necessarily guarantee that a future observation will not contribute new themes [34]. These limitations can impact the validity of the study.

6. Conclusions and Implications

The COVID-19 pandemic has introduced immense stresses on the Canadian healthcare system and the mental health of frontline physicians [1]. This is especially significant for physicians living in under-resourced rural and remote areas, who were already experiencing high pre-pandemic burnout rates [3,4]. For physicians working in rural Ontario, the COVID-19 pandemic has resulted in worsening self-reported overall mental wellbeing due to stressors such as increased workload, risk of infection, limited access to resources and staffing, as well as changing personal relationship dynamics. Current physician wellness resources are often not accessed by burnt-out rural physicians due to difficulty in finding time to focus on self-care. This study recommends establishing a rapid locum supply system, ensuring rural representation at decision-making tables, and taking an organizational approach to support the mental health of rural physicians.

The literature on physician wellness in the rural Canadian context was limited prior to the pandemic. The findings of this study can, thus, be used to guide future research and long-term studies on this important topic in other rural regions of Canada or the world. Healthcare policymakers may also use this data for the development of focused strategies that address rural physician wellness and improve the resilience of the medical workforce. This will, consequently, strengthen the quality of care in rural and remote communities and equip the Canadian healthcare system to effectively support rural physicians during the present and future global public health emergencies.

Author Contributions: Conceptualization of study topic and methodology, A.M. and E.P.; Recruitment, data acquisition (questionnaire and interviews), and data analysis, A.M.; Analysis review, E.P.; Writing—Original Draft Preparation, A.M.; Writing—Review and Editing, A.M. and E.P.; Supervision, E.P.; Project Administration, A.M. and E.P. All authors are accountable for the accuracy and integrity of the work. All authors have read and agreed to the published version of the manuscript.

Funding: This research received no external funding.

Institutional Review Board Statement: Ethics approval was obtained through the Queen's University Health Sciences and Affiliated Teaching Hospitals Research Ethics Board (TRAQ# 6032326, 25 March 2021).

Informed Consent Statement: Written informed consent has been obtained from the participants to publish this paper.

Data Availability Statement: All data are presented in the article.

Acknowledgments: The authors thank the administrative staff of the Society of Rural Physicians of Canada for their assistance in recruitment and processing participant gift cards. The authors also thank Sarah Gower for providing her insight on the recommendations made by the authors, and the Queen's Undergraduate Medical Education student awards committee for recognition with the S.S. Robinson Summer Studentship in Family Medicine in support of this project. Lastly, thank you to the physicians who agreed to share their deeply personal and often difficult stories to shape our understanding.

Conflicts of Interest: The principal investigator is a co-chair of the Society of Rural Physicians of Canada Student Committee.

Appendix A

Interview Guide:

Questions:

(1) When the pandemic started, how did your everyday medical practice and lifestyle change? Are there are any changes unique to your rural context?

Prompts:

- What new measures were put in place in response to the pandemic?
- Did you have access to appropriate types and quantities of PPE?
- How has your workload changed?
- If you needed it, are you able to get time off as easily as before?
- What personal lifestyle changes have you made?

(2) Do you think your mental health state has changed since before the pandemic compared to now? How?

(3) Do you feel like your community was impacted by many/a high number of COVID-19 cases?

If so, please tell me briefly about that.

(4) What are some of the stressors that have impacted your mental health? What are some stressors that are unique to your rural context in your experience?

Prompts:

- How did the risk of infection impact your mental health?
- How did changing information about the virus impact you (both personally and professionally)?
- How did wearing PPE affect you?
- How did all the new infection-control measures affect you?
- Where there any institutional policies that affected your mental health? Were these impacts positive or negative?
- How did quarantining and self-isolation affect you (both personally and professionally)?
- Did you feel more or less isolated than before?
- Was your sleep schedule impacted?
- How did the balance between your career and relationships get impacted?
- How did the public's changing perception of your profession impact your mental health?
- How did the anxiety or response of your patients to the pandemic affect you?
- How did taking care of COVID-19 patients make you feel?
- How did availability of resources (i.e., ventilators, PPE, testing kits, medications, staffing, etc.) in your rural setting impact you?
- Did the pandemic have a financial impact on you?

(5) How has the vaccine rollout in rural Ontario impacted your mental health and practice?

(6) What are some coping methods or things you do to maintain or boost your mental health during the pandemic? What did you find worked or didn't work?

(7) There are several supports that have been made available by the Canadian Medical Association, Ontario Medical Association, Society of Rural Physicians of Canada, and other organizations, including local hospitals. This includes support services, help lines, lists of online resources, etc. Have you ever used any of these resources?

Prompts:

- Do you think the supports are well-tailored to rural physicians like yourself?

- If you did use them, did you find them helpful?/If you didn't use them, what is an ideal resource that you think will help rural physicians during the pandemic?

  Debrief:

(1) Is there anything else you would like to share with me today?
(2) Thank you for sharing your experiences with me today. [Stop recording.] How did you find the interview and questionnaire? Do you have any questions?

Table A1. PSS-4, PHQ-2 and GAD-2 detailed results.

Variable	Response % (*n* = 25)	Variable	Response % (*n* = 25)
PSS-4		**PHQ-2**	
Question 1: In the last month, how often have you felt that you were unable to control the important things in your life?		*Question 1: Over the last 2 weeks, how often have you been bothered by little interest or pleasure in doing things?*	
Never	00.0 (0)	Not at all	48.0 (12)
Almost never	16.0 (4)	Several days	44.0 (11)
Sometimes	56.0 (14)	More than half the days	8.0 (2)
Fairly often	24.0 (6)	Nearly every day	00.0 (0)
Very Often	4.0 (1)		
Question 2: In the last month, how often have you felt confident about your ability to handle your personal problems?		*Question 2: Over the last 2 weeks, how often have you been bothered by feeling down, depressed, or hopeless?*	
Never	00.0 (0)	Not at all	48.0 (12)
Almost never	4.0 (1)	Several days	36.0 (9)
Sometimes	28.0 (7)	More than half the days	16.0 (4)
Fairly often	48.0 (12)	Nearly every day	00.0 (0)
Very Often	20.0 (5)		
		GAD-2	
Question 3: In the last month, how often have you felt that things were going your way?		*Question 1: Over the last 2 weeks, how often have you been bothered by feeling nervous, anxious or on edge?*	
Never	00.0 (0)	Not at all	24.0 (6)
Almost never	4.0 (1)	Several days	32.0 (8)
Sometimes	56.0 (14)	More than half the days	40.0 (10)
Fairly often	32.0 (8)	Nearly every day	4.0 (1)
Very Often	8.0 (2)		
Question 4: In the last month, how often have you felt difficulties were piling up so high that you could not overcome them?		*Question 2: Over the last 2 weeks, how often have you been bothered by not being able to stop or control worrying?*	
Never	4.0 (1)	Not at all	28.0 (7)
Almost never	24.0 (6)	Several days	56.0 (14)
Sometimes	40.0 (10)	More than half the days	16.0 (4)
Fairly often	24.0 (6)	Nearly every day	00.0 (0)
Very Often	8.0 (2)		

Table A2. Quotes from participants about impacts on psychological wellbeing.

Quote Number	Quote
Quote 1	I think that I would honestly say like my mental health, right now, is really quite good. I think that in a lot of ways [. . .] a lot of us have like grown a lot through this and really learned to prioritize [. . .] If you'd asked me this like March of last year, April, May of last year, it would have been a little different. I think that that was a really hard time. Like, I've never been somebody who has insomnia, but I would like, you know, just wake up in a cold sweat, wake up worrying, just really afraid because we had done some math based on some projections of like, if we didn't lock down what would that look like? And like, it was really scary for our town. Like we have four ventilators at the best of times, maybe not . . . maybe only three actually. And we were projecting like you know 300 people would end up in the hospital [. . .] So like, it was stuff of nightmares. Like we were just like projecting that we'd be working hundred-hour weeks and that we'd have no help, and we just didn't know what was coming so it was just . . . it was really scary. (Participant 1)
Quote 2	Yeah, I mean, mental health wise, I would say, I was more stressed by the concerns or responsibilities that were mine and [what] I owned than anything else really. So, again, that sort of morphed throughout the pandemic from the very beginning when I was most worried about my daughter's potential exposures at work, and then quickly realizing that my exposure and my long-term care residents were my biggest responsibility. That was extremely difficult. [. . .] I think I was just exhausted by it, you know. I didn't cry about it. I didn't . . . I couldn't get away from it. It was something that I wasn't used to having that constant stress. You know, [I've had] stressful times at [my] various places of work, but I'd never had this constant, "don't leave it" stress from, you know, last March right through to when . . . our residents didn't get vaccinated until . . . fully vaccinated till the end of February beginning of March. So that was a long haul. (Participant 2)
Quote 3	I would say that, you know, I tend to be pretty calm, [but] I have felt more emotional recently. And you know, that maybe shows up on a zoom meeting or something like that. [. . .] I don't know how much of that to tribute to the pandemic. I think it's because of not having enough physicians to do the job, so we're all working too hard. And so then, being short of sleep. (Participant 3)
Quote 4	[I told myself] "Well, I worked every day for the past two months. Obviously, that wasn't good enough" . . . so just like lots of downplaying the amount of work that I did and kind of self-deprivation that was going on as well. Now, you know, I've come to the realization that you know, things will eventually get better but definitely in a very dark place where it's like, you know, what's even the point sort of thing if, if this is . . . if this is the new the new way of life? (Participant 3)
Quote 5	Just because it's been now a year and a half, and it's not like before that year and a half, everything's peachy right? It was a struggle to access resources prior. And then you have a pandemic that augments that exponentially. Right? And, you know, yeah. So I would say that yes, there definitely has been an effect on one's mental health, for sure. (Participant 10)

Table A3. Quotes from participants about stressors affecting psychological wellbeing of rural physicians.

Stressor	Quote
Excessive Increase in Workload and Responsibilities	Being in a rural area, we are expected to cover the hospital and the emergency department as well as outpatient clinics, so there's increased demand. Again, covering for the acute COVID patients, vaccine clinics, swabbing clinics, all of that falls onto family physicians because we are literally the only physicians in the area, so it all has to come to us. That's probably the biggest thing . . . is having to cover everything and trying to pick up all the pieces of this new pandemic world. (Participant 4) All of a sudden I ended up with all of these leadership responsibilities, a lot more meetings to attend, a lot more just like mental load of like, you know, participating in not only all the planning and the kind of figuring out how we were going to do things locally, but also needing to stay up to date with the torrent, like the fire hose of information that was, you know, coming out daily in terms of like this novel virus and like what do we know about it? [. . .] What kind of measures do we need to take to protect our staff? PPE, like you know all the protocols and treatments, just so much information to filter through. (Participant 5) And then it's like which source do you listen to, you know? Different information coming from, you know . . . not hugely different, but different information coming from different sources. It was overwhelming. Like, I don't find it as bad right now. Like I don't get the same volume of emails, but it was easily like 15 to 20 emails a day. And it has just become like . . . it's just too much to kind of keep track of. (Participant 6)

Table A3. *Cont.*

Stressor	Quote
Management of Risk of Infection	I really felt like the best description that they were asking of us was that we were cannon fodder. That we were going into emergency with absolutely no backup, no hope of coming out and a high risk of getting infected. (Participant 7) It's like watching a scary movie where you know there's no creature that's out, but just the anticipation is almost worse than to see it. So, I found that, you know, we would always be kind of on high alert whenever we had someone with symptoms and there would be a few situations where, you know, I would come in contact with someone that ended up being COVID positive a few days later. And of course, you know, I wore the appropriate PPE, but you're always wondering oh you know, "Did I wash my hands right? Was there a little bit of exposure?" (Participant 8) I was coming home and stripping down in the garage and throwing my clothes directly into the washing machine, going to the shower before I would go near my kids. I still do that. (Participant 4) Like we got vaccines and then all of a sudden, they're sent to Toronto instead. Meanwhile, everybody from Toronto was coming up here to vacation. It was just . . . like that made me feel 100% betrayed, you know. (Participant 9)
Management of Limited Resources in a Rural Setting	I have my N95 [now] if I need it. Like, you know, it just felt like I had some form of, like, protection, because before that it felt like we were just being like lambs to the slaughter—like just go out, work, and you can't wear PPE because we don't have enough. So, it was . . . that was terrifying. (Participant 1) When the pandemic started, we had one ventilator, which is used usually just to transport people to higher level centers. So then of course we were like well what happens if we can't transfer them, right? Like what if the higher-level centers are full and they won't take them? So, then we were having to train our . . . trying to train our nurses on how to care for a ventilated patient, which is not something that they typically do. (Participant 5) . . . And as someone trained in a very urban program, who is now working a 100% as a rural doctor, it feels clearer than ever, especially during the pandemic, that a lot of the people that made policies in urban centres, and a lot of people that are being trained in urban centres and practice medicine in urban centres have no idea what's going on in rural places. (Participant 10) Remember, like everything in an urban center is augmented that much more in a rural center. Like urban centers have much higher physician per capita, right? So, there's more redundancy. So, if someone gets sick you have . . . like there in a rural center, [. . .] you're one person away from your emerg shutting down even before COVID. (Participant 10) The irony is that we're supposed to take care of ourselves as physicians to take care of our patients, but we're not really allowed quote unquote to take any time off. So how do you take care of . . . like how do you take care of yourself when you can't devote any time to your own self care?...There's this weird perception of doctors where we're supposed to be superhuman, and we can just keep on going while denying our own needs and still be excellent all the time for patients, which I find ridiculous, but that's, I just feel like that's kind of an expectation for physicians. (Participant 11)
Impact on Familial, Professional and Public Relationships	We basically spent four months not touching, not hugging, not sleeping in the same bed. That was stressful. It wasn't so much around the decision making, but really around protecting capacity for the local health care system, so that we didn't lose two docs at the same time. (Participant 12) Like you're here either at work working or you're at home parenting. Whereas before there used to be little pockets of time before the kids came home from school or, you know, before you know something else happened. Like so those pockets of time or windows of opportunity for self-care just evaporated and they're still largely gone. [. . .] You just don't have those shared opportunities to kind of indulge in our relationship as a couple. [. . .] I'd sort of say, my partner doesn't feel as fun, but not to any fault of hers or mine. It just feels like, again, it's all work all the time so . . . (Participant 11) I was pretty disillusioned with the public's behavior at large. I found it very difficult. I'm going to have trouble, integrating with some of my neighbors who basically chose to ignore every public health guideline that has been, you know, put out to the public. So, I'm becoming less judgmental of them as time goes by, but it's going to be hard to sort of reintegrate to those. (Participant 2) But I think media, especially social media has been extremely hard on physicians. [. . .] And this gradual transition from being the hero savior who was, you know, the well-loved by the community to being the bad guy who was continuing to try and force public health and Ontario health recommendations for isolation. Feeling that at some point, a portion of the population, not all and probably not the majority, turned against us. (Participant 13)

Table A4. Quotes from participants about coping with the COVID-19 pandemic.

Stressor	Quote
Use of Coping Strategies	Personally, I think that, you know, because there were a lot of work responsibilities and meetings and other kind of work-related and patient care-related events, you know what, I would say that it definitely has eaten into my hobbies and my extracurricular outside of medicine interests, you know. Some of that was because you couldn't do those interests because they weren't, you know, available to do because of public health restrictions, you know. But often because of clinical workload. (Participant 10) My life [outside of work] hasn't changed quite that dramatically in a rural place because there is more space. There's more space, there's more outdoor space, people are more used to doing outdoor things, and like generally have larger homes and patios that you can go to socialize at outdoors. (Participant 14)
Experience Accessing Physician Wellness Resources	In my mind it's [accessing physician wellness services] something that would take even more time. I already have a busy life, which would invariably come either from my sleep or my family life. So, I just have no more time to give, including to myself. (Participant 4) You can kind of imagine sometimes in a smaller community, you might have access to in-person mental health resources but oftentimes you may know the counselor, right? And there may be a bit of familiarity with that relationship, so you may not necessarily want to discuss certain issues with that individual. (Participant 10) The other thing too I sometimes think about is if I reach out and do this and they're concerned about me, is there any repercussions especially like on insurance? [. . .] You know, especially for a younger physician who's still, you know, trying to improve their insurance, you know their disability or whatever, would that have an impact, because if you say no on that questionnaire for "Have you seen a therapist in the last year?" and you lie and they find out somehow that you did and there's reports about it or if it was copied to your family physician, that would have negative repercussions. So, I just felt there's a barrier there, and I never felt like I was bad enough to need it. (Participant 15)

References

1. Galbraith, N.; Boyda, D.; McFeeters, D.; Hassan, T. The mental health of doctors during the COVID-19 pandemic. *BJPsych Bull.* **2021**, *45*, 93–97. [CrossRef]
2. Dove, N. Can international medical graduates help solve Canada's shortage of rural physicians? *Can. J. Rural. Med.* **2009**, *14*, 120–123.
3. Haggerty, T.S.; Fields, S.A.; Selby-Nelson, E.M.; Foley, K.P.; Shrader, C.D. Physician wellness in rural America: A review. *Int. J. Psychiatry Med.* **2013**, *46*, 303–313. [CrossRef]
4. Thommasen, H.V.; Lavanchy, M.; Connelly, I.; Berkowitz, J.; Grzybowski, S. Mental health, job satisfaction, and intention to relocate. Opinions of physicians in rural British Columbia. *Can. Fam. Physician* **2001**, *47*, 737–744.
5. Canadian Medical Association. CMA National Physician Health Survey (NPHS)—A National Snapshot (Rep.). 2018. Available online: https://www.cma.ca/sites/default/files/2018-11/nph-survey-e.pdf (accessed on 15 January 2021).
6. Hogue, A.; Huntington, M.K. Family Physician Burnout Rates in Rural versus Metropolitan Areas: A Pilot Study. *SD Med.* **2019**, *72*, 306–308.
7. Brophy, J.T.; Keith, M.M.; Hurley, M.; McArthur, J.E. Sacrificed: Ontario Healthcare Workers in the Time of COVID-19. *New Solut. J. Environ. Occup. Health Policy* **2020**, *30*, 267–281. [CrossRef]
8. Cabarkapa, S.; Nadjidai, S.E.; Murgier, J.; Ng, C.H. The psychological impact of COVID-19 and other viral epidemics on frontline healthcare workers and ways to address it: A rapid systematic review. *Brain Behav. Immun.-Health* **2020**, *8*, 100144. [CrossRef]
9. Chow, K.M.; Law, B.M.; Ng, M.S.; Chan, D.N.; So, W.K.; Wong, C.L.; Chan, C.W. A review of psychological issues among patients and healthcare staff during two major coronavirus disease outbreaks in China: Contributory factors and management strategies. *Int. J. Environ. Res. Public Health* **2020**, *17*, 6673. [CrossRef]
10. Shreffler, J.; Petrey, J.; Huecker, M. The impact of COVID-19 on healthcare worker wellness: A scoping review. *West. J. Emerg. Med.* **2020**, *21*, 1059. [CrossRef]
11. Chatterjee, S.S.; Bhattacharyya, R.; Bhattacharyya, S.; Gupta, S.; Das, S.; Banerjee, B.B. Attitude, practice, behavior, and mental health impact of COVID-19 on doctors. *Indian J. Psychiatry* **2020**, *62*, 257–265. [CrossRef]
12. Khatun, M.F.; Parvin, M.F.; Rashid, M.M.; Alam, M.S.; Kamrunnahar, M.; Talukder, A.; Rahman Razu, S.; Ward, P.R.; Ali, M. Mental Health of Physicians During COVID-19 Outbreak in Bangladesh: A Web-Based Cross-Sectional Survey. *Front. Public Health* **2021**, *9*, 592058. [CrossRef] [PubMed]

13. Liu, Z.; Wu, J.; Shi, X.; Ma, Y.; Ma, X.; Teng, Z.; You, X.; Zhang, Y.; Zhang, W.; Feng, Z.; et al. Mental Health Status of Healthcare Workers in China for COVID-19 Epidemic. *Ann. Glob. Health* **2020**, *86*, 128. [CrossRef] [PubMed]
14. Robertson, L.J.; Maposa, I.; Somaroo, H.; Johnson, O. Mental health of healthcare workers during the COVID-19 outbreak: A rapid scoping review to inform provincial guidelines in South Africa. *S. Afr. Med. J.* **2020**, *110*, 1010–1019. [CrossRef]
15. De Brier, N.; Stroobants, S.; Vandekerckhove, P.; De Buck, E. Factors affecting mental health of health care workers during coronavirus disease outbreaks (SARS, MERS & COVID-19): A rapid systematic review. *PLoS ONE* **2020**, *15*, e0244052. [CrossRef]
16. Quintana-Domeque, C.; Lee, I.; Zhang, A.; Proto, E.; Battisti, M.; Ho, A. Anxiety and depression among medical doctors in Catalonia, Italy, and the UK during the COVID-19 pandemic. *PLoS ONE* **2021**, *16*, e0259213. [CrossRef]
17. Gold, J.A. COVID-19: Adverse mental health outcomes for healthcare workers. *BMJ Clin. Res. Ed.* **2020**, *369*, m1815. [CrossRef]
18. Di Tella, M.; Romeo, A.; Benfante, A.; Castelli, L. Mental health of healthcare workers during the COVID-19 pandemic in Italy. *J. Eval. Clin. Pract.* **2020**, *26*, 1583–1587. [CrossRef]
19. Shechter, A.; Diaz, F.; Moise, N.; Anstey, D.E.; Ye, S.; Agarwal, S.; Birk, J.L.; Brodie, D.; Cannone, D.E.; Chang, B.; et al. Psychological distress, coping behaviors, and preferences for support among New York healthcare workers during the COVID-19 pandemic. *Gen. Hosp. Psychiatry* **2020**, *66*, 1–8. [CrossRef]
20. Preti, E.; Di Mattei, V.; Perego, G.; Ferrari, F.; Mazzetti, M.; Taranto, P.; Di Pierro, R.; Madeddu, F.; Calati, R. The Psychological Impact of Epidemic and Pandemic Outbreaks on Healthcare Workers: Rapid Review of the Evidence. *Curr. Psychiatry Rep.* **2020**, *22*, 43. [CrossRef]
21. Siyal, F.J.; Shaikh, Z.A.; Ahmed, S.Z.; Shahid, M.A.; Agha, F.; Khoso, M.; Shahani, M.P. Anxiety among COVID-19 Physicians during the Pandemic in the Health Care Center of the Rural Region. *Arch. Pharm. Pract.* **2020**, *11*, 91–93.
22. De Kock, J.H.; Latham, H.A.; Leslie, S.J.; Grindle, M.; Munoz, S.A.; Ellis, L.; Polson, R.; O'Malley, C.M. A rapid review of the impact of COVID-19 on the mental health of healthcare workers: Implications for supporting psychological well-being. *BMC Public Health* **2021**, *21*, 104. [CrossRef] [PubMed]
23. Frontline Workers and COVID-19: Coping with Stress. Available online: http://www.emro.who.int/mnh/news/frontline-workers-and-covid-19-coping-with-stress.html (accessed on 15 January 2021).
24. Penwell-Waines, L.; Ward, W.; Kirkpatrick, H.; Smith, P.; Abouljoud, M. Perspectives on Healthcare Provider Well-Being: Looking Back, Moving Forward. *J. Clin. Psychol. Med. Settings* **2018**, *25*, 295–304. [CrossRef] [PubMed]
25. Physician Wellness+ Initiative Dedicates $15 Million to Support Physicians and Medical Learners. Available online: https://www.cma.ca/news/physician-wellness-initiative-dedicates-15-million-support-physicians-and-medical-learners (accessed on 22 November 2021).
26. COVID-19 Resource Guide. Available online: https://srpc.ca/COVID-19-Resource-Guide (accessed on 17 October 2021).
27. CMPA COVID-19 Hub. CMPA. Available online: https://www.cmpa-acpm.ca/en/covid19 (accessed on 17 October 2021).
28. COVID-19 and Mental Health Resources, Tools: We Are Here for You. Available online: https://psychiatry.queensu.ca/faculty/mental-health-resources (accessed on 20 February 2021).
29. Lieberman, C. Coronavirus: Quebec Doctor's Death by Suicide Sends Shockwaves through Canada's Medical Community. 13 January 2021. Available online: https://globalnews.ca/news/7569318/coronavirus-doctor-karine-dion/ (accessed on 20 February 2021).
30. Canadian Medical Association. One Year into the Pandemic: High Levels of Anxiety and Increased Fatigue among Physicians. Available online: https://www.cma.ca/news-releases-and-statements/one-year-pandemic-high-levels-anxiety-and-increased-fatigue-among (accessed on 17 October 2021).
31. CTVNews.ca. Tracking Every Case of COVID-19 in Canada. 15 January 2021. Available online: https://www.ctvnews.ca/health/coronavirus/tracking-every-case-of-covid-19-in-canada-1.4852102 (accessed on 7 November 2021).
32. Lauzon, A.; Douglas, D.; Caldwell, W.; Ragetlie, N. Ontario—State of Rural Canada. 6 October 2015. Available online: http://sorc.crrf.ca/ontario/ (accessed on 15 January 2021).
33. Statistics Canada. *Rural and Small Town Canada Analysis Bulletins*; Cat. No. 21-006-XIE; Statistics Canada: Ottawa, ON, Canada, 2001.
34. Lowe, A.; Norris, A.C.; Farris, A.J.; Babbage, D.R. Quantifying thematic saturation in qualitative data analysis. *Field Methods* **2018**, *30*, 191–207. [CrossRef]
35. Sapra, A.; Bhandari, P.; Sharma, S.; Chanpura, T.; Lopp, L. Using Generalized Anxiety Disorder-2 (GAD-2) and GAD-7 in a Primary Care Setting. *Cureus* **2020**, *12*, e8224. [CrossRef]
36. Maurer, D.M. Screening for depression. *Am. Fam. Physician* **2012**, *85*, 139–144. [PubMed]
37. Warttig, S.L.; Forshaw, M.J.; South, J.; White, A.K. New, normative, English-sample data for the short form perceived stress scale (PSS-4). *J. Health Psychol.* **2013**, *18*, 1617–1628. [CrossRef] [PubMed]
38. Jason, L.; Glenwick, D. (Eds.) *Handbook of Methodological Approaches to Community-Based Research: Qualitative, Quantitative, and Mixed Methods*; Oxford University Press: Oxford, UK, 2016.
39. Maguire, M.; Delahunt, B. Doing a thematic analysis: A practical, step-by-step guide for learning and teaching scholars. *All Irel. J. High. Educ.* **2017**, *9*, 3.
40. Draaisma, M. Ontario Will No Longer Give AstraZeneca COVID-19 Vaccine as 1st Dose Due to Blood Clot Risk. CBCnews. 12 May 2021. Available online: https://www.cbc.ca/news/canada/toronto/ontario-update-astrazeneca-vaccine-1.6022545 (accessed on 9 November 2021).

41. Government of Ontario. Frequently Asked Questions. Home and Community Care Support Services North East. Available online: http://www.nelhin.on.ca/resources/frequentlyaskedquestions.aspx (accessed on 9 November 2021).
42. Farkas, B.; Manns, B.; Ogundeji, Y.; Wellstead, D.; Clement, F. *Recruitment and Retention of Rural and Remote Physicians—The Role of Alternate Payment Models*; Abridged Report; University of Calgary: Calgary, AB, Canada, 2020.
43. Peckham, C. Medscape National Physician Burnout, Depression & Suicide Report 2019. Medscape. 16 January 2019. Available online: https://www.medscape.com/slideshow/2019-lifestyle-burnout-depression-6011056 (accessed on 21 November 2021).
44. Health Force Ontario. Ontario Physician Locum Programs. Available online: https://www.healthforceontario.ca/en/M4/Ontario_Physician_Locum_Programs (accessed on 17 October 2021).
45. Shapiro, D.E.; Duquette, C.; Abbott, L.M.; Babineau, T.; Pearl, A.; Haidet, P. Beyond burnout: A physician wellness hierarchy designed to prioritize interventions at the systems level. *Am. J. Med.* **2019**, *132*, 556–563. [CrossRef]
46. Canadian Medical Association. Building the Case for Physician Wellness in Canada. Canadian Medical Association. 27 July 2020. Available online: https://www.cma.ca/physician-wellness-hub/content/investing-physician-wellness-in-Canada (accessed on 22 November 2021).
47. Imai, H.; Matsuishi, K.; Ito, A.; Mouri, K.; Kitamura, N.; Akimoto, K.; Mino, K.; Kawazoe, A.; Isobe, M.; Takamiya, S.; et al. Factors associated with motivation and hesitation to work among health professionals during a public crisis: A cross sectional study of hospital workers in Japan during the pandemic (H1N1) 2009. *BMC Public Health* **2010**, *10*, 672. [CrossRef]
48. Damery, S.; Draper, H.; Wilson, S.; Greenfield, S.; Ives, J.; Parry, J.; Petts, J.; Sorell, T. Healthcare workers' perceptions of the duty to work during an influenza pandemic. *J. Med. Ethics* **2010**, *36*, 12–18. [CrossRef]
49. Shanafelt, T.D.; Noseworthy, J.H. Executive Leadership and Physician Well-being: Nine Organizational Strategies to Promote Engagement and Reduce Burnout. *Mayo Clin. Proc.* **2017**, *92*, 129–146. [CrossRef] [PubMed]
50. Canadian Medical Association. Equity and Diversity in Medicine. December 2019. Available online: https://policybase.cma.ca/viewer?file=%2Fmedia%2FPolicyPDF%2FPD20-02.pdf#page=1 (accessed on 21 November 2021).
51. Plain, C. Four Key Priorities for Achieving Rural Health Equity. University of Minnesota. 17 March 2021. Available online: https://www.sph.umn.edu/news/four-key-priorities-for-achieving-rural-health-equity/ (accessed on 21 November 2021).
52. Saunders, B.; Sim, J.; Kingstone, T.; Baker, S.; Waterfield, J.; Bartlam, B.; Burroughs, H.; Jinks, C. Saturation in qualitative research: Exploring its conceptualization and operationalization. *Qual. Quant.* **2018**, *52*, 1893–1907. [CrossRef] [PubMed]

healthcare MDPI

Article

Suicidal Ideation during the COVID-19 Pandemic among A Large-Scale Iranian Sample: The Roles of Generalized Trust, Insomnia, and Fear of COVID-19

Chung-Ying Lin [1], Zainab Alimoradi [2], Narges Ehsani [2], Maurice M. Ohayon [3], Shun-Hua Chen [4,*], Mark D. Griffiths [5] and Amir H. Pakpour [2,6,*]

[1] Institute of Allied Health Sciences, National Cheng Kung University Hospital, College of Medicine, National Cheng Kung University, Tainan 701, Taiwan; cylin36933@gmail.com
[2] Social Determinants of Health Research Center, Research Institute for Prevention of Non-Communicable Diseases, Qazvin University of Medical Sciences, Qazvin 3419759811, Iran; zainabalimoradi@yahoo.com (Z.A.); narges.ehsany@yahoo.com (N.E.)
[3] Stanford Sleep Epidemiology Research Center (SSERC), School of Medicine, Stanford University, Stanford, CA 94305, USA; mohayon@stanford.edu
[4] School of Nursing, Fooyin University, Kaohsiung 83102, Taiwan
[5] International Gaming Research Unit, Psychology Department, Nottingham Trent University, Nottingham NG1 4FQ, UK; mark.griffiths@ntu.ac.uk
[6] Department of Nursing, School of Health and Welfare, Jönköping University, 55111 Jönköping, Sweden
* Correspondence: ft104@fy.edu.tw (S.-H.C.); Amir.pakpour@ju.se (A.H.P.); Tel.: +886-7-7811151-6154 (S.-H.C.); Fax: +886-7-7835112 (S.-H.C.)

Citation: Lin, C.-Y.; Alimoradi, Z.; Ehsani, N.; Ohayon, M.M.; Chen, S.-H.; Griffiths, M.D.; Pakpour, A.H. Suicidal Ideation during the COVID-19 Pandemic among A Large-Scale Iranian Sample: The Roles of Generalized Trust, Insomnia, and Fear of COVID-19. *Healthcare* **2022**, *10*, 93. https://doi.org/10.3390/healthcare10010093

Academic Editors: Yasuhiro Kotera and Elaina Taylor

Received: 18 November 2021
Accepted: 24 December 2021
Published: 4 January 2022

Abstract: The novel 2019 coronavirus disease (COVID-19) is still not under control globally. The pandemic has caused mental health issues among many different cohorts and suicidal ideation in relation to COVID-19 has been reported in a number of recent studies. Therefore, the present study proposed a model to explain the associations between generalized trust, fear of COVID-19, insomnia, and suicidal ideation during the COVID-19 pandemic among a large-scale Iranian sample. Utilizing cluster sampling with multistage stratification, residents from Qazvin province in Iran were invited to participate in the present study. Adults aged over 18 years (n = 10,843; 6751 [62.3%] females) completed 'paper–and-pencil' questionnaires with the assistance of a trained research assistant. Structural equation modeling (SEM) was applied to understand the associations between generalized trust, fear of COVID-19, insomnia, and suicidal ideation. Slightly over one-fifth of the participants (n = 2252; 20.8%) reported suicidal ideation. Moreover, the SEM results indicated that generalized trust was indirectly associated with suicidal ideation via fear of COVID-19 and insomnia. Furthermore, generalized trust was not directly associated with suicidal ideation. The proposed model was invariant across gender groups, age groups, and participants residing in different areas (i.e., urban vs. rural). Generalized trust might reduce individuals' suicidal ideation during the COVID-19 pandemic period via reduced levels of fear of COVID-19 and insomnia. Healthcare providers and policymakers may want to assist individuals in developing their generalized trust, reducing fear of COVID-19, and improving insomnia problems to avoid possible suicidal behaviors.

Keywords: generalized trust; psychological distress; sleep; insomnia; suicide

1. Introduction

The psychological distress induced by the novel coronavirus disease-2019 (COVID-19) has had a large negative impact across different countries and regions worldwide [1–6]. With such long-lasting suffering, it is likely that some individuals develop various severe mental health problems, including suicidal ideation [7]. A recent paper reported that in Bangladesh during the pandemic, the prevalence of suicidal ideation was 5% during 1 to

10 April 2020, 6% during 8 to 25 April 2020, 12.8% during April to May 2020, and 19.0% during July 2020 [8].

Moreover, constructing the relationships between these factors is needed to help build a theoretical model to guide potential treatment directions for target outcomes [9,10]. In the present study, a proposed model was constructed to provide knowledge concerning the underlying psychological mechanisms for healthcare providers to understand how suicidal ideation is generated. In this regard, three important factors (i.e., generalized trust, fear of COVID-19, and insomnia) are proposed to be important factors contributing to suicidal ideation during the COVID-19 pandemic [11–16].

Generalized trust (which is also called 'general trust' or simply 'trust' by some researchers) is defined as "a willingness to be vulnerable to the actions of others" [17] and is found to be an important societal factor in assisting individuals in maintaining daily living [18]. Generalized trust has been found to be a factor that contributes to good social relationships, which may subsequently improve individuals' health via the mutual trust and positive interactions between individuals [19–21]. Consequently, individuals' psychological health can be improved [22–25]. Therefore, it is hypothesized that generalized trust during the COVID-19 pandemic may help reduce individuals' fear of COVID-19, insomnia, and suicidal ideation.

A systematic review shows that the suicidal ideation and behaviors during the COVID-19 pandemic were increased compared with the period before the COVID-19 pandemic [8]. Moreover, evidence has shown that some suicide attempts are related to fear of COVID-19 or worries about COVID-19 [11,16]. For example, the first suicide case in Bangladesh involved a man who killed himself because he experienced COVID-19-like symptoms, although the autopsy showed he was not infected with COVID-19 [11]. In another study, fear of COVID-19 was significantly associated with suicidal ideation among 1013 English-speaking Americans [16]. Similarly, it was found that fear of COVID-19, together with psychological distress, explained 27% of the risk regarding suicidal ideation among 595 Spanish adults [26]. The relationship between fear and suicidal ideation has also been found in another coronavirus (i.e., SARS) [27]. Therefore, in extreme cases, fear of COVID-19 may trigger an individual's irrational thoughts and result in suicide.

Aside from fear of COVID-19, insomnia has been found to be another important factor in contributing to individuals' suicidal ideation. Before the COVID-19 pandemic, there was evidence that insomnia severity was a useful indicator in predicting suicidal ideation [13]. Moreover, prior research, including randomized controlled trials, has shown that tackling insomnia can substantially reduce individuals' suicidal ideation [12,14]. There is also some evidence that insomnia has been associated with suicidal ideation during the pandemic [16]. Moreover, sleep problems during the COVID-19 pandemic have been found to be serious [28,29]. Therefore, insomnia could be a trigger for individuals to develop suicidal ideation during the COVID-19 pandemic.

To the best of the present authors' knowledge, the contemporary evidence has supported the associations between fear of COVID-19, insomnia, and suicidal ideation during the COVID-19 pandemic [16,26]. However, there is a lack of empirical evidence regarding how generalized trust is associated with fear of COVID-19, insomnia, and suicidal ideation during the COVID-19 pandemic period. However, prior to the COVID-19 pandemic, generalized trust was found to be associated with low levels of fear, insomnia, and suicidal ideation. Therefore, the present study proposed a model that adopts generalized trust as the independent variable, fear of COVID-19 and insomnia as mediating variables, and suicidal ideation as the dependent variable. The proposed model was examined among a large-scale Iranian sample to provide evidence regarding the psychological mechanism underlying generalized trust and suicidal ideation. More specifically, the present study investigated whether insomnia and fear of COVID-19 are mediators in the association between generalized trust and suicidal ideation among the Iranian general population during the COVID-19 pandemic.

2. Materials and Methods

2.1. Procedures and Participants

The target population in the present study was the general population from Qazvin province in Iran, including residents who lived in urban or rural areas during the COVID-19 pandemic. The entire population residing in Qazvin, a province located in the central part of Iran, comprised 1,273,761 individuals in 2018 with 51% of the Qazvin population being male. The present study used cluster sampling with multistage stratification to recruit participants. First, 70 strata were stratified according to the administration districts in Qazvin. Second, several health centers in each stratum were randomly selected with the consideration of the population size in the stratum. Third, a list of families in the health centers was provided by each health center and several families were randomly selected. Fourth, interviewers who had received formal training contacted the selected families and introduced the study to them. More specifically, the participants' rights and autonomy were clearly described. Once eligible individuals were willing to participate, an interviewer performed a home visit to describe further study details to the participant. If they provided written informed consent, they were allowed to complete the survey.

There was no obvious pattern of selection bias and there was a relatively high response rate (78%). The present study recruited a sample that was considered to be representative of the general adult population in Qazvin. The survey period was administered between 19 February and 9 April 2021. In order to maximize the representativeness of the general adult population, there were only two inclusion criteria. These were being (i) an adult aged 18 years or older living in Qazvin, and (ii) an adult who provided written informed consent for participation. Guests and tourists in Qazvin were excluded from the study.

2.2. Measures

Suicidal ideation was assessed using Item 9 from the nine-item Patient Health Questionnaire (PHQ-9). The PHQ-9 suicide item has been found to be an effective tool for assessing the prevalence of suicidal ideation. The PHQ-9 suicide item assesses passive thoughts of death and the desire for self-harm among respondents (i.e., "Thoughts that you would be better off dead, or thoughts of hurting yourself in some way?") within the last two weeks. The Iranian version of the PHQ-9 has been found to be valid and reliable [30].

General trust was assessed using the six-item Generalized Trust Scale (GTS) [21,31]. All six items (e.g., "Most people are basically honest") are rated on a five-point scale with a higher score indicating a higher level of generalized trust [21,31]. The GTS was recently translated into Persian for Iranians and demonstrated satisfactory psychometric properties (Cronbach's $\alpha = 0.89$ and is unidimensional among the Iranian general population [18,32]).

Fear of COVID-19 was assessed using the seven-item Fear of COVID-19 Scale (FCV-19S) [33]. All seven items (e.g., "I am most afraid of coronavirus-19") are rated on a five-point scale with a higher score indicating a higher level of COVID-19 fear [34,35]. The FCV-19S was originally developed in Persian among the Iranian general population and the scale demonstrated satisfactory psychometric properties (Cronbach's $\alpha = 0.82$ and is unidimensional among the Iranian general population [33]).

Insomnia was assessed using the seven-item Insomnia Severity Index (ISI) [36]. All seven items (e.g., "How satisfied/dissatisfied are you with your current sleep pattern?") are rated on a five-point scale with a higher score indicating a higher level of insomnia [37,38]. The ISI has been translated into Persian for Iranians and the Persian ISI demonstrated satisfactory psychometric properties (Cronbach's α between 0.82 and 0.87 among Iranian patients with insomnia [39] and is unidimensional among Iranian patients with cancer [37]).

Demographic information was collected using a background information sheet including their age (in years), sex (male or female), educational status (university, diploma, high school, secondary school, primary school, or no formal education), marital status (married or single), and accommodation (city or rural).

2.3. Data Analysis

Descriptive statistics were first carried out to understand the participants' characteristics, including their age, sex, educational status, marital status, accommodation, insomnia, fear of COVID-19, and generalized trust. Moreover, the frequency of suicidal ideation was calculated for the sample and different subgroups of the sample (e.g., male and female). Then, Pearson correlations were conducted to examine the zero-order correlations between the studied variables in the proposed model (i.e., suicidal ideation, insomnia, fear of COVID-19, and generalized trust). The proposed model in the present study was then evaluated using structural equation modeling (SEM) with full information maximum likelihood.

In the SEM, all the instrument scores were summed to be used as an observed variable in the proposed model to fulfill the principle of parsimony. The proposed model was first evaluated using fit indices of nonsignificant χ^2, comparative fit index (CFI), and Tucker–Lewis index (TLI) > 0.9, together with root mean square error of approximation (RMSEA) and standardized root mean square residual (SRMR) < 0.08 [40–42]. When the proposed model shows satisfactory fit, the mediated effects of insomnia and fear of COVID-19 in the association between generalized trust and suicidal ideation were checked using the bootstrapping method. More specifically, 5000 bootstrapping samples were generated. Then, the lower and upper limits in the 95% confidence interval (CI) were calculated and the bootstrapping samples were adopted to evaluate the mediated effects. When both limits of the 95% CI do not include 0 (i.e., both are negative values or both are positive values), the mediated effect is supported [43].

The proposed model was then evaluated for its invariance across different subgroups, including gender group (males vs. females), age group (mean age above 35.54 years vs. mean age below 35.54 years), and accommodation group (urban residents vs. rural residents). Multigroup SEM was used to examine the invariance with four nested models; M1: configural model which did not constrain any parameters in the proposed model; M2: a model based on M1 to constrain path coefficients in the proposed model being equal across subgroups; M3: a model based on M2 to constrain correlation coefficients in the proposed model being equal across subgroups; and M4: a model based on M3 to constrain residuals in the proposed model being equal across subgroups. Invariance was supported when ΔCFI (i.e., the difference of CFI between every two nested models) > −0.01, ΔSRMR (i.e., the difference of SRMR between every two nested models) < 0.03, and ΔRMSEA (i.e., the difference of RMSEA between every two nested models) < 0.015 [44].

2.4. Patient and Public Involvement Statement

Study participants and the public were not involved in the design, conduct, reporting, or dissemination plans of the present study.

2.5. Ethical Consideration

The Ethics Committee of Qazvin University Medical Sciences approved the study procedures (reference number IR.QUMS.REC.1399.418).

3. Results

In the large-scale sample of 10,843 participants (mean age = 35.54 years; SD = 12.00), 6751 were females (62.3%), and 2252 reported suicidal ideation (20.8%). The present sample was well educated with over half of them (n = 6991) having a diploma or higher degree (64.5%). Nearly three-quarters of the participants were married (n = 8092; 74.6%) and slightly more than three-quarters resided in a city (n = 8187; 75.5%). A higher prevalence of suicidal ideation was found among those who (i) were female, (ii) had a lower educational level, (iii) were married, and (iv) were living in a city (Table 1). The participants' levels of insomnia, fear of COVID-19, and generalized trust performances are also presented in Table 1.

Table 1. The demographic characteristics of study participants (n = 10,843).

	Mean ± SD or N (%)	Suicidal Ideation	
		No (n = 8591)	Yes (n = 2252)
Age	35.54 ± 12.00		
Sex			
Female	6751 (62.3%)	5242 (77.6%)	1509 (22.4%)
Male	4092 (37.7%)	3349 (81.8%)	743 (18.2%)
Educational status			
University	4230 (39.0%)	3423 (80.9%)	807 (19.1%)
Diploma	2761 (25.5%)	2180 (79.0%)	581 (21.0%)
High school	974 (9.0%)	768 (78.9%)	206 (21.1%)
Secondary school	1540 (14.2%)	1192 (77.4%)	348 (22.6%)
Primary school	986 (9.1%)	760 (77.1%)	226 (22.9%)
No formal education	352 (3.2%)	268 (76.1%)	84 (23.9%)
Marital status			
Married	8092 (74.6%)	6343 (78.4%)	1749 (21.6%)
Single	2751 (25.4%)	2248 (81.7%)	503 (18.3%)
Accommodation			
City	8187 (75.5%)	6476 (79.1%)	1711 (20.9%)
Rural	2656 (24.5%)	2115 (79.6%)	541 (20.4%)
Insomnia	8.69 ± 5.47		
Fear of COVID-19	21.10 ± 6.95		
Generalized trust	2.81 ± 0.87		

Suicidal ideation was positively and significantly correlated with insomnia (r = 0.327; $p < 0.001$) and fear of COVID-19 (r = 0.353; $p < 0.001$), and negatively and significantly correlated with generalized trust (r = −0.070; $p < 0.001$). Generalized trust was negatively and significantly correlated with insomnia (r = −0.100; $p < 0.001$) and fear of COVID-19 (r = −0.183; $p < 0.001$). Moreover, insomnia and fear of COVID-19 were positively and significantly correlated (r = 0.271; $p < 0.001$).

The proposed model was supported by the fit indices, including CFI (0.986), TLI (0.984), RMSEA (0.032), SRMR (0.028), and the nonsignificant χ^2 test ($p = 0.13$). In addition, generalized trust had direct effects on insomnia (standardized coefficient [β] = −0.100; $p < 0.001$) and fear of COVID-19 (β = −0.183; $p < 0.001$) but not on suicidal ideation (β = 0.007; $p = 0.393$). Insomnia (β = 0.250; $p < 0.001$) and fear of COVID-19 (β = 0.286; $p < 0.001$) both had direct effects on suicidal ideation (Figure 1). The mediated effects of insomnia and fear of COVID-19 in the association between generalized trust and suicidal ideation were supported by the 95% CI of bootstrapping samples (Table 2).

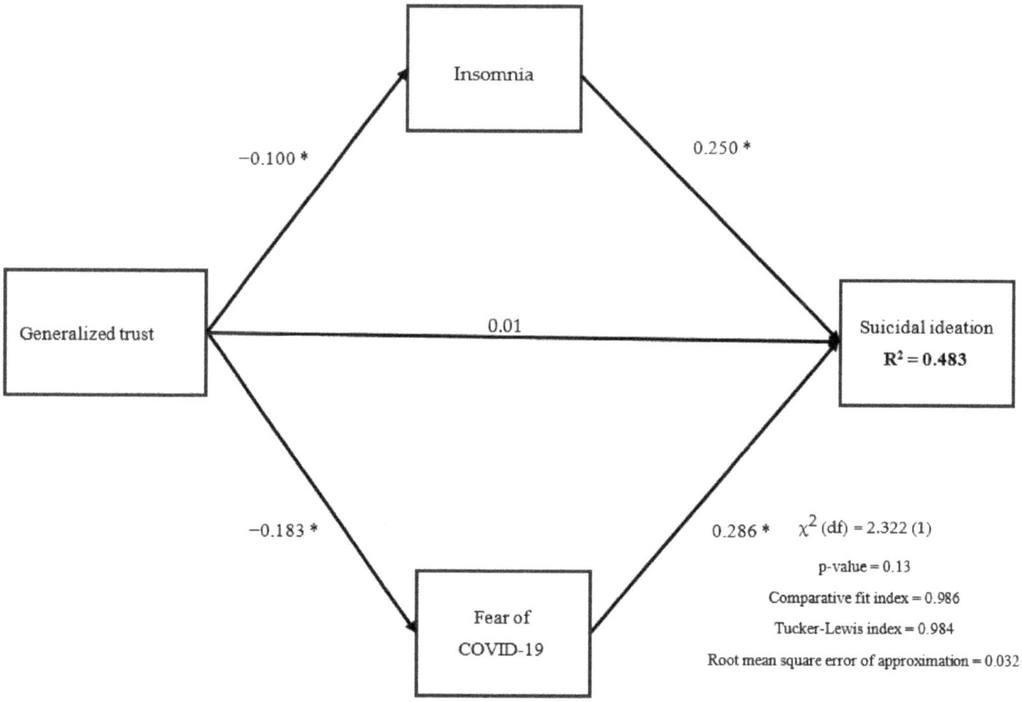

Figure 1. Hypothesized mediation model and findings. * $p < 0.05$.

Table 2. Models that tested mediated effects of fear and insomnia.

	Stand. Coeff.	Unstand. Coeff.	Bootstrapping SE	Bootstrapping LLCI	Bootstrapping ULCI	*p*-Value
Total effect of generalized trust on suicidal ideation	−0.07	−0.032	0.004	−0.041	−0.024	<0.001
Direct effect of generalized trust on suicidal ideation	0.007	0.003	0.004	−0.004	0.011	0.393
Direct effect of generalized trust on mediators						
Fear of COVID-19	−0.183	−1.453	0.082	−1.618	−1.297	<0.001
Insomnia	−0.100	−0.860	0.085	−1.030	−0.694	<0.001
Indirect effect of generalized trust on suicidal ideation						
Total indirect effect	−0.078	−0.036	0.002	−0.041	−0.032	<0.001
Through fear of COVID-19	−0.045	−0.021	0.001	−0.024	−0.018	<0.001
Through insomnia	−0.013	−0.006	0.001	−0.009	−0.004	<0.001

Note: age, gender, education, accommodation, and marital status were adjusted for the model. Unstand. Coeff. = unstandardized coefficient; LLCI = lower limit in 95% confidence interval; ULCI = upper limit in 95% confidence interval.

The proposed model was then tested for its invariance across different subgroups. The multigroup SEM showed that the path coefficients in the proposed model were invariant across gender groups (ΔCFI = 0.002 to 0.006; ΔRMSEA = −0.004 to −0.001; ΔSRMR = −0.002 to 0.000), age groups (ΔCFI = 0.001 to 0.004; ΔRMSEA = −0.006 to −0.003; ΔSRMR = −0.003 to −0.001), and type of residence (ΔCFI = 0.001 to 0.005; ΔRMSEA = −0.002; ΔSRMR = −0.005 to 0.000) in terms of path coefficients, correlations, and residuals (Table 3).

Table 3. Path invariance across age, gender, and living place through multigroup structural equation modeling analysis.

Model and Comparisons	Fit Statistics									
	χ^2 (df)/p	$\Delta\chi^2$ (Δdf)/p	CFI	ΔCFI	TLI	ΔTLI	RMSEA	ΔRMSEA	SRMR	ΔSRMR
Gender (males vs. females)										
M1: Unconstrained	7.40 (2)/0.02	-	0.981	-	0.979	-	0.032	-	0.038	-
M2: Structural weights	13.65 (7)/0.06	6.25 (5)/0.28	0.987	0.006	0.980	0.001	0.028	−0.004	0.037	−0.001
M3: Structural covariances	15.14 (8)/0.06	7.74 (6)/0.26	0.989	0.002	0.982	0.002	0.027	−0.001	0.037	0.000
M4: Structural residuals	18.59 (11)/0.07	11.19 (9)/0.26	0.991	0.002	0.985	0.003	0.024	−0.003	0.035	−0.002
Age (>35.54 years vs. <35.54 years)										
M1: Unconstrained	7.68 (2)/0.02	-	0.980	-	0.978	-	0.033	-	0.039	-
M2: Structural weights	12.81 (7)/0.08	5.13 (5)/0.40	0.984	0.004	0.983	0.005	0.027	−0.006	0.037	−0.002
M3: Structural covariances	13.60 (8)/0.09	5.92 (6)/0.43	0.985	0.001	0.984	0.001	0.022	−0.005	0.036	−0.001
M4: Structural residuals	17.32 (11)/0.10	9.64 (9)/0.38	0.988	0.003	0.986	0.002	0.019	−0.003	0.033	−0.003
Living (city vs. rural)										
M1: Unconstrained	7.01 (2)/0.03	-	0.982	-	0.980	-	0.030	-	0.036	-
M2: Structural weights	13.81 (7)/0.055	6.80 (5)/0.24	0.983	0.001	0.982	0.002	0.028	−0.002	0.036	0.000
M3: Structural covariances	14.18 (8)/0.08	7.17 (6)/0.31	0.984	0.001	0.983	0.001	0.026	−0.002	0.035	−0.001
M4: Structural residuals	16.80 (11)/0.11	9.79 (9)/0.37	0.989	0.005	0.987	0.004	0.024	−0.002	0.030	−0.005

CFI = comparative fit index; TLI = Tucker–Lewis index; RMSEA = root mean square error of approximation; SRMR = standardized root mean square residual.

4. Discussion

The present study used a large sample of the Iranian general public to examine the associations between generalized trust, fear of COVID-19, insomnia severity, and suicidal ideation. The findings indicated that the proposed mediation model (i.e., fear of COVID-19 and insomnia as mediators in the association between generalized trust and suicidal ideation) was supported by the SEM fit indices. Moreover, both mediated effects derived from fear of COVID-19 and insomnia were significant, whereas the direct effect between generalized trust and suicidal ideation was non-significant. The proposed mediation model was further found to be invariant across the three examined subgroups: males vs. females, younger participants vs. older participants, and participants residing in an urban area vs. participants residing in a rural area.

Generalized trust was found to be associated with suicidal ideation indirectly via fear of COVID-19 and insomnia but not directly associated with suicidal ideation. Yamamura (2015) used a survey dataset that was representative of the Japanese general public (i.e., Japanese General Social Survey) and found that high generalized trust can deter suicidal ideation [15]. However, to the best of the present authors' knowledge, it is unclear why generalized trust is associated with reduced suicidal ideation. Therefore, the present study's findings provide the possible mechanism in the association between generalized trust and suicidal ideation during the COVID-19 pandemic. That is, generalized trust may decrease individuals' fear of COVID-19 and insomnia, both of which are important predictors of suicidal ideation [12–14,16]. Moreover, existing evidence indicates that generalized trust can help individuals decrease psychological distress and improve psychological health [22–25]. Therefore, higher levels of generalized trust may reduce individuals' fear of COVID-19 and insomnia during the COVID-19 pandemic, resulting in lower suicidal ideation among individuals.

Fear of COVID-19 and insomnia are potential factors contributing to individuals' suicide attempts. Completed suicides during the COVID-19 pandemic were reported and one of the most salient reasons for these individuals committing suicide was the fear of COVID-19 [11,45]. Moreover, individuals who have been exposed to environments with a high risk of COVID-19 infection are likely to have stronger suicidal ideation. This can be explained by worrying that they might transmit their own COVID-19 infection to their loved ones [45]. Regarding the association between insomnia and suicidal ideation, strong evidence from randomized controlled trials has supported the causal effects between insomnia and suicidal ideation (i.e., when individuals' sleep quality improves, their suicidal thoughts are reduced [12–14]). Therefore, the findings of the present study echo prior literature regarding the relationship between insomnia and suicidal ideation.

Based on the present study's results, there are several clinical implications. Policymakers should implement policies that can improve generalized trust among society members, especially utilizing educational programs. However, given that the increase of generalized trust takes time, other strategies are needed to tackle the emergent needs during the COVID-19 pandemic. Therefore, healthcare providers need to design fear reduction programs or insomnia improvement programs (e.g., cognitive behavioral therapy for insomnia) to reduce the levels of fear of COVID-19 and insomnia severity among individuals [12–14]. Moreover, such programs should be converted into online versions to increase the feasibility of success [46]. Programs on reducing fear may work on pervasive anxiety given that this type of anxiety seems to be an important factor contributing to low generalized trust. In other words, when pervasive anxiety with COVID-19 is reduced in general, it is possible to increase generalized trust in the general population. Although it is postulated that pervasive anxiety is an important factor that may contribute to low generalized trust, the present study does not have any evidence showing the associations between pervasive anxiety and generalized trust. Therefore, future studies are encouraged to further probe this issue.

There are some limitations in the present study. First, the present study used a cross-sectional design to examine the associations between generalized trust, fear of COVID-19, insomnia, and suicidal ideation. Although the relationships between these factors are supported by potential mechanisms and previous empirical research, the present findings only provide empirical evidence concerning associations rather than causality between these factors. Therefore, future studies are warranted to provide stronger evidence (e.g., using a longitudinal study design) to verify the model proposed in the present study. Second, the present sample was recruited from Iran, a country that substantially values harmony and family. Therefore, it is unknown whether the present findings can be applied to a country that values individualism. Third, the severity of the COVID-19 outbreak was different across countries worldwide [47,48]. Therefore, the present findings may not generalize to other countries with different severity levels of COVID-19 than Iran.

5. Conclusions

The present study suggests that generalized trust might reduce levels of fear of COVID-19 and insomnia; then, the reduced levels of fear of COVID-19 and insomnia might decrease suicidal ideation. Therefore, healthcare providers may want to find ways to improve generalized trust to resolve the mental health problems during the COVID-19 pandemic. For example, helping people cope with pervasive anxiety, a potentially important factor influencing individuals' generalized trust, may be a solution.

Author Contributions: Conceptualization, A.H.P.; methodology, C.-Y.L.; validation, A.H.P. and Z.A.; formal analysis, A.H.P. and C.-Y.L.; investigation, Z.A. and N.E.; data curation, Z.A.; writing—original draft preparation, C.-Y.L., S.-H.C. and A.H.P.; writing—review and editing, M.D.G., S.-H.C. and M.M.O. All authors have read and agreed to the published version of the manuscript.

Funding: Shun-Hua Chen received funding from the Ministry of Science and Technology, Taiwan (MOST 110-2314-B-242-003).

Institutional Review Board Statement: The study was conducted according to the guidelines of the Declaration of Helsinki and approved by the Ethics Committee of Qazvin University Medical Sciences (reference number IR.QUMS.REC.1399.418, 20 January 2021).

Informed Consent Statement: Written informed consent has been obtained from the patient(s) to publish this paper.

Data Availability Statement: The datasets used for this research on adolescents cannot be shared with the public as per the privacy policy and regulations of Qazvin University of Medical Sciences.

Acknowledgments: The study was supported by the Research Institute for Prevention of Non-Communicable Diseases, Qazvin University of Medical Sciences.

Conflicts of Interest: The authors declare no conflict of interest.

References

1. Alijanzadeh, M.; Harati, T. The role of social capital in the implementation of social distancing during the COVID-19 pandemic. *Asian J. Soc. Health Behav.* **2021**, *4*, 45. [CrossRef]
2. Ashraf, A.; Ali, I.; Ullah, F. Domestic and gender-Based violence: Pakistan scenario amidst COVID-19. *Asian J. Soc. Health Behav.* **2021**, *4*, 47. [CrossRef]
3. Olashore, A.A.; Akanni, O.O.; Fela-Thomas, A.L.; Khutsafalo, K. The psychological impact of COVID-19 on health-care workers in African countries: A systematic review. *Asian J. Soc. Health Behav.* **2021**, *4*, 85.
4. Patil, S.T.; Datar, M.C.; Shetty, J.V.; Naphade, N.M. "Psychological consequences and coping strategies of patients undergoing treatment for COVID-19 at a tertiary care hospital": A qualitative study. *Asian J. Soc. Health Behav.* **2021**, *4*, 62.
5. Rajabimajd, N.; Alimoradi, Z.; Griffiths, M.D. Impact of COVID-19-related fear and anxiety on job attributes: A systematic review. *Asian J. Soc. Health Behav.* **2021**, *4*, 51–55. [CrossRef]
6. Sharma, R.; Bansal, P.; Chhabra, M.; Bansal, C.; Arora, M. Severe acute respiratory syndrome coronavirus-2-associated perceived stress and anxiety among indian medical students: A cross-sectional study. *Asian J. Soc. Health Behav.* **2021**, *4*, 98.
7. Pramukti, I.; Strong, C.; Sitthimongkol, Y.; Setiawan, A.; Pandin, M.G.R.; Yen, C.-F.; Lin, C.-Y.; Griffiths, M.D.; Ko, N.-Y. Anxiety and suicidal thoughts during the COVID-19 pandemic: Cross-country comparative study among Indonesian, Taiwanese, and Thai University students. *J. Med. Internet Res.* **2020**, *22*, e24487. [CrossRef]
8. Mamun, M.A. Suicide and suicidal behaviors in the context of COVID-19 pandemic in Bangladesh: A systematic review. *Psychol. Res. Behav. Manag.* **2021**, *14*, 695. [CrossRef]
9. Lin, C.-Y.; Scheerman, J.F.; Yaseri, M.; Pakpour, A.H.; Webb, T.L. A cluster randomised controlled trial of an intervention based on the health action process approach for increasing fruit and vegetable consumption in Iranian adolescents. *Psychol. Health* **2017**, *32*, 1449–1468. [CrossRef]
10. Lin, C.-Y.; Strong, C.; Scott, A.J.; Broström, A.; Pakpour, A.H.; Webb, T.L. A cluster randomized controlled trial of a theory-based sleep hygiene intervention for adolescents. *Sleep* **2018**, *41*, zsy170. [CrossRef]
11. Mamun, M.A.; Griffiths, M.D. First COVID-19 suicide case in Bangladesh due to fear of COVID-19 and xenophobia: Possible suicide prevention strategies. *Asian J. Psychiatry* **2020**, *51*, 102073. [CrossRef]
12. McCall, W.V.; Benca, R.M.; Rosenquist, P.B.; Youssef, N.A.; McCloud, L.; Newman, J.C.; Case, D.; Rumble, M.E.; Szabo, S.T.; Phillips, M. Reducing suicidal ideation through insomnia treatment (REST-IT): A randomized clinical trial. *Am. J. Psychiatry* **2019**, *176*, 957–965. [CrossRef]
13. McCall, W.V.; Blocker, J.N.; D'Agostino, R., Jr.; Kimball, J.; Boggs, N.; Lasater, B.; Rosenquist, P.B. Insomnia severity is an indicator of suicidal ideation during a depression clinical trial. *Sleep Med.* **2010**, *11*, 822–827. [CrossRef]
14. Trockel, M.; Karlin, B.E.; Taylor, C.B.; Brown, G.K.; Manber, R. Effects of cognitive behavioral therapy for insomnia on suicidal ideation in veterans. *Sleep* **2015**, *38*, 259–265. [CrossRef]
15. Yamamura, E. Comparison of social Trust's effect on suicide ideation between urban and non-urban areas: The case of Japanese adults in 2006. *Soc. Sci. Med.* **2015**, *140*, 118–126. [CrossRef]
16. Killgore, W.D.; Cloonan, S.A.; Taylor, E.C.; Fernandez, F.; Grandner, M.A.; Dailey, N.S. Suicidal ideation during the COVID-19 pandemic: The role of insomnia. *Psychiatry Res.* **2020**, *290*, 113134. [CrossRef]
17. Jasielska, D.; Rogoza, R.; Zajenkowska, A.; Russa, M.B. General trust scale: Validation in cross-cultural settings. *Curr. Psychol.* **2019**, *290*, 113134. [CrossRef]

18. Lin, C.Y.; Namdar, P.; Griffiths, M.D.; Pakpour, A.H. Mediated roles of generalized trust and perceived social support in the effects of problematic social media use on mental health: A cross-sectional study. *Health Expect.* **2021**, *24*, 165–173. [CrossRef]
19. Ashraf, N.; Bohnet, I.; Piankov, N. Decomposing trust and trustworthiness. *Exp. Econ.* **2006**, *9*, 193–208. [CrossRef]
20. Johnson, N.D.; Mislin, A.A. Trust games: A meta-analysis. *J. Econ. Psychol.* **2011**, *32*, 865–889. [CrossRef]
21. Yamagishi, T.; Yamagishi, M. Trust and commitment in the United States and Japan. *Motiv. Emot.* **1994**, *18*, 129–166. [CrossRef]
22. Abbott, S.; Freeth, D. Social capital and health: Starting to make sense of the role of generalized trust and reciprocity. *J. Health Psychol.* **2008**, *13*, 874–883. [CrossRef]
23. Fahmi, M.; Panjaitan, N.A.; Habibie, I.; Siregar, A.Y.; Amarullah, G.; Sunjaya, D.K. Does your neighborhood protect you from being depressed? A study on social trust and depression in Indonesia. *BMC Public Health* **2019**, *19*, 1371–1378. [CrossRef]
24. Sato, Y.; Aida, J.; Tsuboya, T.; Shirai, K.; Koyama, S.; Matsuyama, Y.; Kondo, K.; Osaka, K. Generalized and particularized trust for health between urban and rural residents in Japan: A cohort study from the JAGES project. *Soc. Sci. Med.* **2018**, *202*, 43–53. [CrossRef]
25. Tokuda, Y.; Fujii, S.; Inoguchi, T. Individual and country-level effects of social trust on happiness: The Asia barometer survey. In *Trust with Asian Characteristics*; Springer: Berlin/Heidelberg, Germany, 2017; pp. 123–139.
26. Soto-Sanz, V.; Falcó, R.; Piqueras, J.A. Socio-emotional strengths against psychopathology and suicidal ideation in fear of COVID-19. *Curr. Psychol.* **2021**. advance online publication. [CrossRef]
27. Leaune, E.; Samuel, M.; Oh, H.; Poulet, E.; Brunelin, J. Suicidal behaviors and ideation during emerging viral disease outbreaks before the COVID-19 pandemic: A systematic rapid review. *Prev. Med.* **2020**, *141*, 106264. [CrossRef]
28. Alimoradi, Z.; Broström, A.; Tsang, H.W.; Griffiths, M.D.; Haghayegh, S.; Ohayon, M.M.; Lin, C.-Y.; Pakpour, A.H. Sleep problems during COVID-19 pandemic and its' association to psychological distress: A systematic review and meta-analysis. *EClinicalMedicine* **2021**, *36*, 100916. [CrossRef]
29. Alimoradi, Z.; Gozal, D.; Tsang, H.W.; Lin, C.Y.; Broström, A.; Ohayon, M.M.; Pakpour, A.H. Gender-specific estimates of sleep problems during the COVID-19 pandemic: Systematic review and meta-analysis. *J. Sleep Res.* **2021**. advance online publication. [CrossRef]
30. Dadfar, M.; Kalibatseva, Z.; Lester, D. Reliability and validity of the Farsi version of the Patient Health Questionnaire-9 (PHQ-9) with Iranian psychiatric outpatients. *Trends Psychiatry Psychother.* **2018**, *40*, 144–151. [CrossRef]
31. Yamagishi, T.; Kosugi, M. Character detection in social exchange. *Cogn. Stud.* **1999**, *6*, 179–190.
32. Ahorsu, D.K.; Lin, C.-Y.; Imani, V.; Carlbring, P.; Nygårdh, A.; Broström, A.; Hamilton, K.; Pakpour, A.H. Testing an app-based intervention to improve insomnia in patients with epilepsy: A randomized controlled trial. *Epilepsy Behav.* **2020**, *112*, 107371. [CrossRef]
33. Ahorsu, D.K.; Lin, C.-Y.; Imani, V.; Saffari, M.; Griffiths, M.D.; Pakpour, A.H. The Fear of COVID-19 Scale: Development and initial validation. *Int. J. Ment. Health Addict.* **2020**. advance online publication. [CrossRef]
34. Chang, K.-C.; Hou, W.-L.; Pakpour, A.H.; Lin, C.-Y.; Griffiths, M.D. Psychometric testing of three COVID-19-related scales among people with mental illness. *Int. J. Ment. Health Addict.* **2020**. advance online publication. [CrossRef]
35. Lin, C.Y.; Hou, W.L.; Mamun, M.A.; Aparecido da Silva, J.; Broche-Pérez, Y.; Ullah, I.; Masuyama, A.; Wakashima, K.; Mailliez, M.; Carre, A. Fear of COVID-19 Scale (FCV-19S) across countries: Measurement invariance issues. *Nurs. Open* **2021**, *8*, 1892–1908. [CrossRef]
36. Bastien, C.H.; Vallières, A.; Morin, C.M. Validation of the Insomnia Severity Index as an outcome measure for insomnia research. *Sleep Med.* **2001**, *2*, 297–307. [CrossRef]
37. Lin, C.; Cheng, A.S.K.; Nejati, B.; Imani, V.; Ulander, M.; Browall, M.; Griffiths, M.D.; Broström, A.; Pakpour, A.H. A thorough psychometric comparison between Athens Insomnia Scale and Insomnia Severity Index among patients with advanced cancer. *J. Sleep Res.* **2020**, *29*, e12891. [CrossRef]
38. Lin, C.-Y.; Cheng, A.S.K.; Imani, V.; Saffari, M.; Ohayon, M.M.; Pakpour, A.H. Advanced psychometric testing on a clinical screening tool to evaluate insomnia: Sleep condition indicator in patients with advanced cancer. *Sleep Biol. Rhythm.* **2020**, *18*, 343–349. [CrossRef]
39. Yazdi, Z.; Sadeghniiat-Haghighi, K.; Zohal, M.A.; Elmizadeh, K. Validity and reliability of the Iranian version of the insomnia severity index. *Malays. J. Med. Sci.* **2012**, *19*, 31–36.
40. Cheng, C.-P.; Luh, W.-M.; Yang, A.-L.; Su, C.-T.; Lin, C.-Y. Agreement of children and parents scores on Chinese version of pediatric quality of life inventory version 4.0: Further psychometric development. *Appl. Res. Qual. Life* **2016**, *11*, 891–906. [CrossRef]
41. Lin, C.Y.; Broström, A.; Griffiths, M.D.; Pakpour, A.H. Psychometric evaluation of the Persian eHealth Literacy Scale (eHEALS) among elder Iranians with heart failure. *Eval. Health Prof.* **2020**, *43*, 222–229. [CrossRef]
42. Nejati, B.; Fan, C.W.; Boone, W.J.; Griffiths, M.D.; Lin, C.Y.; Pakpour, A.H. Validating the Persian Intuitive Eating sSale-2 among breast cancer survivors who are overweight/obese. *Eval. Health Prof.* **2020**. advance online publication. [CrossRef]
43. Lin, C.-Y.; Tsai, M.-C. Effects of family context on adolescents' psychological problems: Moderated by pubertal timing, and mediated by self-esteem and interpersonal relationships. *Appl. Res. Qual. Life* **2016**, *11*, 907–923. [CrossRef]
44. Chen, F.F. Sensitivity of goodness of fit indexes to lack of measurement invariance. *Struct. Equ. Modeling* **2007**, *14*, 464–504. [CrossRef]

45. Rahman, M.E.; Al Zubayer, A.; Bhuiyan, M.R.A.M.; Jobe, M.C.; Khan, M.K.A. Suicidal behaviors and suicide risk among Bangladeshi people during the COVID-19 pandemic: An online cross-sectional survey. *Heliyon* **2021**, *7*, e05937. [CrossRef]
46. Majd, N.R.; Broström, A.; Ulander, M.; Lin, C.-Y.; Griffiths, M.D.; Imani, V.; Ahorsu, D.K.; Ohayon, M.M.; Pakpour, A.H. Efficacy of a theory-based cognitive behavioral technique app-based intervention for patients with insomnia: Randomized controlled trial. *J. Med. Internet Res.* **2020**, *22*, e15841. [CrossRef]
47. Khafaie, M.A.; Rahim, F. Cross-country comparison of case fatality rates of COVID-19/SARS-COV-2. *Osong Public Health Res. Perspect.* **2020**, *11*, 74–80. [CrossRef]
48. Hashim, M.J.; Alsuwaidi, A.R.; Khan, G. Population risk factors for COVID-19 mortality in 93 Countries. *J. Epidemiol. Glob. Health* **2020**, *10*, 204–208. [CrossRef]

healthcare

MDPI

Article

Emotional Contagion in the Online Depression Community

Jingyun Tang, Guang Yu * and Xiaoxu Yao

School of Management, Harbin Institute of Technology, 92 Xidazhi Street, Nangang District, Harbin 150001, China; tang_jing_yun@126.com (J.T.); xiaoxu.yao06@gmail.com (X.Y.)
* Correspondence: yug@hit.edu.cn; Tel.: +86-139-3649-0774

Abstract: Negative emotions are prevalent in the online depression community (ODC), which potentially puts members at risk, according to the theory of emotional contagion. However, emotional contagion in the ODC has not been confirmed. The generalized estimating equation (GEE) was used to verify the extent of emotional contagion using data from 1548 sample users in China's popular ODC. During interaction, the emotional themes were analyzed according to language use. The diurnal patterns of the interaction behaviors were also analyzed. We identified the susceptible groups and analyzed their characteristics. The results confirmed the occurrence of emotional contagion in ODC, that is, the extent to which the user's emotion was affected by the received emotion. Our study also found that when positive emotional contagion occurred, the replies contained more hopefulness, and when negative emotional contagion occurred, the replies contained more hopelessness and fear. Second, positive emotions were easier to spread, and people with higher activity in ODC were more susceptible. In addition, nighttime was an active period for user interaction. The results can help community managers and support groups take measures to promote the spread of positive emotions and reduce the spread of negative emotions.

Keywords: emotional contagion; depression; online depression community; social media; text-mining

Citation: Tang, J.; Yu, G.; Yao, X. Emotional Contagion in the Online Depression Community. *Healthcare* **2021**, *9*, 1609. https://doi.org/10.3390/healthcare9121609

Academic Editors: Yasuhiro Kotera and Elaina Taylor

Received: 13 October 2021
Accepted: 20 November 2021
Published: 23 November 2021

1. Introduction

Human emotions are strongly influenced by social interactions. Previous studies have shown that happiness [1], depression [2], loneliness [3], and other emotions can spread in social networks. Emotional contagion [4] is initially considered to occur between people who are often in close contact, such as families [5] and roommates [6,7]. Emotions also spread in face-to-face communication, such as during workplace interactions [6,8,9], and even in experimental situations [10].

The emergence of social media platforms has changed the traditional ways of communication. Personal emotions are expressed in texts that are posted online [11]. Various recent contributions have advanced the hypothesis that emotions may be spread throughout online social networks [12–21]. For example, a study found that after users update their status with emotional content, their friends are significantly more likely to make a valence-consistent post [22]. However, this study did not distinguish between emotional contagion and homogeneity; that is, users like to make friends and communicate with people with the same emotional state. To solve this problem, Kramer et al. [23] designed an experiment by reducing the amount of emotional content (positive and negative) in the news feed on Facebook and found that people produced fewer positive posts and more negative posts when positive expressions were reduced and vice versa. However, manipulation of the information that users see raises ethical concerns. In an empirical study [24], members of mental health communities were shown to experience significant increases in anxiety, anger, and negative emotions following reports of several celebrity suicides. In another study, it was shown that rainfall directly influenced the emotional content of Facebook users' status messages and also affected the status messages of friends in other cities who were not experiencing rainfall [14]. In short, most of the research on emotional contagion based on social media is realized by observing the emotional expression of users after they see a status update with emotional content. Consequently, some studies compared the spread of different emotions and found some evidence that on social media, people usually share positive emotions more often than negative ones. However, anger is an exception [17]. Some studies show that posts containing anger are more likely to spread among several users [25].

The functioning of social media platforms may even intensify the negative aspects of outrage, such as harassment or potentially anger, instead of turning outrage into a force of social progress [26]. An online community, as a form of social media, helps users establish social networks, enhancing communication and mutual influence [27]. In recent years, an increasing number of online health communities have formed where people with the same diseases gather. Online health communities can increase opportunities for patients to communicate with their peers [28], reduce the stigma of illness [29], and provide social support that is not limited by space and time [30]. Therefore, it is generally believed that online health communities have a positive impact on users [31–35]. Research on the benefits of online health communities highlights psychosocial benefits, such as reduced depression [33,35], anxiety [33,36], stress [33], and negative emotions [37,38]. While these studies reveal the emotional changes of users, they do not consider the relationship between emotional changes and interaction. Furthermore, most of these studies are not based on mental health communities, such as the online depression community (ODC).

The prevalence of depression has been increasing every year of late. According to a WHO report, depression affects more than 350 million people worldwide and is the leading cause of disability [39]. Depressed individuals are reluctant to share their emotional states with family and friends, but they are willing to share with their peers in ODCs, as they may feel it is less risky to disclose their experiences online [40,41]. Negative emotion is a primary characteristic of depressive individuals and is also common in ODCs [42]. The emotional contagion theory [43] suggests that prolonged interactions with depressed individuals and their negative emotions can worsen the symptoms of depression [38]. In ODCs, users frequently communicate with numerous other users [14]. However, emotional contagion in interactions within ODCs has not been confirmed.

In the online community, users engage in text-based interactions, so when we study emotional contagion in an online community, we should consider whether the emotional content of communication among users will affect the emotion of receiving users. Considering the large amount of negative emotional content and emotional interaction in ODCs, it is necessary to study emotional contagion in this community. This study aims to verify the emotional contagion in ODC interactions. Based on the epidemic model, three factors of the emotional transmission process are analyzed: themes of emotional interaction, diurnal pattern of interaction, and characteristics of susceptible groups. The results provide suggestions for the management of ODCs to provide better support for community users.

2. Methods

2.1. Data Collection

In this study, the ODC under consideration comprises a comment thread appearing on Sina Weibo, which is one of the most popular social media platforms in China and is similar to Twitter [44]. In 2012, a Sina Weibo user "zoufan" died by suicide related to depression and posted a farewell posting. An increasing number of people suffering from depression or depression symptoms tend to post comments under the farewell posting to disclose their depression, express their feelings, and communicate with others. At present, over 1 million comments have been posted under the farewell posting, and the number continues to grow. Many studies have used this comment thread as an ODC to study depression-related issues, because it has the largest concentration of people with depressive symptoms and has more posts than other ODCs [45–47]. Moreover, as there is no speech restriction in this ODC, members can express their feelings and thoughts freely, even including thoughts on suicide. As a result, the community has numerous emotional posts, including negative emotions. Therefore, it provides a suitable data source for the study of emotional contagion. We developed a Python program to automatically download postings.

These postings are published anonymously on an open and accessible platform. Therefore, according to the ethical guidelines from Benton et al. [48], an ethical review was not required in this study. In addition, to protect users' anonymity, any information that could be trace to users is hidden.

2.2. Data Selection

In order to verify the emotional contagion in ODC, we must first get the member's longitudinal emotional changes, which requires that the member must stay in the community for a certain period of time. Those users who only stay in the community for a short period were not considered. Second, as emotional contagion is based on interaction, those members who do not interact with others in the community were not included in this study. We identified 1548 sample users for our study who were active in the community for 4 weeks or more and interacted with others at least once. Their original posts (n = 242,788) and interactive posts (n = 67,987) with others were extracted. In addition, their gender was retrieved.

2.3. Emotional Analysis

We built a text classifier to identify emotions in postings. Classification is modeled as a supervised learning process in which a training dataset is required. The coding process for the training dataset was done as follows. First, two psychology students coded the 10,000 postings (randomly selected from the 242,788 original postings and 67,987 interactive posting) as positive, neutral, or negative. Second, where there is a disagreement, they discussed each of the postings until an agreement was reached. Finally, a third researcher who is also experienced in mental health coded 1000 postings randomly sampled from the 10,000 postings to compute the inter-rater reliability. All Kappas are >0.85.

We trained a text classifier using bidirectional encoder representations from transformers (BERT), which is a new language representation model developed by Google in 2018 that obtains new state-of-the-art results on 11 natural language processing tasks [49]. The accuracy of the trained classifier is 84.28%, and the F1 score is 83.54, which shows that the classifier has a good performance.

Each posting after classification has an emotional label. To obtain the user's emotional state in a time period, we need to aggregate user postings to calculate the emotional value. The emotional value EV_{it} for user i in the epoch t is defined as follows:

$$EV_{it} = \frac{TP_{it} - TN_{it}}{T_{it}}$$

where TP_{it} is the total number of positive postings published by user i in the epoch t. TN_{it} is the total number of negative posts published by user i in the epoch t. T_{it} is the total number of posts published by user i in the epoch t. The closer a user's EV_{it} is to 1, the more positive his or her emotion will be. On the contrary, the closer EV_{it} is to -1, the more negative the user's emotion is.

The users' received emotional value REV_{it} from the received replies is calculated as follows:

$$REV_{it} = \frac{TRP_{it} - TRN_{it}}{TR_{it}}$$

where TRP_{it} is the total number of received positive replies by user i in the epoch t. TRN_{it} is the total number of received negative replies by user i in the epoch t. TR_{it} is the total number of all replies received by user i in the epoch t. The closer a user's REV_{it} value is to 1, the more positive the emotion the user receives. On the contrary, the closer a user's REV_{it} value is to -1, the more negative the emotion the user receives.

2.4. Longitudinal Changes

First, we wanted to understand the emotional change of these users during their participation in the community. Therefore, we calculated and clustered the user's longitudinal emotional changes. Specifically, we divided the total time of user's active engagement in the community into three equal time spans: the early, middle, and late periods of participation. Then, the values of EV_{it} in each period were calculated according to the method mentioned in Section 2.3. Thus, we obtained the 1×3 vector of each user's longitudinal emotion change. Finally, we applied the classic k-means clustering algorithm to cluster all the user's longitudinal emotional changes. The optimal number of clusters from the k-means clustering results was estimated using the R package NbClust [50].

2.5. Statistical Analysis of Emotional Contagion

After determining the emotional changes of users during their participation in the community, we wanted to know whether these changes were affected by the interaction with others in the community, that is, whether there is an emotional contagion. We observed the emotional changes of users on a weekly basis and investigated whether their emotions in the current week are affected by the emotions received in the current week. In addition, considering the lag of the impact, the emotions received in the previous week may also affect the users' emotional expressions in the current week. Thus, we hypothesize the following:

Hypothesis 1 (H1). *The emotions received by users in the current week have a positive impact on the users' emotional expressions in the current week.*

Hypothesis 2 (H2). *The emotions received by users in the previous week have a positive impact on the users' emotional expressions in the current week.*

Figure 1 shows the research model of our study. The user's EV_{it} (in the current week) was calculated as the dependent variable. The user's REV_{it} (in the current week) and REV_{it-1} (in the previous week) were calculated as independent variables. In addition, the user's EV_{it-1} (in the previous week) was included as the control variable to eliminate serial correlation in the errors and to control the user's intrinsic and stable emotional states. Then, we obtained the temporal sequences of all variables for each user.

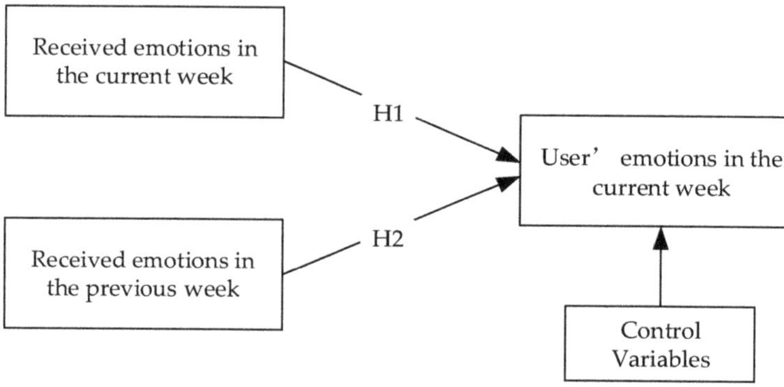

Figure 1. Research model.

Generalized estimating equation (GEE) is a general statistical approach that facilitates the analysis of data collected in longitudinal measures designs [51]. GEE has been popularly applied in clinical trials and biomedical studies [52,53]. Some studies use GEE to verify the spread of depression in real social networks [2]. Therefore, the models were estimated using a GEE. An independent working correlation structure was assumed for the clusters [54].

2.6. Elements of Emotional Contagion

Information and emotions spread in social networks similarly to pathogens [55]. Based on the three elements of epidemic transmission [56]—infection sources, transmission routes, and susceptible population—we divided the process of emotional contagion into three elements—emotional interaction information, interactive behavior, and susceptible group—and then analyzed these elements to promote positive emotional contagion and reduce negative emotional contagion.

2.6.1. Emotional Interaction Information

After verifying the extent of the emotional contagion, we wanted to determine which emotional themes and language use were more likely to cause emotional contagion. Therefore, we calculated the difference of interactive information between the week when the emotional contagion occurred and the week when the emotional contagion did not occur by all sample users.

We extracted the replies in which positive emotional contagion occurred; in other words, the response received by each user extracted when users $REV_{it} > 0$ and $EV_{it} - EV_{it-1} > 0$. For comparison, we extracted the replies when positive emotional contagion did not occur, that is, the replies received by each user extracted when users $REV_{it} > 0$ and $EV_{it} - EV_{it-1} < 0$. All the replies were segmented, and the positive emotion words were extracted from HowNet's emotional dictionary [57]. Finally, the top 100 extracted emotion words were classified and counted. After considering measures of widely recognized basic emotions [58], we grouped the positive emotional words into (1) relaxed, (2) thankful, (3) praise, (4) hopeful, (5) like, (6) happiness, and (7) respect.

Similarly, we extracted the response postings when negative emotional contagion occurred (users $REV_{it} < 0$ and users $EV_{it} - EV_{it-1} < 0$) and did not occur (users $REV_{it} < 0$ and users $EV_{it} - EV_{it-1} > 0$). All of the response postings were segmented, and the top 100 negative emotion words were extracted. Negative emotions include (1) disgust, (2) fear, (3) anxiety, (4) boredom, (5) guilt, (6) hopelessness, (7) sadness, and (8) anger.

In addition, we also assessed the use of pronouns, which are categorized by the use of the first person and the second person, in positive and negative interactions.

2.6.2. Interactive Behavior

To facilitate the monitoring of user interaction behavior, the distribution of positive and negative interactions over 24 h was investigated.

2.6.3. Characteristics of Susceptible Groups

Finally, in order to explore if different users have different susceptibility to emotional contagion, we designed and calculated the user susceptibility index as follows:

$$SI_{pos} = \frac{C_{imp}}{C_{pos}} \text{ and } SI_{neg} = \frac{C_{wor}}{C_{neg}}$$

where SI_{pos} is the positive susceptibility index, C_{imp} is the total number of weeks when users $REV_{it} > 0$ and $EV_{it} - EV_{it-1} > 0$, C_{pos} is the total number of weeks when users $REV_{it} > 0$. SI_{neg} is the negative susceptibility index, C_{wor} is the total number of weeks when users $REV_{it} < 0$ and $EV_{it} - EV_{it-1} < 0$, C_{neg} is the total number of weeks when users $REV_{it} < 0$. Users with $SI_{pos} > 0.85$ or $SI_{neg} > 0.85$ are considered as susceptible groups, while users with probability $SI_{pos} < 0.15$ or $SI_{neg} < 0.15$ are considered as non-susceptible groups. This threshold has been used to divide those that are highly and scarcely susceptible to emotional contagion [59]. For example, if a user has been active in the community for 10 weeks, their $REV_{it} > 0$ in 5 weeks, and in 4 of these 5 weeks, their emotions have improved, then their $SI_{pos} = 0.8$. In addition, if the user's $REV_{it} < 0$ during the other 5 weeks, and in one of these 5 weeks, their emotions worsened, then their $SI_{neg} = 0.2$. Then, the user is considered among both the positive susceptible group and negative non-susceptible group.

We compared the community participation and demographic characteristics (gender) of the susceptible and non-susceptible groups. The users' community activity and span time were defined, that is, the total number of posts and the duration of engagement with the community. The Wilcoxon rank-sum and signed-rank test [60] was adopted to determine whether there was a statistically significant difference in the distributions of activity and span time between the two groups. Pearson's chi-square test was used to determine whether there was a statistically significant difference in gender distribution.

3. Results

3.1. Longitudinal Changes

After obtaining the longitudinal emotional changes for each user by emotion classification and calculation, we clustered the longitudinal emotional changes for these users. The K-means clustering results show that the optimal number of clusters is 2. The clustering results and the proportion of each group are shown in Figure 2. In addition, we plotted the mean and standard deviation of EV for each group at each period in Figure 2. Group 1, with mild negative emotions, accounted for a small proportion of the sample users in the community (23.13%). This group's posts contained fewer negative emotions in the later phase of their participation compared to the earlier period. Group 2, with more negative emotions, accounted for 76.87%. During participation in the community, the emotions of this group are relatively stable, and negative emotions tend to increase in the later period. In other words, only a small number of sample users in the community improved their emotions.

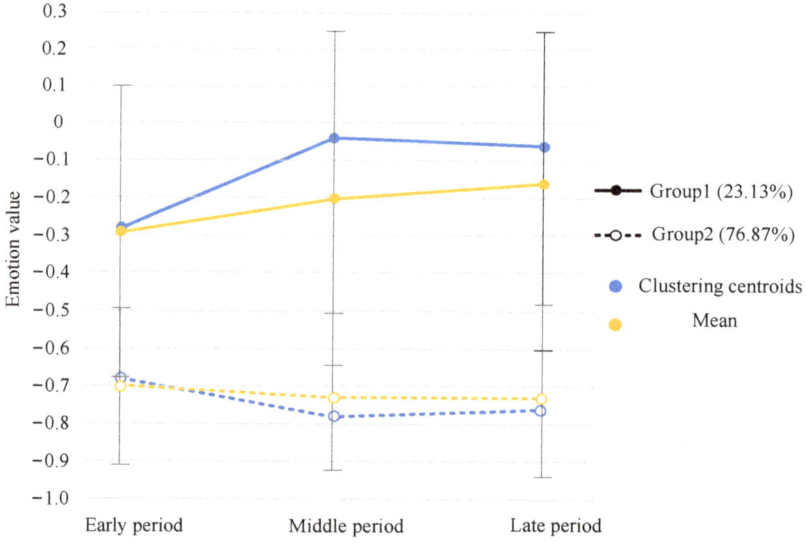

Figure 2. Longitudinal emotional changes for sample users in the ODC.

3.2. Emotional Contagion

The GEE regression models provide parameter estimates in the form of β-coefficients. As shown in Table 1, the results suggest that the user's emotion in the current week is affected by the emotion received in the current week. In addition, the emotion received in the previous week has no significant impact on the user's emotion in the current week. Therefore, it is evident that there is emotional contagion in the ODC and that it occurs on a short time scale. To check for multicollinearity, we measured the variance inflation factor for all variables. All variance inflation factors were far below the value of 2.5.

Table 1. Results of generalized estimation equation for emotional contagion.

	Dependent Variable: Current User's EV		
	Co-Eff	**S.E.**	***p*-Value**
Current EV received	0.031	0.0101	0.001
Previous EV received	−0.017	0.0098	0.087
Previous user's EV	0.500	0.0250	0.000
Constant	−0.248	0.0152	0.000

Note: EV—emotional value.

3.3. *Elements of Emotional Contagion*

3.3.1. Themes of Emotional Interaction Information

Figure 3 shows the probability of emotional interaction themes appearing in the replies. As shown, the common emotional themes in positive interactions were hopefulness, like, and praise. Replies that were more hopeful improved the recipient's emotion. In negative interactions, users expressed more sadness, hopelessness, and fear. Among them, replies containing more hopelessness and fear were more likely to have a negative impact on the recipient. In addition, we also counted pronoun use, categorized by person; the first person was used more frequently in negative interactions, while the second person was used more frequently in positive interactions.

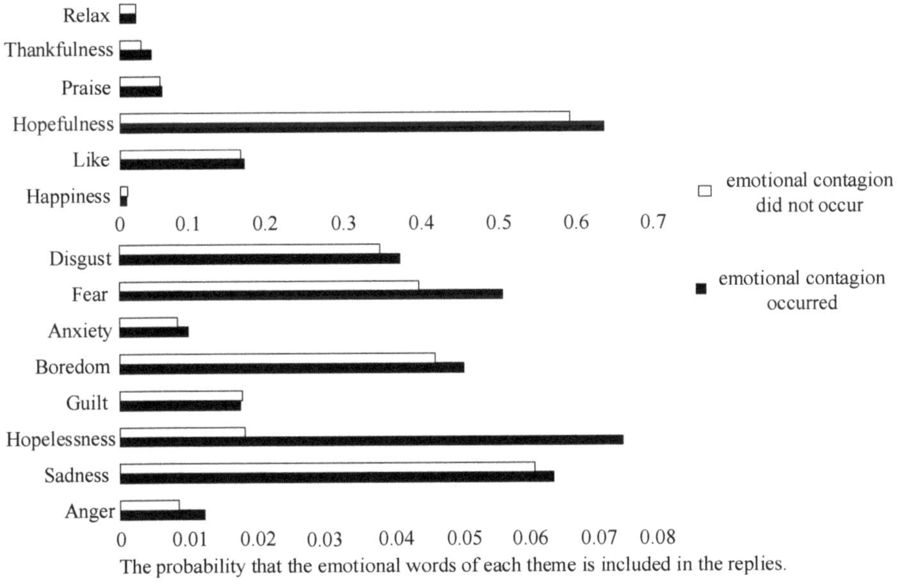

Figure 3. The themes of emotional interaction.

3.3.2. Diurnal Pattern of Interaction

Figure 4 shows the diurnal pattern of positive and negative emotional interactions of users in the community. The period from 22:00 to 02:00 was found to be the frequent time for user interaction; during this period, positive interactions were more common than negative interactions. Second, for the period from 11:00 to 22:00, user interaction was also relatively active, and negative interactions were more frequent than positive interactions. Therefore, interactions during these periods should also be considered.

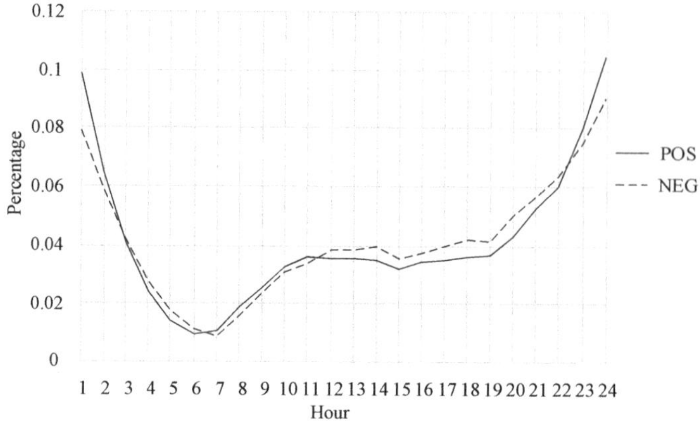

Figure 4. Time distribution of interaction.

3.3.3. Characteristics of Susceptible Groups

Figure 5 shows the complementary cumulative distribution of the user's positive and negative susceptibility indices (SIs). As shown, users were more susceptible to positive emotions; that is, positive emotions were easier to spread than negative emotions.

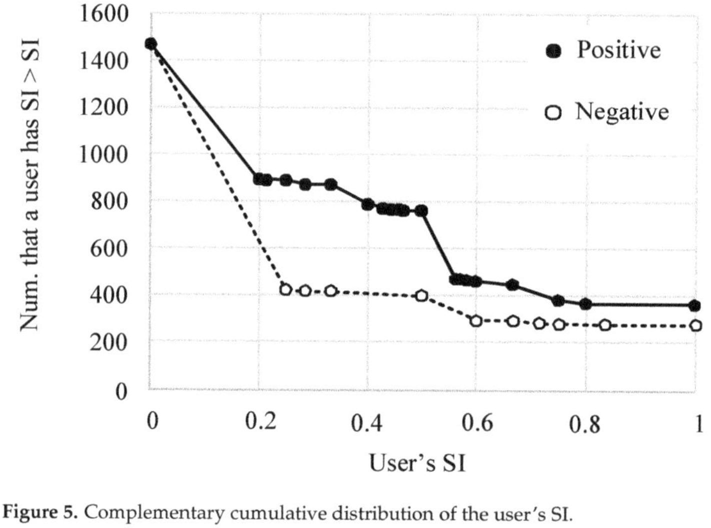

Figure 5. Complementary cumulative distribution of the user's SI.

We examined the online characteristics of susceptible and non-susceptible groups, including gender, activity, and span time. As shown in Table 2, the activity of the susceptible groups was generally higher than that of the non-susceptible groups. Users who posted and interacted frequently in the community were more likely to be affected by emotions from interactions, whether positive or negative. However, there was no significant difference in community time span between the susceptible and non-susceptible groups. In addition, the number of female users was three times that of male users. The results suggest that there was no significant difference in gender distribution between the susceptible and non-susceptible groups. The fact that no significant differences were found may be related to the small size of our sample. Using the power analysis, in order to find significant differences with some probability (power = 0.8, sig. level = 0.05), it would require expanding the data sample to at least 12 times the current size.

Table 2. Characteristics of the susceptible population.

	Positive Emotional Contagion			Negative Emotional Contagion		
	Susceptible Groups: 320	**Non-Susceptible Groups: 603**	*p*	**Susceptible Groups: 278**	**Non-Susceptible Groups: 1045**	*p*
Gender			0.435			0.466
No. of males, n (%),	73, 22.81	153, 25.37		70, 25.18	239, 22.87	
No. of females, n (%)	247, 77.19	450, 74.63		208, 74.82	806, 77.13	
Activity			<0.01			<0.01
Mean	157.55	131.91		221	154	
Median	97	79		117	93	
Loyalty			0.73			0.48
Mean	296.04	306.79		332	291	
Median	205	202		212	205	

Finally, based on the results of this study, a model of users' online behavior was designed to predict users' emotional changes after interactions in ODC, as shown in Figure 6. The model describes the likelihood that users with different levels of susceptibility will improve or worsen their emotions after receiving different emotional messages. The purpose of the model is to help community managers identify individuals who are affected by others' negative emotions to intervene and assist the identified individuals in a timely manner.

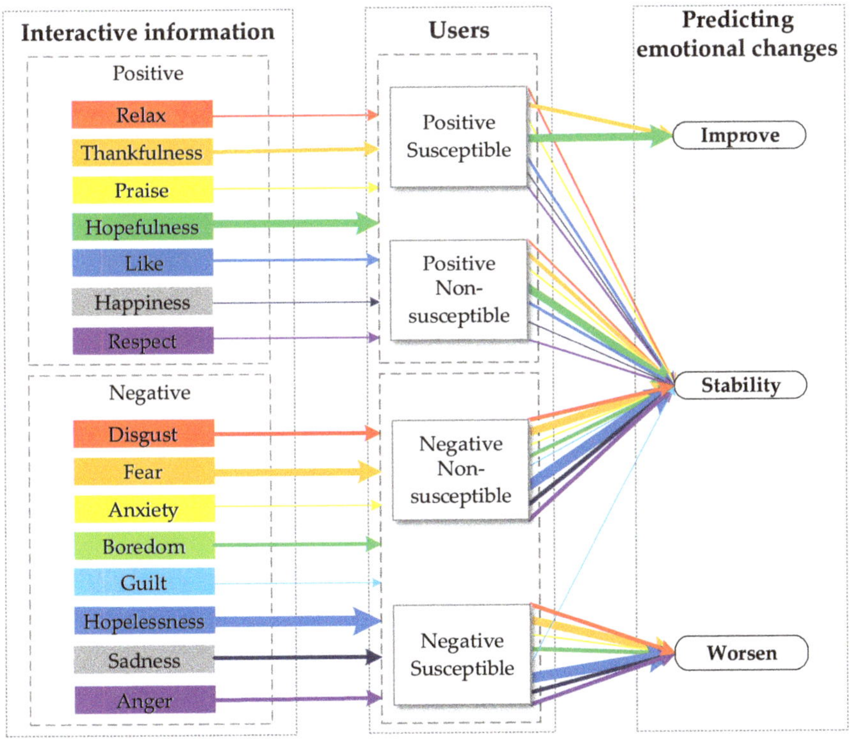

Figure 6. Model of users' online behavior. Note: The thickness of the line represents the degree of influence of different emotional themes.

4. Discussion

This study examined emotional contagion in the ODC, including the three elements of emotional contagion: themes of emotional interaction, diurnal pattern of interaction, and characteristics of the susceptible group.

First, according to the emotional longitudinal change, we found that the sample users of the community were mainly divided into two groups. One was the group with mild negative emotions, which accounted for a small proportion of the sample users. This group generally showed a trend toward emotional improvement during their participation in ODC. The other group had considerable negative emotions; they comprised the main sample users, showing a slight trend toward worsening negative emotions. This emotional longitudinal change was different from results of previous studies based on online healthy communities [33,35,36] that found that online interaction can improve depression and anxiety. Our results showed that the impact of the ODC on users is complex and cannot be generalized. The frequent expression of negative emotions in online communities was associated with higher levels of depression symptoms [61–63]. Therefore, for the second group that had considerable negative emotions, the trend of worsening emotions may be accompanied by the aggravation of depression symptoms, which should be of concern to community managers and support groups.

Second, after understanding the emotional changes of users in ODCs, we verified that the emotions received by users in the interaction will have an impact on users' emotions; that is, there was evidence of emotional contagion in the community. Although studies have shown that the suppression of emotional expressions backfires for depression and accepting one's feelings, even negative ones, is the key to psychological well-being [64], online communities might be a platform that enables individuals to exchange their feelings and thoughts. However, given the massive scale of social networks such as the ODC, even small effects can have significant aggregated consequences. Frequent negative emotion exchange in ODCs will expand the emergence and spread of negative emotions, which is a potential risk. Therefore, community managers need to pay attention to negative communication and their negative impact in ODCs. At the same time, in order to promote the spread of positive emotions, the community should introduce more emotional support groups to encourage community members to actively deal with the disease. In addition, we found that emotional contagion exists on a short time scale, which requires managers to take timely and continuous response measures.

Third, through the themes of emotional interaction and use of language in the process of user interaction, we found that users pay more attention to interactive objects by using the second person in the positive emotional interaction. The common emotional themes in positive interactions were hopefulness, like, and praise. We believe that most of the positive interactions are designed to provide emotional support. Previous studies have shown that emotional support can help patients cope better with depression [53]. However, emotional support can also be divided into many categories. Compared with like and praise, emotional support with more hopefulness can have a more positive impact on users. In addition, in negative interactions, users were more likely to express their feelings by using the first person. Users expressed more sadness, hopelessness, and fear. Among them, hopelessness and fear could significantly affect the emotions of the receiving users. Previous studies have also shown that content that evokes high-arousal emotions (e.g., fear) is more likely to spread than content that evokes low-arousal emotions (e.g., sadness) [65]. Communication with despair and fear often contain a lot of discussions about death and suicide, which can convey wrong health concepts and behaviors in patients. Studies have found that the more frequently users communicate with people with suicidal ideation, the more likely that users would become suicidal [47]. Therefore, we should design a text monitoring system based on the high-frequency emotional words related to hopelessness and fear that we have collected. When such words appear in interactions, the support group can be informed to facilitate a timely intervention.

Fourth, in the diurnal pattern of user interaction, the period of frequent interaction from 22:00 to 02:00 should be focused on for monitoring, as should the afternoon and evening period of 11:00 to 22:00, in which users have relatively more negative interactions. Intervening and helping users when they are most active would be more effective. In addition, users interact frequently at night, indicating that they may have insomnia problems, so the help group should pay attention to the possibility of other complications.

Fifth, by calculating the SI, we found that positive emotions are easier to spread than negative emotions. Therefore, encouraging positive interaction among users could effectively improve the emotional state of the community. By defining and analyzing the susceptible group characteristics, we found that groups susceptible to both positive and negative emotions are more active users in the community. Such groups may be more dependent on the community. Groups susceptible to negative emotions constitute the main category of users that community managers need to pay attention to. Identifying these groups and encouraging them to communicate with positive groups could effectively prevent the spread of negative emotions.

In China, the scarcity of mental health services and resources leads to 90% of individuals with depression not being treated [45,66]. In addition, the influence of traditional culture has deepened the stigma of mental illness, encouraging depressed people to hide their depression in real life [67]. ODCs provide a platform for them to express their thoughts and communicate with each other, as well as provide the opportunity to gain health-related knowledge. However, our findings show that most users of ODCs do not improve their depression during their participation. Long-term exposure to negative emotions in the community may exacerbate levels of depression symptoms. ODCs, which gather a large number of people with depression, are the best platform for implementing online interventions at low cost and should be given high priority by mental health institutions in China.

This study has some limitations that need to be addressed in future research. First, we are not aware of whether users of ODC have depression. Further research is needed to determine the depression levels of the users. In addition, additional personal information about the users, such as their degree of depression, age, or education level, may affect the level of emotional contagion. In the follow-up study, additional user information should be obtained by means of scale and questionnaire surveys, which can be added to the model variables. Finally, we only used data from a single ODC. In a follow-up study, additional ODCs from more platforms will be included to expand our study sample size and verify the research results.

Despite these shortcomings, this study provides evidence of emotional contagion in ODCs. When positive emotional contagion occurred, the replies contained more hopefulness, and when negative emotional contagion occurred, the replies contained more hopelessness and fear. Compared with negative emotions, positive emotions are easier to spread, and people with higher activity were more susceptible. In addition, night time was an active period for user interaction. These results can potentially help community managers and support groups in taking measures to promote the spread of positive emotions and reduce the incidence of negative emotions, responding to signs of distress in a timely manner.

Author Contributions: Conceptualization, J.T. and G.Y.; methodology, J.T.; software, J.T.; validation, J.T., G.Y. and X.Y.; formal analysis, J.T. and X.Y.; investigation, J.T.; resources, J.T.; data curation, J.T.; writing—original draft preparation, J.T.; writing—review and editing, J.T., G.Y. and X.Y.; visualization, J.T.; supervision, G.Y.; project administration, G.Y.; funding acquisition, G.Y. All authors have read and agreed to the published version of the manuscript.

Funding: This study was funded by the National Natural Science Foundation of China (Grant Nos. 71774041). The authors wish to express their sincere appreciation to the editors and reviewers of this paper.

Institutional Review Board Statement: Not applicable.

Informed Consent Statement: Not applicable.

Data Availability Statement: Publicly available datasets were analyzed in this study. This data can be found here: [https://m.weibo.cn/1648007681/3424883176420210] (accessed on 19 November 2021).

Conflicts of Interest: The authors declare no conflict of interest.

References

1. Fowler, J.H.; Christakis, N.A. Dynamic spread of happiness in a large social network: Longitudinal analysis over 20 years in the Framingham Heart Study. *BMJ* **2008**, *337*, a2338. [CrossRef] [PubMed]
2. Rosenquist, J.N.; Fowler, J.H.; Christakis, N.A. Social Network Determinants of Depression. *Mol. Psychiatry* **2011**, *16*, 273–281. [CrossRef]
3. Cacioppo, J.T.; Fowler, J.H.; Christakis, N.A. Alone in the crowd: The structure and spread of loneliness in a large social network. *J. Personal. Soc. Psychol.* **2009**, *97*, 977. [CrossRef]
4. Hatfield, E.; Cacioppo, J.T.; Rapson, R.L. Emotional contagion: Cambridge studies in emotion and social interaction. Cambridge, UK: Cambridge University Press. errors-in-variables regression model when the variances of the measurement errors vary between the observations. *Stat. Med.* **1994**, *21*, 1089–1101.
5. Larson, R.W.; Almeida, D.M. Emotional transmission in the daily lives of families: A new paradigm for studying family process. *J. Marriage Fam.* **1999**, *61*, 5–20. [CrossRef]
6. Barsade, S.G. The ripple effect: Emotional contagion and its influence on group behavior. *Adm. Sci. Q.* **2002**, *47*, 644–675. [CrossRef]
7. Howes, M.J.; Hokanson, J.E.; Loewenstein, D.A. Induction of depressive affect after prolonged exposure to a mildly depressed individual. *J. Personal. Soc. Psychol.* **1985**, *49*, 1110. [CrossRef]
8. Bono, J.E.; Ilies, R. Charisma, positive emotions and mood contagion. *Leadersh. Q.* **2006**, *17*, 317–334. [CrossRef]
9. Van Kleef, G.A.; De Dreu, C.K.; Manstead, A.S. The interpersonal effects of anger and happiness in negotiations. *J. Personal. Soc. Psychol.* **2004**, *86*, 57. [CrossRef] [PubMed]
10. Fowler, J.H.; Christakis, N.A. Cooperative behavior cascades in human social networks. *Proc. Natl. Acad. Sci. USA* **2010**, *107*, 5334–5338. [CrossRef]
11. Gill, A.J.; Gergle, D.; French, R.M.; Oberlander, J. Emotion rating from short blog texts. In Proceedings of the SIGCHI Conference on Human Factors in Computing Systems, Florence, Italy, 5–10 April 2008; pp. 1121–1124.
12. Choi, S.; Kim, E.M. Between Instagram browsing and subjective well-being: Social comparison or emotional contagion? *Media Psychol.* **2020**, *24*, 1–25. [CrossRef]
13. Counts, M.D.C.S.; Gamon, M. Not all moods are created equal! exploring human emotional states in social media. In Proceedings of the International AAAI Conference on Web and Social Media (ICWSM), Dublin, Ireland, 4–7 June 2012; pp. 1–8.
14. Coviello, L.; Fowler, J.H.; Franceschetti, M. Words on the web: Noninvasive detection of emotional contagion in online social networks. *Proc. IEEE* **2014**, *102*, 1911–1921. [CrossRef]
15. Coviello, L.; Sohn, Y.; Kramer, A.D.; Marlow, C.; Franceschetti, M.; Christakis, N.A.; Fowler, J.H. Detecting emotional contagion in massive social networks. *PLoS ONE* **2014**, *9*, e90315. [CrossRef] [PubMed]
16. Garcia, D.; Garas, A.; Schweitzer, F. Positive words carry less information than negative words. *EPJ Data Sci.* **2012**, *1*, 3. [CrossRef]
17. Goldenberg, A.; and Gross, J.J. Digital emotion contagion. *Trends Cogn. Sci.* **2020**, *24*, 316–328. [CrossRef] [PubMed]
18. Golder, S.A.; Macy, M.W. Diurnal and seasonal mood vary with work, sleep, and daylength across diverse cultures. *Science* **2011**, *333*, 1878–1881. [CrossRef]
19. Hancock, J.T.; Gee, K.; Ciaccio, K.; Lin, J.M.H. I'm sad you're sad: Emotional contagion in CMC. In Proceedings of the 2008 ACM Conference on Computer Supported Cooperative Work, San Diego, CA, USA, 8–12 November 2008; pp. 295–298.
20. Harris, R.B.; Paradice, D. An investigation of the computer-mediated communication of emotions. *J. Appl. Sci. Res.* **2007**, *3*, 2081–2090.
21. Mei, Q.; Ling, X.; Wondra, M.; Su, H.; Zhai, C. Topic sentiment mixture: Modeling facets and opinions in weblogs. In Proceedings of the 16th international conference on World Wide Web, Banff, AB, Canada, 8–12 May 2007; pp. 171–180.
22. Kramer, A.D. The spread of emotion via Facebook. In Proceedings of the SIGCHI Conference on Human Factors in Computing Systems, Austin, TX, USA, 5–10 May 2012; pp. 767–770.
23. Kramer, A.D.; Guillory, J.E.; Hancock, J.T. Experimental evidence of massive-scale emotional contagion through social networks. *Proc. Natl. Acad. Sci. USA* **2014**, *111*, 8788–8790. [CrossRef]
24. Dang-Xuan, L.; Stieglitz, S. Impact and Diffusion of Sentiment in Political Communication-An Empirical Analysis of Political Weblogs. In Proceedings of the ICWSM, Dublin, Ireland, 4–7 June 2012.
25. Wang, J.; Wei, L. Fear and hope, bitter and sweet: Emotion sharing of cancer community on twitter. *Soc. Media Soc.* **2020**, *6*. [CrossRef]
26. Brady, W.J.; Crockett, M.J. How Effective Is Online Outrage? *Trends Cogn. Sci.* **2019**, *23*, 1–2. [CrossRef] [PubMed]
27. Ba, S.; Wang, L. Digital health communities: The effect of their motivation mechanisms. *Decis. Support Syst.* **2013**, *55*, 941–947. [CrossRef]

28. Fox, S. *Peer-to-Peer Healthcare. Many People–Especially Those Living with Chronic or Rare Diseases–Use Online Connections to Supplement Professional Medical Advice [Pew Internet & American Life Project]*; Pew Research Center: Washington, DC, USA, 2011.
29. Bargh, J.A.; McKenna, K.Y.; Fitzsimons, G.M. Can you see the real me? Activation and expression of the "true self" on the Internet. *J. Soc. Issues* **2002**, *58*, 33–48. [CrossRef]
30. Kallinikos, J.; Tempini, N. Patient data as medical facts: Social media practices as a foundation for medical knowledge creation. *Inf. Syst. Res.* **2014**, *25*, 817–833. [CrossRef]
31. Bartlett, Y.K.; Coulson, N.S. An investigation into the empowerment effects of using online support groups and how this affects health professional/patient communication. *Patient Educ. Couns.* **2011**, *83*, 113–119. [CrossRef]
32. Griffiths, K.M.; Calear, A.L.; Banfield, M. Systematic review on Internet Support Groups (ISGs) and depression (1): Do ISGs reduce depressive symptoms? *J. Med. Internet Res.* **2009**, *11*, e40. [CrossRef]
33. Setoyama, Y.; Yamazaki, Y.; Namayama, K. Benefits of peer support in online Japanese breast cancer communities: Differences between lurkers and posters. *J. Med. Internet Res.* **2011**, *13*, e122. [CrossRef] [PubMed]
34. van Uden-Kraan, C.F.; Drossaert, C.H.; Taal, E.; Shaw, B.R.; Seydel, E.R.; van de Laar, M.A. Empowering processes and outcomes of participation in online support groups for patients with breast cancer, arthritis, or fibromyalgia. *Qual. Health Res.* **2008**, *18*, 405–417. [CrossRef] [PubMed]
35. van Uden-Kraan, C.F.; Drossaert, C.H.; Taal, E.; Seydel, E.R.; van de Laar, M.A. Participation in online patient support groups endorses patients' empowerment. *Patient Educ. Couns.* **2009**, *74*, 61–69. [CrossRef]
36. Høybye, M.T.; Dalton, S.O.; Deltour, I.; Bidstrup, P.E.; Frederiksen, K.; Johansen, C. Effect of Internet peer-support groups on psychosocial adjustment to cancer: A randomised study. *Br. J. Cancer* **2010**, *102*, 1348–1354. [CrossRef]
37. Shaw, B.R.; Hawkins, R.; McTavish, F.; Pingree, S.; Gustafson, D.H. Effects of insightful disclosure within computer mediated support groups on women with breast cancer. *Health Commun.* **2006**, *19*, 133–142. [CrossRef]
38. Park, A.; Conway, M. Longitudinal changes in psychological states in online health community members: Understanding the long-term effects of participating in an online depression community. *J. Med. Internet Res.* **2017**, *19*, e71. [CrossRef] [PubMed]
39. World Health Organization. *Depression and Other Common Mental Disorders: Global Health Estimates (No. WHO/MSD/MER/2017.2)*; World Health Organization: Geneve, Switzerland, 2017.
40. Barney, L.J.; Griffiths, K.M.; Banfield, M.A. Explicit and implicit information needs of people with depression: A qualitative investigation of problems reported on an online depression support forum. *BMC Psychiatry* **2011**, *11*, 1–11. [CrossRef] [PubMed]
41. De Choudhury, M.; De, S. Mental health discourse on reddit: Self-disclosure, social support, and anonymity. In Proceedings of the Eighth International AAAI Conference on Weblogs and Social Media, Ann Arbor, MI, USA, 1–4 June 2014.
42. Thompson, R.J.; Mata, J.; Jaeggi, S.M.; Buschkuehl, M.; Jonides, J.; Gotlib, I.H. The everyday emotional experience of adults with major depressive disorder: Examining emotional instability, inertia, and reactivity. *J. Abnorm. Psychol.* **2012**, *121*, 819. [CrossRef] [PubMed]
43. Hatfield, E.; Cacioppo, J.T.; Rapson, R.L. Emotional contagion. *Curr. Dir. Psychol. Sci.* **1993**, *2*, 96–100. [CrossRef]
44. Sina Weibo Data Center. *Weibo User Development Report*; Sina Weibo: Beijing, China, 2013.
45. Tang, J.; Yu, G.; Yao, X. A comparative study of online depression communities in china. *Int. J. Environ. Res. Public Health* **2020**, *17*, 5023. [CrossRef]
46. Wang, Z.; Yu, G.; Tian, X.; Tang, J.; Yan, X. A study of users with suicidal ideation on sina weibo. *Telemed. J. E Health* **2018**, *24*, 702–709. [CrossRef]
47. Wang, Z.; Yu, G.; Tian, X. Exploring behavior of people with suicidal ideation in a Chinese online suicidal community. *Int. J. Environ. Res. Public Health.* **2018**, *16*, 54. [CrossRef] [PubMed]
48. Benton, A.; Coppersmith, G.; Dredze, M. Ethical Research Protocols for Social Media Health Research. In Proceedings of the First ACL Workshop on Ethics in Natural Language Processing, Valencia, Spain, 4 April 2017.
49. Devlin, J.; Chang, M.W.; Lee, K.; Toutanova, K. Bert: Pre-training of deep bidirectional transformers for language understanding. *arXiv* **2018**, arXiv:1810.04805.
50. Charrad, M.; Ghazzali, N.; Boiteau, V.; Niknafs, A. Nbclust: An r package for determining the relevant number of clusters in a data set. *J. Statal Softw.* **2014**, *61*, 1–36. [CrossRef]
51. Liang, K.Y.; Zeger, S.L. Longitudinal data analysis using generalized linear models. *Biometrika* **1986**, *73*, 13–22. [CrossRef]
52. Ballinger, G.A. Using generalized estimating equations for longitudinal data analysis. *Organ. Res. Methods* **2004**, *7*, 127–150. [CrossRef]
53. Wang, M. Generalized estimating equations in longitudinal data analysis: A review and recent developments. *Adv. Stat.* **2014**, *2014*, 303728. [CrossRef] [PubMed]
54. Schildcrout, J.S.; Heagerty, P.J. Regression analysis of longitudinal binary data with time-dependent environmental covariates: Bias and efficiency. *Biostatistics* **2005**, *4*, 633–652. [CrossRef] [PubMed]
55. Hodas, N.O.; Lerman, K. The simple rules of social contagion. *Sci. Rep.* **2014**, *4*, 4343. [CrossRef]
56. Hill, A.L.; Rand, D.G.; Nowak, M.A.; Christakis, N.A. Emotions as infectious diseases in a large social network: The SISa model. *Proc. R. Soc. B Biol. Sci.* **2010**, *277*, 3827–3835. [CrossRef] [PubMed]
57. Yan, J.; Bracewell, D.B.; Ren, F.; Kuroiwa, S. The creation of a Chinese emotion ontology based on HowNet. *Eng. Lett.* **2008**, *16*, 166–171.
58. Ekman, P. Facial expression and emotion. *Am. Psychol.* **1993**, *48*, 384. [CrossRef]

59. Ferrara, E.; Yang, Z. Measuring emotional contagion in social media. *PLoS ONE* **2015**, *10*, e0142390. [CrossRef]
60. Woolson, R.F. Wilcoxon signed-rank test. *Wiley Encycl. Clin. Trials* **2007**, 1–3. [CrossRef]
61. Cavazos-Rehg, P.A.; Krauss, M.J.; Sowles, S.; Connolly, S.; Rosas, C.; Bharadwaj, M.; Bierut, L.J. A content analysis of depression-related tweets. *Comput. Hum. Behav.* **2016**, *54*, 351–357. [CrossRef]
62. De Choudhury, M.; Counts, S.; Horvitz, E. Social media as a measurement tool of depression in populations. In Proceedings of the 5th Annual ACM Web Science Conference, Paris, France, 2–4 May 2013; pp. 47–56.
63. Settanni, M.; Marengo, D. Sharing feelings online: Studying emotional well-being via automated text analysis of Facebook posts. *Front. Psychol.* **2015**, *6*, 1045. [CrossRef] [PubMed]
64. Gross, J. Antecedent- and response-focused emotion regulation: Divergent consequences for experience, expression, and physiology. *J. Personal. Soc. Psychol.* **1998**, *74*, 224–237. [CrossRef]
65. Berger, J.; Milkman, K.L. What makes online content viral? *J. Mark. Res.* **2012**, *49*, 192–205. [CrossRef]
66. Phillips, M.R.; Zhang, J.; Shi, Q.; Song, Z.; Ding, Z.; Pang, S.; Li, X.; Zhang, Y.; Wang, Z. Prevalence, treatment, and associated disability of mental disorders in four provinces in China during 2001–2005: An epidemiological survey. *Lancet* **2009**, *373*, 2041–2053. [CrossRef]
67. Yang, L.H.; Kleinman, A. Face and the embodiment of stigma in china: The cases of schizophrenia and aids. *Soc. Sci. Med.* **2008**, *67*, 398–408. [CrossRef]

healthcare

MDPI

Article

Moderating Effect of Nurse's Character on the Relationship between Attitudes toward Nursing Care of the Dying and Performance of Terminal Care in South Korea

Kawoun Seo

Department of Nursing, Joongbu University, Goyang 32713, Korea; kwseo@joongbu.ac.kr

Abstract: This study aimed to investigate the moderating effect of nurses' characteristics on the relationship between attitudes toward nursing care and terminal care performance of hospice nurses. The participants included nurses working in hospice care units in general hospitals in South Korea. Data collected from August 1 to 31, 2020 were analyzed using *t*-test, ANOVA, Pearson's correlation coefficients, and hierarchical multiple regression. The performance of terminal care was positively correlated with attitudes toward nursing care of the dying ($r = 0.45$, $p < 0.001$) and nurses' characteristics ($r = 0.60$, $p < 0.001$). There was also a positive correlation ($r = 0.58$, $p < 0.001$) between attitudes toward nursing care for dying patients and nurses' characteristics. Nurses' characteristics had a significant moderating effect on the relationship between attitudes toward nursing care of the dying and performance of terminal care. This means that the nurses' character had a buffering effect on the relationship between attitudes toward nursing care of the dying and performance of terminal care in hospice care units. These findings suggest that continuous and repetitive educational programs on terminal care need to be enhanced, and strategies to strengthen attitudes toward nursing care of the dying and nurses' character need to be included in these programs.

Keywords: hospices; terminal care; character; attitude toward death; nurses

Citation: Seo, K. Moderating Effect of Nurse's Character on the Relationship between Attitudes toward Nursing Care of the Dying and Performance of Terminal Care in South Korea. *Healthcare* **2021**, *9*, 1195. https://doi.org/10.3390/healthcare9091195

Academic Editors: Yasuhiro Kotera and Elaina Taylor

Received: 8 August 2021
Accepted: 8 September 2021
Published: 10 September 2021

Publisher's Note: MDPI stays neutral with regard to jurisdictional claims in published maps and institutional affiliations.

1. Introduction

Although hospice palliative care commenced in Korea in 1965, progress has been slow owing to the failure to establish a stable financial foundation while conducting hospice palliative care from a religious and private perspective [1]. However, interest in hospice palliative care has been increasing with the growing awareness of the importance of end-of-life care in the context of a growing, aging population and the implementation of the national health insurance hospice benefit system [1,2]. In addition, with the establishment of the Act on the Determination of Lifetime Medical Care for Patients in the Hospice, Palliative Care, and on the Dying Process, the user rate is expected to increase as the service scope expands to include terminal cancer as well as acquired immunodeficiency syndrome, chronic obstructive pulmonary disease, and chronic cirrhosis [1,3].

Hospice palliative care involves the creation of a multidisciplinary team to improve the quality of life of terminally ill patients, along with that of their families. The aim is to provide holistic care, including physical, emotional, spiritual, legal, and financial counseling [4]. Most nurses in hospice palliative care units are responsible for providing nursing care to terminally ill patients. They provide physical care to reduce pain and discomfort caused by the symptoms, as well as psychological or spiritual counseling for those suffering from fear of death. Spiritual nursing also helps patients to improve their quality of life and maintain their dignity [5]. Therefore, for terminally ill patients, end-of-life nursing provided by hospice nurses plays a very important role in improving their quality of life. Among the factors that influence the performance of terminal care by hospice nurses, the most significant is the attitude toward nursing care for dying patients.

Attitudes toward nursing care for dying patients refer to the positive or negative emotions and perceptions of nurses who provide nursing care to these patients [6]. Patients and their families who are assigned nurses with positive attitudes toward nursing care are considered to have a high quality of life and are likely to face death more gracefully [7]. Furthermore, attitudes toward nursing care for dying patients were found to significantly influence the performance of terminal care by intensive care unit nurses [8], and it is also an important influencing factor in the performance of terminal care of cancer ward nurses [9]. However, some of these nurses encounter difficulties in performing terminal care due to the lack of knowledge, experience, or values, a limited understanding of end-of-life patients, or communication problems with other medical personnel [10]. In addition, nurses said that they found terminal care challenging as they experienced spiritual conflicts caused by anxiety or skepticism about death [3]. In particular, the dearth of nurses, workload processes, and lack of communication hinders the performance of terminal care [8]. However, to maintain the quality of life and dignity of individuals who use the hospice ward, it is critical to launch efforts to improve the performance of hospice nurses' terminal care.

Meanwhile, clinical nursing character refers to desirable attitudes and behaviors that active clinical nurses should possess [11], including having a personality that is suitable for caring for a patient. These include a sense of responsibility, a passion for improving work ability, and maintaining composure in situations where emotions are high. Other qualities include coping, empathy, the ability to work smoothly in cooperation with colleagues, behaviors that respect the personality of others, and recognition of the existence or value of others [11]. These qualities enable nurses to overcome barriers to terminal care. Effectively, nurses' competence in terms of their end-of-life nursing attitude is expressed through empathy [12], as well as their ability to maintain their composure despite the stress arising from terminal care or to change one's perception of death through their experiences to realize the meaning of care for end-of-life patients [13]. Considering the above, it can be predicted that a nurse's character, particularly having a positive attitude toward nursing care of the dying, can serve as a buffer and help in overcoming obstacles to performing effective terminal care. However, there are no comprehensive studies on the relationship between attitudes toward nursing care of the dying and performance of terminal care among hospice nurses in Korea, while studies on nurses' characteristics are limited.

Therefore, this study aimed to examine whether a nurse's character has a moderating effect on the relationship between attitudes toward nursing care of the dying and the performance of hospice nurses' terminal care. The results of this study can be used as basis data for developing an intervention program to enhance the performance of hospice nurses' terminal care in Korea.

2. Materials and Methods

2.1. Design

This study explores the moderating effect of a nurse's character on the relationship between attitudes toward nursing care of the dying and terminal care performance of hospice nurses.

2.2. Participants and Data Collection

The participants of this study included 158 hospice nurses who worked in hospice wards of general hospitals located in D city and C and K provinces for more than three months. They understood the purpose of the study and voluntarily agreed to participate.

The data for this study were collected using a self-report questionnaire consisting of topics such as general characteristics, clinical career, attitudes toward nursing care of the dying, performance of terminal care, and the character of nurses. After explaining the purpose of the study and the contents of the survey, the questionnaires were distributed to hospice nurses in the nursing department of each hospital. The data collection period was from 1 to 31 August 2020.

2.3. Measurements

The questionnaire consists of eight questions regarding the general characteristics and career, 20 items that evaluate attitudes toward nursing care of the dying, 21 items that measure the performance of terminal care, and 53 items that evaluate the nurse's character, for a total of 102 items.

2.3.1. Attitudes toward Nursing Care of the Dying (ANCD)

To evaluate attitudes toward nursing care of the dying, the Frommel Attitudes Toward Nursing Care of the Dying Scale (FATCOD) developed by Frommelt and translated by Cho and Kim was used [14,15]. The original tool consisted of 20 questions regarding nurses' attitudes toward patients and 10 questions on family care for terminally ill patients. However, only 20 questions on nurses' attitudes were used in this study. Each question was answered on a 4-point Likert scale. At the time of tool development [14], the reliability of the tool was Cronbach's $\alpha = 0.98$, 0.94 in the study by Cho and Kim [15], and in this study, the reliability of the tool was Cronbach's $\alpha = 0.79$.

2.3.2. Performance of Terminal Care

The performance of terminal care was measured using the performance of the terminal care tool developed by Park and Choi and revised by Chung [16,17]. This tool has 21 questions overall: eight questions each for the physical and psychological domains, and five questions regarding spiritual aspects. Responses were answered using a 4-point Likert scale. At the time of tool development [16], the reliability of the tool was Cronbach's $\alpha = 0.96$, 0.80 in the study by Chung [17], and, in this study, the reliability of the tool was Cronbach's $\alpha = 0.89$.

2.3.3. Nurse's Character (ND)

The nurses' character was measured using a nurse's character tool developed by Park [11]. This tool consists of 53 questions. Each question was answered using a 5-point Likert scale, and the higher the total score, the better the personality of the nurse. In Park's study [11], the reliability was Cronbach's $\alpha = 0.94$; in this study, the reliability of the tool was Cronbach's $\alpha = 0.95$.

2.4. Data Analysis

The collected data were analyzed using the SPSS/WIN 24.0. Descriptive statistics were used for general characteristics of the study participants, and Pearson's correlation coefficients were used to determine the correlation between ANCD, performance of terminal care, and ND. The moderating effect of NC on the relationship between ANCD and performance of terminal care of hospice nurses was analyzed through hierarchical multiple regression. Reliability was measured using Cronbach's α.

3. Results

3.1. Participants' Characteristics

The general characteristics of the study participants are listed in Table 1. The average age of the participants in this study was 39.4 years ($\pm$10.8) with those in their 30s or younger accounting for 30.4% of all participants. Among the participants, 47.5% had attained a bachelor's degree, while 39.9% had attained a master's degree. Among the participants, 68.4% engaged in religious activities, 48.1% were married, and 22.2% lived alone. The average hospice nurse's career was 3.98 ($\pm$4.16) years, whereas the majority of the nurses (33.5%) had less than one year of experience. Only 24.1% of the nurses were hospice specialist nurses, and 66.5% said they were satisfied with their jobs.

The mean score of attitudes toward nursing care of the dying in this study was 3.22 ($\pm$0.31) points. Meanwhile, the mean score for the performance of terminal care was 3.18 ($\pm$0.33) points. The mean score for nurses' character was 4.06 ($\pm$0.40) points.

Table 1. General characteristics of participants (n = 158).

Characteristics	Categories	M ± SD or n (%)	MIN–MAX
Age (yr)		39.4 ± 10.8	24.0–68.0
	≤30	48 (30.4)	
	31–40	45 (28.5)	
	41–50	38 (24.1)	
	≥50	27 (17.1)	
Educational level	3-year nursing school	20 (12.7)	
	Bachelor's degree	119 (75.3)	
	Master's degree or higher	19 (12.0)	
Having religion	Yes	108 (68.4)	
	No	50 (31.6)	
Marital status	Yes	76 (48.1)	
	No	82 (51.9)	
Living alone	Yes	48 (22.2)	
	No	123 (77.8)	
Career of hospice nurse (yr)		3.98 ± 4.16	0–24.0
	≤1	53 (33.5)	
	2–3	42 (26.6)	
	4–10	41 (26.0)	
	≥11	22 (13.9)	
Hospice specialist nurse	Yes	38 (24.1)	
	No	120 (75.9)	
Job satisfaction	dissatisfaction	7 (4.4)	
	moderate	46 (29.1)	
	satisfaction	105 (66.5)	
Attitudes toward nursing care of the dying		3.22 ± 0.31	2.40–3.80
Performance of terminal care		3.18 ± 0.33	2.57–4.00
Nurse's character		4.06 ± 0.40	2.87–4.97

3.2. *Differences in the Performance of Terminal Care According to General Characteristics*

There were significant differences in the performance of terminal care based on the general characteristics of the participants according to age (F = 7.04, $p < 0.001$), religion (t = 4.26, $p < 0.001$), marital status (t = 7.04, $p < 0.001$), career of hospice nurses (F = 4.95, p = 0.003), hospice specialist nurses (t = 2.06, p = 0.004), and job satisfaction (F = 5.09, p = 0.007) (Table 2).

Table 2. Differences in performance of terminal care according to general characteristics (N = 158).

Characteristics	Categories	Performance of Terminal Care	
		M ± SD	t or F (p)
Age (yr)	≤30 [a]	3.06 ± 0.31	7.04 (<0.001) a < b < c,d
	31–40 [b]	3.13 ± 0.31	
	41–50 [c]	3.27 ± 0.30	
	≥51 [d]	3.36 ± 0.33	
Educational level	Diploma	3.20 ± 0.30	0.36 (0.697)
	Bachelor's degree	3.19 ± 0.34	
	Master's degree	3.12 ± 0.29	
Having religion	Yes	3.25 ± 0.33	4.26 (<0.001)
	No	3.02 ± 0.27	
Marital status	Yes	3.28 ± 0.33	3.79 (<0.001)
	No	3.09 ± 0.30	

Table 2. *Cont.*

Characteristics	Categories	Performance of Terminal Care	
		M ± SD	t or F (p)
Living alone	Yes No	3.20 ± 0.33 3.10 ± 0.32	0.81 (0.439)
Career of hospice nurse (yr)	≤1 [a] 2–3 [b] 4–10 [c] ≥11 [d]	3.14 ± 0.39 3.07 ± 0.26 3.07 ± 0.34 3.27 ± 0.32	4.95 (0.003) b,c < d
Hospice specialist nurse	Yes No	3.28 ± 0.32 3.28 ± 0.33	2.06 (0.054)
Job satisfaction	Dissatisfaction [a] Moderate [b] Satisfaction [c]	2.91 ± 0.37 3.10 ± 0.28 3.23 ± 0.33	5.09 (0.007) a < c

The differences and order among groups in the post-hoc comparison are indicated by superscript letters (a, b, c, d).

3.3. Correlation between Independent Variables

ANCD had a positive correlation with the performance of terminal care (r = 0.45, $p < 0.001$) and NC (r = 0.58, $p < 0.001$). There was also a positive correlation between the performance of terminal care and NC (r = 0.60, $p < 0.001$).

3.4. Moderating Effects of Nurse's Character in Relation to Attitude toward Nursing Care of the Dying and Performance of Terminal Care

A multi-collinearity test was conducted before hierarchical regression analysis to confirm the moderating effects of NC on the relationship between ANCD and the performance of terminal care. The Durbin–Watson value was 2.019, and there was no autocorrelation of error. The limit of tolerance was 0.58–0.95, all above 0.1, and the variance expansion coefficients of all variables ranged from 1.06–1.73, not exceeding 10. Therefore, there was no multicollinearity among the variables.

The explanatory power of Model 1, in which only control variables were inputted with performance of terminal care as a dependent variable, was 14.8%, and none of the variables had a significant influence (F = 4.88, $p < 0.001$). In Model 2, ANCD was added as an independent variable to the control variable of Model 1 and to understand its effect. The explanatory power increased to 24.3%, and ANCD ($\beta = 0.35$, $p < 0.001$) had a significant effect on the performance of terminal care (F = 7.25, $p < 0.001$). In Model 3, NC was added to the independent variable in Model 2 to determine its moderating effect. The explanatory power of Model 3 was 37.2%, and only NC ($\beta = 0.49$, $p < 0.001$) had a statistically significant effect (F = 11.25, $p < 0.001$). Finally, in order to examine the moderating effect, the interaction variable between ANCD and NC was added to the independent variable of Model 3. The explanatory power of Model 4 increased to 41.3%, and among the independent variables, religion ($\beta = 0.15$, $p = 0.037$), ANCD ($\beta = 0.18$, $p = 0.032$), NC ($\beta = 0.51$, $p < 0.001$), and the interaction between ANCD and NC ($\beta = 0.22$, $p < 0.001$) were found to be statistically significant variables (F = 11.99, $p < 0.001$). As a result of testing the moderating effect of NC on the relationship between ANCD's effect on performance of terminal care, it was found that explanatory power gradually increased to 24.3% in Model 2, 37.2% in Model 3, and 41.3% in Model 4. Since the p-value is also less than 0.05 in all models, the moderating effect can be confirmed. In conclusion, NC was found to have a moderating effect between the ANCD and performance of terminal care. In other words, with regard to the relationship between ANCD and performance of terminal care, it was found that the higher the NC, the greater the effect of the ANCD on the performance of terminal care. Therefore, the NC greatly amplified the attitude toward nursing care for dying patients (Table 3).

Table 3. Moderating effect of NC on the relationship between ANCD and performance of terminal care (n = 158).

Categories	1			2			3			4		
	β	t	p	β	t	p	β	t	p	β	t	p
Age (yr)	0.17	10.33	0.185	0.14	1.19	0.237	0.08	0.68	0.495	0.06	0.53	0.599
Having religion (ref: no)	0.17	10.92	0.057	0.13	1.54	0.127	0.14	10.80	0.074	0.15	20.10	0.037
Marital status (ref: no)	0.09	0.92	0.360	0.04	0.45	0.656	−0.04	−0.50	0.620	−0.03	−0.37	0.713
Career of hospice nurse (yr)	0.00	0.05	0.959	0.05	0.44	0.662	0.01	0.11	0.914	0.02	0.16	0.876
Hospice specialist nurse (ref: no)	0.07	0.93	0.353	0.03	0.39	0.694	0.05	0.80	0.427	0.05	0.76	0.451
Job satisfaction (ref: dissatisfaction)												
moderate	0.12	0.65	0.517	0.04	0.23	0.823	−0.08	−0.49	0.623	−0.05	−0.34	0.734
satisfaction	0.23	10.21	0.229	0.03	0.16	0.871	0.12	−0.69	0.494	−0.08	−0.47	0.640
ANCD				0.35	4.43	<0.001	0.13	10.56	0.122	0.18	20.17	0.032
NC							0.49	50.60	<0.001	0.51	60.06	<0.001
ANCD × NC										0.22	3.83	0.001
R^2 (ΔR^2)	0.148			0.243 (0.095)			0.372 (0.129)			0.413 (0.041)		
F	4.88			7.25			11.25			11.99		
p	(<0.001)			(<0.001)			(<0.001)			(<0.001)		

ANCD: Attitudes toward Nursing Care of the Dying; NC: Nurse's Character.

4. Discussion

The terminal care performance of hospice nurses is critical in enhancing the quality of life of hospice patients, alleviating their physical and psychological distress, and ensuring a peaceful death. Therefore, this study was conducted to confirm the moderating effect of NC, an individual competency, on the relationship between ANCD and the performance of terminal care by hospice nurses.

The mean score of hospice nurses for ANCD was 3.22 points (range 1–4 points), which is higher than that reported by Frommelt, who stated an average of 2 points at the time of tool development [14]. In addition, this score was higher than the 2.83 points found in a study of intensive care unit nurses [8] and the 2.86 points found in a study of general hospital nurses [12]. A possible reason is that a relatively higher proportion of patients require terminal care in hospice wards than in general wards. In addition, the goal of the hospice palliative unit is to improve the quality of life of patients while aiding them in achieving a peaceful death as compared to the intensive care unit, which focuses more on the recovery and resuscitation of the patient. However, in a previous study targeting hospice palliative ward nurses [18], the level of ANCD was similar to that of a study targeting intensive care unit nurses. In this study, unlike previous studies [18], there is a possibility that the score was relatively high because the nurses' attitudes toward the patient's family were excluded from among the research tools. Since most nurses provide patient-centered care, there is a possibility that the proportion given to the patient's family is relatively lower than that to the patient. However, considering that the patient's family also faces various difficulties [19], further research on ANCD, including the patient's family, should be conducted in subsequent studies.

The mean score for NC was 4.06 points (range 1–5 points), which was higher than the 3.67 points found in a study targeting general hospital nurses [20] and the 3.84 points found in a study targeting clinical nurses [21]. Considering the subdomains, in this study, the nurses scored higher in the interpersonal domain than in the self-related domain, and they scored highest in terms of civility, followed by reliability and interactional justice. These results are similar to those of a study on general hospital nurses. These areas pertain to respecting end-of-life patients and recognizing their value so that they can maintain their quality of life until their death. This is considered very important for hospice nurses, who play a role in helping patients maintain their dignity [5]. Good character is a quality that all nurses who provide nursing care to patients, as well as hospice nurses, should possess as a basic qualification, and it is a very important factor that affects the quality of nursing service received by patients [20]. In addition, NC is related to nurses' anxiety about the death of others and affects care behaviors [22]. However, research on NC is incomplete; therefore, further studies should focus on various aspects of a clinical nurse's character.

The mean score for terminal care performance in this study was 3.18 points (range 1–4 points), similar to the mean score of 3.17 points found in a study of long-term care hospital nurses [23]. However, in a study of intensive care unit nurses [8], the mean score of performance of terminal care was 2.60 points, which was higher than that in this study. The performance of hospice nurses' terminal care was found to be influenced by age, educational level, religion, marital status, career of hospice nurses, whether the nurse is a hospice specialist, and job satisfaction. This result differs from that in the study of intensive care unit nurses [8], where there was no difference in the performance of terminal care according to general characteristics. In addition, considering the subdomains, among the areas of terminal care, hospice nurses focused most on psychological nursing, followed by physical and spiritual nursing. However, clinical nurses and intensive care unit nurses focused most on physical, psychological [8,12], then spiritual nursing, demonstrating that hospice nurses provided more psychological nursing than nurses in other fields. This can be attributed to the fact that the goal of a hospice unit is different from that of other units. When the level of nursing for each area was compared, the level of spiritual nursing was found to be lower than that of clinical nurses [12]. This result could be explained by the results of a qualitative study dealing with the spiritual conflict of hospice nurses, who found that hospice nurses felt vulnerable when facing the death of a patient and experienced immense spiritual conflict as they were disturbed by the state of affairs before death [3]. These spiritual conflicts likely resulted in burnout among hospice nurses. Another study found that compassion fatigue caused by hospice nurses while caring for patients and families creates psychological stress [24]. For this reason, hospice nurses emphasized the need for a program to address the burnout and distress caused by compassion fatigue [24]. However, these results may be due to the characteristics of the participants in this study. In this study, nurses' performance in terminal care was found to be higher among nurses who were older, married, and had more than 10 years of experience in the hospice ward. However, more than 50% of the participants of this study had less than three years of experience in the hospice ward. Therefore, in future research, further studies on the relationship between the experience of the hospice ward and the level of spiritual nursing is needed.

Terminal care performance was positively correlated with ANCD and NC. Religion, job satisfaction, ANCD, and NC were strong factors that influenced the performance of terminal care. In addition, NC was found to have a moderating effect on the relationship between ANCD and terminal care performance. In other words, since ANCD and NC interact, even if ANCD is low, NC buffers and improves the performance of terminal care. A previous study explored the factors influencing the performance of terminal care among intensive care unit nurses and found that stress during end-of-life nursing [8] and the stress associated with end-of-life nursing can be caused by negative attitudes or burdens of the patient or caregiver toward hospice care, interpersonal problems, or lack of knowledge and perception of hospice care [8,24,25]. Thus, even if nurses have a positive attitude toward nursing care of the dying, it is difficult for their attitudes to lead to enhanced performance unless they establish a trusting relationship with the patient's family. At this time, NC affects the formation of smooth interpersonal relationships and resolution of conflicts in the clinical field [20]. This is because trust, listening, and responsibility, which are necessary elements for forming good interpersonal relationships, are the main components of a nurse's clinical nursing character [11]. Therefore, a nurse with good clinical nursing character gives trust to the patient, even if the ANCD is rather poor, and has the capacity to listen to their story and practice responsible actions. These competencies will soon lead to terminal care performance. In addition, it is difficult to provide appropriate terminal nursing if there is not enough time or accurate communication between the doctor, peer nurse, or hospice team. In this situation, nurses do their best, striving to make up for their deficiencies, treating the patient sincerely and with kindness, respect, and empathy, while cooperating with other medical personnel. It is believed that, if they are polite and can be fair toward the weak, this positive attitude toward nursing care of the dying can lead to better performance of terminal care. These capabilities make up a nurse's character [11]. In a qualitative study on the adaptation of hospice nurses to their roles, nurses assigned to the hospice ward who faced difficulties naturally and positively accepted their perception of death, understood and empathized with the patients and their families, and practiced respectful nursing [10]. It is said that they adapted to their own roles while playing the role of coordinator in their relationship with medical personnel. Therefore, to enhance the performance of terminal care by hospice nurses, it is imperative to introduce educational programs regarding end-of-life care so that nurses can develop a positive attitude toward nursing care for the dying. Additionally, when developing such a program, it is necessary to seek ways to improve clinical nursing characteristics.

Despite these results, this study has several limitations. First, since this study selected participants through a convenience sample, the results of this study cannot be generalized to all hospice nurses. Second, since these study data were collected during the COVID-19 period, these situations may have affected the results of this study. However, since studies on NC for hospice nurses are not diverse, the results of this study emphasize the importance of education on nurses' characteristics and provide important fundamental data for the development of policies and programs to enhance the performance of terminal care.

5. Conclusions

This study attempted to provide fundamental data for strategies to improve the performance of terminal care for hospice patients by examining the moderating effect of nurses' characteristics on the relationship between the attitudes toward nursing care of the dying and the performance of hospice nurses' terminal care. It was found that religion, job satisfaction, attitudes toward nursing care of the dying, and nurses' characteristics affected the performance of hospice nurses' terminal care. In addition, it was confirmed that there is a moderating effect of nurses' character on the relationship between attitudes toward nursing care of the dying and performance of terminal care. Based on this, we suggest the following: first, it is necessary to include elements to improve nurses' character in the educational program to improve the performance of hospice nurses' terminal care from a clinical point of view. Second, in terms of education, education should be provided so that nursing students can have the right knowledge and awareness of hospice nursing, and, at the same time, education should be provided to improve clinical nursing characteristics. Third, in terms of research, repeated studies that expand the scope of investigation are needed, and in-depth follow-up studies on nurses' hospice ward careers and spiritual nursing should be conducted.

Funding: This research was supported by the Joongbu University Research & Development Fund in 2020.

Institutional Review Board Statement: The study was conducted in accordance with the guidelines of the Declaration of Helsinki and approved by the Institutional Review Board of Joongbu University (IRB No: JIRB-2020080303-01-200901).

Informed Consent Statement: Informed consent was obtained from all participants involved in the study.

Data Availability Statement: The data presented in this study are available upon request from the corresponding author. The data are not publicly available because of privacy concerns.

Conflicts of Interest: The author declare no conflict of interest.

References

1. Kim, C.G. The history of hospice and palliative care in Korea. *Korean J. Hosp. Palliat. Care.* **2019**, *22*, 1–7. [CrossRef]
2. Oh, J.H.; Lee, D.H.; Lim, J.W.; Shin, Y.J.; Park, D.H.; Yoo, H.L. Hospice palliative medical service system improvement plan–Focusing on reinforcing linkages between types and promoting patient-centered integrated use. In *Policy Proposal Report*; Health Insurance Review & Assessment Service: Gangwon, Korea, 2020.

3. Lee, B.S.; Kwak, S.Y. Experience of spiritual conflict in hospice nurses: A phenomenological study. *J. Korean Acad. Nurs.* **2017**, *47*, 98–109. [CrossRef]
4. Kelley, A.S.; Morrison, R.S. Palliative care for the seriously ill. *N. Engl. J. Med.* **2015**, *373*, 747–755. [CrossRef]
5. Kim, B.H.; Choe, S.O.; Chung, B.Y.; Yoo, Y.S.; Kim, H.S.; Kang, K.A.; Yu, S.J. Job analysis for role identification of general hospice palliative nurse. *Korean J. Hosp. Palliat. Care.* **2010**, *13*, 13–23. [CrossRef]
6. Yang, S.A. Convergence analysis of the factors influencing terminal care attitude. *J. Korea Converg. Soc.* **2015**, *6*, 73–88. [CrossRef]
7. Ali, W.G.M.; Ayou, N.S. Nurses' attitudes toward caring for dying patient in Mansoura university hospital. *J. Med. Biomed. Sci.* **2010**, *1*, 16–23.
8. Ko, M.J.; Moon, S.H. Factors influencing performance of end-of life care by ICU care. *J. Korean Acad. Psychiatry Ment. Health Nurs.* **2016**, *25*, 327–337. [CrossRef]
9. Noh, S.S.; Lee, C.K.; Sung, Y.H. Predictors of terminal care performance of clinical nurses for cancer patients. *J. Korean Crit. Care Nurs.* **2016**, *9*, 61–70.
10. Kwak, S.Y.; Lee, B.S. Role adaptation process of hospice nurses. *J. Korean Acad. Nurs. Adm.* **2018**, *24*, 149–160. [CrossRef]
11. Park, J.H. Development and validation of nurse's character scale for care in clinical settings. *J. Korean Acad. Soc. Nurs. Educ.* **2016**, *22*, 137–151. [CrossRef]
12. Kim, E.H.; Lim, Y.M. The mediating effect of compassionate competence on the relationship between nurses' spirituality and attitude toward care of dying patients. *J. East-West Nurs. Res.* **2019**, *25*, 166–173. [CrossRef]
13. Seol, E.M.; Koh, C.K. Experiences of critical care nurses caring for dying patients. *J. Korean Crit. Care Nurs.* **2018**, *11*, 1–10.
14. Frommelt, K.H.M. The effects of death education on nurses' attitudes toward caring for terminally ill persons and their families. *Am. J. Hosp. Palliat. Med.* **1991**, *8*, 37–43. [CrossRef]
15. Cho, H.J.; Kim, E.S. The effect of the death education program on the death anxiety and attitudes toward nursing care of the dying patients of nursing student. *Korean J. Hosp. Palliat. Care.* **2005**, *18*, 163–172.
16. Park, S.J.; Choi, S.H. A study on the degrees of death orientation and terminal care performance of nurses. *J. Korean Acad. Fundam. Nurs.* **1996**, *3*, 285–297.
17. Chung, S.Y. A Study of Terminal Care Stress and Terminal Care Performance of ICU Nurse. Master's Thesis, Hanyang University, Seoul, Korea, 2013.
18. Kim, S.J. Death anxiety and terminal care attitudes of nurses in hospice palliative units. *Health Nurs.* **2015**, *27*, 23–34.
19. Washington, K.T.; Parker Oliver, D.; Smith, J.B.; McCrae, C.S.; Balchandani, S.M.; Demiris, G. Sleep problems, anxiety, and global self-rated health among hospice family caregivers. *Am. J. Hosp. Palliat. Med.* **2018**, *35*, 244–249. [CrossRef] [PubMed]
20. Yom, E.Y.; Seo, K. Influences of interpersonal problems and character of nurses on quality of nursing service. *J. Korean Acad. Nurs. Adm.* **2018**, *24*, 445–454. [CrossRef]
21. Han, S.; Oh, J.; Huh, B.; Kim, H. Mediating effects of resilience between nurse's character and happiness for nurses in general hospitals. *J. Korean Acad. Nurs. Adm.* **2020**, *26*, 22–30. [CrossRef]
22. Shim, J.Y.; Lee, S. Death anxiety and needs of interpersonal care behaviors of nurses in hospice wards: Focused on types of personality. *J. Korea Contents Assoc.* **2021**, *21*, 737–745. [CrossRef]
23. Lee, L.J.; Park, H.S. Death anxiety and terminal care performance of nurses at long term care hospitals. *Korean J. Hosp. Palliat. Care.* **2017**, *20*, 37–45. [CrossRef]
24. Jason, T.H. Mindful self-care and secondary traumatic stress mediate a relationship between compassion satisfaction and burnout risk among hospice care professionals. *Am. J. Hosp. Palliat. Med.* **2018**, *35*, 1099–1108. [CrossRef]
25. Park, E.; Bang, K.D. The relationship between terminal care stress and knowledge and perception of hospice-palliative care among pediatric nurses. *Perspect. Nurs. Sci.* **2019**, *16*, 55–64. [CrossRef]

healthcare MDPI

Article

Comparison of Academic Motivation between Business and Healthcare Students in Online Learning: A Concurrent Nested Mixed-Method Study

Yasuhiro Kotera [1,*], Valentina Gorchakova [2], Sarah Maybury [3], Ann-Marie Edwards [3] and Hiromasa Kotera [4]

[1] School of Health Sciences, University of Nottingham, Nottingham NG7 2TU, UK
[2] College of Business, Law and Social Sciences, University of Derby, Derby DE22 1GB, UK
[3] College of Health, Psychology and Social Care, University of Derby, Derby DE22 1GB, UK
[4] Department Linguistics, University of Potsdam, 14476 Potsdam, Germany
* Correspondence: yasuhiro.kotera@nottingham.ac.uk

Abstract: While the demand for online education and the diversity of online students have been increasing worldwide, how online students motivate themselves to continuously engage in learning remains to be appraised. Research in the face-to-face contexts reports that academic motivation is central to student success and wellbeing, and the type of motivation can differ by subject. In particular, the motivation of business students and healthcare students can differ considerably. This study aimed to understand the motivation of online students, and compare them between business and healthcare students using a concurrent nested mixed-method design with correlation and thematic analyses. A survey regarding motivation, learning enjoyment, and study willingness was responded to by 120 online students (61 business and 59 healthcare). Business students were associated with extrinsic motivation, whereas healthcare students were associated with intrinsic motivation. While students in both groups enjoyed the pursuit of knowledge, healthcare students valued the process and accomplishment, whereas business students regarded education as steppingstones in their careers. Findings can help educators develop effective motivational support for these student groups.

Keywords: academic motivation; online students; intrinsic motivation; extrinsic motivation; mixed-method

Citation: Kotera, Y.; Gorchakova, V.; Maybury, S.; Edwards, A.-M.; Kotera, H. Comparison of Academic Motivation between Business and Healthcare Students in Online Learning: A Concurrent Nested Mixed-Method Study. *Healthcare* **2022**, *10*, 1580. https://doi.org/10.3390/healthcare10081580

Academic Editor: Florin Oprescu

Received: 25 July 2022
Accepted: 18 August 2022
Published: 19 August 2022

1. Introduction

Online learning has gained significant prominence in higher education institutions (HEIs) over the past two decades, with many incorporating this into their long-term strategies [1,2]. Several advantages are associated with online learning for students, namely reduced time and geographical constraints, increased control over class pace and engagement, and fewer costs associated with them, i.e., on commuting or relocation, altogether providing a more flexible and attractive format for many [3,4]. The nature of online learning requires students to possess efficient time management skills and the ability to self-direct their learning [5]. These benefits extend to HEIs, who can reach a broader population of students and those who would otherwise be unable to access traditional face-to-face higher education [2,6].

Despite the noted unique advantages, some challenges have been reported. The reduced peer and teacher interaction can cause motivation issues [7–9]. In addition, as noted by Curasi and Burkhalter [10], the profile of online students is becoming increasingly untraditional, including married students with children, and students who are employed full-time. These students are less able and less motivated to dedicate additional time to course materials to complete the course, as their outside lives tend to be prioritised over their studies [10]. Additionally, a lack of belongingness to the university or a learning community sabotages their motivation [2], accompanied by a strong sense of loneliness [11].

These factors could help explain the poor performance and high attrition rates associated with online learning [12].

More specifically, online business students and healthcare students face motivational challenges. Pre-COVID studies reported that although many business students appreciated the self-paced nature of online learning (e.g., fitting studies in their work life), a lack of personal support experienced in online learning relative to face-to-face learning was challenging to business students especially those who were new to online learning [13,14]. During the COVID-19 pandemic, business studies were relatively unaffected by the forced transition to online, because many components of their curriculum were adaptable to online learning [15]. However, a lack of the social aspect of learning (e.g., discussion and conversation with lecturers and peers) was demotivating for many business students [16] Likewise, in pre-COVID research, online healthcare students reported their motivation was compromised in group work where good engagement from all group members was required [17]. To counter this problem, educators implemented measures such as mandatory attendance to online lectures and activities. However, these measures could counteract the self-paced nature of online learning, disadvantaging those with a job with irregular working hours, career duties, and living in a different time zone [18]. Unsurprisingly online students' intrinsic motivation for asynchronous learning was higher than for synchronous learning, relating to a sense of autonomy [19]. These challenges were intensified during the COVID-19 pandemic: many students' duties outside the curriculum have increased (e.g., deployment to hospital, obscured boundary between study and life [20], making it hard for them to stay motivated to study [21,22]. The forced transition to online learning was experienced positively by healthcare students who were undertaking desk-based learning [23], whereas those in clinical practice experienced it negatively as their practice was interrupted [24]

Understanding academic motivation continues to be of importance to educators, as a necessary function of maintaining student engagement and retention, and to encourage positive academic outcomes [25]. This has become even more pronounced in the context of the global COVID-19 pandemic, which continues to disrupt conventional teaching methods, forcing many HEIs to provide online instruction, even for those who did not originally enroll in online learning courses [5]. The demand for online learning has clearly increased worldwide.

One well-established theory of motivation is Deci and Ryan [26] Self-Determination Theory (SDT), a model that seeks to explain human needs, motivation, and wellness in various social contexts [12]. SDT argues that autonomy, competence, and relatedness are critical components of motivation [27]. It distinguishes between three types of motivation: (1) intrinsic motivation, carrying out an activity based on pleasure or enjoyment, (2) extrinsic motivation, behaviours undertaken for reasons beyond innate satisfaction, and (3) amotivation, the absence of intentionality [28]. Within the context of education, the focus of SDT is on promoting student self-confidence, the value of education, and a desire to learn [29]. Both intrinsic and extrinsic motivation act as a mediating role for academic performance through online learning behaviours [30]. Those are essential components of education [29], indicating the relevance and importance of motivation in education.

The findings of Griffin et al. suggest that the level of intrinsic motivation demonstrated by students constitutes the most powerful indicator of student success. Students with a high level of intrinsic motivation tend to exhibit increased engagement in classes, experience positive self-esteem and a sense of subjective wellbeing, and excel academically [31]. In contrast, the predominance of extrinsic motivation can result in negative outcomes such as boredom, poor class attendance, and academic performance, and amotivation is negatively associated with engagement, learning potential, and wellbeing [28,32]. This is important as many students have adopted a consumerist approach to higher education, where extrinsic motivators (i.e., grades) take precedence, fostering an environment of shallow cognitive engagement [29,33]. Not only can motivation be a powerful predictor of academic achievement, but it also can impact decisions to commence and continue specific learning behaviours, helping students become successful professionals [34]. Moreover, intrinsic motivation is a known predictor of student psychological well-being, a significant determinant of the decision to remain enrolled in their course [35,36]. The positive impacts of intrinsic motivation are diverse.

Educators and researchers have begun to pay attention to interventions to support students' intrinsic motivation. However, this has encountered several challenges due to educational goals and motivation types that differ across academic disciplines [37,38]. Those disciplines that are perceived as 'hard', such as business, management, and technology, place more emphasis on the application of what is learned versus the knowledge itself, attracting students with high levels of extraversion, linked to high extrinsic motivation [37,39]. For example, although some shift towards intrinsic motivators such as meaning is reported among business students [40], a business degree is highly sought due to the reputation of high earnings potential and elevated social image [39,41]. Additionally, it is thought that intrinsic motivation is stimulated in some contexts in business education [42]. Cheng et al. [43] highlighted that in the context of business students pursuing personal certifications, the prevalent use of rewards (i.e., monetary) acts as reinforcers of extrinsic motivation and overshadows the potential for intrinsic motivators, such as a sense of belonging and ability. Furthermore, in an investigation of motivation in business students in India undertaking an MBA, the brand name of an institution was a main motivator based on its relevance to career prospects [44]. These studies share a common aspect whereby the pursuit of a business degree is perceived as a necessary step to reach the desired goal of a career and life in the corporate world, rather than the standalone merits that a business education may provide.

On the other hand, while some extrinsic rewards do act as an attractor [45], many healthcare students, categorised as 'soft' disciplines, are inspired by intrinsic motivators such as an altruistic desire to help others [46,47]. For example, nursing is often seen as a 'calling' rather than a job, where the opportunity to care for others, being part of a team, and a sense of pride in the services provided are cited as primary drivers of pursuing a nursing degree and subsequent career [48,49]. When examining motivation in students taking nutrition classes, Maurer et al. [29] found these students to be principally driven by intrinsic motivation, specifically the contribution that the course would have to their understanding of their own health status. In a study examining career choice and motivation of medical students and the potential impact on their academic performance, almost 75% of participants cited reasons that related to intrinsic motivation [35]. These reasons predominantly included intellectual curiosity, and experience of illness or death either personally or within their family Kim et al. [35], highlighting the relevance of intrinsic motivation in healthcare students.

These intrinsic drivers may become threatened when more external rewards are used [50]. External rewards such as money 'crowd out' intrinsic motivation, shifting the emphasis from engaging in a task for internal feelings of self-determination, to obtaining the reward itself [51,52]. Furthermore, while intrinsic motivation and extrinsic motivation are not dichotomous (i.e., one individual can have high levels of both intrinsic and extrinsic motivation), often a larger external reward could spoil intrinsic motivation, leaving extrinsic motivation as the dominant driver of behaviour [53]. The implications of the predominance of extrinsic motivation in students extend beyond academic achievement and success. In a study investigating approaches to recruiting peer mentors, Anghelcev and Eighmey [54] found that students receiving no monetary reward were the most willing to become peer mentors, followed by those receiving a low reward, and those receiving the highest monetary value were the least willing to participate. Moreover, the use of external rewards such as a desire for recognition has implications for enjoyment. Children who are attached to parents' praises may lose a sense of enjoyment in an originally enjoyed behaviour [55], whereas nurturing a sense of competence and autonomy (components of intrinsic motivation) in students can lead to elevated feelings of enjoyment towards a certain task and may promote positive cognitive and behavioural outcomes [56]. Despite these findings, researchers remain divided over the impact of external rewards [57]. Therefore, it is important to evaluate whether a) students enjoy studying the subject (enjoyment), and b) students would still study in the programme if they were not to receive a grade (willingness), and how those are related to each type of motivation.

Considering the significance of furthering our understanding of motivation in learning, it is crucial to explore the academic motivations of students across different academic disciplines, particularly in the context of online learning. Accordingly, this study sought to add to the literature on the similarities and differences between the types of motivation among healthcare and business online students. Four research questions were established.

RQ1: Are the levels of each type of motivation different between business and healthcare students?

RQ2: How is each type of motivation related to each other?

RQ3: How is each type of motivation related to learning enjoyment and willingness to study without any external reward?

RQ4: What are the similarities and differences in key academic motivations and goals of business and healthcare students?

2. Materials and Methods

2.1. Study Design

This study employed a concurrent nested mixed-method design [58] Qualitative findings (secondary focus) were used to complement the quantitative findings (primary focus), to gain a better understanding of the students' motivation.

2.2. Study Context

The study was conducted at an online department of a university in the United Kingdom, where almost all students are working professionals. Undergraduate business and healthcare students were recruited from January to June 2021 using an online survey link embedded in the programme announcements delivered by an academic who was not a co-author of this article. The survey included quantitative and qualitative questions. About 300 students were approached in each discipline.

2.3. Participants and Recruitment

One hundred and twenty students completed the survey in total. Sixty-one business students (36 females and 25 males; Age M = 33.59, SD = 10.56, Range 18–57 years) consisted of 31 UK students and 30 international students (11 other Europeans, 6 Asians, 5 Africans, and 1 Oceanian: 7 did not report). All business students were enrolled in a business and management programme. Fifty-nine healthcare students (48 females and 11 males; Age M = 45.28, SD = 11.24, Range 25–70 years) consisted of 38 UK students and 21 international students (7 other Europeans, 4 Africans, 2 Asians, and 1 Oceanian: 7 did not report). Thirty-six students were in counselling, 17 in nursing, 4 in social care and 2 in perioperative practice programmes. As reliable data were not available, the representativeness of our samples with the general populations of business and healthcare students was not certain, however, our samples' demographics were roughly comparable with the entire cohorts [59,60]. Demographic data were summarised in Table 1.

2.4. Data Collection

The research ethics committee at the university approved this study. Consent was received from all participants before participation. Participants were informed at the beginning of the study that (a) the participation was arbitrary, (b) they were allowed to withdraw at any time just by closing the browser, (c) reasons for withdrawal would not be asked, and (d) all responses would be anonymised, and (e) the study might be published or presented for academic purposes. Should a participant feel distressed while completing the questionnaire, contact details of the student wellbeing services in the university were introduced.

Table 1. Demographic data of participants.

Demographics	Business Students (*n* = 61)	Healthcare Students (*n* = 59)
Gender: *n* (%)		
Female	36 (59%)	48 (81%)
Male	25 (41%)	11 (19%)
Age		
Mean	33.59	45.28
SD	10.56	11.24
Range	18–57	25–70
Residence: *n* (%)		
UK (Home)	31 (51%)	38 (64%)
Other Europe	11 (18%)	7 (12%)
Asian	6 (10%)	2 (3%)
African	5 (8%)	4 (7%)
Oceanian	1 (2%)	1 (2%)
Unanswered	7 (11%)	7 (12%)
Programme: *n* (%)		
Business and Management	61 (100%)	
Counselling		36 (61%)
Nursing		17 (29%)
Social Care		4 (7%)
Perioperative		2 (3%)

2.5. *Instruments*

2.5.1. Quantitative

The quantitative component of the survey consisted of the Academic Motivation Scale (AMS; [61]) and two additional yes/no questions.

AMS is a 28-item scale that evaluates the levels of three types of motivation, which are further categorised into seven subtypes measured by four items each: (A) amotivation, (B) extrinsic motivation (external, introjected, and identified regulation), and (C) intrinsic motivation (to know, to accomplish, and to experience stimulation). Example items include 'I can't see why I go to college and frankly, I couldn't care less' for amotivation, 'In order to have a better salary later on' for extrinsic motivation, and 'For the pleasure I experience when I discover new things never seen before' for intrinsic motivation, which are responded on a seven-point Likert scale (from 1 = 'Does not correspond at all' to 7 = 'Corresponds exactly'). All of the subscales have adequate to high reliability with Cronbach's alphas between 0.62 and 0.91 [61].

Based on the assumptions of SDT, the Academic Motivation Scale (AMS) was formulated by Vallerand et al. (1992) [61] as an instrument to measure intrinsic motivation, extrinsic motivation, and amotivation in students. While the AMS has been extensively researched to investigate motivation in high school and college generally, more work can be conducted to apply the AMS to specific disciplines [29]. Therefore, the AMS was used in the current study to assess motivation levels among participants.

The two yes/no questions were 'Do you enjoy the subject by itself?' and 'If no rewards are given (just pure joy of learning experience), would you still study?'. In the analysis, 'yes' was coded as '1', and 'no' as '0'.

2.5.2. Qualitative

The seven items in the qualitative component were (1) 'Why did you decide to study online?', (2) 'What motivates you to engage in academic activities?', (3) 'What is/are your main goal(s) of study?', (4) 'What do you want to achieve in this online programme?'. (5) 'What motivates you to engage in learning?', (6) 'What hinders you from engaging in learning?', and (7) 'Is there anything else that you feel I should have asked, or that you would like to add?'.

2.6. *Analyses*

To respond to RQ1–3, a correlation analysis was conducted. The point biserial correlation coefficient was used for the students' discipline (0 = Healthcare, 1 = Business) to address RQ1. For the qualitative data (RQ4), thematic analysis was conducted to capture the themes in the participants' text responses. Following Braun and Clarke's guidelines [62,63], the text responses were read several times, and codes were generated (Table 2), which led to the creation of themes (Table 3). Themes were reviewed by all co-authors and reached an agreement.

Table 2. Generating initial codes (examples).

Focus Area	Initial Codes
Healthcare Students	Self-improvement Pleasure of learning Challenging oneself Intellectual curiosity Staying current Sense of discovery Empowerment Adapting to change Self-reflection Professional development Unlimited accomplishment Deeper understanding
Business Students	Increased success International recognition Perceived competence Screening for employers Enhanced credibility Pay rises Sense of satisfaction Contingency plan Professional qualifications Foundation for career Potential promotion Social capital

Table 3. Themes and examples of participant comments.

Themes	Example of Participant Comment
T1: Striving for the Pursuit of Knowledge	I have not been rewarded for my learning because I do this for myself. I would still study without rewards, after all studying itself is rewarding (HCS). Absolutely. Learning is a gift. Knowledge is power and the one thing that no one can take away. There is pride and joy in learning (BS).
T2: Learning Process and Second Chance in Healthcare Students	It feels really good to learn something that is useful, it doesn't cause me to feel more important than people who don't have qualifications: it is more I am fulfilling my own potential no matter how old I am (HCS).
T3: Education as Steppingstone for Business Students	I felt inferior to all of my friends and colleagues who are university graduates (BS).

HCS = Healthcare Students; BS = Business Students.

3. Results

3.1. Quantitative Analysis

No outliers were identified. P-p plots demonstrated that the quantitative data were normally distributed. The levels of each type of motivation were summarised in Table 4. The reliability of each motivation type was high in both student groups (α = 0.75–0.88 in business, α = 0.72–0.84 in healthcare). Pearson correlation was performed (Table 5).

Table 4. Levels of each type of motivation in business students (n = 61) and healthcare students (n = 59).

	Business Students			Healthcare Students		
	M	**SD**	α	**M**	**SD**	α
IMK	5.45	1.17	0.88	5.99	0.94	0.79
IMA	5.11	1.46	0.87	5.29	1.18	0.75
IMS	3.73	1.40	0.86	4.42	1.28	0.72
EMID	5.16	1.33	0.75	5.10	1.45	0.75
EMINT	4.72	1.63	0.85	4.50	1.73	0.82
EMER	4.73	1.73	0.86	4.00	1.74	0.84
AM	1.39	.86	0.82	1.31	0.73	0.78

IMK = Intrinsic motivation to know. IMA = Intrinsic motivation to accomplish. IMS = Intrinsic motivation to experience stimulation. EMID = Extrinsic motivation for identified regulation. EMINT = Extrinsic motivation for introjected regulation. EMER = Extrinsic motivation for external regulation. AM = Amotivation.

Table 5. Correlation between the discipline, gender, age, nationality, types of motivation, learning enjoyment, and willingness to study without reward among online students (n = 120).

		1	2	3	4	5	6	7	8	9	10	11	12	13
1	Discipline (0 = HC, 1 = Biz)	–												
2	GN (0 = F, 1 = M)	0.23 *	–											
3	Age	−0.46 **	−0.15	–										
4	Nationality (0 = UK, 1 = Overseas)	0.14	0.24 **	−0.17	–									
5	IMK	−0.24 **	−0.13	0.24 **	−0.13	–								
6	IMA	−0.07	−0.21 *	0.19 *	−0.22 *	0.67 **	–							
7	IMS	−0.25 **	−0.10	0.29 **	−0.02	0.61 **	0.61 **	–						
8	EMID	0.04	0.05	−0.23 *	0.11	0.21 *	0.18 *	0.20 *	–					
9	EMINT	0.033	−0.10	−0.04	−0.23 *	0.22 *	0.56 **	0.36 **	0.30 **	–				
10	EMER	0.19 *	0.03	−0.31 **	0.15	−0.02	0.19 *	0.12	0.76 **	0.44 **	–			
11	AM	0.05	0.27 **	−0.14	0.12	−0.29 **	−0.29 **	−0.02	0.02	−0.07	0.08	–		
12	Enjoy (0 = No, 1 = Yes)	−0.07	0.05	0.06	0.01	0.24 *	0.10	0.07	0.03	0.03	−0.07	−0.06	–	
13	Willingness (0 = No, 1 = Yes)	−0.30 **	−0.15	0.12	0.07	0.38 **	0.20 *	0.23 *	−0.14	−0.13	−0.23 *	−0.09	0.20 *	–

HC = Healthcare. Biz = Business. IMK = Intrinsic motivation to know. IMA = Intrinsic motivation to accomplish. IMS = Intrinsic motivation to experience stimulation. EMID = Extrinsic motivation for identified regulation. EMINT = Extrinsic motivation for introjected regulation. EMER = Extrinsic motivation for external regulation. AM = Amotivation. Enjoy = Whether they enjoy learning the subject. Willingness = Whether they would be still willing to study the subject if there were no reward given for completing the programme. * $p < 0.05$, ** $p < 0.01$.

Students' discipline was negatively associated with intrinsic motivation to know (IMK), intrinsic motivation to experience stimulation (IMS) and no reward, while positively associated with extrinsic motivation for external regulation (EMER), suggesting that healthcare students had significantly higher levels of IMK, IMS and willingness to study without reward, and lower EMER than business students (RQ1). Overall, intrinsic motivation and extrinsic motivation were positively interrelated, whereas extrinsic motivation and amotivation were not (RQ2). Lastly, only IMK was associated with learning enjoyment among all types of motivation. All three types of intrinsic motivation (to know, to accomplish, and to experience stimulation) were positively associated, and EMER was negatively associated with willingness to study without reward (RQ3).

3.2. Qualitative Analysis

Thematic analysis on responses of business and healthcare students has revealed three prominent themes, based on one similarity and two differences in the motivating drivers to engage in academic studies and in learning in general (Table 2). The motivations shared by students of both disciplines included the following.

Striving for the Pursuit of Knowledge (T1)

For many healthcare students, the opportunity to learn more and enhance their understanding of certain subjects, i.e., intrinsic rewards, was both a major motivating factor in engaging with course tasks, and their desired outcome: The prospect of obtaining a reward was perceived as secondary.

> *HCS (Healthcare Student): I have not been rewarded for my learning because I do this for myself. I would still study without rewards, after all studying itself is rewarding.*

This was also the case for quite a few business students, one expressing with certainty their willingness to study without the prospect of reward or recognition.

> *BS (Business Student): Absolutely. Learning is a gift. Knowledge is power and the one thing that no one can take away. There is pride and joy in learning.*

One healthcare student expressed a degree of opposition towards the concept of receiving rewards for their studies, viewing this as a hindrance to the quality of their learning experience.

> *HCS: In some ways, I find it distracting to have to study with a specific agenda in mind, rather than just being able to explore the materials at my leisure and to discover for myself the parts that are relevant and interesting to me.*

These comments suggest that strong orientation towards intrinsic motivation in healthcare students. On the other hand, business students, who were employed, were often enrolled in the course solely on the request of their employers who were providing the funds, and therefore had conditions attached to them. Moreover, some business students noted that consideration for certain jobs and promotions were contingent on obtaining specific degrees, which can lead to high extrinsic motivation.

> *BS: My employer is paying my fees which motivates me.*
>
> *BS: I did it as a part-sponsored degree within my work to help me achieve a promotion and better prospects for the future.*

Contrary to healthcare students, these comments by business students indicate a strong connection between their degree and employment. That said, there were some responses from healthcare students which demonstrated clear extrinsic motivation, whether to continue their professional development, or to work towards a successful career.

> *HCS: To supplement my existing undergraduate, post-graduate, and master's level qualifications with a further accreditation relevant to my chosen profession.*
>
> *HCS: Future career development, to be more competent, to be internationally recognised, for more salary, etc.*

These responses indicate that many business- and healthcare-related degrees are structured for the purpose of employment, and to fit the needs of employers. When combined with the costs of obtaining a degree, the prevalence of extrinsic motivation in participant responses can somewhat be expected.

Responses provided by both business and healthcare students also suggest the fluidity of motivation: intrinsic and extrinsic motivation can change. While some students enrolled in their course specifically for practical (extrinsic) considerations, they noted that over time they began to value the course for the content alone, and the opportunity to learn.

> *HCS: It was initially to get promotion as BSc Nursing was needed to be promoted to level 13, in the place I work some have been in level 12 for 11–12 years... But now I found academics very interesting though can be difficult and challenging.*

Moreover, another participant highlighted a contextual impact on motivation. Their motivation was shaped by their current circumstances and responsibilities, noting that once they were in a better position, they would study for the passion of learning itself.

> *BS: I do enjoy learning new things, but at the moment my focus is on obtaining the end reward of a bachelor's degree. However, in the future, once I am in a comfortable working position and have achieved what I wanted, I would continue learning for the learning experience.*

Learning Process and Accomplishment in Healthcare Students (T2)

Many healthcare students noted that the process of learning and a sense of accomplishment as key motivators. Not only is the knowledge gained from a university education a main motivator for students, but the different steps that comprise the learning process was another major theme for healthcare students.

> *HCS: I am not only exercising my physical muscles but also my mental and cognitive abilities give me an immeasurable amount of satisfaction.*

> *HCS: I feel a sense of achievement in the process [of] taking part and pushing my comfort zones.*

These comments suggest that the opportunity to be in an environment where they can challenge themselves while developing their mental acuity is their major motivator. The learning process is personal, acting as a boost to their self-esteem, rather than an instrument to prove their value over others.

> *HCS: It feels really good to learn something that is useful, it doesn't cause me to feel more important than people who don't have qualifications: it is more I am fulfilling my own potential no matter how old I am.*

The majority of online healthcare students at our institution are registered professionals in employment, and many of them have other non-academic commitments (e.g., childcaring). Their life context is often more restricted than typical young students who have just graduated from high school to afford an education. Therefore, online healthcare students reported a sense of accomplishment to manage their studies in online learning.

Their degree provides them with a sense of satisfaction. Their motivation to engage with their course and desired outcome were intrinsically oriented.

> *HCS: Feelings of validation in my abilities [are my motivation to study].*

> *HCS: To fulfill my potential which did not happen while I was at school. To be a "good" role model to my two daughters.*

Taken together, many healthcare students enjoy the process of learning to expand their knowledge and skills, as well as a sense of accomplishment derived from progressing their studies (i.e., intrinsic motivators) while managing other duties.

Education as Steppingstone for Business Students (T3)

Many business students perceived university education as a steppingstone to advancing their careers. While they acknowledged the importance of the learning process, they also noted the social value attached to getting a university education.

A university education was treated as an element of social capital, providing a sense of membership and common understanding based on shared experience. Being a university graduate comes with baseline assumptions in relation to intelligence level, organisational and time management skills, interpersonal skills, and to some degree socioeconomic background. Therefore, those who do not fall under the category of university graduate, may be made to feel like an outsider.

> *BS: [I] felt inferior [to] all of my friends and colleagues [who] are university graduates.*

> *BS: My full degree to enable me to apply for certain roles that require minimum 2:1.*

> *BS: To have a job where I can earn a good salary.*

> *BS: Get the necessary UK qualification to enter/access a new work sector.*

Online business students, who already worked in business at the time of the study, expressed feelings of inadequacy because, unlike their colleagues, they do not have a university degree. Therefore, the motivation at the forefront for them appears to be utilising their degree as a resource to achieve a level playing field with their social circle. Similarly, other business students noted a desire to demonstrate their ability by obtaining a degree. In response to the question around the primary motivation for enrolling in the course and engaging in their unit activities, two business students provided the following answers.

BS: To achieve higher grades and show my work to my peers.

BS: Hmm . . . like the letters after my name to be honest.

These comments suggest that business students regard university education as a steppingstone to enhance their credibility, and being more accepted in their business community.

3.3. Synthesis

The findings in our quantitative analyses and qualitative analyses were roughly consistent with each other: healthcare students were, in general, more intrinsically oriented than business students, and intrinsic motivation was associated with learning enjoyment. Our qualitative analyses deepened these findings that both groups were motivated to pursue knowledge, however healthcare students enjoyed the learning process whereas business students somewhat regard education as a steppingstone in their career.

4. Discussion

The purpose of this study was to evaluate the academic motivation of online business and healthcare students. Our results show that (RQ1) healthcare students had a higher level of intrinsic motivation, whereas business students had a higher level of extrinsic motivation, (RQ2) intrinsic motivation and extrinsic motivation were positively inter-related, (RQ3) intrinsic motivation was associated with both enjoyment and intention to study, and (RQ4) while students in both groups enjoyed the pursuit of knowledge, healthcare students valued the learning process and accomplishment, whereas business students regarded education as a steppingstone.

One novel finding of our mixed-method analyses may be that although the demographic data were different between online students and face-to-face students (e.g., age and life stage), their motivational characteristics were similar. Previous face-to-face research reported a high level of extrinsic motivation in business students, and a high level of intrinsic motivation in healthcare students [36,64–66]. We found the same motivation characteristics were present in online students. Moreover, we further identified their perceptions towards learning in relation to motivation: healthcare students were process-oriented, whereas business students were more focused on their career steps. These findings have several implications. For example, to help maintain academic motivation, educators can focus on the learning process in healthcare programmes whereas educators in business can link their curriculum to student career steps.

High intrinsic motivation was associated with healthcare students, whereas high extrinsic motivation was associated with business students. Healthcare students enjoyed the learning process and accomplishment, both of which are rather intrinsic rewards. On the other hand, business students consider education as a steppingstone, an extrinsic reward. In line with the face-to-face research [39], academic motivation was different between these two groups in the online context: in the face-to-face setting, business students were more associated with extrinsic motivation, whereas healthcare students were more associated with intrinsic motivation as well. Moreover, this trend is present when these students have become professionals/employees [52,67–69]. These findings suggest that securing a prestigious and highly salaried position is the primary driver for choosing a business education programme [70], whereas healthcare students are more likely to study a healthcare subject out of personal interest or enjoyment [35], leading to improved learning outcomes. These characteristics were deemed similar (a) between online and face-to-face education, and (b) between pre-graduation and post-graduation. Indeed, many healthcare students see working in the healthcare setting as a vocation rather than a job, with a sense of calling to a particular healthcare profession, the basis of patient-centred and compassionate care [71]. A study conducted by Wu et al. (2020) [72] found that intrinsic motivation was significantly and positively associated with self-efficacy, learning engagement, and academic performance in medical students. This suggests that engaging with student/employee motivation should be conducted over time, including when they are preparing to work, and when they are working.

Considering these positive associations of intrinsic motivation, business students could benefit from shifting their extrinsic motivation to an intrinsic one. Cognitive evaluation theory informs that incorporating engaging learning activities such as reading stimulating materials and self-awareness diaries helps students focus on their grades to their learning [73]. Moreover, enhancing self-compassion may be helpful too as self-compassion assists the extrinsic to intrinsic shift [52,73]. These are consistent with the significant interrelations between intrinsic and extrinsic motivation. Business educators need to embed this training and education in their curriculum to prevent business students from experiencing the negative consequences of extrinsic motivation. It is also essential to mention that intrinsic motivation can be harmful to students if their passion is obsessive (based on uncontrollable compulsion) instead of harmonious (balanced engagement; [65,74]). Self-awareness needs to be encouraged among healthcare students [75,76]. Moreover, during the pandemic, a sense of loneliness was identified as a debilitating factor for healthcare students' well-being and motivation [77]. To improve motivation (i.e., to cope with amotivation) psycho-social support was recommended to augment a sense of connection and belongingness for healthcare students [78,79]. Post-traumatic growth, which many healthcare students are related to, due to traumatic events in their practice, is associated with intrinsic motivation, and career support was recommended to support their intrinsic motivation [80–82]. In sum, both extrinsic motivation and intrinsic motivation can contribute to academic performance, and these types of motivation can change across different phases of the curriculum influenced by various intrapersonal and interpersonal factors [83,84]. Further understanding of motivation among university staff is recommended.

Significant associations between intrinsic motivation and enjoyment, and willingness (and no significant associations between extrinsic motivation and enjoyment, and willingness) were consistent with the SDT that students who are passionate about the subject tend to enjoy studying and be more willing to study without external rewards. Especially, the willingness to study was significantly associated with healthcare students, which can be further explained using the qualitative findings.

Our qualitative analyses helped to understand our quantitative findings more holistically. Both student groups enjoyed learning, which is related to the intrinsic motivation to maintain the desire to continue the learning process [85]. Students who are intrinsically motivated are self-driven, challenged, and find enjoyment in their studies [86]. Student involvement in the learning process increases when students find study engaging, effective, and valuable [87]. Online professional students who chose to engage in a further degree have intrinsic motivation to enjoy the pursuit of knowledge.

The different perspectives on learning between healthcare and business students deepened our quantitative findings: healthcare students focused on the process and a sense of accomplishment from learning, whereas business students considered their career ahead and regarded education as a steppingstone. Generally speaking, the process of and accomplishment in learning are pertinent to intrinsic motivation, whereas a focus on further career steps is more pertinent to extrinsic motivation [28]. However, healthcare educators may need to be mindful that, as noted above, an obsessive need for accomplishment can trigger extrinsic motivation. If the standard for accomplishment is set extrinsically, social comparison can take place, leading to a more negative evaluation of oneself [88]. Self-awareness training may be helpful for healthcare students to remain intrinsic towards their learning [89]. On the other hand, for business students, how to bring their external focus inward would be more important. Interventions such as life crafting or ikigai (a Japanese concept roughly translated as a meaning in life) approaches may be helpful as these ask students about the intrinsic meaning of their career goals [90,91]. Future research needs to evaluate the impact of those approaches on students' motivation and wellbeing.

While this research provides valuable insight into the motivation of business and healthcare students, it is important to keep the limitations in mind when interpreting the results. First, in addition to the modest size, our sample was recruited through convenience sampling from one academic institution in the UK, limiting our findings' generalisability. Moreover, considering the wide impact of COVID-19 on academic motivation, students in other subjects should be evaluated. Secondly, a self-report survey was used, thus response biases might have been present [92]. Additionally, a shorter version of the Academic Motivation Scale [93] could have been used to reduce the workload of participants, improving the accuracy and response rate. Moreover, in the qualitative part, participants' emotional tone was not captured in this study. Other possible motivational factors such as the employment status [94] were not considered in this study. Relatedly, the impact of COVID-19 was not assessed. These two factors could have a meaningful impact on our findings. For example, an unstable employment status negatively affected people's psycho-social well-being [95]. During the pandemic, university students' academic motivation was compromised, as their sense of belonging declined [96]. Future research needs to consider these factors. Lastly, the study does not examine whether students had specifically chosen to study online or whether e-learning was the only option available to them [97]. Future research needs to address these limitations to identify more robust evidence for online students' motivation.

5. Conclusions

Academic motivation is essential for student success, however, it remains to be appraised in online education, which has been gaining importance today. In particular, business and healthcare students learning online encounter some motivational challenges. Our study identified that business students were extrinsically motivated to regard education as a steppingstone, whereas healthcare students were intrinsically motivated to focus on the process and accomplishment. Self-compassion and meaning-oriented interventions may be helpful for business students, and self-awareness training, psycho-social support and career guidance may be useful for healthcare students to support their wellbeing. Our findings will help educators and well-being services develop effectively, and tailor mental health care in those student groups.

Author Contributions: Conceptualization, Y.K. and V.G.; methodology, Y.K. and H.K.; software, Y.K.; validation, all authors; formal analysis, Y.K. and S.M.; investigation, all authors; resources, Y.K. and V.G.; data curation, Y.K. and V.G.; writing—original draft preparation, Y.K., V.G., S.M. and A.-M.E.; writing—review and editing, all authors; project administration, Y.K. All authors have read and agreed to the published version of the manuscript.

Funding: This research received no external funding.

Institutional Review Board Statement: The study was conducted in accordance with the Declaration of Helsinki and approved by the Ethics Committee of the University of Derby, reference no: 271118YKVG, approved on 16 December 2021.

Informed Consent Statement: Informed consent was obtained from all subjects involved in the study.

Data Availability Statement: The data that support the findings of this study are available on request from the corresponding author. The data are not publicly available due to privacy or ethical restrictions.

Conflicts of Interest: The authors declare no conflict of interest.

References

1. Vlachopoulos, D.; Makri, A. Quality Teaching in Online Higher Education: The Perspectives of 250 Online Tutors on Technology and Pedagogy. *Int. J. Emerg. Technol. Learn. (iJET)* **2021**, *16*, 40–56. [CrossRef]
2. Kotera, Y.; Cockerill, V.; Green, P.; Hutchinson, L.; Shaw, P.; Bowskill, N. Towards Another Kind of Borderlessness: Online Students with Disabilities. *Distance Educ.* **2019**, *40*, 170–186. [CrossRef]
3. Bambara, C.S.; Bambara, C.S.; Harbour, C.P.; Davies, T.G.; Athey, S. Delicate Engagement: The Lived Experience of Community College Students. *Community Coll. Rev.* **2009**, *36*, 219–238. [CrossRef]
4. El-Mansour, B.; Davison, M. Students' Positive and Negative Experiences in Hybrid and Online Classes. *Coll. Stud. J.* **2007**, *41*, 242–248.
5. Mashifana, T. Students' Preferred Mode of Teaching: Blended vs. Traditional Face to Face Instruction. In Proceedings of the 2020 IFEES World Engineering Education Forum—Global Engineering Deans Council, WEEF-GEDC 2020, Cape Town, South Africa, 16–19 November 2020. [CrossRef]
6. Shonfeld, M.; Ronen, I. Online Learning for Students from Diverse Backgrounds: Learning Disability Students, Excellent Students and Average Students. *IAFOR J. Educ.* **2015**, *3*, 13–29. [CrossRef]
7. Freeman, T.M.; Anderman, L.H.; Jensen, J.M. Sense of Belonging in College Freshmen at the Classroom and Campus Levels. *J. Exp. Educ.* **2007**, *75*, 203–220. [CrossRef]
8. Jaggars, S.S. Choosing Between Online and Face-to-Face Courses: Community College Student Voices. *Am. J. Distance Educ.* **2014**, *28*, 27–38. [CrossRef]
9. Yardimci, F.; Bektaş, M.; Özkütük, N.; Muslu, G.K.; Gerçeker, G.Ö.; Başbakkal, Z. A Study of the Relationship between the Study Process, Motivation Resources, and Motivation Problems of Nursing Students in Different Educational Systems. *Nurse Educ. Today* **2017**, *48*, 13–18. [CrossRef]
10. Curasi, C.; Burkhalter, J. Student Attitudes Toward Coursework: An Examination of the Motivation of University Business Students. Available online: https://www.researchgate.net/publication/260517085_Student_Attitudes_Toward_Coursework_An_Examination_of_the_Motivation_of_University_Business_Students (accessed on 10 June 2010).
11. Kotera, Y.; Chircop, J.; Hutchinson, L.; Rhodes, C.; Green, P.; Jones, R.M.; Kaluzeviciute, G.; Garip, G. Loneliness in Online Students with Disabilities: Qualitative Investigation for Experience, Understanding and Solutions. *Int. J. Educ. Technol. High. Educ.* **2021**, *18*, 64. [CrossRef]
12. Chen, K.-C.; Jang, S.-J. Motivation in Online Learning: Testing a Model of Self-Determination Theory. *Comput. Hum. Behav.* **2010**, *26*, 741–752. [CrossRef]
13. Landrum, B.; Bannister, J.; Garza, G.; Rhame, S. A Class of One: Students' Satisfaction with Online Learning. *J. Educ. Bus.* **2021**, *96*, 82–88. [CrossRef]
14. Tseng, H.; Yi, X.; Yeh, H.-T. Learning-Related Soft Skills among Online Business Students in Higher Education: Grade Level and Managerial Role Differences in Self-Regulation, Motivation, and Social Skill. *Comput. Hum. Behav.* **2019**, *95*, 179–186. [CrossRef]
15. Szopiński, T.; Bachnik, K. Student Evaluation of Online Learning during the COVID-19 Pandemic. *Technol. Soc. Chang.* **2022**, *174*, 121203. [CrossRef] [PubMed]
16. Zizka, L.; Probst, G. Learning during (or despite) COVID-19: Business Students' Perceptions of Online Learning. *Qual. Assur. Educ.* 2022. [CrossRef]
17. Moule, P.; Ward, R.; Lockyer, L. Nursing and Healthcare Students' Experiences and Use of e-Learning in Higher Education. *J. Adv. Nurs.* **2010**, *66*, 2785–2795. [CrossRef]
18. Kotera, Y.; Rawson, R.; Okere, U.; Ucheaga, M.B.; Williams, A.; Chircop, J.; Lyte, G.; Spink, R.; Green, P. Teaching Sustainability to Online Healthcare Students: A Viewpoint. *Int. J. High. Educ. Sustain.* **2022**, *4*, 38. [CrossRef]
19. Bailey, D.; Almusharraf, N.; Hatcher, R. Finding Satisfaction: Intrinsic Motivation for Synchronous and Asynchronous Communication in the Online Language Learning Context. *Educ. Inf. Technol.* **2021**, *26*, 2563–2583. [CrossRef]
20. Kotera, Y.; Spink, R.; Brooks-Ucheaga, M.; Green, P.; Rawson, R.; Rhodes, C.; Chircop, J.; Williams, A.; Okere, U.; Lyte, G. Teaching Healthcare Professional Students in Online Learning during COVID-19: Reflection of University Lecturers. *J. Concurr. Disord.* **2020**, *3*, 193–201. [CrossRef]
21. Suzuki, T.; Murayama, A.; Kotera, Y.; Bhandari, D.; Senoo, Y.; Tani, Y.; Harada, K.; Kawamoto, A.; Sato, S.; Sawano, T.; et al. Cross-Country Student Perceptions about Online Medical Education during the COVID-19 Pandemic. *Int. J. Environ. Res. Public Health* **2022**, *19*, 2840. [CrossRef]
22. Mcfadden, S.; Id, S.G.; Daly-Lynn, J.; O'neill, B.; Marley, J.; Hanratty, C.; Shepherd, P.; Ramsey, L.; Breen, C.; Duffy, O.; et al. Academic, Clinical and Personal Experiences of Undergraduate Healthcare Students during the COVID-19 Pandemic: A Prospective Cohort Study. *PLoS ONE* **2022**, *17*, e0271873. [CrossRef]
23. Bianchi, S.; Gatto, R.; Fabiani, L. Effects of the SARS-CoV-2 Pandemic on Medical Education in Italy: Considerations and Tips. *Euromediterranean Biomed. J.* **2020**, *15*, 100–101. [CrossRef]
24. Vijayan, R. Teaching and Learning during the COVID-19 Pandemic: A Topic Modeling Study. *Educ. Sci.* **2021**, *11*, 347. [CrossRef]
25. Griffin, R.; MacKewn, A.; Moser, E.; VanVuren, K.W. Learning Skills and Motivation: Correlates To Superior Academic Performance. *Bus. Educ. Accredit.* **2013**, *50*, 53–65.
26. Deci, E.L.; Ryan, R.M. *Intrinsic Motivation and Self-Determination in Human Behavior*; Plenum Press: New York, NY, USA, 1985. [CrossRef]

27. Ryan, R.M.; Deci, E.L. *Self-determination Theory: Basic Psychological Needs in Motivation, Development and Wellness*; Guilford Press: New York, NY, USA, 2017.
28. Ryan, R.M.; Deci, E.L. Intrinsic and Extrinsic Motivation from a Self-Determination Theory Perspective: Definitions, Theory, Practices, and Future Directions. *Contemp. Educ. Psychol.* **2020**, *61*, 1–11. [CrossRef]
29. Maurer, T.W.; Allen, D.; Gatch, D.B.; Shankar, P.; Sturges, D. A Comparison of Student Academic Motivations across Three Course Disciplines. *J. Scholarsh. Teach. Learn.* **2013**, *13*, 77–89.
30. Meng, X.; Hu, Z. The Relationship between Student Motivation and Academic Performance: The Mediating Role of Online Learning Behavior. *Qual. Assur. Educ.* 2022. [CrossRef]
31. White, R.L.; Bennie, A.; Vasconcellos, D.; Cinelli, R.; Hilland, T.; Owen, K.B.; Lonsdale, C. Self-Determination Theory in Physical Education: A Systematic Review of Qualitative Studies. *Teach. Teach. Educ.* **2021**, *99*, 103247. [CrossRef]
32. Moore, S.; Armstrong, C.; Pearson, J. Lecture Absenteeism among Students in Higher Education: A Valuable Route to Understanding Student Motivation. *J. High. Educ. Policy Manag.* **2008**, *30*, 15–24. [CrossRef]
33. Rose, S. Academic Success of Nursing Students: Does Motivation Matter? *Teach. Learn. Nurs.* **2011**, *6*, 181–184. [CrossRef]
34. Campos-Sánchez, A.; López-Núñez, J.A.; Carriel, V.; Martín-Piedra, M.Á.; Sola, T.; Alaminos, M. Motivational Component Profiles in University Students Learning Histology: A Comparative Study between Genders and Different Health Science Curricula. *BMC Med. Educ.* **2014**, *14*, 46. [CrossRef]
35. Kim, K.J.; Hwang, J.Y.; Kwon, B.S. Differences in Medical Students' Academic Interest and Performance across Career Choice Motivations. *Int. J. Med. Educ.* **2016**, *7*, 52. [CrossRef] [PubMed]
36. Kotera, Y.; Cockerill, V.; Chircop, J.; Kaluzeviciute, G.; Dyson, S. Predicting Self-Compassion in UK Nursing Students: Relationships with Resilience, Engagement, Motivation, and Mental Wellbeing. *Nurse Educ. Pract.* **2021**, *51*, 1–7. [CrossRef] [PubMed]
37. Breen, R.; Lindsay, R. Different Disciplines Require Different Motivations for Student Success. *Res. High. Educ.* **2002**, *43*, 693–725. [CrossRef]
38. Neumann, R. Disciplinary Differences and University Teaching. *Stud. High. Educ.* **2001**, *26*, 135–146. [CrossRef]
39. Kotera, Y.; Conway, E.; van Gordon, W. Mental Health of UK University Business Students: Relationship with Shame, Motivation and Self-Compassion. *J. Educ. Bus.* **2019**, *94*, 11–20. [CrossRef]
40. Hurst, A.; Pearce, A.; Parish, S.; Erickson, C.; Schnidman, A.; Vesty, L.; Garlinghouse, M.; Pavela, A. *Purpose at Work: The Largest Global Study on the Role of Purpose in the Workforce*; Imperative: Seattle, WA, USA, 2016.
41. Downey, J.P.; McGaughey, R.; Roach, D. Attitudes and Influences toward Choosing a Business Major: The Case of Information Systems. *J. Inf. Technol. Educ. Res.* **2011**, *10*, 231–251. [CrossRef]
42. Brahm, T.; Jenert, T.; Wagner, D. The Crucial First Year: A Longitudinal Study of Students' Motivational Development at a Swiss Business School. *High. Educ.* **2017**, *73*, 459–478. [CrossRef]
43. Cheng, P.Y.; Lin, M.L.; Su, C.K. Attitudes and Motivations of Students Taking Professional Certificate Examinations. *Soc. Behav. Personal.* **2011**, *39*, 1303–1314. [CrossRef]
44. Sahoo, F.M.; Sahoo, K.; Jena, L.K. Management of Students' Motivation in Business Schools: A Test of an Indigenous Model. *Int. J. Manag. Concepts Philos.* **2019**, *12*, 117–132. [CrossRef]
45. Nilsson, K.E.L.; Stomberg, W.; Margareta, I. Nursing Students Motivation toward Their Studies—A Survey Study. *BMC Nurs.* **2008**, *7*, 6. [CrossRef]
46. Boughn, S.; Lentini, A. Why Do Women Choose Nursing? *J. Nurs. Educ.* **1999**, *38*, 156–161. [CrossRef] [PubMed]
47. Law, W.; Arthur, D. What Factors Influence Hong Kong School Students in Their Choice of a Career in Nursing? *Int. J. Nurs. Stud.* **2003**, *40*, 23–32. [CrossRef]
48. Eley, D.; Eley, R.; Bertello, M.; Rogers-Clark, C. Why Did I Become a Nurse? Personality Traits and Reasons for Entering Nursing. *J. Adv. Nurs.* **2012**, *68*, 1546–1555. [CrossRef] [PubMed]
49. Kotera, Y.; Cockerill, V.; Chircop, J.G.E.; Forman, D. Mental Health Shame, Self-Compassion and Sleep in UK Nursing Students: Complete Mediation of Self-Compassion in Sleep and Mental Health. *Nurs. Open* **2021**, *8*, 1325–1335. [CrossRef] [PubMed]
50. Daniel, T.L.; Esser, J.K. Intrinsic Motivation as Influenced by Rewards, Task Interest, and Task Structure. *J. Appl. Psychol.* **1980**, *65*, 566–573. [CrossRef]
51. Deci, E.L.; Benware, C.; Landy, D. The Attribution of Motivation as a Function of Output and Rewards. *J. Personal.* **1974**, *42*, 652–667. [CrossRef]
52. Kotera, Y.; Taylor, E.; Fido, D.; Williams, D.; Tsuda-McCaie, F. Motivation of UK Graduate Students in Education: Self-Compassion Moderates Pathway from Extrinsic Motivation to Intrinsic Motivation. *Curr. Psychol.* **2021**, *1*, 1–14. [CrossRef]
53. Deci, E.L. Effects of Externally Mediated Rewards on Intrinsic Motivation. *J. Personal. Soc. Psychol.* **1971**, *18*, 105–115. [CrossRef]
54. Anghelcev, G.; Eighmey, J. The Impact of Extrinsic Incentives on Students' Willingness to Volunteer as Peer Mentors: Implications for Advertising Education. *J. Advert. Educ.* **2013**, *17*, 5–16. [CrossRef]
55. Kohn, A. *Unconditional Parenting: Moving from Rewards and Punishments to Love and Reason*; Atria Paperback: New York, NY, USA, 2005; ISBN 13: 9780743487481.
56. Wang, C.K.J.; Liu, W.C. Promoting Enjoyment in Girls' Physical Education: The Impact of Goals, Beliefs, and Self-Determination. *Eur. Phys. Educ. Rev.* **2007**, *13*, 145–164. [CrossRef]

57. Cameron, J.; Banko, K.M.; Pierce, W.D. Pervasive Negative Effects of Rewards on Intrinsic Motivation: The Myth Continues. *Behav. Anal.* **2017**, *24*, 1–44. [CrossRef] [PubMed]
58. Aultman, J. Utilizing a Concurrent Nested Design for Strengthening Mixed Methods Research. In *SAGE Research Methods Cases: Medicine and Health*; SAGE Publications, Ltd.: Thousand Oaks, CA, USA. [CrossRef]
59. Kotera, Y.; Green, P.; Sheffield, D. Mental Health of Therapeutic Students: Relationships with Attitudes, Self-Criticism, Self-Compassion, and Caregiver Identity. *Br. J. Guid. Couns.* **2021**, *49*, 701–712. [CrossRef]
60. Kotera, Y.; Conway, E.; van Gordon, W. Ethical Judgement in UK Business Students: Relationship with Motivation, Self-Compassion and Mental Health. *Int. J. Ment. Health Addict.* **2019**, *17*, 1132–1146. [CrossRef]
61. Vallerand, R.J.; Pelletier, L.G.; Blais, M.R.; Briere, N.M.; Senecal, C.; Vallieres, E.F. The Academic Motivation Scale: A Measure of Intrinsic, Extrinsic, and Amotivation in Education. *Educ. Psychol. Meas.* **1992**, *52*, 1003–1017. [CrossRef]
62. Clarke, V.; Braun, V. Thematic Analysis. *J. Posit. Psychol.* **2017**, *12*, 297–298. [CrossRef]
63. Braun, V.; Clarke, V. Using Thematic Analysis in Psychology. *Qual. Res. Psychol.* **2006**, *3*, 77–101. [CrossRef]
64. Kotera, Y.; Green, P.; Sheffield, D. Positive Psychology for Mental Wellbeing of UK Therapeutic Students: Relationships with Engagement, Motivation, Resilience and Self-Compassion. *Int. J. Ment. Health Addict.* **2021**, *20*, 1611–1626. [CrossRef]
65. Kotera, Y.; Green, P.; Sheffield, D. Roles of Positive Psychology for Mental Health in UK Social Work Students: Self-Compassion as a Predictor of Better Mental Health. *Br. J. Soc. Work* **2020**, *50*, 2002–2021. [CrossRef]
66. Conway, E.; Kotera, Y. Ethical Judgement and Intent in Business School Students: The Role of the Psyche? *Int. J. Ethics Educ.* **2020**, *5*, 151–186. [CrossRef]
67. Kotera, Y.; Adhikari, P.; van Gordon, W. Motivation Types and Mental Health of UK Hospitality Workers. *Int. J. Ment. Health Addict.* **2018**, *16*, 751–763. [CrossRef]
68. Kotera, Y.; Adhikari, P.; van Gordon, W. The Relationship between Work Motivation and Worker Profile in UK Hospitality Workers. *Int. J. Educ. Psychol. Couns.* **2017**, *2*, 231–243.
69. Kotera, Y.; Ozaki, A.; Miyatake, H.; Tsunetoshi, C.; Nishikawa, Y.; Kosaka, M.; Tanimoto, T. Qualitative Investigation into the Mental Health of Healthcare Workers in Japan during the COVID-19 Pandemic. *Int. J. Environ. Res. Public Health* **2022**, *19*, 568. [CrossRef] [PubMed]
70. Bengtsson, E.; Teleman, B. Motivation, Learning Strategies and Performance Among Business Undergraduates at University Colleges in Sweden. *Bus. Manag. Educ.* **2019**, *17*, 111–133. [CrossRef]
71. Puchalski, C.M.; Guenther, M. Restoration and Re-Creation: Spirituality in the Lives of Healthcare Professionals. *Curr. Opin. Supportive Palliat. Care* **2012**, *6*, 254–258. [CrossRef]
72. Wu, H.; Li, S.; Zheng, J.; Guo, J. Medical Students' Motivation and Academic Performance: The Mediating Roles of Self-Efficacy and Learning Engagement. *Med. Educ. Online* **2020**, *25*, 1742964. [CrossRef]
73. McEvoy, G.M. Increasing Intrinsic Motivation to Learn in Organizational Behavior Classes. *J. Manag. Educ.* **2011**, *35*, 468–503. [CrossRef]
74. Lalande, D.; Vallerand, R.J.; Lafrenière, M.A.K.; Verner-Filion, J.; Laurent, F.A.; Forest, J.; Paquet, Y. Obsessive Passion: A Compensatory Response to Unsatisfied Needs. *J. Personal.* **2017**, *85*, 163–178. [CrossRef]
75. Healy, D.; Sharry, P.M. Promoting Self Awareness in Undergraduate Nursing Students in Relation to Their Health Status and Personal Behaviours. *Nurse Educ. Pract.* **2011**, *11*, 228–233. [CrossRef]
76. Rasheed, S.P.; Younas, A.; Sundus, A. Self-Awareness in Nursing: A Scoping Review. *J. Clin. Nurs.* **2019**, *28*, 762–774. [CrossRef]
77. Tahara, M.; Mashizume, Y.; Takahashi, K. Mental Health Crisis and Stress Coping among Healthcare College Students Momentarily Displaced from Their Campus Community Because of COVID-19 Restrictions in Japan. *Int. J. Environ. Res. Public Health* **2021**, *18*, 7245. [CrossRef]
78. Baker, S.R. Intrinsic, Extrinsic, and Amotivational Orientations: Their Role in University Adjustment, Stress, Well-Being, and Subsequent Academic Performance. *Curr. Psychol.* **2004**, *23*, 189–202. [CrossRef]
79. Kotera, Y.; Fido, D. Effects of Shinrin-Yoku Retreat on Mental Health: A Pilot Study in Fukushima, Japan. *Int. J. Ment. Health Addict.* **2021**, 1–13. [CrossRef]
80. Yun, M.R.; Lim, E.J.; Yu, B.; Choi, S. Effects of Academic Motivation on Clinical Practice-Related Post-Traumatic Growth among Nursing Students in South Korea: Mediating Effect of Resilience. *Int. J. Environ. Res. Public Health* **2020**, *17*, 4901. [CrossRef] [PubMed]
81. Kotera, Y. A Qualitative Investigation into the Experience of Neuro-Linguistic Programming Certification Training among Japanese Career Consultants. *Br. J. Guid. Couns.* **2018**, *46*, 39–50. [CrossRef]
82. Kotera, Y.; Sheffield, D. Disney Strategy for Japanese University Students' Career Guidance: A Mixed Methods Pilot Study. *J. Natl. Inst. Career Educ. Couns.* **2017**, *38*, 52–61. [CrossRef]
83. Lemos, M.S.; Veríssimo, L. The Relationships between Intrinsic Motivation, Extrinsic Motivation, and Achievement, Along Elementary School. *Procedia Soc. Behav. Sci.* **2014**, *112*, 930–938. [CrossRef]
84. Orsini, C.; Binnie, V.I.; Wilson, S.L. Determinants and Outcomes of Motivation in Health Professions Education: A Systematic Review Based on Self-Determination Theory. *J. Educ. Eval. Health Prof.* **2016**, *13*, 19. [CrossRef]
85. Reeve, J. The Interest-Enjoyment Distinction in Intrinsic Motivation. *Motiv. Emot.* **1989**, *13*, 83–103. [CrossRef]
86. Afzal, H.; Ali, I.; Aslam Khan, M.; Hamid, K. A Study of University Students' Motivation and Its Relationship with Their Academic Performance. *Int. J. Bus. Manag.* **2010**, *5*, 4. [CrossRef]

87. Ly, P.; Nyoman Sudana Degeng, I.; Setyosari, P. Relationship between Achievement Motivation and Learning Outcomes on Land Law Course vy Student of PPKN Nusa Cendana University. In Proceedings of the International Conference on Education (ICE2) 2018: Education and Innovation in Science in the Digital Era, Bandung, Indonesia, 9 October 2018; pp. 699–705.
88. Swallow, S.R.; Kuiper, N.A. Social Comparison and Negative Self-Evaluations: An Application to Depression. *Clin. Psychol. Rev.* **1988**, *8*, 55–76. [CrossRef]
89. Benbassat, J.; Baumal, R. Enhancing Self-Awareness in Medical Students: An Overview of Teaching Approaches. *Acad. Med.* **2005**, *80*, 156–161. [CrossRef] [PubMed]
90. Kotera, Y.; Fido, D. *Ikigai: Towards a Psychological Understanding of a Life Worth Living*; Concurrent Disorders Society Publishing: Toronto, ON, Canada, 2021; ISBN 978-1-7778018-0.
91. Schippers, M.C.; Ziegler, N. Life Crafting as a Way to Find Purpose and Meaning in Life. *Front. Psychol.* **2019**, *10*, 2778. [CrossRef] [PubMed]
92. Kotera, Y.; van Laethem, M.; Ohshima, R. Cross-Cultural Comparison of Mental Health between Japanese and Dutch Workers: Relationships with Mental Health Shame, Self-Compassion, Work Engagement and Motivation. *Cross Cult. Strateg. Manag.* **2020**, *27*, 511–530. [CrossRef]
93. Kotera, Y.; Conway, E.; Green, P. Construction and Factorial Validation of a Short Version of the Academic Motivation Scale. *Br. J. Guid. Couns.* **2021**, 1–10. [CrossRef]
94. Wang, C.; Pan, R.; Wan, X.; Tan, Y.; Xu, L.; Ho, C.S.; Ho, R.C. Immediate Psychological Responses and Associated Factors during the Initial Stage of the 2019 Coronavirus Disease (COVID-19) Epidemic among the General Population in China. *Int. J. Environ. Res. Public Health* **2020**, *17*, 1729. [CrossRef]
95. Bhandari, D.; Kotera, Y.; Ozaki, A.; Abeysinghe, S.; Kosaka, M.; Tanimoto, T. COVID-19: Challenges Faced by Nepalese Migrants Living in Japan. *BMC Public Health* **2021**, *21*, 752. [CrossRef]
96. Marler, E.K.; Bruce, M.J.; Abaoud, A.; Henrichsen, C.; Suksatan, W.; Homvisetvongsa, S.; Matsuo, H. The Impact of COVID-19 on University Students' Academic Motivation, Social Connection, and Psychological Well-Being. *Scholarsh. Teach. Learn. Psychol.* **2021**. [CrossRef]
97. Stark, E. Examining the Role of Motivation and Learning Strategies in the Success of Online vs. Face-to-Face Students. *Online Learn.* **2019**, *23*, 234–251. [CrossRef]

healthcare

MDPI

Article

Mental Health Shame, Caregiver Identity, and Self-Compassion in UK Education Students

Yasuhiro Kotera [1,*], Freya Tsuda-McCaie [2], Ann-Marie Edwards [2], Divya Bhandari [3], Dan Williams [4] and Siobhan Neary [4]

[1] School of Health Sciences, University of Nottingham, Nottingham NG7 2HA, UK
[2] College of Health, Psychology of Social Care, University of Derby, Derby DE22 1GB, UK; freya.mccaie@gmail.com (F.T.-M.); annm.edwards@icloud.com (A.-M.E.)
[3] Medical Governance Research Institute, Tokyo 108-0074, Japan; rayordeal3@gmail.com
[4] College of Arts, Humanities and Education, University of Derby, Derby DE22 1GB, UK; d.williams@derby.ac.uk (D.W.); s.neary@derby.ac.uk (S.N.)
* Correspondence: yasuhiro.kotera@nottingham.ac.uk

Abstract: Although students in education have high rates of mental health problems, many of them do not ask for help, which can exacerbate their symptoms. One reason for their low help-seeking is shame associated with mental health problems. As education students aspire to provide care for children, they may feel ashamed to care for themselves, as the role identity theory suggests. Self-compassion is reported to reduce shame and mental health problems. This study explored the relationships between mental health problems, mental health shame, self-compassion, and caregiver identity among UK education students. One hundred and nine postgraduate students completed four self-report scales regarding those constructs. Correlation and regression analyses were performed. Mental health problems were positively associated with shame and identity, and negatively associated with self-compassion. Self-compassion was the only significant predictor of mental health problems. Findings will help educators and education students to develop effective approaches for their mental health problems.

Keywords: education students; mental health; mental health shame; caregiver identity; self-compassion

Citation: Kotera, Y.; Tsuda-McCaie, F.; Edwards, A.-M.; Bhandari, D.; Williams, D.; Neary, S. Mental Health Shame, Caregiver Identity, and Self-Compassion in UK Education Students. *Healthcare* **2022**, *10*, 584. https://doi.org/10.3390/healthcare10030584

Academic Editor: Alyx Taylor

Received: 10 February 2022
Accepted: 17 March 2022
Published: 21 March 2022

Publisher's Note: MDPI stays neutral with regard to jurisdictional claims in published maps and institutional affiliations.

1. Introduction

1.1. Mental Health of UK Postgraduate Education Students

Mental health problems among postgraduate students in the UK are concerningly prevalent [1]. Globally, graduate students are six times more likely to experience anxiety and depression than the general population [2]. Postgraduate education students, including trainee teachers, may be especially vulnerable to mental health problems, because of the stress and workload associated with managing work placements and academic work [3]. Additionally, the emotional impact of undergoing teacher training is significant, and students experience anxiety, self-doubt, and fatalistic feelings [4]. Indeed, psychological distress is prevalent among trainee teachers in the UK [5]. Poor mental health is associated with reduced academic achievement and attrition in tertiary education [6,7]. Despite this, research on the mental health of education postgraduate students is scarce [5], and no studies have investigated the relationships between mental health shame, caregiver identity, and self-compassion on mental health symptoms in this population.

1.2. Role Identity

Role identity has been used to explore helping professionals' reluctance to seek help for mental health problems. Role identity theory argues that individuals' perceptions of their personal and social identities, and societal and individual associations with those identities, shape role identity. Role identity is an idealised version of self, which provides a schema for appraisal and determination of thoughts and action, and creates expectations of self [8]. Siebert and Siebert [9] applied role identity theory to helping professionals' reluctance to seek help for mental health issues. They theorised that the association between caregiver identity and lack of help-seeking may occur because social workers overlook their mental health problems because of an unwillingness to acknowledge that they have similar problems to their clients, as such an acknowledgment would challenge and destabilise their idealised role identity [9]. A caregiver identity is viewing oneself and/or believing others view one as a helper or caregiver either professionally or personally (or both). Role identity as a caregiver was associated with burnout, depression, and not seeking help among social workers [9], and, in healthcare students, was predictive of mental health problems [10–12].

Caregiver identity may be prevalent among education students, and exacerbate mental health challenges by decreasing help-seeking behaviour. Education students' social identities, influenced by their professional and academic roles, may include the identity of 'teacher' [13]. Given the responsibilities that comprise teachers' work (i.e., nurturing, safeguarding, and developing young people), postgraduate education students may associate the role of 'teacher' with 'caregiver', and thus identify as caregivers. Indeed, teaching is classified as one of the caring professions [14]. Furthermore, popular social discourses depict 'good' teachers as selfless, heroic [15], and caring [16]; that is, 'good' teachers foreground and care for their pupils' wellbeing. These associations with the teacher identity may impact education students' expectations of themselves in the same way that idealised notions of 'helping professionals' impacted social workers' self-expectations [9]. Thus, teachers may face difficulties in reconciling their role with that of someone in need of support, thereby inhibiting acknowledgment of, and seeking help for, mental health problems. Indeed, the British Education Research Association argue that that challenges around the teacher identity may contribute to mental health problems among postgraduate education students [3]. Yet, caregiver identity and its relationship with mental health has not been explored with postgraduate education students.

1.3. Mental Health Shame

Mental health shame is feeling ashamed for having a mental health problem (Kotera et al. 2019d), and has been associated with poor mental health in business, psychotherapy, occupational therapy, social work, and nursing students in the UK [10,11,17,18]. One explanation for this association is that mental health shame reduces the likelihood that an individual will seek help for their mental health problems [19]. Indeed, mental health shame was associated with reduced help-seeking among young adults with mental health problems [20], and UK social work students [21].

Mental health shame may be especially prevalent among postgraduate education students because of incompatibility between stigmatised notions of people with mental health problems and idealised depictions of educators. Stigmatised attitudes may be internalised, and result in feelings of shame [22]. Stigmatised beliefs about people with mental health problems include that they are weak, irresponsible, childlike, and unable to care for themselves [23]. Identities associated with teachers include 'expert' and 'authority' [24], and 'manager' or 'professional' [25]. These two identities may seem incompatible. Additionally, within the teaching profession, the dominant morality foregrounds institutional wellbeing over individual wellbeing, promoting resilience as the solution to work-based challenges [26]. Trainee teachers may internalise this morality, and interpret mental health problems as reflecting a lack of personal resilience. Thus, they may feel their mental health problems are indicative of weakness and their poor suitability as an educator—exacerbating mental health shame, and preventing help-seeking. Relatedly, caregiver identity may heighten mental health shame. Indeed, among social work students, mental health shame was associated with caregiver identity [10].

1.4. Self-Compassion

Self-compassion is associated with improved wellbeing and decreased mental health problems [27]. The construct has three central components: self-kindness, common humanity, and mindfulness [28]. Self-kindness describes an understanding and sympathetic attitude towards oneself in response to failure or pain (rather than self-criticism and judgment). Common humanity is the feeling that one's experiences, including painful or difficult ones, are shared aspects of the human condition, and thus are unifying rather than isolating. Mindfulness is a state in which painful thoughts and feelings can be acknowledged and held, but not overly identified with or equated with the self [28]. Neff posits that self-compassion conceptualises a healthy relationship to oneself, and is protective against the negative impacts of rumination, self-judgement, and isolation [28]. Research supports the protective mechanism of self-compassion, and it has been negatively associated with mental health symptoms among UK social work, counselling, occupational therapy, and business students [10–12,17] with burnout; reduced wellbeing and self-criticism among UK midwifery students [29]; and with lower distress among university students more broadly [30].

1.5. Three Affect Regulation Systems

Gilbert theories that imbalances between the three affect regulation systems can lead to mental health problems. The three systems regulating affect are: (i) the threat and protection system; (ii) the drive, excitement, and resource-seeking system; and (iii) the contentment, soothing, and safety system [31]. Imbalances—especially an overdeveloped and dominant threat system, and an under-activated soothing system—are a source of psychopathology and poor wellbeing [31]. Individuals with high levels of shame and self-criticism have dominant threat systems, and struggle to activate their soothing systems [32]. Thus, self-criticism and shame, in combination with low levels of self-reassurance, leads to an activated threat system, resulting in lowered wellbeing and an increased risk of mental health problems [31]. In contrast, self-compassion activates the soothing system, creating balance between the three systems [31]. Indeed, compassionate mind training reduced participants' levels of self-criticism and shame [33].

The interplay of affect regulation systems may explain the protective role of self-compassion, and the negative impact of caregiver identity and mental health shame on mental health problems. Mental health shame may result in an activated threat system, increasing risk of mental health problems. Caregiver identity, being positively associated with mental health shame, may also result in an activated threat system. Additionally, caregiver identity may result in heightened self-criticism (associated with the threat system), as individuals feel a sense of failure or inadequacy around experiencing mental health challenges. In activating the soothing system, self-compassion may counter the negative impacts of mental health shame and caregiver identity on mental health problems. However, no study has investigated the relationship between self-compassion, mental health shame, and mental health problems in postgraduate education students. Understanding the prevalence and functions of caregiver identity, self-compassion, and mental health shame in the development of mental health problems among postgraduate education students may help universities and health services develop targeted, ameliorative strategies to prevent or address mental health problems, and improve students' wellbeing.

1.6. Aims

Therefore, this study aimed to explore relationships among mental health problems, mental health shame, self-compassion, and caregiver identity in UK education students. Mental health was evaluated in terms of depression, anxiety, and stress, as anxiety and depression are the most common mental health disorders in the general population [34], and stress is common among trainee teachers [5]. Three research questions were considered:

RQ1. How are mental health shame, self-compassion, and caregiver identity related to each mental health problem (depression, anxiety, and stress)?

RQ2. How do mental health shame, self-compassion, and caregiver identity predict each mental health problem (depression, anxiety, and stress)?
RQ3. How do mental health shame and caregiver identity predict self-compassion?

2. Materials and Methods

2.1. Participants

Participants had to be 18 years old or older, and studying in an education programme at a UK university at the time of the study: students taking a study break were excluded. Participants were recruited via convenience sampling using a paper-based survey distributed by programme tutors instead of the researchers to avoid response biases. Of 120 part-time graduate students who were introduced to the study, 109 (91%; 70 females, 39 males; Age 27.39 ± 7.94 years old, range 21–55 years old; 104 British, 2 other Europeans, and 1 Asian) completed four mental health scales, satisfying the required sample size calculated by power analysis (84: two tails, p H1 = 0.30, α = 0.05, Power = 0.80, p H0 = 0) [35]. Coloured paper was prepared for students with visual impairments, and one student used it. Paper data were digitised by a research assistant, who was not a co-author of this study. Compared with the general population of UK education students, which is 78% [36], our sample recruited slightly fewer females (64%). No compensation was awarded for completing the survey. We did not ask for a reason for withdrawal to the 11 participants, adhering to the ethical guidelines: no reason nor complaint was received. Ethics approval was obtained from the University Research Ethics Committee.

2.2. Instruments

The Depression Anxiety and Stress Scale (DASS21), a shortened version of DASS42 [37], was used to assess mental health problems. DASS21 consists of 21 items on a four-point Likert scale (0 = did not apply to me at all to 3 = applied to me very much or most of the time; a higher score indicates poorer mental health) divided into three subscales, seven items each: depression (e.g., 'I found it difficult to work up the initiative to do things'), anxiety (e.g., 'I was worried about situations in which I might panic and make a fool of myself'), and stress (e.g., 'I found myself getting agitated'). These subscales had good reliability: α = 0.87–0.94 [38].

Mental health shame was assessed using the Attitudes Towards Mental Health Problems (ATMHP), consisting of 35 four-point Likert items (0 = do not agree at all to 3 = completely agree; a higher score indicates stronger shame). ATMHP involves four sections: (i) their community's and family's attitudes towards mental health problems (community and family attitudes, e.g., 'My community/family sees mental health problems as a personal weakness'); (ii) their views on how their community and family would see them if they had a mental health problem (community and family external shame, e.g., 'I think my community/family would see me as inferior'); (iii) how they view themselves if they had a mental health problem (internal shame, e.g., 'I would see myself as inadequate'); and (iv) how their family would be perceived if they had a mental health problem (family-reflected shame, e.g., 'My family would be seen as inadequate'), and how much they worry about themselves when a close relative had a mental health problem (self-reflected shame, e.g., 'I would worry that others would not wish to be associated with me'). All of the subscales had good Cronbach's alphas of between 0.85 and 0.97 [39].

Self-compassion was evaluated using the Self-Compassion Scale-Short Form [40]. This self-report measure is a shortened version of the original 26-item Self-Compassion Scale [28], comprising 12 items (e.g., 'When something painful happens I try to take a balanced view of the situation.') on a five-point Likert scale (1 = almost never to 5 = almost always; a higher score indicates more self-compassion). Cronbach's alpha was found to be 0.86 [40].

Lastly, the Role Identity Scale (RIS) was used to assess caregiver identity. This eight-item measure considers (a) how participants view themselves as a caregiver, and (b) how they perceive how others view themselves as a caregiver [9]. Participants respond to how much they agree to each item (e.g., 'I have heard I am a natural helper or caregiver') on a five-point Likert scale (1 = strongly disagree to 5 = strongly agree; a higher store indicates a stronger caregiver identity). The reliability of RIS was high ($\alpha = 0.78$).

2.3. Procedure

First the collected data were screened for outliers and the assumptions of parametric tests. Second, correlations between their mental health problems, mental health shame, self-compassion, and caregiver identity were examined. Third, multiple regression analyses were conducted to identify significant predictors for each mental health problem. Finally, another multiple regression analysis was conducted to identify predictors for self-compassion.

3. Results

Analyses were conducted using IBM SPSS version 25.0. No outliers were identified. All variables demonstrated good internal reliability ($\alpha = 0.87$–0.96; Table 1).

Table 1. Descriptive statistics: mental health problems, mental health shame, self-compassion, and caregiver identity in UK education students ($n = 109$).

Scale	Construct (Range)	M	SD	α
Depression Anxiety and Stress Scale-21	Mental Health Problems			
	Depression (0–42)	13.72	11.92	0.93
	Anxiety (0–42)	12.07	9.13	0.77
	Stress (0–42)	19.88	10.52	0.84
Attitudes Towards Mental Health Problems	Mental Health Shame			
	Negative Mental Health Attitudes (0–24)	5.20	5.36	0.91
	External Shame (0–30)	7.22	8.36	0.96
	Internal Shame (0–15)	8.68	5.24	0.97
	Reflective Shame (0–36)	7.17	7.72	0.90
Self-Compassion Scale-Short Form	Self-Compassion (1–5)	2.73	0.77	0.86
Role Identity Scale	Caregiver Identity (8–40)	30.13	5.11	0.75

3.1. Relationships among Mental Health, Mental Health Shame, Self-Compassion, and Sleep

Because all variables apart from self-compassion were not normally distributed (Shapiro–Wilk test, $p < 0.05$), data were square-root-transformed to satisfy the assumption of normality [41]. Pearson's correlation was calculated (Table 2).

Table 2. Correlations among mental health problems, mental health shame, self-compassion, and caregiver identity in UK education students ($n = 109$).

		1	2	3	4	5	6	7	8	9	10	11
1	Gender (0 = M, 1 = F)	-										
2	Age	0.06	-									
3	Depression	0.11	0.06	-								
4	Anxiety	0.19 *	0.02	0.69 **	-							
5	Stress	0.23 *	0.04	0.74 **	0.70 **	-						
6	Negative Mental Health Attitudes	0.06	0.15	0.38 **	0.25 **	0.41 **	-					
7	External Shame	0.20 *	0.10	0.49 **	0.38 **	0.51 **	0.72 **	-				
8	Internal Shame	0.08	0.06	0.40 **	0.28 **	0.39 **	0.41 **	0.53 **	-			
9	Reflected Shame	0.03	0.01	0.34 **	0.24 *	0.34 **	0.48 **	0.59 **	0.62 **	-		
10	Self-Compassion	−0.13	−0.03	−0.66 **	−0.53 **	−0.66 **	−0.37 **	−0.46 **	−0.46 **	−0.27 **	-	
11	Caregiver Identity	0.20 *	0.06	0.29 **	0.26 **	0.23 *	0.22 *	0.28 **	0.32 **	0.19 *	−0.24 *	-

* $p < 0.05$, ** $p < 0.01$. For gender (0 = M, 1 = F), point-biserial coefficients are reported.

Mental health problems were positively associated with all the mental health shame subscales (external shame being the strongest positive correlate) and caregiver identity, and negatively associated with self-compassion (RQ1). Furthermore, self-compassion was negatively related to all the mental health shame subscales (again, external shame was the strongest) and caregiver identity.

3.2. Predictors of Mental Health Problems

Multiple regression analyses were conducted to explore the relative contribution of mental health shame, self-compassion, and caregiver identity to each mental health problem (Table 3). Mental health shame was calculated by summing all of the subscale scores [11]. First, gender and age were entered to statistically adjust for their effects (step one), and then mental health shame, self-compassion, and sleep were entered (step two). Adjusted coefficients of determination (Adj. R^2) were reported. Multicollinearity was not a concern (VIF < 10). Mental health shame, self-compassion, and caregiver identity accounted for 26–49% of the variance in each mental health problem, indicating a large effect size [42]. Self-compassion was the only significant predictor ($p < 0.001$), negatively predicting all mental health problems (RQ2): Depression B = −4.22, Anxiety B = −2.99, and Stress B = −3.22.

Table 3. Multiple regression: mental health shame, self-compassion, and caregiver identity to mental health problems among education students (n = 109).

	Depression			**Anxiety**			**Stress**		
		95% CI			**95% CI**			**95% CI**	
	B	**Lower**	**Upper**	**B**	**Lower**	**Upper**	**B**	**Lower**	**Upper**
Step 1									
Gender (0 = M, 1 = F)	0.39	−0.34	1.13	0.63 *	0.02	1.24	0.69 *	0.13	1.26
Age	0.01	−0.03	0.06	<0.001	−0.04	0.04	0.01	−0.03	0.04
Step 2									
Gender (0 = M, 1 = F)	−0.10	−0.66	0.47	0.28	−0.27	0.83	0.37	−0.07	0.82
Age	0.01	−0.03	0.04	<0.001	−0.03	0.03	<0.001	−0.03	0.03
Negative Mental Health Attitudes	−0.02	−0.28	0.25	−0.08	−0.33	0.18	0.06	−0.15	0.26
External Shame	0.19	−0.05	0.43	0.15	−0.09	0.38	0.13	−0.06	0.32
Internal Shame	−0.05	−0.35	0.24	−0.11	−0.40	0.18	−0.05	−0.28	0.18
Reflected Shame	0.11	−0.13	0.35	0.08	−0.15	0.32	0.09	−0.10	0.28
Self-Compassion	−4.22 ***	−5.51	−2.94	−2.99 ***	−4.24	−1.74	−3.22 ***	−4.23	−2.22
Caregiver Identity	0.41	−0.17	0.99	0.38	−0.19	0.94	0.04	−0.41	0.50
Adj R^2 Δ		49%			26%			43%	

B = unstandardised regression coefficient. * $p < 0.05$; *** $p < 0.001$.

3.3. Predictors of Self-Compassion

Lastly, another multiple regression analysis was performed to appraise the relative contribution of mental health shame and caregiver identity to self-compassion (Table 4). Again, multicollinearity was not a concern (VIF < 10). Mental health shame and caregiver identity accounted for 25% of the variance in self-compassion, indicating a large effect size [42]. External shame ($p < 0.05$, B = −0.04) and internal shame ($p < 0.001$, B = −0.07) were significant predictors for self-compassion (RQ3).

Table 4. Multiple regression: mental health shame and caregiver identity to self-compassion among education students (n = 109).

	Self-Compassion		
		95% CI	
	B	**Lower**	**Upper**
Step 1	1.75	1.54	1.97
Gender (0 = M, 1 = F)	−0.06	−0.16	0.03
Age	<0.001	−0.01	0.01
Step 2	2.02	1.54	2.50
Gender (0 = M, 1 = F)	−0.02	−0.10	0.07
Age	<0.001	<0.001	0.01
Negative Mental Health Attitudes	−0.02	−0.06	0.03
External Shame	−0.04 *	−0.07	<0.001
Internal Shame	−0.07 ***	−0.11	−0.03
Reflected Shame	0.03	−0.01	0.06
Caregiver Identity	−0.03	−0.12	0.06
Adj R^2 Δ		25%	

B = unstandardised regression coefficient. * $p < 0.05$; *** $p < 0.001$.

4. Discussion

There is little evidence in the literature on the mental health of postgraduate education students, despite a prevalence of mental health problems in this particular group [43]. To the best of our knowledge, this is the first study investigating the relationship between self-compassion, caregiver identity, mental health shame, and mental health problems in postgraduate education students. Regarding RQ1, all mental health problems were indeed positively associated with mental health shame and caregiver identity, and negatively associated with self-compassion. In addition, external shame was the strongest correlate with self-compassion. Moreover, our regression analysis revealed that all mental health problems were only predicted by self-compassion (RQ2). Lastly, another regression identified that external and internal shame were significant predictors of self-compassion (RQ3). We discuss each finding in detail below.

In the present study, a positive relationship between self-compassion, mental health shame, and caregiver identity were noted, supported by previous findings showing the impact of self-compassion and caregiver identity on mental health problems [44,45]. The findings of this study demonstrate that self-compassion can result in long-lasting improvements in mental health symptoms, such as decreased depression and greater happiness [46], which would be particularly helpful for postgraduate education students, such as trainee teachers. Siebert and Siebert [9] were the first to research role identity and distress. They found that strong caregiver role identity is significantly associated with psychological and occupational distress, and negatively correlated with help-seeking [9]. These findings are consistent with [47], who highlight the importance of cultivating self-compassion in order to reduce the negative consequences associated with external shame. This has implications for the role of self-compassion-based interventions in shame reduction, which warrants further investigation [48]. The results also demonstrate the need to understand the specific challenges faced by postgraduate education students, and how practising self-compassion can prove beneficial. Counselling and other mental health care interventions may be necessary to help improve students' health and wellbeing [49].

Though mental health shame, self-compassion, and caregiver identity predicted 26–49% of variance in each mental health problem, analyses found that self-compassion was a significant predictor, negatively predicting each mental health problem. Findings indicate that individuals with higher levels of self-compassion may be more resilient [50], thus lowering levels of anxiety, depression, and stress-related illnesses [51,52]. However, our study found that external and internal shame negatively predicted self-compassion in postgraduate education students, which is consistent with previous research on self-compassion and mental health shame [11,18]. Shame is linked to psychopathology in various ways, evoking negative feelings across a range of mental health issues, such as anxiety and depression [48,53]. Education students could benefit from Compassionate Mind Training (CMT), which has proven effective in increasing levels of self-compassion in caring professions, such as psychotherapy [29] and education [54]. Developed by Gilbert [31], CMT is a therapeutic approach designed to promote and develop self-compassion for self and others, and is regarded as an important tool to improve mood and wellbeing [46]. For example, 31 nursing, social work, counselling, psychology, and teaching students were tested to see if Mindfulness-Based Stress Reduction (MBSR) could help them avoid stress and increase self-compassion. It was found that practising mindfulness increased self-compassion levels, implying the importance of raising awareness of self-compassion, and its benefits for the mental health of students [55,56]. Indeed, mindfulness is recommended for trainee teachers [57]; however, self-compassion has not been assessed in depth in this population. The findings have implications for further research on self-compassion and its relationship with mental health problems, particularly in postgraduate education students. Moreover, incorporating wellness programmes into the university student experience [58], as well as mindfulness-based intervention in the graduate curriculum, may yield mental health benefits [59], such as enhancing the cognitive and emotional capacity of trainees to cope with the demands of caring professions [60]. Wellbeing education was effective for reducing stress among students [61]. Furthermore, if proven to be effective in these contexts, these interventions could be beneficial in other occupational and international settings. For example, how 'self-compassion' is perceived is different by culture and sector [62]. The ability to improve psychological wellbeing following self-compassion training is both pragmatic and beneficial; hence, it is worthy of wider applications. Further research is therefore warranted.

Limitations

The present study has several limitations. First, the lack of ethnically diverse students is likely to limit the generalisability of those underrepresented groups. Moreover, this study did not examine the mental health of postgraduate students in other countries. Considering the cross-cultural differences observed in mental health problems and shame [63], the international generalisability of our findings needs to be assessed. Second, this study relied on self-reported measures. Results are often influenced by self-reported surveys when individuals report their own experiences, thus limiting the accuracy of the findings. Alternative data collection methods would enhance future research. Lastly, because participants were recruited from a single UK academic institution, the findings may not represent other higher education institutions, which may have implications for the mental health of enrolled postgraduate students.

5. Conclusions

The findings of our study revealed that mental health problems were positively associated with mental health shame and caregiver identity. Similarly, those experiencing internal and external mental health shame were less likely to have self-compassion. However, students with higher self-compassion were found less likely to face mental health problems, which highlights self-compassion as a protective factor against mental health problems. Therefore, interventions (such as CMT and MSBR) aimed to improve self-compassion, and reduce mental health shame, should be designed and promoted to enhance the overall wellbeing of the students.

Author Contributions: Conceptualization, Y.K.; methodology, Y.K.; software, Y.K.; validation, all authors; formal analysis, Y.K.; investigation, Y.K. and D.W.; resources, Y.K. and D.W.; data curation, Y.K. and D.W.; writing—original draft preparation, Y.K., F.T.-M., A.-M.E. and D.B.; writing—review and editing, Y.K., A.-M.E. and S.N. All authors have read and agreed to the published version of the manuscript.

Funding: This research received no external funding.

Institutional Review Board Statement: The study was conducted in accordance with the Declaration of Helsinki, and approved by the Ethics Committee of University of Derby (protocol code 011017YK and 1 October 2017).

Informed Consent Statement: Informed consent was obtained from all subjects involved in the study.

Data Availability Statement: The data that support the findings of this study are available from the corresponding author, Y.K., upon reasonable request. All procedures followed were in accordance with the ethical standards of the responsible committee on human experimentation (institutional and national), and with the Helsinki Declaration of 1975, as revised in 2000 (5). Informed consent was obtained from all patients for being included in the study. No animal or human studies were carried out by the authors for this article.

Conflicts of Interest: The authors declare no conflict of interest.

References

1. Murray, S. Mental health: More support needed for postgraduate students. Postgraduates. *The Guardian*, 16 February 2021.
2. Evans, T.M.; Bira, L.; Gastelum, J.B.; Weiss, L.T.; Vanderford, N.L. Evidence for a mental health crisis in graduate education. *Nat. Biotechnol.* **2018**, *36*, 282–284. [CrossRef]

3. Quickfall, A.; Clarke, E.; Thompson, S. *The PGCE Journey: Wellbeing and Workload*; British Educational Research Association: London, UK, 2020; p. 16.
4. Hayes, J.-A. Values-based nursing and fitness to practice issues. *Key Concepts Issues Nurs. Ethics* **2017**, *106*, 175–189. [CrossRef]
5. Chaplain, R.P. Stress and psychological distress among trainee secondary teachers in England. *Educ. Psychol.* **2008**, *28*, 195–209. [CrossRef]
6. Andrews, B.; Wilding, J.M. The relation of depression and anxiety to life-stress and achievement in students. *Br. J. Psychol.* **2004**, *95*, 509–521. [CrossRef]
7. Poh Keong, P.; Chee Sern, L.; Foong, M.; Ibrahim, C. The Relationship between Mental Health and Academic Achievement among University Students—A Literature Review. *Glob. Publ.* **2013**, *6*, 318–323.
8. McCall, G.J.; Simmons, J.L. *Identities and Interactions: An Examination of Human Associations in Everyday Life*; Free Press: New York, NY, USA, 1966; ISBN 0029206308.
9. Siebert, D.C.; Siebert, C.F. The Caregiver Role Identity scale a validation study. *Res. Soc. Work Pract.* **2005**, *15*, 204–212. [CrossRef]
10. Kotera, Y.; Green, P.; Sheffield, D. Mental Health Attitudes, Self-Criticism, Compassion and Role Identity among UK Social Work Students. *Br. J. Soc. Work* **2019**, *49*, 351–370. [CrossRef]
11. Kotera, Y.; Green, P.; Sheffield, D. Mental health of therapeutic students: Relationships with attitudes, self-criticism, self-compassion, and caregiver identity. *Br. J. Guid. Couns.* **2019**, *49*, 701–712. [CrossRef]
12. Kotera, Y.; Green, P.; Sheffield, D. Mental health shame of UK construction workers: Relationship with masculinity, work motivation, and self-compassion. *Rev. Psicol. Trab. Organ.* **2019**, *35*, 135–143. [CrossRef]
13. Fitzmaurice, M. Constructing professional identity as a new academic: A moral endeavour. *Stud. High. Educ.* **2013**, *38*, 613–622. [CrossRef]
14. Hugman, R. *New Approaches in Ethics for the Caring Professions*; Palgrave Macmillan: London, UK, 2005; ISBN 1403914710.
15. Alsup, J. *Teacher Identity Discourses: Negotiating Personal and Professional Spaces*; Routledge: Abingdon, UK, 2006; ISBN 9781410617286.
16. Moore, A. *The Good Teacher: Dominant Discourses in Teaching and Teacher Education*; RoutledgeFalmer: Abingdon, UK, 2004; ISBN 0415335655.
17. Kotera, Y.; Conway, E.; Van Gordon, W. Mental health of UK university business students: Relationship with shame, motivation and self-compassion. *J. Educ. Bus.* **2019**, *94*, 11–20. [CrossRef]
18. Kotera, Y.; Cockerill, V.; Chircop, J.; Kaluzeviciute, G.; Dyson, S. Predicting self-compassion in UK nursing students: Relationships with resilience, engagement, motivation, and mental wellbeing. *Nurse Educ. Pract.* **2021**, *51*, 102989. [CrossRef] [PubMed]
19. Cage, E.; Stock, M.; Sharpington, A.; Pitman, E.; Batchelor, R. Barriers to accessing support for mental health issues at university. *Stud. High. Educ.* **2018**, *45*, 1637–1649. [CrossRef]
20. Rüsch, N.; Müller, M.; Ajdacic-Gross, V.; Rodgers, S.; Corrigan, P.W.; Rössler, W. Shame, perceived knowledge and satisfaction associated with mental health as predictors of attitude patterns towards help-seeking. *Epidemiol. Psychiatr. Sci.* **2014**, *23*, 177–187. [CrossRef]
21. Ting, L. Depressive symptoms in a sample of social work students and reasons preventing students from using mental health services: An exploratory study. *J. Soc. Work Educ.* **2011**, *47*, 253–268. [CrossRef]
22. Byrne, P. Stigma of mental illness and ways of diminishing it. *Adv. Psychiatr. Treat.* **2000**, *6*, 65–72. [CrossRef]
23. Corrigan, P.W.; Edwards, A.B.; Green, A.; Diwan, S.L.; Penn, D.L. Prejudice, Social Distance, and Familiarity with Mental Illness. *Schizophr. Bull.* **2001**, *27*, 219–225. [CrossRef]
24. Rahimi, A.; Bigdeli, R.A. The dynamicity and flexibility of EFL teachers' role identities in language institutes. *ABAC J.* **2014**, *34*, 46–55.
25. Aghaei, P.; Bavali, M.; Behjat, F. An in-depth qualitative study of teachers' role identities: A case of Iranian EFL teachers. *Int. J. Instr.* **2020**, *13*, 601–620. [CrossRef]
26. Margolis, J.; Hodge, A.; Alexandrou, A. The teacher educator's role in promoting institutional versus individual teacher well-being. *J. Educ. Teach.* **2014**, *40*, 391–408. [CrossRef]
27. Kotera, Y.; Green, P.; Sheffield, D. Roles of positive psychology for mental health in UK social work students: Self-compassion as a predictor of better mental health. *Br. J. Soc. Work* **2019**, *50*, 2002–2021. [CrossRef]
28. Neff, K. Self-Compassion: An Alternative Conceptualization of a Healthy Attitude Toward Oneself. *Self Identity* **2003**, *2*, 85–101. [CrossRef]
29. Beaumont, E.; Rayner, G.; Durkin, M.; Bowling, G. The effects of compassionate mind training on student psychotherapists. *J. Ment. Health Train. Educ. Pract.* **2017**, *12*, 300–312. [CrossRef]
30. Fong, M.; Loi, N.M. The Mediating Role of Self-compassion in Student Psychological Health. *Aust. Psychol.* **2016**, *51*, 431–441. [CrossRef]
31. Gilbert, P. *The Compassionate Mind*; Robinson: London, UK, 2009; ISBN 978-1-84529-713-8.
32. Gilbert, P. An introduction to Compassion Focused Therapy in Cognitive Behavior Therapy. *Int. J. Cogn. Ther.* **2010**, *3*, 97–112. [CrossRef]
33. Gilbert, P.; Procter, S. Compassionate Mind Training for People with High Shame and Self-Criticism: Overview and Pilot Study of a Group Therapy Approach. *Clin. Psychol. Psychother.* **2006**, *13*, 353–379. [CrossRef]
34. Organisation for Economic Co-Operation and Development. *Health at a Glance: Europe 2018*; OECD Publications: Paris, France, 2018. [CrossRef]

35. Faul, F.; Erdfelder, E.; Buchner, A.; Lang, A.-G. Statistical power analyses using G*Power 3.1: Tests for correlation and regression analyses. *Behav. Res. Methods* **2009**, *41*, 1149–1160. [CrossRef]
36. Higher Education Statistics Agency. *Higher Education Student Statistics: UK, 2016/17—Subjects Studied*; HESA: Cheltenham, UK, 2018.
37. Lovibond, S.; Lovibond, P. *Manual for the Depression Anxiety Stress Scales*, 2nd ed.; Psychology Foundation of Australia: Sydney, Australia, 1995; ISBN 7334-1423-0.
38. Antony, M.M.; Bieling, P.J.; Cox, B.J.; Enns, M.W.; Swinson, R.P. Psychometric properties of the 42-item and 21-item versions of the Depression Anxiety Stress Scales in clinical groups and a community sample. *Psychol. Assess.* **1998**, *10*, 176–181. [CrossRef]
39. Gilbert, P.; Bhundia, R.; Mitra, R.; McEwan, K.; Irons, C.; Sanghera, J. Cultural differences in shame-focused attitudes towards mental health problems in Asian and non-Asian student women. *Ment. Health Relig. Cult.* **2007**, *10*, 127–141. [CrossRef]
40. Raes, F.; Pommier, E.; Neff, K.D.; Van Gucht, D. Construction and factorial validation of a short form of the Self-Compassion Scale. *Clin. Psychol. Psychother.* **2011**, *18*, 250–255. [CrossRef]
41. Field, A. *Discovering Statistics Using IBM SPSS Statistics*, 5th ed.; SAGE Publications Ltd.: Thousand Oaks, CA, USA, 2017; ISBN 9781526419521.
42. Cohen, J. *Statistical Power Analysis for the Behavioral Sciences*; Lawrence Erlbaum Associates: Mahwah, NJ, USA, 1988; ISBN 0805802835.
43. Knapstad, M.; Sivertsen, B.; Knudsen, A.K.; Smith, O.R.F.; Aaro, L.E.; Lonning, K.J.; Skogen, J.C. Trends in self-reported psychological distress among college and university students from 2010 to 2018. *Psychol. Med.* **2021**, *51*, 470–478. [CrossRef]
44. Wu, C.; Pooler, D. Social workers' caregiver identity and distress: Examining the moderating role of self-esteem and social support. *Soc. Work Res.* **2014**, *38*, 237–249. [CrossRef]
45. Sager, K.M. *An Examination of the Roles of Distress, Self-Stigma, Caregiver Role Identity, and Self-Compassion in Clergy Help-Seeking Attitudes*; University of Kansas: Lawrence, KS, USA, 2019.
46. Shapira, L.B.; Mongrain, M. The benefits of self-compassion and optimism exercises for individuals vulnerable to depression. *J. Posit. Psychol.* **2010**, *5*, 377–389. [CrossRef]
47. Ferreira, C.; Pinto-Gouveia, J.; Duarte, C. Self-compassion in the face of shame and body image dissatisfaction: Implications for eating disorders. *Eat. Behav.* **2013**, *14*, 207–210. [CrossRef]
48. Woods, H.; Proeve, M. Relationships of mindfulness, self-compassion, and meditation experience with shame-proneness. *J. Cogn. Psychother.* **2014**, *28*, 20–33. [CrossRef]
49. Cosgrove, F.L.; Corrie, S.; Wolever, R.Q. An exploration of personal benefits reported by students of a health and wellness coach training programme. *Coach. Int. J. Theory Res. Pract.* **2021**, 1–17. [CrossRef]
50. Raes, F. The Effect of Self-Compassion on the Development of Depression Symptoms in a Non-clinical Sample. *Mindfulness* **2011**, *2*, 33–36. [CrossRef]
51. Gerber, Z.; Anaki, D. The Role of Self-compassion, Concern for Others, and Basic Psychological Needs in the Reduction of Caregiving Burnout. *Mindfulness* **2021**, *12*, 741–750. [CrossRef]
52. López, A.; Sanderman, R.; Ranchor, A.V.; Schroevers, M.J. Compassion for Others and Self-Compassion: Levels, Correlates, and Relationship with Psychological Well-being. *Mindfulness* **2018**, *9*, 325–331. [CrossRef]
53. Bilevicius, E.; Single, A.; Bristow, L.A.; Foot, M.; Ellery, M.; Keough, M.T.; Johnson, E.A. Shame mediates the relationship between depression and addictive behaviours. *Addict. Behav.* **2018**, *82*, 94–100. [CrossRef]
54. Maratos, F.A.; Montague, J.; Ashra, H.; Welford, M.; Wood, W.; Barnes, C.; Sheffield, D.; Gilbert, P. Evaluation of a Compassionate Mind Training Intervention with School Teachers and Support Staff. *Mindfulness* **2019**, *10*, 2245–2258. [CrossRef]
55. Birnie, K.; Speca, M.; Carlson, L.E. Exploring Self-compassion and Empathy in the Context of Mindfulness-based Stress Reduction (MBSR). *Stress Health* **2010**, *26*, 359–371. [CrossRef]
56. Newsome, S.; Waldo, M.; Gruszka, C. Mindfulness Group Work: Preventing Stress and Increasing Self-Compassion Among Helping Professionals in Training. *J. Spec. Group Work* **2012**, *37*, 297–311. [CrossRef]
57. Hartigan, B.F. Mindfulness in teacher education: A constructivist approach to stress reduction for teacher candidates and their students. *Child. Educ.* **2017**, *93*, 153–158. [CrossRef]
58. Pasic, M.; Eleftheriades, R.; Fiala, C. The challenges and mental health issues of academic trainees. *F1000Research* **2020**, *9*, 104. [CrossRef]
59. Shapiro, S.L.; Brown, K.W.; Biegel, G.M. Teaching self-care to caregivers: Effects of mindfulness-based stress reduction on the mental health of therapists in training. *Train. Educ. Prof. Psychol.* **2007**, *1*, 105–115. [CrossRef]
60. Tandler, N.; Kirkcaldy, B.; Petersen, L.E.; Athanasou, J. Is there a role for mindfulness and self-compassion in reducing stress in the teaching profession? *Minerva Psichiatr.* **2019**, *60*, 51–59. [CrossRef]
61. Young, T.; Macinnes, S.; Jarden, A.; Colla, R. The impact of a wellbeing program imbedded in university classes: The importance of valuing happiness, baseline wellbeing and practice frequency. *Stud. High. Educ.* **2020**, 1–20. [CrossRef]
62. Kotera, Y.; Ozaki, A.; Miyatake, H.; Tsunetoshi, C.; Nishikawa, Y.; Kosaka, M.; Tanimoto, T. Qualitative Investigation into the Mental Health of Healthcare Workers in Japan during the COVID-19 Pandemic. *Int. J. Environ. Res. Public Health* **2022**, *19*, 568. [CrossRef]
63. Kotera, Y.; Tsuda-McCaie, F.; Maughan, G.; Green, P. Cross-cultural comparison of mental health in social work students between UK and Ireland: Mental health shame and self-compassion. *Br. J. Soc. Work* **2021**, bcab240. [CrossRef]